Lecture Notes in Networks and Systems

Volume 239

The series "Lecture Notes in Networks and Systems" publishes the latest developments in Networks and Systems—quickly, informally and with high quality. Original research reported in proceedings and post-proceedings represents the core of LNNS.

Volumes published in LNNS embrace all aspects and subfields of, as well as new challenges in, Networks and Systems.

The series contains proceedings and edited volumes in systems and networks, spanning the areas of Cyber-Physical Systems, Autonomous Systems, Sensor Networks, Control Systems, Energy Systems, Automotive Systems, Biological Systems, Vehicular Networking and Connected Vehicles, Aerospace Systems, Automation, Manufacturing, Smart Grids, Nonlinear Systems, Power Systems, Robotics, Social Systems, Economic Systems and other. Of particular value to both the contributors and the readership are the short publication timeframe and the world-wide distribution and exposure which enable both a wide and rapid dissemination of research output.

The series covers the theory, applications, and perspectives on the state of the art and future developments relevant to systems and networks, decision making, control, complex processes and related areas, as embedded in the fields of interdisciplinary and applied sciences, engineering, computer science, physics, economics, social, and life sciences, as well as the paradigms and methodologies behind them.

Indexed by SCOPUS, INSPEC, WTI Frankfurt eG, zbMATH, SCImago.

All books published in the series are submitted for consideration in Web of Science.

More information about this series at http://www.springer.com/series/15179

Abdalmuttaleb M. A. Musleh Al-Sartawi ·
Anjum Razzaque · Muhammad Mustafa Kamal
Editors

Artificial Intelligence Systems and the Internet of Things in the Digital Era

Proceedings of EAMMIS 2021

Springer

Editors
Abdalmuttaleb M. A. Musleh Al-Sartawi
Accounting, Finance and Banking
Department
Ahlia University
Manama, Bahrain

Anjum Razzaque
College of Business and Finance
Ahlia University
Manama, Bahrain

Muhammad Mustafa Kamal
School of Strategy and Leadership
Coventry University
Coventry, UK

ISSN 2367-3370 ISSN 2367-3389 (electronic)
Lecture Notes in Networks and Systems
ISBN 978-3-030-77245-1 ISBN 978-3-030-77246-8 (eBook)
https://doi.org/10.1007/978-3-030-77246-8

This Springer imprint is published by the registered company Springer Nature Switzerland AG
The registered company address is: Gewerbestrasse 11, 6330 Cham, Switzerland

Foreword

Sustainable development is one of the greatest challenges of our time. Yet, it has become one of the most sought-after mission of businesses, economies, governments, and societies around the world. Nevertheless, without technology, societies would disintegrate and sustainability would be out of reach. Despite technology creating many of the problems faced by societies, they are also the answer to solving them. As technology is intrinsically ingrained in our lives, any advancement creates economic and social transformation, case in point the First Industrial Revolution to the current, and Fourth Revolution. Today, we are on a brink of a newer and faster revolution which involves sustainable technologies such as the Internet of Things (IoT) and artificial intelligence (AI).

One way or another, the term Internet of Things (IoT) has become unavoidable in the current digital and data-driven economy; especially in the last couple of years where IoT has garnered the interests of scholars, academics, and practitioners. IoT solutions are virtually extending to all areas of everyday life including education, health care, manufacturing, government services, and the financial service industry just to name a few. Although IoT seems futuristic, we use it daily through smartphones, smart tablets, and smartwatches. People, especially, the younger generation are fascinated with the word smart as it is used in concoction with technology in all its forms. The question is how smart technology really is? Another futuristic term, which has never been more attainable and approachable than today, is artificial intelligence (AI). AI is all around us in the form of virtual assistants, drones, and self-driving cars. Despite the advancement in the fields of sustainable technologies such as AI and IoT, only certain tasks have been fully enhanced by technology. However, we need to continuously strive to aim for more than that. The challenges and opportunities in AI and IoT are both interesting and difficult. Scientists, researchers, governments, and Higher Education Institutions are working with reverence and determination to provide new and improved solutions to humankind in the digital era that would support sustainable development.

This book provides a valuable outlook on various concepts and explores novel themes related to AI, IoT, innovation in the digital era, Industry 4.0, sustainable technologies, utilization of blockchain, big data, fintech, cybersecurity, and finally the impending Fifth Industrial Revolution. It looks further into the implications of technology on education, sustainability, and finance. This book is a good step in the direction of offering holistic knowledge to practitioners, scholars, educational institutions, professionals, students, and governments.

Small steps make big changes.

Arafat Salih Aydiner
Abdalmuttaleb M. A. Musleh Al- Sartawi
Anjum Razzaque
Ahmet Faruk Aysan
Conference Co-chairs

Preface

EAMMIS 2021 is being organized by the Bridges Foundation in cooperation with the Istanbul Medeniyet University. EAMMIS 2021 probes a deeper academic and scholarly interpretation through its theme: artificial intelligence systems and the Internet of Things in the digital era. The EAMMIS provides an international platform for researchers, academicians, practitioners, and industry professionals, to present research findings offering deeper insight into novel theories that embrace theoretical, managerial, and practical implications within the fields of management and information systems.

In extensive consultation with the EAMMIS executive committee and the faculty of the organizing universities, it was decided that it is best not to hold only a physical EAMMIS 2021, given the uncertainty of current circumstances. Our organizing committee volunteers have already invested so much effort into EAMMIS 2021, and many of them are already dealing with additional workloads because of COVID-19. This is an event for both scientific and social interactions. We believe we can achieve such a conference through a hybrid conference platform. The conferences call for papers asked for submissions in full research papers and, as a result, attracted many submissions from around the world. Submissions were reviewed in a double-blind process by at least two specialized academics in the relevant fields. The 2021 conference received a total of 111 submissions, and overall, 48 papers were accepted with an acceptance rate of 43.2%. The final set of proceedings have been grouped according to track, each outlined below.

The book is organized into forty-seven chapters and covers various topics categorized according to the following tracks: Track 1: Innovation in the digital era, Track 2: Industry 4.0 management for sustainable technologies, Track 3: Utilization of Blockchain, DLT, and Fintech for AI and IoT in the digital eras, Track 4: Role of Cybersecurity and privacy facilitated by AI and IoT, and Track 5: Fifth Industrial Revolution: The Internet of Ideas.

The papers in Track 1 address the key innovations in the current digital era, Industrial Revolution 4.0. The works here stem from innovative technologies including artificial intelligence and Internet of Things which provide solutions in business processes. Furthermore, different types of clustering algorithms and software-defined networking algorithms are examined. Additionally, the papers tackle various topics in social media usage during COVID-19 pandemic.

The papers in Track 2 explore Industry 4.0 management for sustainable technologies. The works here deal with machine learning for sustainability and a review on Industry 4.0 management for sustainable technologies. Moreover, the papers investigate sustainable innovations such as novel hydro renewable power platform and posture of robo-advisor. Finally, Chapter 18 discusses the methods of increasing productivity and quality with IoT technologies in industrial treatment systems

The papers in Track 3 address a very critical topic which is the utilization of blockchain, DLT, and fintech for AI and IoT in the digital eras. Papers investigate knowledge management in the digital era in relation to Higher Education Institutions. Another paper explores artificial intelligence opportunities and challenges to Higher Education Sector. Finally, several papers discuss blockchain technologies for sustainable development.

The works in Track 4 address governance, cybersecurity, and privacy in AI and IoT. Papers examined, in detail, financial statement and fintech fraud and measures taken to prevent unethical practices through detailed literature review. Additionally, the track explores corporate governance and sustainability of financial industry and the impact of women entrepreneurship on sustainable economies.

Finally, Track 5 addresses the Fifth Industrial Revolution and the Internet of ideas. Papers discuss novel ideas such as smart cities and technology readiness and technology acceptance model.

Nevertheless, in addition to the submissions received and papers accepted, the success of the EAMMIS 2021 conference on artificial intelligence systems and the Internet of Things in the digital era which was held on19–20 March 2021 in Istanbul, Turkey, depended on the efforts of various individuals, committees, and organizations, mainly the Bridges Foundation and Istanbul Medenyet University.

Abdalmuttaleb M. A. Musleh Al- Sartawi
Anjum Razzaque
Muhammad Mustafa Kamal

Acknowledgements

This book offers a discussion platform for researchers, scholars, students, and academics to provide insights and explore the opportunities and challenges of sustainable technologies including artificial intelligence (AI) and Internet of Things (IOT) in the digital era.

This book is inspired by **Prof. Abdulla Yusuf Al Hawaj**, Founding President of Ahlia University, the Kingdom of Bahrain, due to his immense support and efforts towards sustainable development and technology.

I would like to take this opportunity to express my gratitude to all the contributing authors for their valuable chapter submissions and their dedication to join the discussion on sustainable technologies, the Fourth Industrial Revolution and their impact on education, information systems, economy, and the financial industry. Moreover, I would like to thank the reviewers for reviewing the chapters and the authors who served as reviewers.

Special thanks go to the efforts and support of various individuals, committees, and organizations who contributed towards the success of the European, Asian, Middle Eastern, North African Conference on Management & Information Systems 'EAMMIS 2021' which was held on 19–20 March 2021 in Istanbul, Turkey, namely the Bridges Foundation and Istanbul Medenyet University.

Finally, I would like to extend my thanks to the book Series Editor **Prof. Dr. Janusz Kacprzyk** and the Springer editorial team including Editorial Director **Dr. Thomas Ditzinger**, and Senior Editorial Assistant **Mr. Holger Schaepe** for their support and acceptance of this valuable offering to students, practitioners, and professionals in the form of state-of-the-art knowledge.

Abdalmuttaleb M.A. Musleh Al-Sartawi
Anjum Razzaque
Muhammad Mustafa Kamal

Acknowledgments

Conference Committees

Conference Chairs

Anjum Razzaque	Ahlia University, Bahrain
Abdalmuttaleb Al-Sartawi	Ahlia University, Bahrain

Conference Co-chairs

Arafat Salih Aydiner	Istanbul Medeniyet University, Turkey
Muhammad M. Kamal	Coventry University, UK

Conference Executive Committee

Khaled Hussainey	University of Portsmouth, Portsmouth, UK
Michael Stankosky	George Washington University, USA
Vincent Ribiere	Bangkok University, Thailand
Birasnav Muthuraj	New York Institute of Technology, USA
Magdalena Karolak	Zayed University, United Arab Emirates
Mahdi H. Miraz	Xiamen University, Malaysia
Abdulnaser Nour	An-Najah National University, Nablus, Palestine
Nadia Mansour	University of Sousse, Tunisia, and Visiting Researcher, University of Salamanca, Spain
Arafat Salih Aydiner	Istanbul Medeniyet University, Turkey
Ahmet Faruk Aysan	Hamad Bin Khalifa University, Qatar

Programme Committee

Ahmed Alhayky	Alasala University, Kingdom of Saudi Arabia
Alyona Oboznaya	Dean Kyiv National University of Culture and Arts, Ukraine
Amitabh Anand	Université Côte d'Azur, France

Vincent Ribiere	Bangkok University, Thailand
Wafi Al-Karaghouli	Snr. L/C, Brunel University, UK
Yalçın Kirdar	Maltepe University, Istanbul, Turkey
Zorah Abu Kassim	Arab Open University, Bahrain
Pavel Laczko	L/C, Brunel University, UK
Abdul Rahim Sabouni	President and CEO, Emirates College of Technology, United Arab Emirates
Abdulnaser Nour	An-Najah National University: Nablus, Nablus- Palestine
Ahmed El-Masry	Coventry University, Coventry, UK
Ahmed Said	British University in Egypt, Cairo, Egypt
Ahmet Faruk Aysan	Hamad Bin Khalifa University, Qatar
Charles Wankel	St. John's University, New York, USA
Cornelia Ilie	Malmö University, Stockholm, Sweden
Khaled Dahawy	The American University in Cairo, Egypt
Khaled Hussainey	University of Portsmouth, UK
Maria Manuela Cruz-Cunha	Polytechnic Institute of Cávado and Ave, Portugal
Muhteşem Baran	İstanbul Üniversitesi, Istanbul, Turkey
Neelambar Hatti	Lund University, Sweden
Qasim Zureiga	Sulaiman AlRajhi University, Saudi Arabia
Rong Zhang	Nishinippon Institute of Technology, Japan
Sergii Kavun	Interregional Academy of Personnel Management, Kyiv, Ukraine
Talal Al-Hayale	University of Windsor, Canada
Tord Häversjö	Nordic International Management Institute, Stockholm County, Sweden
Waldemar Pfoertsch	International Bussiness Mannheim Business School, Germany
Willy Legrand	IUBH International University of Applied Sciences, Germany
Zahir Irani	University of Bradford, UK
Zenonas Turskis	Professor at Vilnius Gediminas Technical University, Lithuania

EAMMIS 2021 Conference Organization Committee

Sercan Pekel (Research Assistant)	Istanbul Medeniyet University, Turkey
Zehra Ozsurunc (Research Assistant)	Istanbul Medeniyet University, Turkey
Ismail Canoz (Research Assistant)	Istanbul Medeniyet University, Turkey

Contents

Fifth Industrial Revolution: The Internet of Ideas

Innovation in the Digital Era

Model-Based Validation of Business Processes: Case Study of Logistic Sector

Firna Muninggar Kistianti(✉), Muharman Lubis, and Asti Amalian Nur Fajrillah

Telkom University, Bandung, Indonesia
firnamuninggar@student.telkomuniversity.ac.id, {muharmanlubis, astiamalia}@telkomuniversity.ac.id

Abstract. A company or organization has the ultimate goal to achieve their target that it requires a business strategy to manage the process in a company running properly. Business process is the knowledge that becomes a key of asset for the organization that must be maintained by reason of the accuracy in designing business processes to maintain the smooth running of an activity synergistically. Workflow is a representation of a business process consist of a set of activities that are structured to be related to each other so as to produce useful result for the company. But still business processes cannot be said to be correct and allows failure in providing the functions modeled for it. Therefore, the business process that is modeled in graphic notation needs semantic validation to prove the suitability and accuracy of the model according to the business strategy. The purpose of this study is to introduce validation for business processes by modeling business processes to the Petri net supported by analytical tool called WoPeD. which can show defined workflow have consistently meets the specified requirements. According to the theme of the conference, Petri net is a method like artificial intelligence in helping to solve problems that occur in business processes to be able to develop correspond to business needs so that it supports the logistics sector in order to compete in the midst of further business competition.

Keywords: Business process · Validation · Petri net

1 Introduction

In Order to achieve the company business goal, strong coherence between business and IT has become an important factor of competition on all markets and almost all sectors entering the industrial revolution 4.0 [32]. To develop and advance the field of business process knowledge management is artificial intelligent, which is neglected by many practitioner and knowledge management theorists [33]. Artificial Intelligence (AI) technology can be a very effective means of implementing a smart design environment that provides a wide range of information for fast and intelligent decision making throughout the design and planning of business processes to be developed to assist designers [34]. Many organizations apply modeling process as an important point of documenting, analyzing, and redesigning those business processes and supporting them in the appropriate

A. M. A. Musleh Al-Sartawi et al. (Eds.): EAMMIS 2021, LNNS 239, pp. 3–12, 2021.
https://doi.org/10.1007/978-3-030-77246-8_1

information systems [1]. An effective modeling process does not require the use of a lot of elements in modeling but to make use of elements with an unambiguous meaning and help modelers to deliver messages [2]. The use of modeling a business process requires that activities be carried out systematically to increase productivity, efficiency and consistency [3]. Poorly designed business processes can result in delays in response limits, worsening service levels, and imbalance in resource utilization, angry customers, and repeated logging [5, 6]. Business process modeling language such as BPMN do not have a feature to determine whether the business process model described is optimal and error-free, so additional analysis is needed for the business process mode [35].

Therefore, in modeled business process modeling is an activity that is prone to errors, the above case requires a validation technique: checking for completeness of semantics to make sure a workflow processes behave as well as predictable in all scenarios, to ensure that the business process model makes sense is required to perform validation [4, 7]. There is a modeling language that can run validation technique called Petri net with offered by business process modeling tools [23]. Petri net shows that the transition can take place smoothly where the business process model can be simulated and validated before the implementation stage is visually described with the workflow [8]. This study aims to try to respond the drivers of validation analysis of proper business process model by using Petri net because the modeling language to describe the business process can help to model system behavior in terms of workflow and it is supported by modeling tools to automate model validation. So that the use if AI based in Petri net knowledge helps the logistics sector increase efficiency to meet the needs of business process modeling, planning and simulation before being implemented for the company.

2 Literature Review

2.1 Business Process

A business process consists of a collection of activities as well-defines tasks and order for their execution [37]. Process modeling becomes an instrument to overcome the complexity of planning and process control so that process model design is often become problematic which has a spontaneous influence on the economic utilization of the project associated with the underlying processes [9]. It is an effective method for designing, analyzing, simulating, and improving the current process [10] thus affecting the correctness of the resulting process model. There are two levels of correctness [29] first, the process model must respect the syntax/modeling rules defined by the process modeling language/standards used. Second, the model must be semantically correct/accurate from a business point of view [30].

Modeling language like Petri net, BPMN, UML are essentially flow chart, they represent a business process as a specific type of graph with nodes that capture activity and arcs that defining the control flow [36]. In modeling a business process which can also be called a modeling process it is necessary to use a workflow that represents a business process to reveal the composition of each tasks that is compiled together to explain the entire business process from various perspectives [4]. Workflow nets are Petri nets with two special sets of nodes, called initial and final nodes [42]. There are many anomalies that can lead to incorrect business process represented by a workflow, such as deadlocks, lack of synchronization and infinite cycles [17]. In this case soundness properties described in [17] as criteria that must be met when handling workflow processes are very important to represent the final state of a system that demonstrates compliance with the requirements can be achieved. For process model in [17] and [1] behavior semantics and methods for analyzing soundness using Petri net.

2.2 Validation Model

Validation can be used to test semantic completeness to make sure a designed workflow behaves according to requirements in all scenarios [4]. By doing validation it can detect that the modeled business process model allows it to produce semantic errors or carelessness that can interfere with interpretation by the modeler and require model refinement [19]. Therefore, methods and techniques are needed to help support this increasing modern business process modeling to analyze business processes that have problems with the complexity of the design and work process of business processes [29] (Fig. 1).

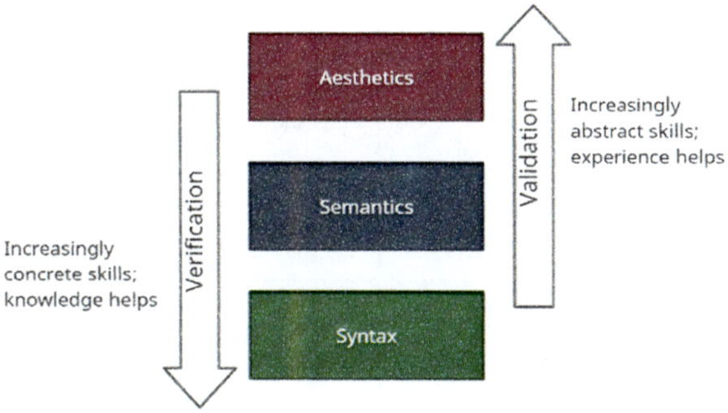

Fig. 1 Quality and V&V of models (Unhelkar 2005)

There are verification and validation techniques that have been developed. V&V is a quality technique intended to prevent and detect errors, inconsistencies, and incompleteness consists of activities requiring re-checking to make sure that the model is accurate, the consistency of the model in the business process being modeled is validated by creating a series of examination lists related to semantics checks [11]. Validation of a good business process model is very important for business process design, because it can evaluate the process designed with the aim of radically revising it, in order to accommodate the improvement needs or new organizational goals [39].

2.3 Petri Net

Petri nets facilitate workflow-based specifications, tools for system modeling on conceptual stage, enabling a seamless transition from the concept state to implementation testing allowing the analysis and validation process to perform researched simulations before proceeding to the business process stage to be applied [18]. Petri nets are a suitable tool for depicting true concurrency and a Petri net based approach has been presented mainly focusing on the health and accuracy of workflows [40]. There are two types of nodes in the Petri net, called place and transition, which have initial marking as initial state [20]. Directional marking between the transition and places called arcs [15]. Generally, places are expressed with circular shapes and transition expressed with rectangles, Petri net depict graphs that can divide two sets that can be reconnected, which means that using an arc into a net the marking is done by connecting one place to the transition or one transition connected to the place, the arc is not allowed to connect with fellow places or fellow transition [31]. Petri nets can be understood easily by a graphical representation of the system being modeled [41] (Fig. 2).

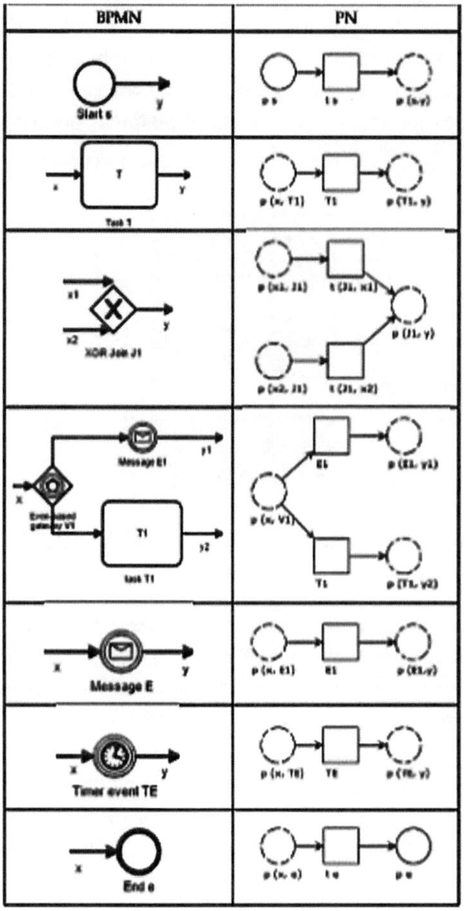

Fig. 2 Business process element converted to Petri nets (Ramautar 2017)

Processing of activities that is carried out by transition on the Petri net is indicated by the input place as marking the start of the task and the output marking the end of the task [26]. A system with dynamic behavior can be characterized by means of a token, which is illustrated as a small black circle in places [27]. Petri net analysis is classified as behavioral and structural analysis, the behavioral properties shows that we can analyze two properties, boundedness and liveness [21]. Typically, abstractions of the workflows used to check the soundness properties which ensure that behavioral process modelling in the absence of live-locks, deadlocks, and other anomalies [38].

3 Research Methodology

In finding problems for testing the validation of business process models, Petri net supporting tools are needed that help to represent and analyze process performance in detail. User can also use Petri net tools as a graphic editor and code generator [12]. A validation abstract description and a specification of model process can be described by doing a Petri net analysis technique [22, 27]. In process modeling guide helps to achieve good quality which has simple rules that define the model properties desired [23] (Table 1).

Table 1 List of process model guideline for business process

Criteria	Description	References
Apply symmetric modeling	Model should as structure as possible. Symmetric structures increase understandability of business process model	[6, 16, 25]
Make use of the model in using a few elements as possible	Model should minimize concurrency levels which means reducing the role of parallel gateway operations and sub-process	[6, 16]
Activity description	Model in describing each activity must provide brief clarity	[6]
Explicit usage of beginning and ending event	The use of start and end events is needed to represent the various state that start and finish the modeled process	[6, 16]
Explicit usage of gateways	Model must divide or merge sequence flows using gateway. Gateway act as either join nodes or split node and use the same type of gateway	[6, 16]
Kind of relation	Model must be able to identify the types of relationships between activities in the business process model	[6]
Labeling event	All events must have a label that represents the state of the process (start and end event, message event, or parallel gateway event)	[6, 25]
Pattern	Models should use modeling patterns according to notation and workflow	[6, 22]

Process model guideline identifies desirable property that can be used as a guide when converting a process model into a model that is equivalent to a behavior, but easier to understand [16]. After the business process model is modeled, it requires to simulate the model before the implementation stage [26]. In fact, the main objective of qualitative analysis which concentrates on property is to prove that the model is valid [27] (Table 2).

Table 2 List of guideline for sound Petri net

Criteria/Property	Description	References
Reachability	Property that represent the active token situation can reach the destination of the sequence of activities or all nodes in the business process model to reach the final state	[6, 20, 22]
Liveness	Property states certain situations can always be reached and follow the appropriate route through workflow-net	[6, 20]
Boundedness	Any place where the number of token must not exceed the number for each mark reached	[15, 20, 21]

In this study, Petri net tools were used to simulate business process models and check the model using WoPeD. WoPeD is used as a simulation tool [13]. WoPeD (Workflow Petri Net Designer) is a simple open source software for modeling, simulating and analysis of workflow processes and decrypting resources using workflow net, part of the Petri net class [14]. The semantical analysis consists of a wizard mode and an expert mode [15].

4 Discussion

We used two different business process model which are the case studies of this research in the logistics sector with the errors found. The first step before validating a business process is to transform business process elements into a workflow net that is modeled according to the Petri net element. Furthermore, to find out the workflow net is said to be sound, the business process must go through the WoPeD analysis tool by doing testing that that provides soundness criteria through the Petri net analysis method. The aim is to prove a business process that has been modeled is validated to make sense. Business process in its transformation to Petri net show every single place and one transition describes an activity or task that occurs in a business process in carrying out the process.

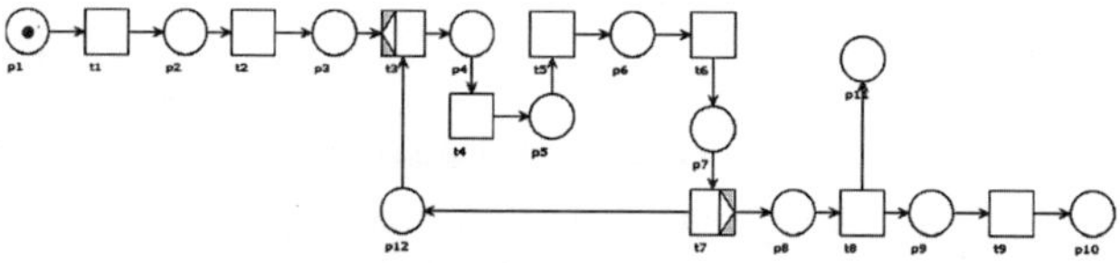

Fig. 3 Petri net model with unsatisfied workflow-net property (Developed by authors)

On the Fig. 3 Petri net that represent the business process model have initial marking at place *p1* as a node which marks the start of a single token being placed to simulate until it reaches the end point. There is *t7* as an activity breaker transition which is represented by an XOR-split operator which shows that the model carries out the selected traffic lane activity and has *t3* as a transition for the selected traffic lane of the operator, namely XOR-join, this condition can be said that the operator usage rule fulfilled. Based on the model above, the error occurs because it does not meet the requirements as a workflow-net. This model causes confusion because two sink place are found, that happen the token split into two parts at place *p10* and *p11* and end up in those two places. The process model can't be said to be well structures which causes the workflow not to work properly. So there needs to be an improvement by using the concept of process collaboration by adding a transition with activity have event label like message after transition *t8* which can make workflow-net can be modeled better.

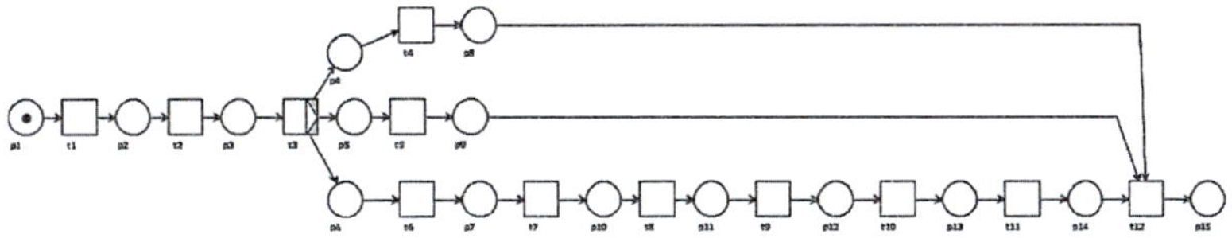

Fig. 4 Petri net model with a liveness error (Developed by authors)

Petri net model on Fig. 4 shows a successful case workflow-net starting by placing the token on *p1* as the source place and the ending place of token on *p15* as the sink place. There is a *t3* transition as an activity breaker called operator XOR-split. When it comes to the *t3* parallel gateway that release more than one arch, the token will split into three path to the place that consist *t4, t5* and *t6* which will be selected as token traffic acting to reach the destination which will be recombined to *t12*. This model will be simulated by WoPeD analysis tools to find out whether the model depicted on the Petri net meets the soundness criteria or an unfulfilled property and shows an error in the model.

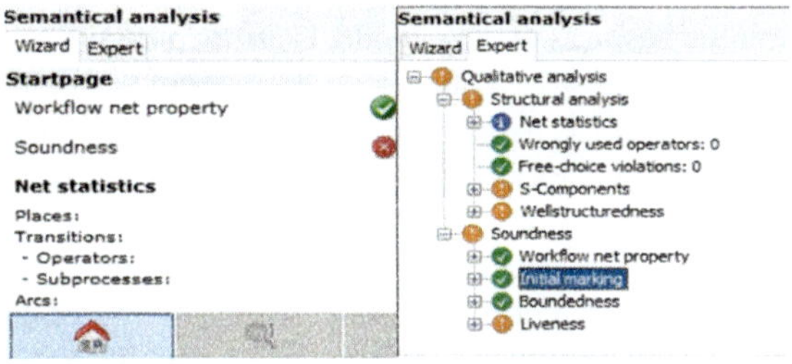

Fig. 5 Simulation test result (Interpretation of Fig. 3. (Developed by authors)

The semantical analysis simulation results are shown in Fig. 5 with WoPeD tools to show modeling errors. These results conclude the modeling of business processes is not sound because there is a liveness error in the soundness property causes the model is not clearly structured and all nodes in place are not covered by the active token. This happens because there is a dead transition on tokens that have passed *t3* that cannot be active after being in place*p8, p9,* and *p14* caused by the use of XOR-split operator without XOR-join as its partner. Thus, as a result its non-live because the workflow cannot be reached by all tokens which causes an invalid business process model.

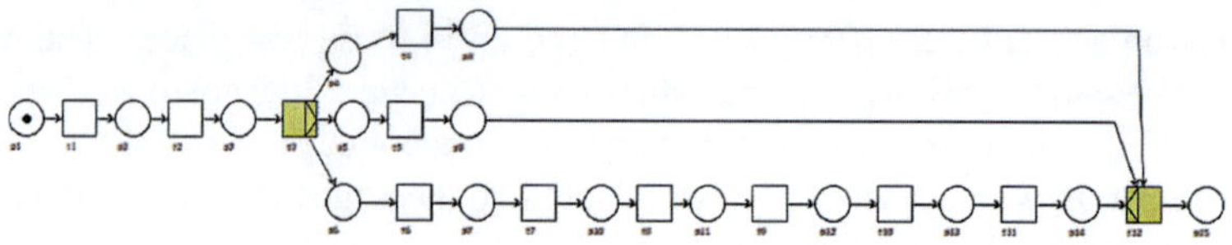

Fig. 6 Soundness Petri net model (Developed by authors)

The conclusion that the Petri net model on Fig. 6 shows there is a structured and obvious workflow so that the soundness property is fulfilled. Thus, it is necessary to add an XOR-join operator in transition *t12* as a pair of XOR-split operators in transition *t3*. Matching parallel gateway operators are paired with same color mark when the model is analyzed using WoPeD. Analysis of the petri net model can be said to be correct if a workflow net conditions are structured and all places are passed by the token. Validation analysis can be said that the model is bounded, if the process finds the ending place when there is a bounded on the place that the token passes and it is necessary to know on a workflow net, the token marking on the source place must be exactly one token. Validation analysis can say that the model is live if there are no deadlocks because all transitions on the workflow-net can be passed by tokens.

5 Conclusion

In this paper, Petri net provides good inside knowledge when checking errors to validate whether business process model is in appropriate with the scenario. Petri net provides property analysis that can help business expert in validating the business processes being modeled to find out if the business process model is error-free, Petri net uses workflow-net as a Petri net class to demonstrate solutions to business process model errors. There are results from semantic analysis to validate the business process model by workflow-net, it proves that the case of deadlocks in the workflow of the business process modeling is usually caused by an incorrect use of activity-breaking operators, namely XOR split-join nor AND split-join and ensure that all nodes must pass the token while token is firing. When a modeled business process meets the criteria of liveness and boundedness, the soundness property can be fulfilled which indicates that business process is valid. For the future research suggestions, it is possible to practice quality techniques from the validation model as a whole so that it can help improve business processes to better support the company's business goals.

References

1. Mendling, J., Leopold, H., Pittke, F.: 25 challenges of semantic process modeling. Int. J. Inf. Syst. Softw. Eng. Big Companies (IJISEBC) **1**(1), 78–94 (2014)

2. Haseeb, J., Ahmad, N., Malik, S., Anjum, A.: Application of formal methods to modelling and analysis aspects of business process reengineering. Business Process Manage. J. **26**(2), 548–569 (2010)
3. Mutarraf, U., Barkaoui, K., Li, Z., Wu, N., Qu, T.: Transformation of business process model and notation models onto petri nets and their analysis. Adv. Mech. Eng. **10**(1), 1–21 (2018)
4. Pranevičius, H., Misevičienė, R.: Verification of business process workflows. Technol. Econ. Dev. Economy **18**(4), 623–635 (2012)
5. van der Aalst, W.: Business process simulation survival guide. In: van der Aalst, W. (ed.) Handbook on Business Process Management, vol. 1, pp. 337–370. W.M.P, Netherlands (2014)
6. Re, B., Polini, A., Gnesi, S., Ferrari, A., Fornari, F., Spagnolo, G., Corradini, F.: Quality assessment strategies for BP models. FP7 ICT STREP Project (2015)
7. Weber, I., Hoffmann, J., Mendling, J.: Semantic business process validation. In: Conference Paper (2008)
8. Adam, N.R., Atluri, V., Huang, W.K.: Modeling and analysis of workflows using petri nets. J. Intell. Inf. Syst. **10**, 131–158 (1988)
9. Becker, J., Rosemann, M., Von Uthmann, C.: Guidelines of business process modeling. In: van der Aalst, W. et al. (eds.) Business Process Management LNCS. Springer, Heidelberg, vol. 1806, pp. 30–49 (2000)
10. Ramadan, M., Elmongui, H.G., Hassan, R.: BPMN formalization using coloured petri nets. In: 2nd Annual International Conference on Software Engineering & Applications (2011)
11. Unhelkar, B.: Verification and Validation for Quality of UML 2.0 Models. A Wiley-Interscience Publication, Hoboken (2005)
12. Jie, T.W., Ameedeen, M.A.: A survey of petri net tools. In: Lecture Notes in Electrical Engineering, vol. 315, no. 8, pp. 537–551 (2015)
13. Wang, Y.Y., Lai, A.-F., Shen, R.-K., Yang, C.Y., Shen, V.R., Chu, Y.H.: Modeling and verification of an intelligent tutoring system based on petri net theory. Math. Biosci. Eng. **16**(5), 4947–4975 (2019)
14. Ge, X., Paige , R.F., McDermid, J.A.: Model-based failure analysis of business process. In: 13th International Conference on Enterprise Information Systems, pp. 388–391. SCITEPRESS, UK (2011)
15. Ramautar, V.: Analysis of BPMN collaboration diagrams using petri nets. Leiden Institute of Advanced Computer Science (LIACS) (2017)
16. Mendling, J., Reijers, H., van der Aalst, W.: Seven process modeling guidelines (7PMG) . Inf. Softw. Technol. **52**(2), 127–136 (2010)
17. Tao, X., Liu, G., Yang, B., Yan, C., Jiang, C.: Workflow nets with tables and their soundness. IEEE Trans. Ind. Inf. **16**(3), 1–1 (2019)
18. Adam, N.R., Atluri, V., Huang, W.K.: Modeling and analysis of workflow using petri nets. J. Intell. Inf. Syst. **10**, 131–158 (1998)
19. Chapurlat, V., Braesch, C.: Verification, validation, qualification and certification of enterprise models: statements and opportunities. Comput. Ind. **59**(7), 711–721 (2008)
20. Murata, T.: Petri nets: properties, analysis and applkations. Proc. IEEE **77**(4), 541–580 (1989)
21. Shen, V., Gibran: Petri net analysis on path protection of a multipath system witk bounded buffers. In: IEEE International Conference on Consumer Electronics-Taiwan, pp. 1–2. IEEE, Taiwan (2014)
22. van der Aalst, W., Stahl, C.: Modeling Business Processes A Petri Net-Oriented Approach, 1st edn. The MIT Press, Cambrige (2011)
23. Avila, D.T., Santos, R.I., Mendling, J., Thom, L.H.: A systematic literature review of process modeling guidelines and their empirical support. Business Process Manage. J. (2020)
24. Lohmann, N., Verbeek, E., Dijkman, R.: Petri net transformations for business processes – a survey. In: Jensen, K., van der Aalst, W. (eds.) Lecture Notes in Computer Science. Springer, Heidelberg, pp. 46–63 (2009)

25. Groner, G., Boskovic, M., Parreiras, F.S., Gasevic, D.: Modeling and validation of business process families. Inf. Syst. **38**(5), 709–726 (2013)
26. Ou-Yang, C., Lin, Y.D.: BPMN-based business process model feasibility analysis: a petri net approach. Int. J. Prod. Res. **46**(14), 3763–3781 (2008)
27. Salimifard, K., Wright, M.: Petri net-based modeling of workflow system: an overview. Eur. J. Oper. Res. **134**(3), 664–676 (2000)
28. Bollen, P.: BPMN: a meta model for the happy path. METEOR, Maastricht, no. 003 (2010)
29. van der Aalst, W., Hee, V.: Business process redesign: a petri net-based approach. Comput. Ind. **29**, 15–26 (1996)
30. Radulian, R.: Rethinking BPMN: how to build good process models using BPMN. Cprime (2020)
31. Dijkman, R.M., Dumas, M., Ouyang, C.: Semantics and analysis of business process models in BPMN. Inf. Softw. Technol. **50**(12), 1281–1294 (2008)
32. Paschek, D., Luminosu, C.T., Draghici, A.: Automated business process management – in times of digital transformation using machine learning or artificial intelligence. MATEC Web Conf. **121**(04007), 1–8 (2017)
33. Liebowitz, J.: Knowledge management and its link to artificial intelligence. Expert Syst. Appl. **20**, 1–6 (2001)
34. Huang, G.Y.: Analysis of artificial intelligence based petri net approach to intelligent integration of design. In: Proceedings of the Fifth International Conference on Machine Learning and Cybernetics, pp. 13–16 (2006)
35. Anggrainingsih, R., Yohanes, S.P., Salamah, U.: Analisis Dan Verifikasi Workflow Mengggunakan Petri (Studi kasus; Proses Bisnis di Universitas Sebelas Maret). SEMINAR NASIONAL TEKNOLOGI INFORMASI & KOMUNIKASI TERAPAN (2014)
36. Ottensooser, A., Fekete, A., Reijers, H.A., Mendling, J., Menictas, C.: Making sense of business process descriptions: an experimental comparison of graphical and textual notations. J. Syst. Softw. **85**(3), 596–606 (2012)
37. Liu, C., Zeng, Q., Duan, H., Wang, L., Tan, J., Ren, C., Yu, W.: Petri net based data-flow error detection and correction strategy for business processes. IEEE Access **8**, 43265–43276 (2020)
38. van der Aalst, W., van Hee, K., Hofstede, A., Sidorova, N., Verbeek, H., Wynn, M.: Soundness of workflow nets: classification, decidability, and analysis. Formal Aspects Comput. **23**(3), 333–363 (2010)
39. Clempner, J.: Verifying soundness of business processes: a decision process Petri nets approach. Expert Syst. Appl. **41**, 5030–5040 (2014)
40. Yu, W., Ding, Z., Wang, X., Crossley, R.D.: Petri net-based methods for analyzing structural security in e-commerce business processes. Future Gener. Comput. Syst. **109**, 611–620 (2018)
41. Yu, W.Y., Yan, C.G., Ding, Z.J., Jiang, C.J., Zhou, M.C.: Modeling and validating e-commerce business process based on petri nets. IEEE Trans. Syst. Man Cybernet.: Syst. **44**(3), 327–341 (2014)
42. Hee, K., Sidorova, N., van der Werf, J.E.: Business process modeling using petri nets. In: Transactions on Petri Nets and Other Models of Concurrency, vol. 7, pp. 116–161 (2013)

COVID-19 Lung Image Classification Based on Logistic Regression and Support Vector Machine

Nazish(✉), Syed Irfan Ullah, Abdus Salam, Wajid Ullah, and Muhammad Imad

Department of Computing and Technology, Abasyn University, Peshawar, Pakistan
{syed.irfanullah,dr.salam,arbabwajid.ullah}@abasyn.edu.pk

Abstract. COVID-19 is an infectious disease caused by SARS-Cov2 that has spread rapidly worldwide. According to the World Health Organization (WHO), the total cases of 4374783839 are reported from different countries. In this consequence, it is necessary to diagnose automatically COVID-19, which helps in prevention during spreading among people. In this study, we have used machine learning techniques to diagnose and classify the COVID-19 and normal patients from chest X-ray images using a machine learning technique. The proposed system involves pre-processing, feature extraction, and classification. In the pre-processing, the image is to enhance and improve the contrast. In the feature extraction, the Histogram of Oriented Gradients has been applied to extract the image's feature. Finally, in classification two different machine learning techniques (Support Vector Machine and Logistic Regression) have been used to classify COVID-19 and normal patients. The result analysis shows that the SVM achieved the highest accuracy of 96% and provide a better result than logistic regression (92% accuracy).

Keywords: COVID-19 · Machine Learning · Classification · SVM · Logistic Regression

1 Introduction

The coronavirus or Novel COVID-19 firstly appears in the Chinese province Hubei in Dec 2019, which is rapidly spread further in multiple countries around the world [1]. The virus that caused COVID-19 disease is called Severe Acute Respiratory Syndrome CoronaVirus-2 (SARS-COV-2) [2]. Men are more infected than women, and there is no death ratio in children between 0–9 years [3]. The COVID-19 virus is rapidly spread from person to person through respiratory transmission. The symptoms of Corona-virus take 14 days to appear. Symptoms of respiratory are fever, cough, and dyspnea can lead to an extreme acute respiratory syndrome, influenza, septic shock, multi-organ failure, and death. But their fatality rate is 2% [4]. The virus is caused by cold such as Middle East Respiratory Syndrome (MERS-COV) and infected from dromedary to human. In contrast, the second virus is Severe Acute Respiratory Syndrome (SARS-COV), infected from musk cats to humans, but COVID-19 is recognized as a virus [4, 5].

A. M. A. Musleh Al-Sartawi et al. (Eds.): EAMMIS 2021, LNNS 239, pp. 13–23, 2021.
https://doi.org/10.1007/978-3-030-77246-8_2

The X-Ray in the chest became the first medical examination used to help and diagnosed the COVID-19 diseases. Chest X-Ray has a very significant role in the medical sector to address the critical health care crisis. The doctor uses X-Rays of the chest to monitor care for conditions such as pneumonia, lung cancer, fluid, or accumulation of air around the lungs and other medical conditions. According to WHO, 'CO' stands for Corona, 'VI' stands for a Virus, and 'D' for the disease [6]. COVID-19 spreads between people through mouth and nose secretions via direct, indirect, or near contact with infected people [7].

The COVID-19 virus infects all ages of people; however, two groups of people at high risk for COVID-19. The one group consists of older people over 60 ages and the other group, which are underlying medical conditions with different diseases [8].

Machine learning technology has been used to improve the performance accuracy for predicting COVID-19 infection and non-infection diseases. Machine learning show impressive performance in medical image processing in the different field. ML supports various pandemic outbreaks and communicable and non-communicable diseases such as SARS, EBOLA, HIV, COVID-19, cancer, diabetes, and heart diseases. Several machine learning techniques have been used for the classification of COVID-19 based on three statistical features. The texture of lung images is classified between malignant and benign lung images using SVM [27].

WHO has provided advice to all communities and community partners to ensure that they are safe from COVID-19 without alienation, stigmatization, or access to services and social care. According to WHO, the global situations of the Coronavirus cases worldwide, as of 9 July 2020, the total confirmed cases are 11,874,226, and deaths are 545,481. The highest cases were reported in America; the total confirmed cases are 6125802, and death cases are 272606. The lowest cases were reported in the Western Pacific, where confirmed cases are 234815, and the death cases are 7515 [9]. The other region cases, according to WHO are presented in Fig. 1.

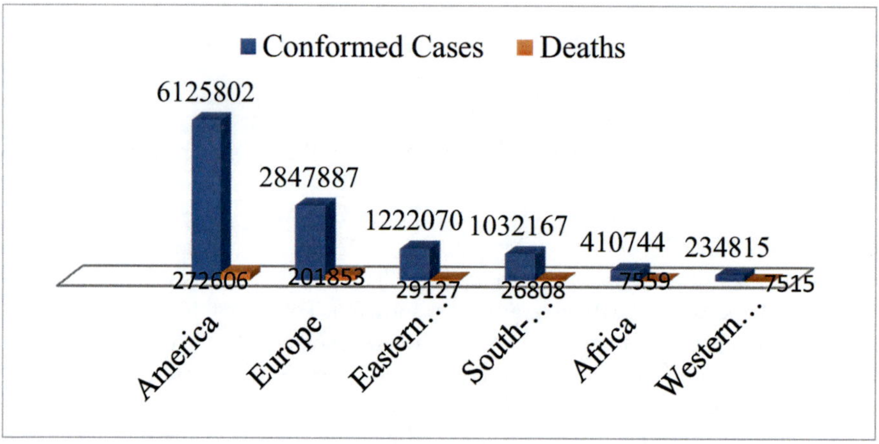

Fig. 1. Region-wise COVID19 infected areas, as of 9 July 2020

According to WHO Situation by Country, as of 9 July 2020. The highest cases were reported in the USA, the total infected patients from a COVID-19 are 2973695, and the death cases are 130893 reported. Brazil, India, Russia also have high highly infected cases were reported from COVID-19. The lowest cases infected by COVID-19 were reported in Pakistan (total confirmed cases 240848), and death cases are 4983) [9] (Fig. 2).

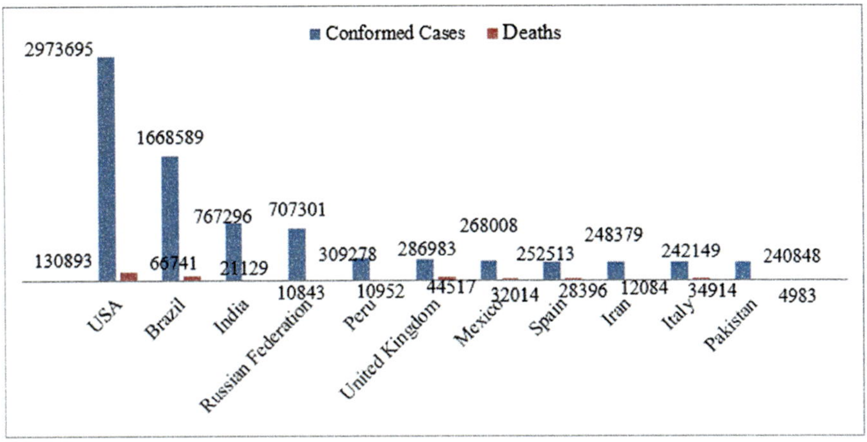

Fig. 2. Country-wise infected area, as of 9 July 2020

The COVID-19 pandemic impacts all aspects related to society on their mental and physical health. The machine learning model will help to identify, control, track and manage the disease, prediction of outbreaks during data analysis and decision-making process [26].

The paper is organized as follows: A literature review for COVID-19 is expressed in detail in Sect. 2. Section 3 discusses the flowchart and the dataset descriptions. Section 4 describes the experimental setup on X-ray datasets using machine learning techniques. In Sect. 5, we discuss our findings. Finally, in Sect. 6, the conclusion and future work are summarized.

2 Literature Review

Narinder Singh Punn proposed machine learning and deep learning algorithms (SVR, PR, DNN, LSTM) to analyze the epidemic of COVID-19. The aim of the article to understand exponential behaviour in the future of COVID-19 by utilizing using real-time information. The Johns Hopkins university dataset has been used, consisting of daily case reports and daily time series summary for death, recovery, and confirmed cases of COVID-19. The result shows that PR (polynomial regression) produced a minimum root mean square error among other approaches [10].

Akib Mohi Ud Din Khanday proposed a technique to classified textual clinical reports using an ensemble machine learning technique. The feature (frequency/inverse document frequency, a bag of words, and report length) was used and provided to machine learning classifiers. The Logistic Regression and multinomial Naïve Bayes provide better results than other classification algorithms [11].

Milind Yadav presents a novel technique for detecting coronavirus based on chest X-Ray using machine learning techniques such as Support Vector Regression, Simple Linear Regression and Polynomial Regression. The result analysis shows that SVR provides better performance among other regression techniques [12].

Prabira Kumar Sethy presented a deep-running technique that is based on X-Ray images. The supported vector machine-classified the X-Ray imaged from deep features. The proposed classification model of the resnet50 plus SVM achieved an accuracy of 95%. The data set is downloaded from the online repository of GitHub, Kaggle site [13].

Mucahid Barstugan has been presented a technique for early phase detection of COVID-19 from the CT images named by WHO. The detection of COVID-19 from four different datasets with different patch sizes from 150 chest X-Ray images. The feature is extracted from the Grey Level Co-occurrence Matrix, Local Directional Pattern, Grey Level Run Length Matrix, Grey level Size Zone Matrix and Discrete Wavelet Transforms algorithms for classification. A support vector machine has been used to classify the COVID-19 images from the extracted features during the classification process. 2-fold, 5-fold, and 10-fold cross-validations methods have been used during the classification process. The result analysis shows that the GLSZM provides better performance with a 10-fold cross-validation method [14].

Chuansheng Zheng develops a deep learning model to detect COVID-19 on the chest CT dataset. Each patient's lung region has been segmented based on a pre-trained model (UNet) and fed into a deep neural network to predict the COVID-19 infection.499 volumes were collected for training, and 131 volumes were used for testing purposes. The deep learning technique obtained 95% ROC accuracy and 97% PR accuracy. While the sensitivity, specificity is 90% and 91%, respectively. When the probability threshold value kept 0.5 for COVID-19 positive and negative, the algorithm obtained 90% of accuracy, the positive predictive value is 84%, and the negative predictive value is 98% [15].

Davide Ferrari discussed the severe respiratory failure, lack of ventilation in the hospital during COVID-19. The dataset was collected from the admitted patients in the hospital of COVID-19, including medical history, clinical data from the electronic chart, demographics. The dataset has been used to train and test the model by using a machine learning framework. The performance accuracy of the presented techniques is 84% without failing the FN rate [16].

Parnian Afshar presented an alternative model for the identification of COVID-19 infected patients. The CNN based model has been used for the diagnosis of COVID-19 from X-Ray images. The proposed model achieved 95% accuracy, 90% sensitivity, 95.8% specificity [17].

Wei Tse Li presented techniques to diagnose COVID-19 based on symptoms and expected test results using machine learning techniques, also identify the correlation between the clinical variable, cluster COVID-19 subtypes. The model classifies between COVID-19 and influenza patients based on a clinical variable. The result shows that the XGBoost model achieved 92% accuracy during COVID-19 and influenza patients [18].

Imran Razzak [19] develops a pre-trained model to differentiate it from viral pneumonia, bacterial pneumonia and improve the diagnostic performance using the transfer learning technique. CT imaging is efficiently used to diagnose COVID-19 infected patients. The performance of different CNN architecture has been compared. The result analysis using a 10-fold cross-validation method better results in accuracy.

Md. Martuza Ahamad presenting a technique for the prediction of COVID-19 disease from the feature examined of individuals concerned such as age, gender, fever observation, travel history, and clinical detail such as cough and lung infection. A different technique of machine learning has been applied to collect data, where the XGBoost algorithm provides the highest accuracy of (>85) during prediction and correctly indicates COVID-19 from all the age groups. While the statistical analysis shows that the prediction of fever symptoms is 41.1%, cough is 30.3%, lung infection is 13.1%, and the runny nose 8.43%. At the same time, 54.4% didn't have any symptoms for the diagnosis of COVID19 [20].

Gergo Pinter proposes a hybrid machine learning technique for the prediction of COVID19 using the Hungary dataset. The adaptive network-based fuzzy inference system and multi-layered perceptron imperialist competitive algorithm has been used in a hybrid method to predict the infected persons and their mortality rate. The validation is completed in 9 days, which provides the model accuracy with promising results [21].

R. Sujth suggests a method to predict the spreading COVID19 from the Kaggle dataset. The linear regression, multilayer perceptron, and vector regression method have been used to identify the affected rate in India. The VR model shows that the confirmed death and recovered cases are 17289.29,492.20,1241.68, confirmed by the forecast [22].

COVID19 is one of the main challenging problems which face all over worldwide. Amir Ahmad presents a review of the prediction of COVID19 using machine learning to improve performance accuracy. The four research themes identified are traditional machine learning regression, deep learning regression, network analysis, social media, and search query database method [23].

WHO identified the coronavirus is a global pandemic issue. Muhammad Imad proposes different machine learning techniques for the classification of COVID-19 from chest X-Ray datasets. Three main steps have been used, which are pre-processing, feature extraction, and classification. In the pre-processing, the noise has been removing and improves the contrast by using a median and average filter. In the feature extraction, the HOG algorithm has been used to extract the discriminant feature from the images. In classification, four different machine learning classifiers (SVM, K-Nearest Neighbors, Random Forest, Naïve Bayes algorithm, and Decision Tree) has been applied. The result shows that the SVM provided better results than other classifiers [24].

3 Research Methodology

In this section, we discussed in detail related to the flow diagram and the dataset description.

3.1 Dataset Description

The dataset is gathered from the Kaggle repository via online resources with Covid19 infected patients and Normal patients to train and test our proposed model. 70% of the dataset is used for the training phase, and 30% of the dataset is used for the testing phase [25]. Figure 3 present the chest X-ray images of Normal and COVID-19 patients are given respectively.

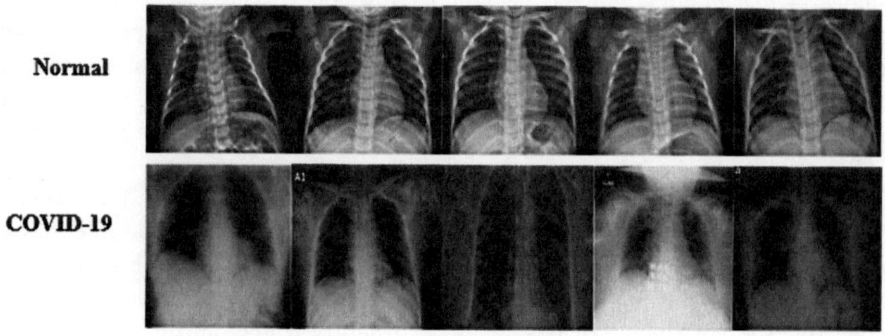

Fig. 3. Example of COVID-19 and Normal Images

3.2 Methods

This section provides the details of the proposed methodology for COVID-19 classification from chest X-ray Images. The model has been comprised of pre-processing, feature extraction, and classification. The detailed overview of the flow diagram through a different component of the method is discussed in Fig. 4.

3.2.1 Pre-processing:

In image pre-processing, the image data has been pre-processed to improve the contrast and the quality of data in the image and remove the unwanted distortion and resize the image and thereby improve the prediction results. De-noising can be done through filtering techniques. The median filter is also known as a non-linear filter, used to eliminate the outlier, decrease the sharpness, and widely used to remove the salt and pepper noise. An average filter is a simple technique used for smoothing the image and reducing the amount of intensity variation between the pixels.

3.2.2 Feature Extraction

Histogram oriented gradient is a feature descriptor technique that is used to extract the feature from the datasets. The gradient directions of the target image are computed by dividing them into small regions (cells) to implement these descriptors. The algorithm is divided into multiple stages while the gradient values are calculated in the first stage, and then a cell histogram is created. The HOG descriptor concatenates the components of

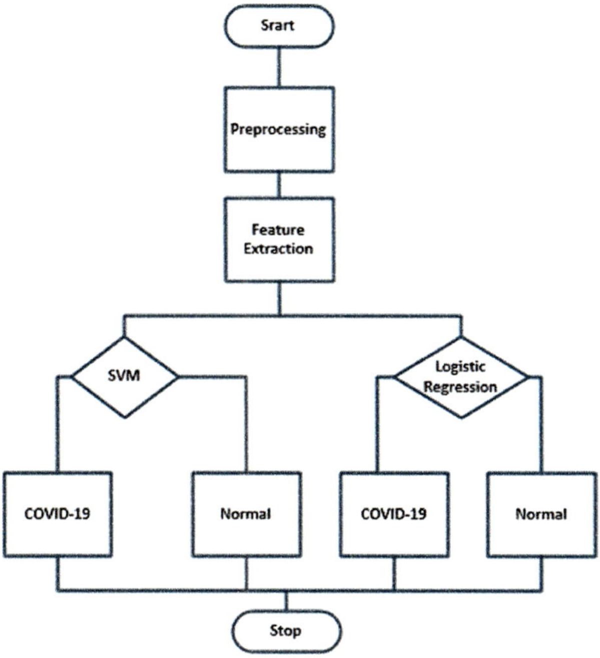

Fig. 4. Flowchart of The Proposed Methodology

the normalized histograms as a vector from all the block regions. Several parameters are considered in HOG: the number of orientation bins, the size of the cell in pixels, and the number of cells per block. HOG features are comprised of gradient angles and distribution of magnitude. The distribution of the orientation-based channel has spanned from 0° to 180°. These cells are categorized into normalized locally and spatially connected larger blocks to provide invariance in fluctuations in contrast and illumination. The HOG feature vector is produced by concatenation of these normalized cell histograms.

3.2.3 Classification

After the feature extraction, two different classification techniques (Logistic Regression, Support Vector Machine) were used to classify the dataset into COVID-19 and Normal from Chest X-Ray images.

Support Vector Machines. A support vector machine is a supervised learning technique used in machine learning for classification and regression problems. The data has been separated by hyper-plane with an N-dimensional space that classifies the data points. The support vector is the data points closer to the hyperplane, which maximizes the margin between classifiers. The position of the support vectors depends upon these hyperplanes.

Logistic Regression. Logistic regression is a supervised learning technique based on probability function and used for classification. In the regression task, the output value is a numerical form that exists on a continuous scale. The relationship between an independent variable(non-linear) is used to determine the dependent variable's output (discrete variable). The sigmoid function is used for a more complex cost function, assigning probabilities between 0 and 1.

4 Experimental Results

4.1 Performance Analysis

Five different metrics were used to evaluate the performance of the machine learning technique. These metrics are accuracy, precision, recall, sensitivity, F1-Score, which is presented in Eq. 1–5.TP, TN, FP, FN represent the value of True Positive, True Negative, False Positive, and False Negative, respectively.

$$\text{Accuracy} = (\text{TN} + \text{TP})/(\text{TN} + \text{TP} + \text{FN} + \text{FP}) \tag{1}$$

$$\text{Recall} = \text{TP}/(\text{TP} + \text{FN}) \tag{2}$$

$$\text{Specificity} = \text{TN}/(\text{TN} + \text{FP}) \tag{3}$$

$$\text{Precision} = \text{TP}/(\text{TP} + \text{FP}) \tag{4}$$

$$\text{F1 - Score} = 2 \times ((\text{Precision} \times \text{Recall})/(\text{Precision} + \text{Recall})) \tag{5}$$

In Eq. 1–5, TP represents that the positive (COVID-19) correctly identified as COVID-19, while FP presents the negative image (Normal) is wrongly identified as positive (COVID-19).TN presented the negative (Normal) images are correctly identified as normal, while the FN incorrectly identified the positive (COVD-19) as negative(normal).

5 Result and Discussion

In this study, the Chest X-ray dataset has been used to predict COVID-19 and normal patients. Two different machine learning models, such as SVM and logistic regression, have been trained to and test on the chest X-ray dataset. The comparison of 2 machine learning techniques using test data is illustrated in Table 1. The highest accuracy has been achieved by SVM of 96%, and their precision is 96%, recall 96%, specificity 96%, and F1-score is 96%. Simultaneously, the Logistic Regression provides 93% accuracy, precision 92%, recall of 92%, a specificity of 92%, and F1-score is 92%, which is the lowest performance than SVM.

Moreover, Fig. 5 present the graphical representation of performance evaluation of the two machine learning techniques (SVM, Logistic Regression) with accuracy, precision, recall, specificity, and F1-score.

Table 1. Comparison of support vector machine and logistic regression

Machine learning algorithm	Feature extraction	TP	FP	TN	FN	Accu	Prec	Rec	Spec	F1-Score
Support vector machine	HOG	266	5	134	10	96%	96%	96%	96%	96%
Logistic regression	HOG	116	19	266	14	92%	92%	92%'	92%	92%

TP = True Positive, FN = False Negative, TN = True Negative, FN = False Negative.
Accu = Accuracy, Prec = Precision, Rec = Recall, Spec = Specificity.

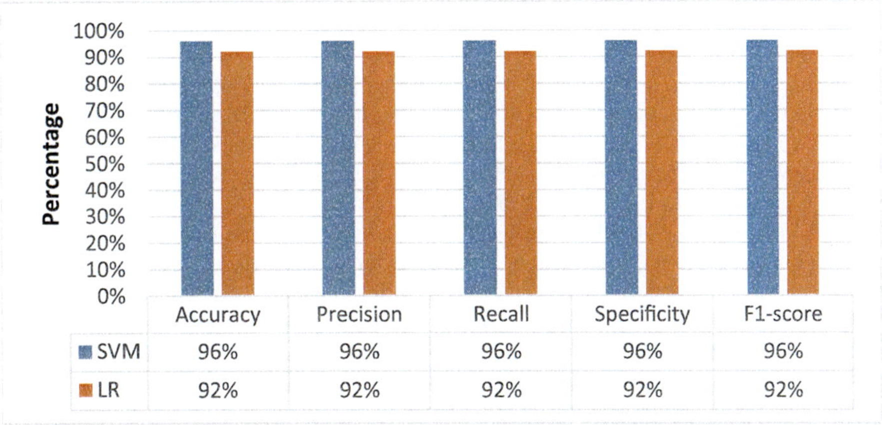

Fig. 5. Performance of SVM and LR in terms of Accuracy, Precision, Recall, Specificity and F1-Score

6 Conclusion

In this research study, we proposed a machine learning technique (Support vector machine and logistic regression) to predict COVID-19 and normal patients using chest X-ray images. The main steps for the diagnosing of COVID-19 are pre-processing, feature extraction, and classification. The image is pre-processed and improve the contrast by using the median and average filtering technique. After the pre-processing, feature extraction has been applied to extract the feature of COVID-19 and normal from the image based on Histogram of Oriented Gradients. Then the image has been classified by using two different machine learning techniques (SVM and logistic regression). The performance result shows that the support vector machine achieved the highest accuracy of 96% and provide better performance than logistic regression (92%). The machine learning technique help to doctor, healthcare and scholars to identify, control and manage the disease during prediction and decision-making process. In the future, our aims are to used other feature extraction with machine learning and deep learning techniques to provide better accuracy and improve performance.

References

1. Yan, L., Zhang, H.T., Xiao, Y., Wang, M., Guo, Y., Sun, C., Tang, X., Jing, L., Li, S., Zhang, M., Xiao, Y., Cao, H., Chen, Y., Ren, T., Jin, J., Wang, F., Xiao, Y., Huang, S., Tan, X., Huang, N., Jiao, B., Zhang, Y., Luo, A., Cao, Z., Xu, H. Yuan, Y.: Prediction of criticality in patients with severe Covid-19 infection using three clinical features: a machine learning-based prognostic model with clinical data in Wuhan. MedRxiv (2020)
2. Lupia, T., Scabini, S., Pinna, S.M., Di Perri, G., De Rosa, F.G., Corcione, S.: 2019-novel coronavirus outbreak: a new challenge. J. Global Antimicrob. Resist. **21**, 22–27 (2020)
3. Here's what coronavirus does to the body. https://www.nationalgeographic.com/science/2020/02/here-is-what-coronavirus-does-to-the-body/ (2021). Retrieved 19 Jan 2021
4. Coronavirus. https://www.who.int/health-topics/coronavirus/coronavirus (2021). Retrieved 19 Jan 2021
5. Huang, C., Wang, Y., Li, X., Ren, L., Zhao, J., Hu, Y., Zhang, J., Hu, J., Zhang, L., Fan, G., Xu, J., Gu, X., Cheng, Z., Yu, T., Cao, B.: Clinical features of patients infected with 2019 novel coronavirus in Wuhan. China. The Lancet **395**(10223), 497–506 (2020)
6. Coronavirus Disease 2019 (COVID-19). https://www.cdc.gov/coronavirus/2019-ncov/cdcresponse/about-COVID-19.html (2021). Retrieved 19 Jan 2021
7. Q&A: How is COVID-19 transmitted? https://www.who.int/vietnam/news/detail/14-07-2020-q-a-how-is-covid-19-transmitted (2021). Retrieved 19 Jan 2021
8. https://www.who.int/docs/default-source/coronaviruse/situation-reports/20200311-sitrep-51-covid-19.pdf?sfvrsn=1ba62e57_10 (2021). Retrieved 19 Jan 2021
9. WHO Coronavirus Disease (COVID-19) Dashboard data explorer. https://covid19.who.int/explorer/ (2021). Retrieved 19 Jan 2021
10. Punn, N.S., Sonbhadra, S.K., Agarwal, S.: COVID-19 epidemic analysis using machine learning and deep learning algorithms. medRxiv (2020)
11. Khanday, A.M.U.D., Rabani, S.T., Khan, Q.R., Rouf, N., Din, M.M.U.: Machine learning based approaches for detecting COVID-19 using clinical text data. Int. J. Inf. Technol. **12**(3), 731–739 (2020)
12. Yadav, M., Perumal, M., Srinivas, M.: Analysis on novel coronavirus (COVID-19) using machine learning methods. Chaos, Solitons Fractals **139**, 110050 (2020)
13. Sethy, P.K., Behera, S.K.: Detection of coronavirus disease (covid-19) based on deep features. Preprints, 2020030300, 2020 (2020)
14. Barstugan, M., Ozkaya, U., Ozturk, S. Coronavirus (covid-19) classification using ct images by machine learning methods. arXiv preprint arXiv:2003.09424 (2020)
15. Zheng, C., Deng, X., Fu, Q., Zhou, Q., Feng, J., Ma, H., Wang, X.: Deep learning-based detection for COVID-19 from chest CT using weak label. medRxiv (2020)
16. Ferrari, D., Milic, J., Tonelli, R., Ghinelli, F., Meschiari, M., Volpi, S., Guaraldi, G.: Machine learning in predicting respiratory failure in patients with COVID-19 pneumonia-challenges, strengths, and opportunities in a global health emergency. medRxiv (2020)
17. Afshar, P., Heidarian, S., Naderkhani, F., Oikonomou, A., Plataniotis, K.N., Mohammadi, A.: Covid-caps: a capsule network-based framework for identification of covid-19 cases from x-ray images. arXiv preprint arXiv:2004.02696 (2020)
18. Li, W.T., Ma, J., Shende, N., Castaneda, G., Chakladar, J., Tsai, J.C., Ongkeko, W.M.: Using machine learning of clinical data to diagnose COVID-19: a systematic review and meta-analysis. BMC Med. Inform. Decision Mak. **20**(1), 1–13 (2020)

19. Razzak, I., Naz, S., Rehman, A., Khan, A., Zaib, A.: Improving Coronavirus (COVID-19) diagnosis using deep transfer learning. medRxiv (2020)
20. Ahamad, M.M., Aktar, S., Rashed-Al-Mahfuz, M., Uddin, S., Liò, P., Xu, H., Moni, M.A.: A machine learning model to identify early-stage symptoms of SARS-Cov-2 infected patients. Expert Syst. Appl. **160**, 113661 (2020)
21. Pinter, G., Felde, I., Mosavi, A., Ghamisi, P., Gloaguen, R.: COVID-19 pandemic prediction for hungary; a hybrid machine learning approach. Mathematics **8**(6), 890 (2020)
22. Sujath, R., Chatterjee, J.M., Hassanien, A.E.: A machine learning forecasting model for COVID-19 pandemic in India. Stoch. Env. Res. Risk Assess. **34**, 959–972 (2020)
23. Ahmad, A., Garhwal, S., Ray, S.K., Kumar, G., Malebary, S.J., Barukab, O.M.: The number of confirmed cases of covid-19 by using machine learning: methods and challenges. Arch. Comput. Methods Eng., 1–9 (2020)
24. Khan, N., Ullah, F., Hassan, M.A., Hussain, A.: COVID-19 classification based on Chest X-Ray images using machine learning techniques. J. Comput. Sci. Technol. Stud. 01–11 (2020)
25. Search | Kaggle. https://www.kaggle.com/search?q=covid+19+dataset+datasetFilTypes%3Apng+datasetFileTypes%3Ajpg+datasetFileTypes%3Ajpeg (2021). Retrieved 19 Jan 2021
26. Chettri, S., Debnath, D., Devi, P.: Leveraging digital tools and technologies to alleviate COVID-19 pandemic. SSRN Electron. J. (2020). https://doi.org/10.2139/ssrn.3626092
27. Abbas, A., Abdelsamea, M., Gaber, M.: Classification of COVID-19 in chest X-ray images using DeTraC deep convolutional neural network. Appl. Intell. **51**(2), 854–864 (2020). https://doi.org/10.1007/s10489-020-01829-7

Performance Analysis of POX and RYU Based on Dijkstra's Algorithm for Software Defined Networking

Naimullah(✉), Syed Irfan Ullah, Arbab Wajid Ullah, Abdus Salam, Muhammad Imad, and Farhat Ullah

Department of Computing, Abasyn University, Peshawar, Pakistan
{syed.irfanullah,arbabwajid.ullah,dr.salam}@abasyn.edu.pk

Abstract. Conventional network architectures are not appropriate for the needs of current businesses, carriers, and end-users. Consequently, new evolving network architecture, Software Defined Network can be used to solve a problem as it is more adaptive, dynamic, manageable, and programmable. The SDN architecture control and data planes are isolated, network information and the state are legalized, and the essential network infrastructure is excluded from the applications. However, a network may achieve a point where the computer or network resources restrict the data flow that is controlled according to the bandwidth. In this paper, the custom network topology is created. To observe the performance of Dijkstra's shortest route algorithm in the SDN open-source controllers: RYU and POX to find out the shortest route between source and the destination nodes. The controller's performance is calculated based on the quality-of-service measures, containing throughput and packet delivery ratio in custom network topology under the various workload. The Mininet emulator tool is used to construct the custom network topology. The results are obtained by simulating different QoS requirements in customized topology in Mininet.

Keywords: SDN · Dijkstra's algorithm · Open flow controller · MININET · POX · RYU · Throughput · Packet delivery ratio

1 Introduction

The volume of media, in general, has been increasing in recent years. There is an increasing request for network flexibility and development, which permit changes. The principle works to minimize system organization and administration costs dramatically [1]. The conventional network problems are solved by SDN, which takes a separate control plane for each network device in the conventional network, containing switches and routers. There is no centralized visibility for network devices for conventional networks, and the Network Management System (NMS) needs to be configured for maximum visibility. As a consequence, Software Defined Networking (SDN) is one of the prevailing paradigms that seek to overcome constraints in conventional computer network architectures in order to meet the dynamic network needs of today [2]. The growth of the

A. M. A. Musleh Al-Sartawi et al. (Eds.): EAMMIS 2021, LNNS 239, pp. 24–35, 2021.
https://doi.org/10.1007/978-3-030-77246-8_3

SDN has gained the attention of the research community in recent years. Especially it provides a versatile centralized management model, which promotes the development of solutions according to the client's request for the network. The SDN concept separates the control plane (the brain of the network) and the data plane (forwarding plane). These are a fundamental element of every network architecture. Control Plane relates to the logic of forwarding and controlling behavior. Its key functions include monitoring topology changes, installing forwarding rules, service provisioning, route computing, etc. Another fundamental aspect is the management plane, sometimes seen as a part of the control plane. In general, the management strategy applies to the tasks responsible for configuring, controlling, and delivering management resources to all levels of the network stack and other areas of the infrastructure. On the other hand, the data plane is relevant to the network portion that forward user traffic. Forwarding is based on the control plane's rules. Another function of the data plane refers to buffering, filtering, packet measurement, etc. in SDN, the control plane is clustered, manages the distributed forwarding plane, and can be fully integrated into the software and install on hardware. An SDN may also be defined as a programmable network. For several years, the concept of programmable networks has been there. It is relevant to the network in which the network device's manner is controlled by software [3].

SDN architecture has two distinct interfaces, specifically the southbound interface and northbound interface. SDN southbound API (Application Program Interface) also connects the control plane with the data plane [4]. This API contains well-known protocols such as Open Flow (i.e., a communication interface that provide access to the data plane switches, routers, etc.) and Open Virtual Switch Database (OVSDB). The SDN northbound interface is the communication channel in the middle of the application layer and control layer [5]. Figure 1 shows the SDN architecture.

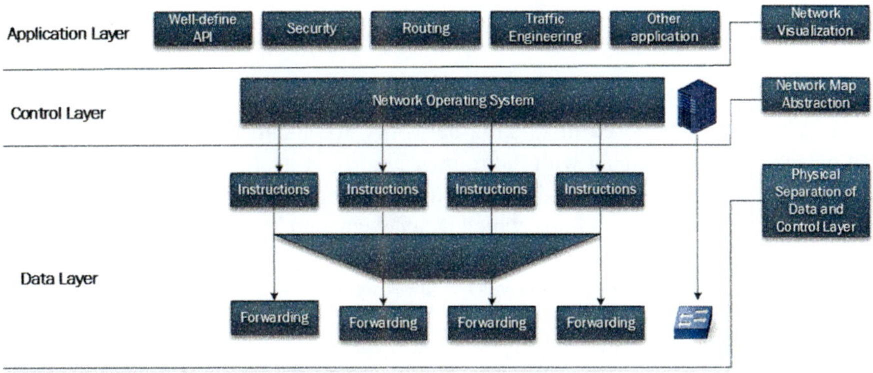

Fig. 1. SDN architecture [23].

The controller is the core of an SDN that controls the traffic (network flows) by controlling the switch and router's flow table. There are several SDN open-source controllers for designing Software-Defined Network platforms, such as RYU, NOX, POX, Floodlight, and Open Daylight [6].

1.1 RYU Controller

RYU is a Japanese word that means flow. It's pronounced as "ree-yooh". It is also a Python based OpenFlow controller with a powerful API that enables developers to program, monitor, and manage an application. Network devices can be designed based on RYU applications via OpenFlow open-source protocol, and its code is available with an Apache License [18].

1.2 POX Controller

POX is a python based open-source OpenFlow controller and is known as one of the most widely used in the research community due to its ease of programmability. It presents a forum for the prototyping and rapid creation of network systems and control apps on the OpenFlow network. POX can be remotely connected to Mininet. Various applications such as firewall, load balancer, intrusion detection and prevention, routing, and switching can be configured on the underlying Open Flow switches. In Mininet, POX is integrated and can also be installed from its GitHub repository [19]. The Table 1 display the various characteristic of POX and RYU controllers.

Table 1. SDN controllers.

Controller name	Programming language	Architecture	Interface	License provider	Supported platform
POX [23]	Python	Centralized	CLI, GUI	Apache	Linux, macOS, Windows
RYU [24]	Python	Centralized	CLI, Web UI	GPL	Linux

Most of the issues posed in existing networks that hamper obtaining appropriate load balancing are associated with the routing algorithm itself. The routing process is currently based on the shortest route algorithm in which every packet searches for a route with the least number of hops, which it can reach, and it is the same for all packets, even if other routes are higher but much faster. This reality reduces usable features of the network, such as the congestion over the shortest route link, the overuse of the same connections, while others are not often used in the network [7].

This research paper evaluates Dijkstra's algorithm implementation over SDN open-source controllers (POX, RYU), which allows for the acquisition and transmission of packets between end devices within the network of the shortest and/or least load routes. In section two, we study some of the relevant work. In section three, the methods and elements required to conduct the SDN performance assessment with the shortest route algorithm are discussed. In section four, the achieved outcome is discussed and the observation and analysis found in this work. Finally, section five describes the conclusion of this work.

2 Literature Review

The majority of comparative studies on SDN controllers relate to performance. The majority of them use bench-marking or other Cbench tools to evaluate metrics related to latency and throughput. In most stance, well-liked Open Flow controllers have selected since they are open source and helpful documentation have existence. What needs to be remembered is that performance and scalability are similar in certain literature situations. Thus, some of the observations below also relate to scalability, which will be further analyzed in the following paragraph.

The performance assessment analysis of five well-known controllers (ONOS, Open Daylight, POX, libfluid, and RYU) has been performed using Iperf, and Ping commands in terms of End-to-End (ETE) throughput and delay. The test was conducted with linear topology and designed with different switches in the Mininet emulator. Although the results show that libfluid provides desired outcomes and POX gives the best delay, the controller selection should base on a number of parameters, according to the user's requirements. In delay calculations, the delay rates increase with the increasing number of network switches and hosts. The libfluid controller had the highest delay than other controllers and the lowest delay values for the ONOS controller [8].

The performance of POX, RYU, ONOS, and OpenDaylight controllers was evaluated based on bandwidth and end-to-end delay parameters. A fixed four-level tree topology comprising 16 hosts was used as a test environment for controllers' performance. The analysis concluded that RYU would have the least end-to-end latency and that ONOS would have the maximum bandwidth among these four controllers. The best-suited controllers were often adopted depending on the priorities or objectives of the result. POX was adopted as one of the most acceptable environments with setup convenience as the highest priority, while the POX performance is not equivalent to OpenDaylight, RYU, and ONOS controllers. This research was limited to performance parameters at all levels and static network topology, thereby contributing to a non-general conclusion regarding controllers in all the other network types with different priority parameters [9].

The performance of four NOX, Beacon, Maestro, NOx, and NOX-MT open-source controllers was measured, presented as an optimized NOX controller by the authors. They addressed that these controllers' true performance may be significantly higher than previously predicted in an integrated network environment. They also developed Cbench to be a benchmarking method primarily designed to test those basic performance metrics for OpenFlow switch emulations. NOX-MT is a multi-threading variant of NOX that uses optimization techniques (e.g., I/O batching) for improvements in baseline performance and has been shown to outperform NOX in overall performance by a factor of 33. This research is outdated because newer versions of these controllers have been introduced, and other significant controllers such as ONOS, Floodlight, and OpenDaylight have also been left out of the evaluation [10].

Round Robin Load-Balancing Algorithm (RRLBA) has been implemented and compared to the Random algorithm based on various load utilizing measuring such as response time, Open Load, and transaction per second. The results indicate that Round-Robin strategies were better than Random. The other beneficial quality in Round Robin is that it distributes the load reasonably, even though Random is not. Their work disadvantage is that they do not check code on real hardware that could produce various outcomes. Another constraint is that code has been checked in the POX controller. The performance of the SDN application is also dependent upon controller output [11].

The POX controller with Python and the Floodlight controller with Java were analyzed as representations of all controllers created in these two computer languages. They carried out a performance comparison on throughput and delay between them. In Mininet, the experiments were conducted using different network topologies. This analysis has also been restricted to two controllers that do not generally include any other python or java-developed controllers; limited network variables have also been calculated [12].

The implementation of the two most common SDN controllers, such as Floodlight and Open Daylight, compared it in terms of accessible delay and packet forwarding bandwidth. It used simulation modeling based on a network flow where the shortest or lowest path algorithm was also implemented. The implementation of the load-balancing algorithm made it possible to enhance the behavior of a Software-Defined Networking in terms of QoS, maximize bandwidth, minimize response times, and optimally disperse the load from the links. As a result, the proposed load-balancing algorithm significantly improves the Open Daylight-based controller's performance in terms of QoS offered [13].

The performance and functionality of POX and Floodlight controllers were examined to determine a well-known SDN controller. There are several open-source controllers. For research purposes, they have selected the Python-based POX and Java-based Floodlight controller. The main parameters, including delay and throughput, compared the POX and Floodlight controller in this article. The parameters have been tested in various typologies. It is noted that the floodlight results are much faster (31 times faster) in all topologies, in terms of throughput, when compared with POX [14].

Different load-balancing algorithms such as Round Robin (RR), Implementation Server Load Balancing (ISLL), Flow statistic (FS), Least Time Based Weighted Load Balancing Algorithm (LTBWLB) compared and implemented in SDN, based on the following parameters such as throughput, response time, and transaction rate to explore withiest load balancing algorithm in SDN controllers. They selected the POX controller supporting the Python programming language and used the Mininet emulation tool to create network topology with the same virtual hardware setting as in a realistic environment. You can use the RYN control to achieve better results and using several controllers rather than just a single controller instead of the POX control [15].

The proactive and reactive SDN schemes were evaluated using the Open Daylight controller and Mininet, simulation of application scenarios from the specific networks and data centers. By describing the limitations of conventional networking architecture, they discussed SDN's background and significance in connection failure recovery. Then the three SDN planes and their interaction mechanisms were identified, along with the importance of SDN to recover link failures. The speed of recovery of failures depends on the time taken to diagnose failures. They identified the state-of-the-art approaches for recognizing connection failures with their pros and cons. A differentiation was accomplished between the proactive and reactive strategy in of scalability, latency, flow operation matching, and robustness to back up the failures, routing information access, and the overheads of routing, controller, and switches [16].

3 Proposed Methodology

In software-defined networks, the controller is liable for picking the most suitable path through the routing algorithms about the network connections load.

Dijkstra's shortest route algorithm is responsible for selecting the most appropriate shortest route between a source and destination node to deliver the data within the specified amount of time. This algorithm performs a vital role in the transmission and collection of the data to the destination port address. The pseudo-code for the Dijkstra algorithm is shown below.

Algorithm 1. Shortest Path using Dijkstra Algorithm

```
function dijkstra (G, S)
      for each vertex V in G
            dist [V] <- infinite
            prev[V] <- NULL
            if V ! = S, add V to priority Queue Q
      dist [S] <- 0
      While Q != empty
            U <- Extract MIN from Q
            For each unvisited neighbour V of U
                  TDistance <- dist [U] + edge_weight (U, V)
                  if TDistance < dist [V]
                        Dist [v] <- TDistance
                        Prev[V] <- U
      Return dist [], prev []
```

As well, Dijkstra's shortest path algorithm was implemented in the POX and RYU controllers and a custom topology created using Mininet emulator with various parameters was run. After that, performance analyses for Throughput and Packet Delivery Ratio have performed on them. So, our simulation topology is shown in Fig. 2.

4 Environment Setup

The RYU and POX controller is executed in the virtual environment made by the VMware Workstation Pro. In the Virtual Box, Ubuntu 18.04 is installed to create the operating environment for the simulation. The network simulation uses a Mininet simulator, which is capable of constructing the network environment and the corresponding simulation within the virtual environment scope. For comparative, we implement a Python-based

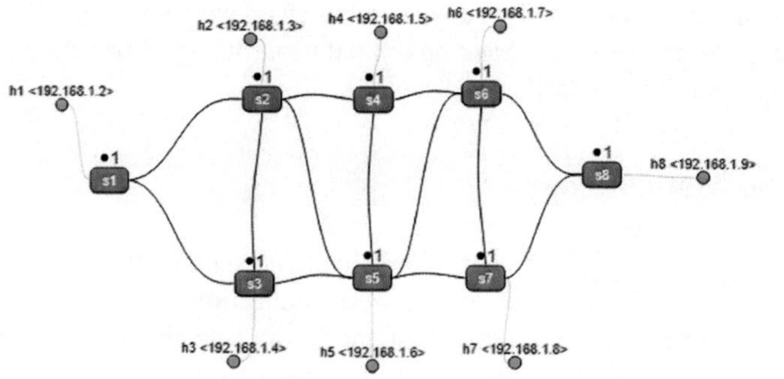

Fig. 2. Custom topology

POX, RYU controller. The measurements of the available throughput and packet delivery ratio of ICMP requests operate for a scenario in which the shortest route algorithm is using [17].

4.1 Performance Evaluation

This section illustrates the outcomes obtained by RYU and POX controllers in the context of the shortest path algorithms described in the previous section. It is worthiness to point out that both SDN controllers are operating over the same custom network topology.

4.1.1 Throughput

Quantity of data that is being sent in a unit of time. Measured bite per second (kbps). Mathematically the throughput can be computed as the number of bits processed per unit time. It is denoted as [25]:

$$Throughput = (\Sigma receivedpacketsize)/time(Kbps)$$

The iperf tool is a popular network testing tool for bandwidth and network connection measurement. The tool can build TCP and UDP data streams and calculate the network throughput through which these streams are carried. The iperf tool performs both client and server features and can measure unidirectionally or bi-directionally throughput between the two end-hosts [20]. The network throughput is calculated using the iperf real-time method. An average throughput between source and destination nodes in custom topology is measured and displayed in the below figures.

The consequences of observations are shown; the test has been carried out for the RYU and POX SDN open-source controller. Figures 3, 4 and 5 display the average throughput values for the RYU and POX controller (Gbp/s). The comparison of the average network throughput of these controllers illustrated in Fig. 5. As a result, the POX controller shows the highest throughput value compared to RYU, a controller over custom network topology. At all, the various number of switches evaluates when network overload POX controller gives better results than RYU. In comparison, the RYU controller displays the lowest throughput value.

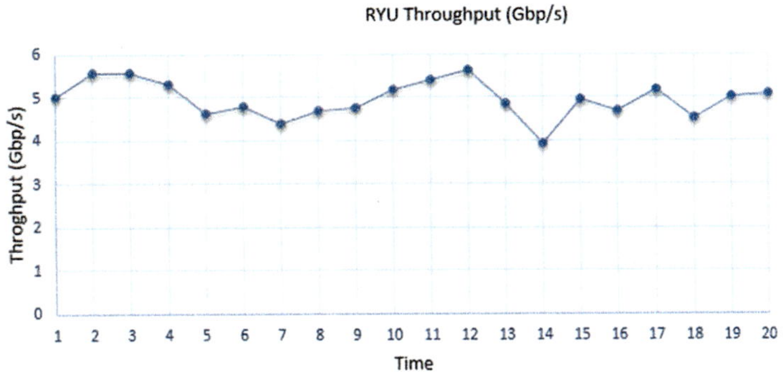

Fig. 3. An average throughput of SDN RYU controller.

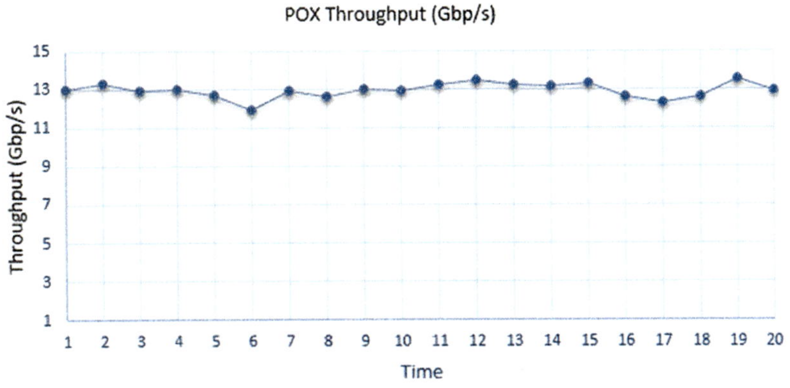

Fig. 4. An average throughput of SDN POX controller.

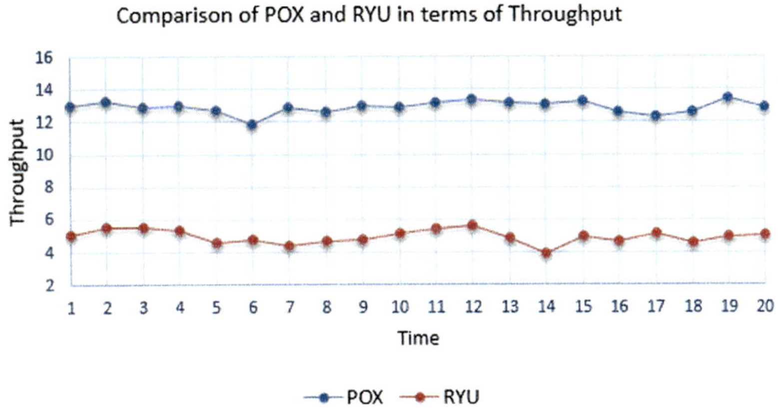

Fig. 5. Average throughput of RYU vs. POX controllers.

4.1.2 Packet Delivery Ratio

Packet delivery ratio is a very important factor to measure the performance of routing protocol in any network. Packet Delivery Ratio (PDR) is an extremely significant feature to measure routing protocol's performance in any network. The protocol performance depends on different parameters selected for simulation. The packet delivery ratio can be achieved from the total number of data packets arrived at destinations divided by the total data packets sent from sources. The performance is better when there is a high packet delivery ratio. It can be seen mathematically as an equation [21, 25].

$$Packet\ Delivery\ Ratio = 100 * \frac{No.\ of\ packets\ Delivered}{Total\ packets\ sent}$$

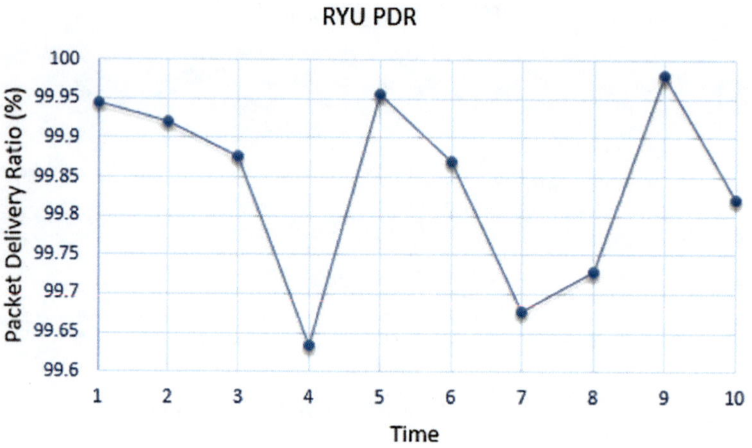

Fig. 6. RYU controller packet delivery ratio

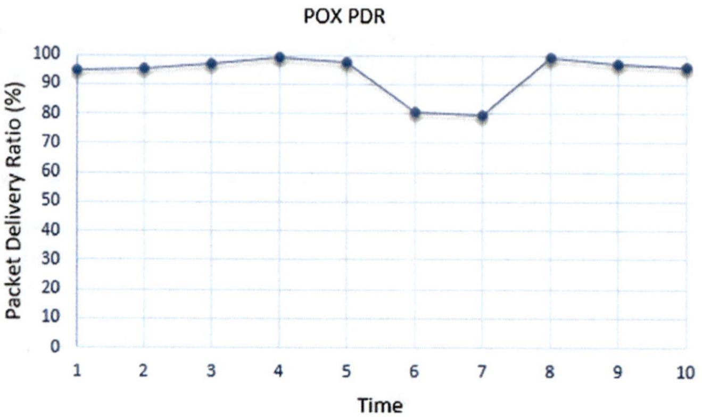

Fig. 7. POX controller packet delivery ratio.

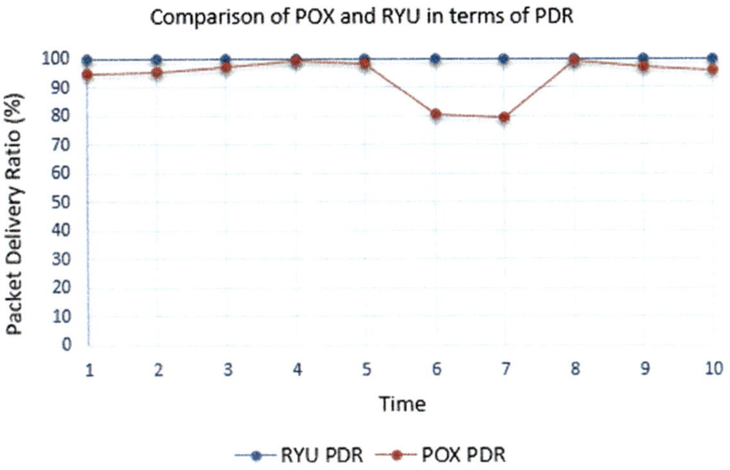

Fig. 8. Comparison of RYU PDR and POX PDR

Figures 6, 7 and 8 demonstrate the Packet Delivery Ratio (PDR) in percentage. The comparison of these two controllers illustrated in Fig. 8. Finally, the Packet Delivery Ratio performance of the RYU controller is better than a POX controller.

5 Conclusion

Software-Defined Networking (SDN) or programmable networks have many traffic management controllers in the network. Each of them differs from the other because of its complexity and the support provided by its developer. The implementation of Dijkstra's shortest path algorithm in SDN controllers such as RYU and POX allowed improving SDN behavior in terms of quality of service, improving throughput, and finding the best optimal shortest path for the packets. All the assessment has been completed using Mininet emulator. The performance evaluation of POX and RYU shows that the POX controller provides a better result in terms of throughput. In the packet delivery ratio, the RYU controller provides better performance. In the future, we aim to use other controllers such as Open daylight, Trema, Floodlight to check the performance and use different parameters like End-To-End Delay, Packet Loss, and Jitter.

References

1. Benzekki, K., El Fergougui, A., Elbelrhiti Elalaoui, A.: Software-defined networking (SDN): a survey. Secur. Commun. Netw. **9**(18), 5803–5833 (2016)
2. Hu, F., Hao, Q., Bao, K.: A survey on software-defined network and openflow: from concept to implementation. IEEE Commun. Surv. Tutor. **16**(4), 2181–2206 (2014)
3. Farhady, H., Lee, H., Nakao, A.: Software-defined networking: a survey. Comput. Netw. **81**, 79–95 (2015)
4. Latif, Z., Sharif, K., Li, F., Karim, M.M., Biswas, S., Wang, Y.: A comprehensive survey of interface protocols for software defined networks. J. Netw. Comput. Appl. **156**, 102563 (2020)

5. Schaller, S., Hood, D.: Software defined networking architecture standardization. Comput. Stand. Interfaces **54**, 197–202 (2017)
6. Blial, O., Ben Mamoun, M., Benaini, R.: An overview on SDN architectures with multiple controllers. J. Comput. Netw. Commun. (2016)
7. Sufiev, H., Haddad, Y.: A dynamic load balancing architecture for SDN. In: 2016 IEEE International Conference on the Science of Electrical Engineering (ICSEE), pp. 1–3. IEEE, November 2016
8. Abdullah, M.Z., Al-Awad, N.A., Hussein, F.W.: Performance evaluation and comparison of software defined networks controllers. Int. J. Sci. Eng. Sci. **2**(11), 45–50 (2018)
9. Stancu, A.L., Halunga, S., Vulpe, A., Suciu, G., Fratu, O., Popovici, E.C.: A comparison between several software defined networking controllers. In: 2015 12th International Conference on Telecommunication in Modern Satellite, Cable and Broadcasting Services (TELSIKS), pp. 223–226. IEEE, October 2015
10. Tootoonchian, A., Gorbunov, S., Ganjali, Y., Casado, M., Sherwood, R.: On controller performance in software-defined networks. In: 2nd {USENIX} Workshop on Hot Topics in Management of Internet, Cloud, and Enterprise Networks and Services (Hot-ICE 12) (2012)
11. Kaur, S., Kumar, K., Singh, J., Ghumman, N.S.: Round-robin based load balancing in software defined networking. In: 2015 2nd International Conference on Computing for Sustainable Global Development (INDIACom), pp. 2136–2139. IEEE March 2015
12. Darianian, M., Williamson, C., Haque, I.: Experimental evaluation of two openflow controllers. In: 2017 IEEE 25th International Conference on Network Protocols (ICNP), pp. 1–6. IEEE October 2017
13. Duque, J.P., Beltrán, D.D., Leguizamón, G.P.: OpenDaylight vs. Floodlight: comparative analysis of a load balancing algorithm for software defined networking. Int. J. Commun. Netw. Inf. Secur. **10**(2), 348–357 (2018)
14. Fancy, C., Pushpalatha, M.: Performance evaluation of SDN controllers POX and floodlight in mininet emulation environment. In: 2017 International Conference on Intelligent Sustainable Systems (ICISS), pp. 695–699. IEEE December 2017
15. Joshi, N., Gupta, D.: A comparative study on load balancing algorithms in software defined networking. In: International Conference on Ubiquitous Communications and Network Computing, pp. 142–150. Springer, Cham February 2019
16. Ali, J., Lee, G.M., Roh, B.H., Ryu, D.K., Park, G.: Software-defined networking approaches for link failure recovery: a survey. Sustainability **12**(10), 4255 (2020)
17. Zhu, L., Karim, M.M., Sharif, K., Li, F., Du, X., Guizani, M.: SDN controllers: benchmarking & performance evaluation. arXiv preprint arXiv:1902.04491 (2019)
18. Ryu, S.D.N.: Framework (sd). Récupéré sur. https://www.osrg.github.io/ryu
19. Bholebawa, I.Z., Dalal, U.D.: Performance analysis of sdn/openflow controllers: pox versus floodlight. Wirel. Pers. Commun. **98**(2), 1679–1699 (2018)
20. Kolahi, S.S., Narayan, S., Nguyen, D.D., Sunarto, Y.: Performance monitoring of various network traffic generators. In: 2011 UkSim 13th International Conference on Computer Modelling and Simulation, pp. 501–506. IEEE March 2011
21. Badlishah, A., Rahman, M.Z.A., Yaakob, N., Amir, A., Ahmad, R., Yoon, S.K., Abd Halim, A.H.: Performance analysis of congestion control mechanism in Software Defined Network (SDN) (2017)
22. Bonfim, M.S., Dias, K.L., Fernandes, S.F.: Integrated NFV/SDN architectures: a systematic literature review. ACM Comput. Surv. (CSUR) **51**(6), 1–39 (2019)

23. Kaur, S., Singh, J., Ghumman, N.S.: Network programmability using POX controller. In: Proceedings of the International Conference on Communication, Computing Systems (ICCCS), vol. 138, pp. 134–138 (2014)
24. Islam, M.T., Islam, N., Al Refat, M.: Node to node performance evaluation through RYU SDN controller. Wirel. Pers. Commun., 1–16 (2020)
25. Khan, I.U., Qureshi, I.M., Aziz, M.A., Cheema, T.A., Shah, S.B.H.: Smart IoT control-based nature inspired energy efficient routing protocol for flying ad hoc network (FANET). IEEE Access **8**, 56371–56378 (2020)

Priority Based Power Delivery System for Electric Vehicle Charging

Noaima Bari(✉), Ahteshamul Haque, Gaurav Ahuja, and V. S. Bharath Kurukuru

Department of Electrical Engineering, Jamia Millia Islamia, New Delhi, India
ahaque@jmi.ac.in

Abstract. With the rapid increase in the development of electric vehicles (EVs), a priority-based power delivery system for EV charging (EVC) has become a significant necessity. This helps in prioritizing the loads with distributed generation systems connected with the grid and prevents negative impacts such as overloading on the power grid. In this paper, an Adaptive Neuro Fuzzy Inference System (ANFIS) based methodology for priority determination of electric vehicles' charging is developed considering the battery's state of charge (SOC), time of stay, distance to be travelled and battery capacity. The proposed approach aims at enhancing the power delivering capability of the utility and overcome the drawback of unbalanced loading while performing EVC. To assess the performance of the proposed approach, the control and simulation of a grid connected EVC is done using 3 level converters. The results depict a smooth ripple free DC voltage which charges the vehicle with the highest priority.

Keywords: Electric vehicle charging · Adaptive neuro fuzzy inference system · Voltage oriented control · Charging prioritization · Battery state of charge

1 Introduction

The depletion of conventional sources of energy such as fossil fuels along with their alarming environmental threats is paving way to look for greener and more efficient sources of energy for the future. In this regard, electric vehicles have garnered great appreciation and attention in the past few years due to their reduced emissions. Moreover, owing to an increase in oil prices, electric vehicles are beginning to compete with the existing conventional Internal Combustion Engine (ICE) run vehicles. Further, increasing research and development in the field of electric vehicles has resulted in better battery technologies along with a reduction in battery prices gradually paving the way to mainstream use of electric vehicles in the near future [1, 2].

With this rapid increase in the numbers, optimal integration of large number of electric vehicles with the grid for charging and discharging purposes has become an extremely important concern in order to prevent negative impacts on the power system operation. Uncoordinated charging, i.e. charging without any strategy at the instance of arrival lead to the most severe impacts on distribution networks such as voltage imbalances [1] and a sharp increase in demand during peak hours. In [3], the impact of

A. M. A. Musleh Al-Sartawi et al. (Eds.): EAMMIS 2021, LNNS 239, pp. 36–45, 2021.
https://doi.org/10.1007/978-3-030-77246-8_4

different charging strategies on two real distribution networks is analyzed to show how it affects investment costs and energy losses in the distribution networks. Smart charging strategies for optimal integration of EVs with the electric power grid are hence, the need of the future.

In [2], a smart charging strategy for shifting load from peak hours to off-peak hours (valley filling) is presented. It uses a Fuzzy controller on a simulated distribution network with electricity prices and State of Charge (SOC) as inputs and charging power as output for optimal charging. In [4], an intelligent charge scheduling scenario for aggregators is addressed by using a fuzzy inference system which takes a PEVs SOC and departure time into account. It assigns priorities according to the fuzzy expert system scoring in which the higher priority scored, and not necessarily the first comer, is charged first considering their final SOC required and the system operational limits. S. Chen et al. in [6] apply an online pre-emptive job scheduling problem with deadlines and commitment to the optimal charge scheduling of PEVs. The research is aimed at maximizing the profit by a trade-off between accepting more customers and reducing penalties due to unfinished jobs. In [5], the impact of different priority criteria on the Time Coordinated Charing (TCC) and Power Coordinated Charging (PCC) EV charging methods is studied. It evaluates the performance of various combinations of SOC, slack time and allotted time on EVC using data-driven simulations and talks about a weighted priority criterion.

While the fuzzy based methods used in the conventional approaches perform well in representing linguistic and structured knowledge with the help of fuzzy sets, they often rely on the knowledge of domain experts [9]. Neural networks, on the other hand, are more efficient in capturing the non-linearities of a system and provide more adaptability. Hence, an integration of a Fuzzy Inference System (FIS) and Artificial Neural Networks (ANN) would produce improved results by incorporating the merits of both methods.

In this work, instead of self-assigning weights for different priority criteria such as SOC, Time of Stay, Distance to be travelled and Battery Capacity, we utilize the benefits of an Adaptive Neuro Fuzzy Inference System (ANFIS) to learn custom weights with the help of neural network trained by backpropagation. The PWM converters which reduce the content of lower harmonics and improve the power factor are used for modelling the electric vehicle charging station connected to a grid. A voltage-oriented control is simulated on the grid side and a combination of constant current and constant voltage control strategies have been used for battery charging and discharging on the EV side producing a smooth dc ripple free voltage for testing charging purposes.

The remaining sections of the paper are arranged as follows: Sect. 2 discusses the development of EVC in a grid connected environment. In Sect. 3 the methodology corresponding to the development of ANFIS controller for prioritized EVC is discussed. The experimental analysis and results are identified in Sect. 4 and the conclusion is provided in Sect. 5.

2 EVC Station

The grid-connected EVC station consists of three main parts: a range of EV chargers, a grid-interface power converter and a system integration controller. Figure 1 shows the configuration of the grid connected charging station.

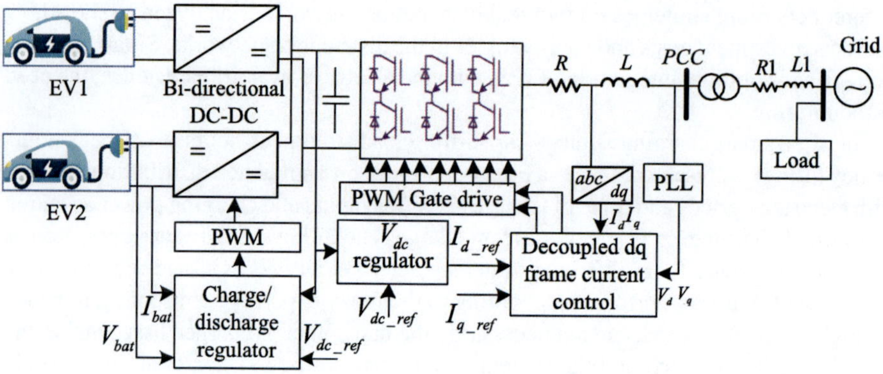

Fig. 1. Configuration of grid-connected charging station

2.1 EV Side Converter Control

A combination of the constant current and constant voltage control mechanisms is utilized for the control of the bidirectional PWM DC/DC converter on the EV side, in order to meet the battery charging and discharging requirements. These requirements are specified in the form of a maximum discharge current and a maximum 30-s discharge pulse current. The former specifies the maximum current at which the battery can be charged or discharged continuously, and the latter specifies the maximum current at which the battery can be charged or discharged for pulses up to 30 s [8]. The charging scheme is chosen so as to ensure a prolonged battery life and high performance.

Initially, assuming that the battery is not pre-charged, a CC charging scheme is utilized until the SOC reaches a threshold of 70%. After this, the control mechanism is switched to CV mode. A nested control structure is implemented using a fast-inner current loop, and a slow-outer voltage loop. The current flowing into the battery is assumed to be positive. The battery voltage V_{bat} is compared with the reference battery voltage V_{bat}^{*} and the result is passed to the first PI controller.

Case 1. If the reference current $I_{ref} < 0$, battery is operating in discharge mode, and the switch is programmed to lie at the bottom position in order to discharge the battery at a constant current.

Case 2. If $I_{ref} > 0$ and $SOC < 70\%$, the switch is shifted to the bottom position and the battery is charged at a constant current.

Case 3. If $I_{ref} > 0$ and $SOC > 70\%$, the switch is shifted to the top position and battery is charged at a constant voltage.

The voltage controller produces a current reference which is used by the current loop controller to regulate the battery charging at the desired current reference. Hence, to prevent this input reference current from exceeding the maximum permissible limit, a saturation block is used having a current limit function as:

$$I_{ref\,(sat)} = f(V_{in}) = \frac{I_{ref\,(max)}(V_{in} - V_1)}{(V_2 - V_1)} \tag{1}$$

Here, V_{in} is the DC input voltage; $I_{ref\,(max)}$ is the maximum current permitted to pass to the EV battery; V_1 is the lower limit of DC bus voltage; V_2 is the DC link voltage setpoint where the power converter reaches its maximum power limit [7].

The output of the current controller is a duty ratio which is constrained to lie between 0 and 1 to prevent non-linear modulation condition of the converter.

$$d = d_{old} + K_{pi}\left(I_{ref} - I_{bat}\right) + K_{ii}\int\left(I_{ref} - I_{bat}\right)dt \tag{2}$$

$$d = d_{old} + K_{pv}(V_{bat}^{*} - V_{bat}) + K_{iv}\int(V_{bat}^{*} - V_{bat})dt \tag{3}$$

Here, d and d_{old} are the recent and previous duty ratios used to control the EV-side converter. K_{pv}, K_{iv} and K_{pi}, K_{ii} are the proportional and integral gains of the outer and inner loop PI Controllers respectively.

2.2 Voltage Oriented Control for GSC

The voltage-oriented control (VOC) strategy which is used to control the bidirectional AC/DC converter has been simulated on the grid side. The VOC technique for GSC is used as it produces low harmonic distortions in the grid and also provides a fast dynamic and static performance due to the inner current loop.

The operation takes place in $\alpha\beta0$ and dq domains. The AC side quantities are first transformed to DC signals using Clarke's and Park's transformation matrices as follows:

$$\begin{bmatrix} v_{S\alpha} \\ v_{S\beta} \\ v_0 \end{bmatrix} = \sqrt{\frac{2}{3}}\begin{bmatrix} 1 & -\frac{1}{2} & -\frac{1}{2} \\ 0 & \frac{\sqrt{3}}{2} & -\frac{\sqrt{3}}{2} \\ \frac{1}{\sqrt{2}} & \frac{1}{\sqrt{2}} & \frac{1}{\sqrt{2}} \end{bmatrix}\begin{bmatrix} v_{Sa} \\ v_{Sb} \\ v_{Sc} \end{bmatrix} \tag{4}$$

$$\begin{bmatrix} v_d \\ v_q \end{bmatrix} = \begin{bmatrix} sin\theta & cos\theta \\ -cos\theta & sin\theta \end{bmatrix}\begin{bmatrix} v_{S\alpha} \\ v_{S\beta} \end{bmatrix} \tag{5}$$

Here v_a, v_b, v_c, are the voltages in *abc* domain; v_α, v_β, v_0 are the transformed voltages in $\alpha\beta0$ domain; v_d and v_q are the transformed voltages in $dq0$ domains and θ is operating phase of the power system. The three phase current quantities i_a, i_b, i_c are also similarly transformed.

The decoupled active and reactive components i_q and i_d are controlled using the decoupled controller with the objective of minimizing the errors between their desired value and measured value. For i_d, a DC-link voltage control approach is used which aims to achieve an active power flow balance in the system. I_{qref} is set to 0 with the aim of unity power factor operation.

$$i_{d_{ref}} = K_{p1}\left(V_{dc_{ref}} - V_{dc}\right) + K_{i1}\int\left(V_{dc_{ref}} - V_{dc}\right)dt \tag{6}$$

$$v_{dc_{ref}} = v_d + 2\pi f L_s i_q - (K_{p2}\left(i_{d_{ref}} - i_d\right) + K_{i2}\int\left(i_{d_{ref}} - i_d\right)dt) \tag{7}$$

$$v_{q_{ref}} = v_q - 2\pi f L_s i_d - (K_{p3}(0 - i_q) + K_{i3}\int(0 - i_q)dt) \tag{8}$$

Here, K_{p1} and K_{i1} are the proportional and integral gains of the PI voltage controller, respectively; K_{p2} and K_{i2} are the gain values for the first PI current controller; K_{p3} and K_{i3} are the gain values for the second PI current controller; and Ls is the source inductance value.

The Park inverse transformation is then applied to the obtained reference values v_{dref} and v_{qref}. Finally, with the help of the PWM technique, gate switching pulses S_{abc} are obtained which control the operation of the AC/DC converter.

$$\begin{bmatrix} v_{\alpha,ref} \\ v_{\beta,ref} \end{bmatrix} = \begin{bmatrix} sin\theta & -cos\theta \\ cos\theta & sin\theta \end{bmatrix} \begin{bmatrix} v_{d,ref} \\ v_{q,ref} \end{bmatrix} \tag{9}$$

3 Methodology

The proposed methodology with adaptive neuro fuzzy inference system (ANFIS) architecture for the priority determination of electric vehicles' charging is shown in Fig. 2. The ANFIS combines the advantages of FIS and artificial neural networks into one single hybrid system. In this method, an initial fuzzy model along with its inputs is derived with the help of rules extracted from the input output data of the system that is being modelled. The neural network is then used to fine tune the rules of the initial fuzzy model to produce the final ANFIS model of the system [10]. The Sugeno fuzzy model is used in this research which has a fuzzy antecedent and a functional consequent. For simplicity of illustration, we have considered a 2-input system. The rules may be stated as follows:

$$\begin{aligned} &Rule1: If\ x\ is\ A1\ and\ y\ is\ B1\ then\ f1 = p1x + q1y + r1 \\ &Rule2: If\ x\ isA2\ and\ y\ is\ B2\ then\ f2 = p2x + q2y + r2 \\ &\vdots \\ &Rulek: If\ x\ is\ Ak\ and\ y\ is\ Bk\ then\ fk = pkx + qky + rk \end{aligned}$$

Here x and y are the crisp inputs; $A1$, $A2$, $A3$, … and $B1$, $B2$, $B3$, … are the linguistic variables (small, medium, etc.) associated with the inputs x and y respectively; f corresponds to the weighted average.

In this paper four linguistic variables (inputs) have been taken into account: Battery's State of Charge, SOC (I_1), Time of Stay, TOS (I_2), Distance to be Travelled (I_3) and Battery's Capacity (I_4), and for each linguistic variable, 3 linguistic values have been considered: Low, Medium and High. A generalized bell shape membership function has been chosen to provide an adequate representation for each of the linguistic values. Twelve nodes are present in each of the layers 1 to 4. The final layer outputs a single priority value 'p' with a value ranging between 0 and 1 for a particular EV.

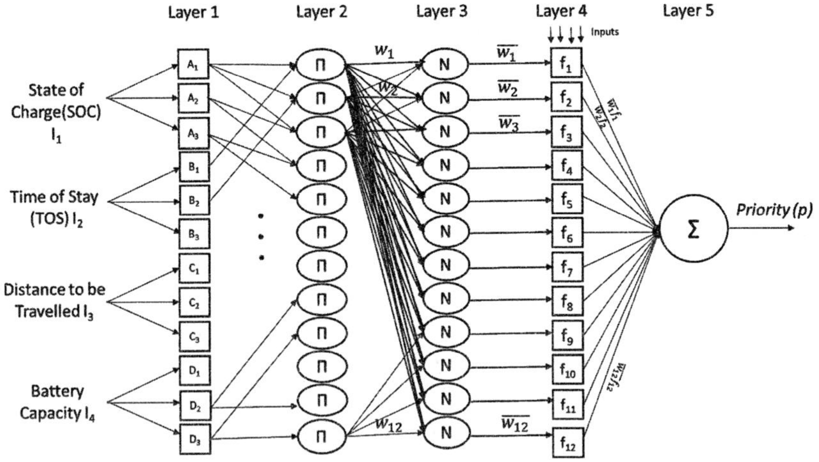

Fig. 2. ANFIS architecture for priority determination for EV charging

Table 1. ANFIS modelling criterion for prioritized EV charging

ANFIS parameters	Value
Number of inputs	4
Membership function	Generalized bell
Number of MFs	3
Algorithm	Hybrid learning algorithm
Number of iterations	50
Fuzzy system	Sugeno type first order system
Output type	Linear
Step size	Initial 1.1 decreased to 0.9
Nodes	12
Data parameter ration	≈84

Initially, a fuzzy model along with its inputs is derived with the help of rules extracted from the input output data of the system. ANFIS is then used to fine tune the rules of the initial fuzzy model to obtain the optimized premise and consequent parameters. Once the premise parameters are obtained, the overall output can be expressed as a linear combination of the consequent parameters. Table 1. presents the details of the ANFIS parameters and their values as used in the proposed methodology.

For the purpose of training, a Hybrid Training Algorithm from MATLAB's Neuro Fuzzy Designer has been utilized. In the forward pass, consequent parameters are first optimized by the least squares estimate, keeping the premise parameters fixed. Once the consequent parameters are obtained, error signals, which are the derivatives of the squared error with respect to each node output, propagate backward from the output layer to the input layer in the backward pass. Premise parameters are updated along the way by using a gradient descent algorithm. The error function used here is the Root Mean Squared Error (RMSE) given by:

$$RMSE = \sqrt{\frac{\sum_{i=1}^{n}(p_i - \widehat{p_i})^2}{n}} \tag{10}$$

Here, n is number of samples in the training batch; p is the value of the estimated priority and $\widehat{p_i}$ is the value of the target priority for the ith data point in the training batch.

The training process is carried out until either the total number of epochs is reached or the desired minimized value of error is obtained.

Each vehicle's priority is determined using the proposed modified ANFIS model architecture. After the comparison between the priority values of each of the EVs, the vehicle with the highest priority is determined in order to get charged first. The converters and their control strategies defined in the previous sections are utilized for charging the EV with a smooth DC ripple free voltage from an optimally integrated grid connected EV charging system. As the vehicles charge, the values of their linguistic variables change resulting in switching of EVs from time to time during the charging process. In order to prevent very rapid switching between EVs, a delay element has also been incorporated which allows the determination of priority only after a delay of 40s.

4 Experiment

The Voltage Oriented control was designed in Simulink and the ANFIS controller is applied with the EVC setup developed in Fig. 1. On starting the simulation, the State of Charge, Time of Stay, Battery capacity and Distance to be travelled is taken into consideration. The increasing SoC implies battery is charging successfully. When Battery's State of Charge is less than 70% the Constant Current charging is done and when battery State of Charge is more than 70% constant voltage charging is done. The battery State of Charge is continuously monitored as shown in Fig. 3 and fed to controller which controls the Switch which decides whether to charge in CC or CV mode.

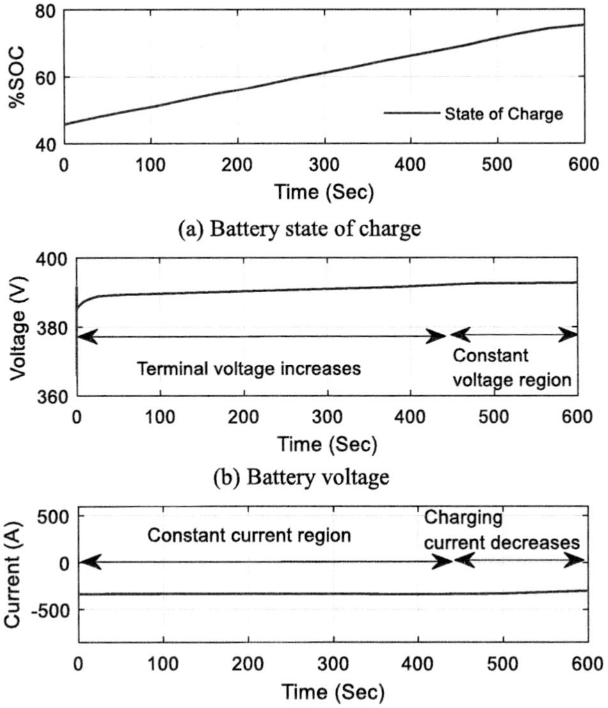

(b) Battery current

Fig. 3. Graphs showing CC and CV region of charging

Further, the action of the developed approach for prioritising the EVC and the corresponding battery characteristic for the three EVs are shown in Fig. 4. The final priority which is determined varies between 0 and 1, when it is 0 eV isn't charging and when it is 1 eV charges.

From the above results, it is identified that the Voltage profile shows very less variation which is a good indicator for VOC operation. Besides, the bidirectional DC-DC converter is working to produce ripple free DC voltage and charging the EV with a constant voltage. The developed ANFIS controller charges battery according to assigned priority. When $0 < T < 60$ EV 1 charge, $60 < T < 80$ EV 2 charges, $80 < T < 120$ EV 3 charges, $120 < T < 140$ EV 2 chargers and so on until all EVs are charged to 100%. In light of the results obtained, it can be said that the voltage-oriented control and bidirectional DC-DC converter have been successfully designed. The controller seems to be correct and stable. It charges the EV with ripple free DC voltage.

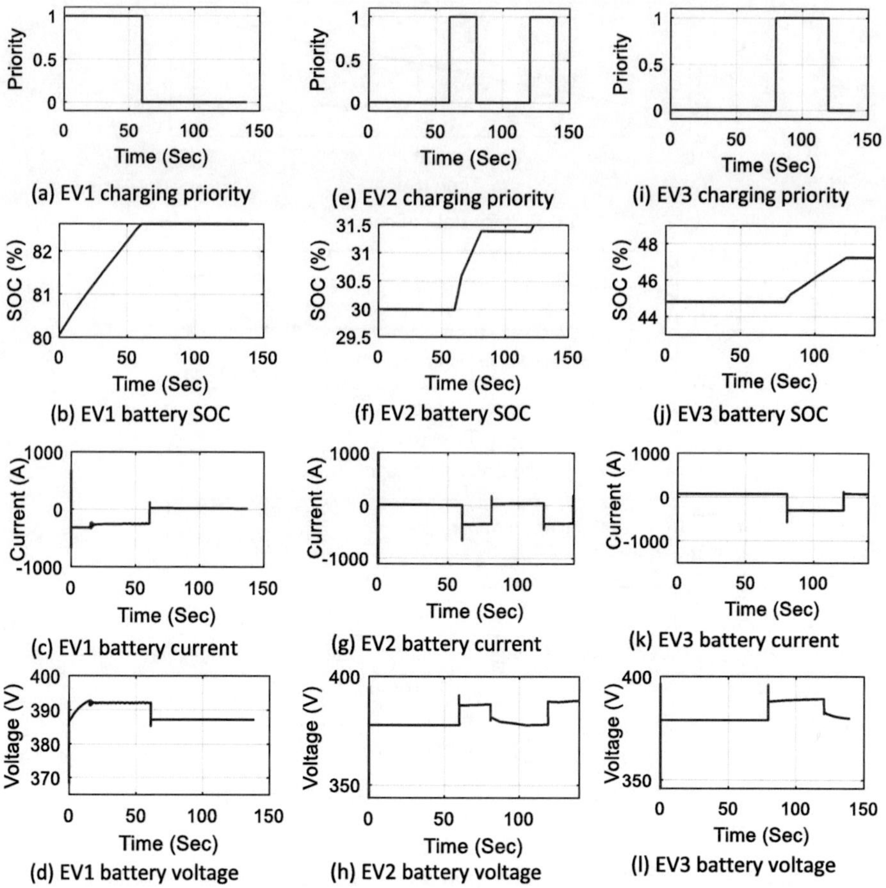

Fig. 4. EV charging characteristics

5 Conclusion

This paper developed an adaptive neuro fuzzy inference system for electric vehicles charging based on their priority. A neuro fuzzy system determines the prioritization of electric vehicle charging based on vehicles battery SoC, Time of Stay, Distance to be travelled and Battery Capacity. It receives an input from battery and user inputs the Time of Stay, Distance to be travelled and Battery Capacity which is utilized to determine priority. Further, a case of three EVs is considered to perform the simulations. The developed approach allocates the charging service to electric vehicles with higher priorities so that only one EV is charged at a time. The priority determination rules are calculated according to arrival time, departure time, and current state-of-charge (SoC), Distance to be Travelled and Battery Capacity.

References

1. Ul-Haq, A., Cecati, C., Strunz, K., et al.: Impact of electric vehicle charging on voltage unbalance in an urban distribution network. Intell. Ind. Syst. **1**, 51–60 (2015)
2. Nour, M., Said, S.M., Ali, A., Farkas, C.: Smart charging of electric vehicles according to electricity price. In: Proceedings of ITCE, pp. 432–437 (2019)
3. Fernandez, L., Roman, T., Cossent, R., et al.: Assessment of the impact of plug-in electric vehicles on distribution networks. IEEE Trans. Power Syst. **26**(1), 206–213 (2011)
4. E. Akhavan-Rezai, M.F., Shaaban, E.F., El-Saadany, F., Karray, F.: Priority-based charging coordination of plug-in electric vehicles in smart parking lots. In: 2014 IEEE PES on Innovative Smart Grid Technologies Conference (ISGT), pp.1–5, 19–22 February 2014
5. Kumar, K.N., Sivaneasan, B., So, P.L.: Impact of priority criteria on electric vehicle charge scheduling. IEEE Trans. Transp. Electrif. **1**, 200–210 (2015)
6. Chen, S., He, T., Tong, L.: Optimal deadline scheduling with commitment. Allerton Conference on Communication, Control, and Computing (2011)
7. Sayed, K., Gabbar, H.A.: Electric vehicle to power grid integration using three-phase three-level AC/DC converter and pi-fuzzy controller. Energies. **9**, 532 (2016). https://doi.org/10.3390/en9070532
8. Bao, K., Li, S., Zheng, H.: Battery charge and discharge control for energy management in EV and utility integration. In: IEEE Power and Energy Society General Meeting, pp. 1–8, July 2012
9. Chaturvedi, D.K.: Soft Computing. Springer Berlin Heidelberg, Berlin, Heidelberg (2008)
10. Buragohain, M.: Adaptive Network Based Fuzzy Inference System (ANFIS) as a Tool for System Identification with Special Emphasis on Training Data Minimization. Ph.D. thesis, Indian Institute of Technology, Guwahati, India (2008)

Clustering Algorithms as a Tool for Odour Classifications in Enose Developments

Yousif Albastaki(✉)

Ahlia University, Manama, Bahrain

Abstract. Odor classifications and identifications aims to identify an odor category a new volatile belongs to from a collection of preidentified categories. Odor classification is critical in the development of many applications, including food spoilage, disease diagnosing, drug smuggling detection and many other implementations. Electronic nose or simple Enose, which utilizes an array of gas sensors and pattern recognition algorithms such as machine learning algorithms are used in classifying and identifying odors. This paper is an attempt to utilize machine learning concepts algorithms to classify odors which can be used in electronic nose developments. The paper describes the concepts of ENoses and their applications. The paper then uses K-means clustering algorithm with R environment software to classify and identify different types of odors. Odors for three different items are collected using OMX-GR odor-meter sensor and the data fed to a clustering machine learning algorithm for classification. The result of the classification showed high level of accuracy. In this research we used three different items for the experiment part: coffee (C), a white board marker (M) and a traditional home-made perfume (P).

1 Introduction

For decades, researchers from different disciplines have been investigating odor classification methods aiming at achieving a perceptual space and promote odor transmission. Unfortunately, none of the proposed methods achieved widespread acceptances or provided an empirical evidence. Odor classification methods deals with solving the problem of recognizing a set of odor types by identifying a new volatile sample it belongs to. This classification of odors plays a significant role in developing ENoses, which is used in many applications like food spoilage detection and food control distribution, medical illness and diseases detection and diagnosis, drugs smuggling and explosive detection.

Human olfactory system [1] or sense of smell, is the sensory system used for classifying smells and identifying different type of odors. Many conducted research claims that of all the senses of human, the smell is the least understood. Despite decades of research, science can still offer no satisfying bases theory for research as to why a specific substance smells the way it does. Also, we do not understand or explain in any depth how we can differentiate the scent of an apple from that of an orange, or how a particular smell can produce long-forgotten memories of a distant time or place. This complicated system is depicted in Fig. 1. Human can smell through specialized sensory

A. M. A. Musleh Al-Sartawi et al. (Eds.): EAMMIS 2021, LNNS 239, pp. 46–56, 2021.
https://doi.org/10.1007/978-3-030-77246-8_5

cells, called olfactory sensory neurons, located high within the nose in a tiny patch of tissue which are directly attached to the brain. There is one odor receptor in each olfactory neuron and these receptors are activated by microscopical molecules released by substances around us.

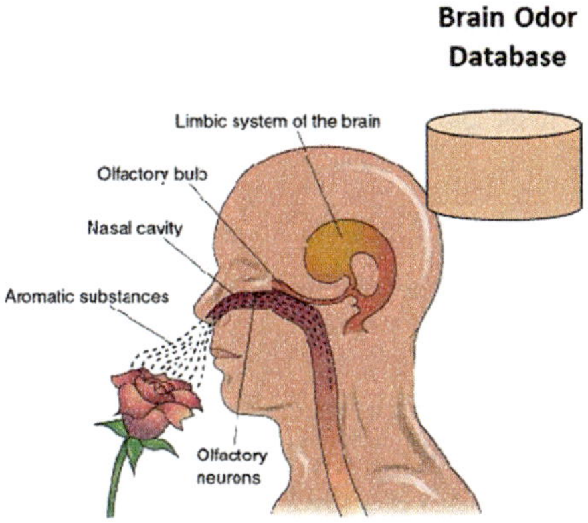

Fig. 1. Human olfactory system

The brain receives signals from the neurons indicating the detection of molecules which leads the smell identification. The number of scents in the atmosphere are much more than the number of receptors, and a mixture of receptors can be activated by any given molecule, producing a special representation in the brain. The brain stores or registers these images as a specific smell in a brain odor database.

Electronic noses or simply ENoses is an attempt to mimic a human olfactory system and the can be designed by utilizing an array of gas sensors supported by an algorithm used for pattern recognition.

This paper is an attempt to utilize machine learning clustering algorithm as a tool for odor classification which can be used in designing electronic noses. The remaining of the paper describes the K-mean algorithm, data collection method and implementation of clustering algorithm for classifying different type of odors.

2 Related Work and Theoretical Background

Advancements in materials, sensors and machine learning technology have contributed to a rapid expansion of research topics related to the Enose [2–6] with diverse applications over the last two decades. The following section will attempt to conduct a critical literature review aiming at identifying research gaps and learning from another research works in the field of Enose design.

Shen and Tao [7] in their article "Applying Electronic Nose Based on Odor Classification and Identification Technology in Detecting the Shelf Life of Fresh Fruits" described the process of boosting the market competitiveness and food safety of fresh fruit products using Electronic Nose techniques through odor classification means. The paper investigated the non-destructive testing (NDT) method for the shelf life of fresh fruits, it adapted the methods of literature analysis, comparative analysis, and experimental detection, as well as the Enose technology focused on odor classification and recognition. The research used an optimized e-nose sensor array, a support vector machine (SVM), a BP neural network model, and a chlorophyll prediction model to render spinach freshness grade discrimination and chlorophyll quantitative prediction using spinach as an example. Finally, the results of the experiments of Shen and Tao demonstrated that e-nose technology focused on odor classification and recognition can enhance fresh fruit shelf-life detection.

Mahat, et al. [8] in their research work "Artificial Odour Classification System" tackled the concept of odor classification by demonstrating multicollinearity in electronic nose (e-nose) classification and looks at some methodological options for dealing with the issue of classification. This research work used the feature extraction method to handle multicollinearity except for the variance-based criterion used to derive raw features which may not be sufficient for the ultimate objective of classification accuracy. As a result, the authors suggest adapting the feature selection approach as it selects only the most desirable functions and Wilk's Lambda and bounded Mahalanobis distance are two distance-based metrics used to evaluate the right features for classification. Clearly, Bounded Mahalanobis distance-based classification performs statistically better than Wilk's Lambda-based classification and as a result, the classification of e-nose with feature selection, according to this research work, is a good option for limiting the cost of experiments while maintaining good classification efficiency.

A recent published research work by Karakaya [9], "Electronic Nose and Its Applications: A Survey", the paper offers a detailed survey of Enose technology and its broad variety of application fields, based on a thorough review of algorithms proposed in the literature, as well as exploitation of relevant domains and potential future research topics. An important conclusion of this research article was the claim that despite the numerous advantages of Enose technology, more improvements in both hardware and software components are needed before ENoses can be used in operation.

Although much research conducted in the file of Enose and odor classification [10–14], it is obvious that there exists a research gape for reliably comparing and identifying odors or estimating their degree of similarity with precision and candidate for such achievement may exist in the utilization of machine learning algorithms such as clustering algorithms.

3 K-Means Clustering Algorithm

Clustering is generally the use of unsupervised techniques to group similar objects. Unsupervised refers to the problem of finding hidden structure within unlabeled data in machine learning. Clustering techniques are unsupervised in the sense that we do not determine the cluster labels in advance. The data structure characterizes the objects of interest and determines how the objects are best grouped.

K-Means [15, 16] is one of the most used unsupervised clustering machine learning algorithms to divide a given set of data into a set of k groups. Where k is the number of groups the analyst has pre-specified. It classifies objects in multiple groups (i.e., odor groups in our case), so that objects within the same cluster are as similar as possible, while objects from different clusters are as dissimilar as possible.

Different implementations of k-means algorithms are available today. The one used in this research work is based on the standard algorithm known as Hartigan-Wong algorithm [17], which defines the total variation within the cluster as the sum of square distances as follows:

$$W(C_k) = \sum_{x_i \epsilon C_k} (x_i - \mu k)^2$$

Where x_i design a data point belonging to the cluster C_k and μ_k is the mean value of the points assigned to the cluster C_k.

The complete square amount within the cluster measures the clustering's compactness and we want it to be as low as appropriate. The total within cluster can be defined as follows:

$$Total\ Within\ Cluster = \sum_{k=1}^{k} w(C_k)$$

$$TWC = \sum_{k=1}^{k} \sum_{x_i \in c_k} (x_i - \mu k)^2$$

Selecting the Value of K in K-means Clustering: The K-means algorithm is a popular data clustering algorithm. Using it requires pre-specifying the number of clusters in the data. Finding the suitable number of clusters for a specified information set is usually a trial-and-error method that makes it harder to decide what constitutes' right' clustering. The following paragraph tackles three different methods used in selecting the value of k in the k-means clustering algorithm [18].

Selecting the Value of k Using Elbow Method Algorithm: The fundamental concept of the elbow rule is to use a matrix of range to offer a sequence of K values between the sample points in each cluster and the cluster centroid. The sum of the squared error amount (SSE) is used as an indicator of efficiency. Iterate the value of K and calculate the value of SSE. Smaller values show the convergence of each cluster [19].

Selecting the Value of k Using the Gap Statistic Algorithm: This algorithm is used to determine the number of clusters of data sets with unknown classification numbers. The fundamental concept of Gap Statistics is to introduce measurements of reference and to calculate the amount of the Euclidean squares between two measurements in each season. The clustering outcomes of the built zero-mean distribution reference are compared to the optimum number of clusters in the information set [20].

Selecting the Value of k Using the Silhouette Coefficient Algorithm: The fundamental concept of this algorithm is based on combining the two factors of cohesion and resolution. The algorithm describes the cohesion to be the similarity between the object and the cluster. This comparison is accomplished by the Silhouette value in the range 1–1. If value of the Silhouette is near to 1, this suggest that the object and the cluster have a close relationship. The model is appropriate and acceptable if a data cluster is produced in a model with a comparatively elevated Silhouette value [21].

This research work will test all the three algorithms to find the appropriate value of k to be used in the k-mean algorithm.

4 Methodology and Data Collection

In this research we used three different items for the experiment part: coffee (C), a white board marker (M) and a traditional home-made perfume (P). We collected 900 samples for the three items, each 300 samples in n interval of 5 min for each item using the OMX-GR odor-meter sensor. OMX-GR handheld Odor Meter is a popular simplified tool for odor collection and analysis, which enables to indicate the relative strength and odor classification numerically. Real-time relative odor strength level can be measured continuously with a pair of semiconductor gas sensors provided to measure odor classification. The sensor supports measuring odor strength in the range between 0 to 999 and odor classifications in the range between 0 to 99. Figure 2 illustrates a small portion of the sample data collected for this research work.

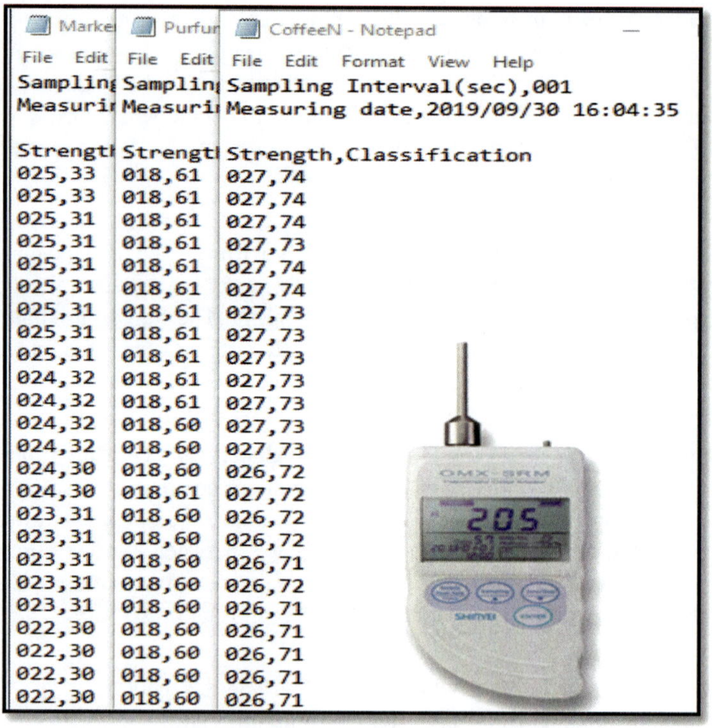

Fig. 2. Sample odor collect for the experiment.

5 Analysis and Results

In this part the readings for three different odors representing three different items will be analyzed. A total of 900 readings were taken for the items (300 samples for each) over 5 sessions of 60 readings each. Each reading has a classification or ID and strength. Figure 3 shows the summary generated by R software for the collected samples for each item as follows:

a) The classification values for the first item (Coffee) odor range between 61 and 75 whereas the strength values range between 22 and 49. The average reading for coffee classification was 70 and for the strength was 30.
b) The classification values for the second item (Home Made Perfume) odor range between 45 and 56 whereas the strength values range between 8 and 18. The average reading for this homemade perfume classification was 51.71 and for the strength was 13.88.
c) The classification values for the third item (White Board Marker) odor range between 13 and 33 whereas the strength values range between 13 and 25. The average reading for this homemade perfume classification was 21.69 and for the strength was 17.67.
d) The classification odor values for all the items together range between 13 and 75 whereas the strength values range between 8 and 49. The average reading for classification of all the items together was 47.84 and the average for the strength was 13.54.

a) Summary Data Collected for Coffee

```
Classification    Strength
Min.   :61.0   Min.   :22.00
1st Qu.:66.0   1st Qu.:24.00
Median :71.0   Median :30.00
Mean   :70.1   Mean   :30.07
3rd Qu.:74.0   3rd Qu.:36.00
Max.   :75.0   Max.   :49.00
```

b) Summary Data Collected for Perfume

```
Classification      Strength
Min.   :45.00   Min.   : 8.00
1st Qu.:48.00   1st Qu.:11.00
Median :53.00   Median :14.00
Mean   :51.71   Mean   :13.88
3rd Qu.:55.00   3rd Qu.:17.00
Max.   :56.00   Max.   :18.00
```

c) Summary Data Collected for Marker

```
Classification      Strength
Min.   :13.00   Min.   :13.00
1st Qu.:18.00   1st Qu.:16.00
Median :21.00   Median :17.00
Mean   :21.69   Mean   :17.67
3rd Qu.:24.00   3rd Qu.:19.00
Max.   :33.00   Max.   :25.00
```

d) Overall Collected Data Summary

```
     Time         Classification      Strength      Class
Min.   :  1.00   Min.   :13.00   Min.   : 8.00   C:300
1st Qu.: 75.75   1st Qu.:24.00   1st Qu.:15.00   M:300
Median :150.50   Median :53.00   Median :18.00   P:300
Mean   :150.50   Mean   :47.84   Mean   :20.54
3rd Qu.:225.25   3rd Qu.:66.00   3rd Qu.:24.00
Max.   :300.00   Max.   :75.00   Max.   :49.00
```

Fig. 3. Summary report for collected samples

Figure 4 shows the distribution of the collected sample of odors for each item with respect to time. The red color represents the first item (Coffee), the blue color represents the second item (White Board Marker) and green color represents the third item (Home Made Perfume).

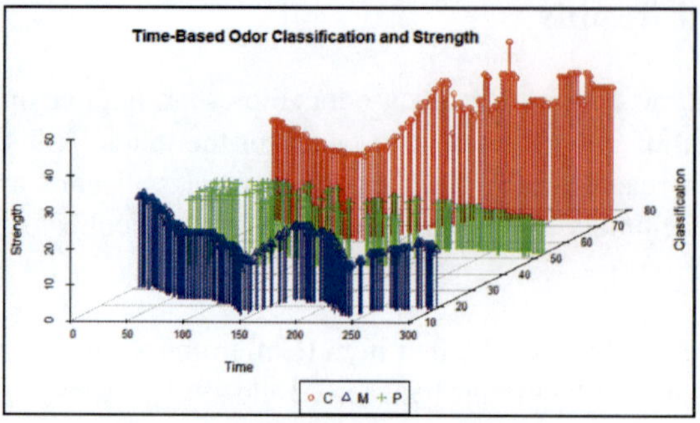

Fig. 4. Distribution of the sampled odor

6 Finding the Appropriate Value of K

To implement the k-means algorithm, we need to find the appropriate value of k to be used in the algorithm to decide how many clusters to use for segmenting the data. If too many clusters are specified, this may lead to too specific groups of data to be useful; conversely, specifying too few may result in heterogeneous grouping. Therefore, to solve this problem, this research work attempts to test and analyze three K-value selection algorithms described earlier: Elbow Method, Gap Statistic, and Silhouette Coefficient by and using the collected data set. R open-source software is used to test each algorithm as follows:

Figure 5 shows the result of using the elbow method with the R code used in top right side of the figure and the suggested k value is 4.

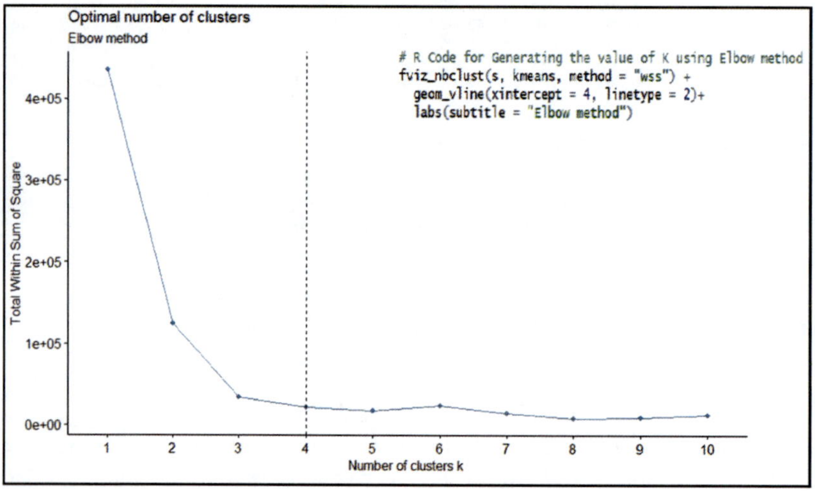

Fig. 5. Value of K using the elbow method.

Figure 6 shows the result of using the Gap Statistics method with the R code used shown in the figure and the suggested k value is 4.

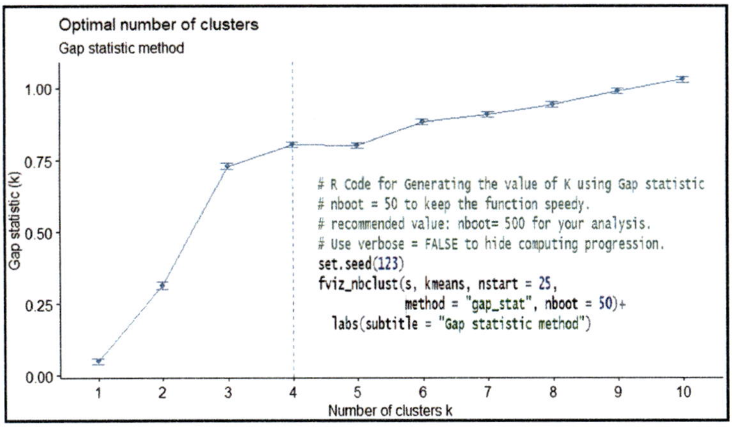

Fig. 6. Value of K using gap statistic method.

Figure 7 shows the result of using the Silhouette method with the R code used shown n the in the figure and the suggested k value is 3.

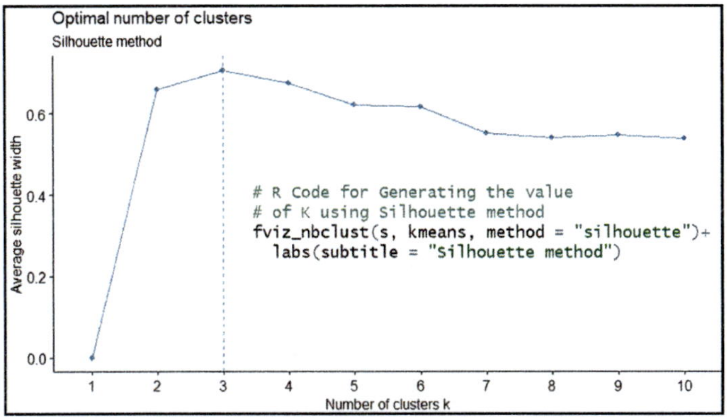

Fig. 7. Value of K using Silhouette method.

In the previous section, three methods of k-value selection algorithms, such as Elbow Method, Gap Statistic and Silhouette Coefficient are used to cluster the data set to obtain the K value and the clustering result of the data set. Table 1 shows the proposed value of k when implemented each algorithm.

The following section with attempt to test and analysing the values of k obtained from each algorithm. The test will be conducted for two values of k, 3 and 4.

Table 1. Values of k proposed by different algorithms.

Algorithm	Suggested value for k
Elbow method	4
Gap statistic method	4
Silhouette method	3

7 Testing the Algorithm for Different Values of k

The aim of clustering analysis is to identify patterns in our odor dataset and create groups according to those patterns. Therefore, if two odor samples have similar characteristics, that means they have the same pattern and consequently, they belong to the same group. By doing clustering analysis we should be able to check what features usually appear together and see what characterizes a group.

Testing the algorithm for k = 4 is illustrated in Fig. 8. It is obvious that there exists an overlap in cluster 1 and 2 which indicates that the algorithm could not classify the sample odors correctly.

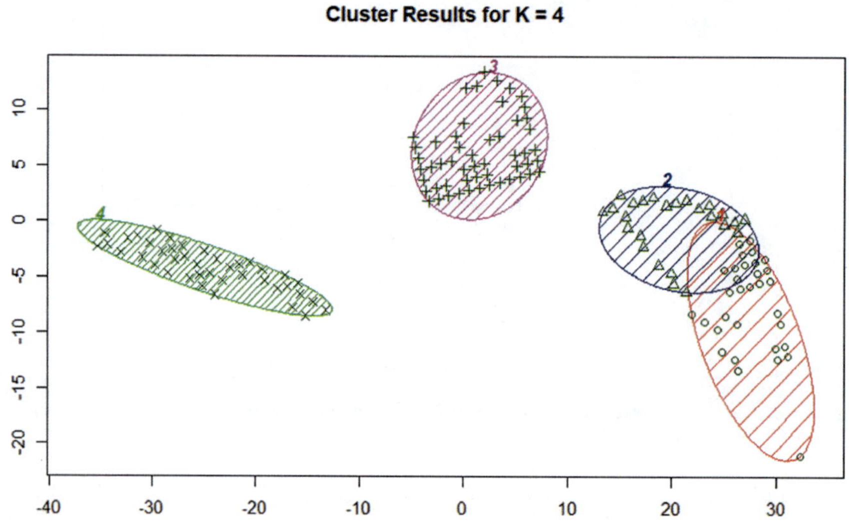

	1	2	3	4
C	149	151	0	0
M	0	0	0	300
P	0	0	300	0

Fig. 8. Cluster results for K = 4

Testing the algorithm for k = 3 is illustrated in Fig. 9. The results for testing value of k = 3 demonstrates 100% accuracy in classifying the sample odors into three groups.

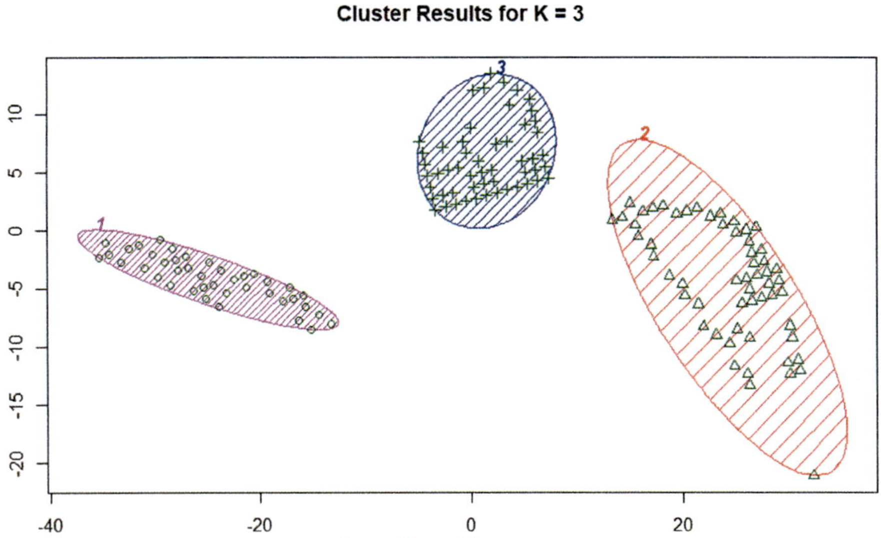

	1	2	3
C	0	300	0
M	300	0	0
P	0	0	300

Fig. 9. Cluster results for K = 3

8 Conclusion

Different methods and algorithms are used by many researchers to attempt in collecting odors and classifying them for many applications. This classification methods of odors, plays a significant role in developing ENoses, which are used in many applications like food spoilage detection and food control distribution, medical illness and diseases detection and diagnosis, drugs smuggling and explosive detection. This research work used three different items for the experiment part: coffee (C), a white board marker (M) and a traditional home-made perfume (P). The collected odors were approximately 900 samples for the three items, each 300 samples in n interval of 5 min for each item using the OMX-GR odor-meter sensor. K-means machine learning algorithm was used for the classifying purposes and implemented in R environment. The results demonstrated very high level of accuracy in terms of classifying and identifying different types of odors.

References

1. Benigus, V., Prah, J.: Olfaction: anatomy, physiology and behavior. Environ. Heal. Perspect. **44**, 15–21 (1982)
2. Schleif, F.M., Hammer, B., Monroy, J., Gonzalez-Jimenez, J., Blanco, L., Biehl, M., Petkov, N.: Odor recognition in robotics applications by discriminative time-series modeling. Pattern Anal. Appl. **19**(1), 207–220 (2016)
3. Gonzalez-Jimenez, J., Monroy, J., Blanco, L.: The Multi-chamber electronic nose–an improved olfaction sensor for mobile robotics. Sensors **11**(6), 6145–6164 (2011)

4. Monroy, J., Gonzalez-Jimenez, J.: Real-time odor classification through sequential bayesian filtering. In: 16th International Symposium on Olfaction and Electronic Nose (ISOEN), Dijon, Burgundy, France (2015)
5. Al-Bastaki, Y., Al-Mutawah, K.: ANN based approach to integrate smell sense in multimedia systems. IJTD **2**(3), 14–23 (2011)
6. Karlik, B., AL-Bastaki, Y.: Real time monitoring odor sensing system using OMX-GR sensor and neural network. WSEAS Trans. Comput. February, 482–175 (2004)
7. Shen, H., Tao, J.: Applying electronic nose based on odour classification and identification technology in detecting the shelf life of fresh fruits. Chem. Eng. Trans. 68. The Italian Association of Chemical Engineering (2018)
8. Mahat, N., Masnan, M., Shakaff, A., Muhd, A., Abdul Kadir, K.: Artificial odour classification system. In: Albastaki, Y., Albalooshi, F., (eds.) Book Chapter. Electronic Nose Technologies and Advances in Machine Olfaction. IGI Global (2018)
9. Karakaya, D., Ulucan, O., Turkan, M.: Electronic nose and its applications: a survey. Int. J. Autom. Comput. **17**(2), 179–209 (2020)
10. Montuschi, P., Mores, N., Trove, A., Mondino, C., Barnes, P.: "The electronic nose in respiratory medicine. Respiration **85**(1), 72–84 (2013). https://doi.org/10.1159/000340044
11. Saraoglu, H., Edin, B.: E-nose system for anesthetic dose level detection using artificial neural network. J. Med. Syst. **31**(6), 475–482 (2007). https://doi.org/10.1007/s10916-007-9087-7
12. Wijaya, D., Sarno, R., Zulaika, E., Sabila, S.: Development of mobile electronic nose for beef quality monitoring. Proc. Comput. Sci. **124**, 728–735 (2017). https://doi.org/10.1016/j.procs.2017.12.211
13. Siadat, M., Losson, E., Ghasemi-Varnamkhasti, M., Mohtasebi. S.: Application of electronic nose to beer recognition using supervised artificial neural networks. In: Proceedings of International Conference on Control, Decision and Information Technologies, IEEE, Metz, France, pp. 640–645 (2014). DOI: https://doi.org/10.1109/CoDIT.6996971.
14. Peng, P., Zhao, X., Pan, X., Ye, W.: Gas classification using deep convolutional neural networks. Sensors 18(1), Article number 157 (2018). DOI: https://doi.org/10.3390/s18010157.
15. Lloyd, S.P.: Least squares quantization in PCM. Technical report RR-5497, Bell Lab, September (1957)
16. MacQueen, J.B.: Some methods for classification and analysis of multivariate observations. In: Le Cam, L.M., Neyman, J., (eds.) Proceedings of the fifth Berkeley symposium on mathematical statistics and probability, vol. 1, pp. 281–297. University of California Press, California (1967)
17. Hartigan, J.A., Wong, M.A.: A k-means clustering algorithm. Appl. Statist. **28**, 100–108 (1979)
18. Yuan, C., Yang, H.: Research on k-value selection method of k-means clustering algorithm. J. Multidiscip. Sci. J. (ISSN 2571–8800) (2019). (Published by MDPI)
19. Bholowalia, P., Kumar, A.: EBK-means: a clustering technique based on elbow method and k-means in WSN. Int. J. Comput. Appl. (0975–8887) 105(9) (2014)
20. Tibshirani, R., Walther, G., Hastie, T.: Estimating the number of clusters in a data set via the gap statistic. J. R. Statist. Soc. Ser. B (Statist. Methodol.) **63**, 411–423 (2001)
21. Kaufmn, I., Rousseeuw, P.J.: Finding Groups in Data an Introduction to Cluster Analysis. John Wiley & Sons, Hoboken (1990)

Bidirectional Octonion Long-Short Term Memory Recurrent Neural Networks for Speech Recognition

Shobarani Salvadi[1(✉)], D. Nagendra Rao[2], and Srinivasan Vatsal[3]

[1] Computer and Engineering, Vardhaman College of Engineering, Shamshabad, Telangana, India
s.shobarani@vardhaman.org
[2] Electronics and Communication Engineering, Hitech College of Engineering, Moinabad, Telangana, India
[3] Electrical and Electronics Engineering, Moinabad, Telangana, India

Abstract. Over the few decades onwards all the researchers got marvelous attention on machine learning due to a lot of applications in different fields like image processing, speech processing, etc. Automatic Speech Recognition (ASR) is an application of speech processing that achieved tremendous results due to the usage of recurrent neural networks (RNN). A unique type of recurrent neural network is long short-term memory (LSTM), to keeps away from the problem of long term dependencies. In this paper, multidimensional octonion algebra is used to process the input entities with multidimensions efficiently, when compared to real-valued models octonion numbers and octonion neural networks solve many tasks with less learning parameters. We propose a new octonion value long-short term memory (OVLSTM) to efficiently represent long-term dependencies among features in speech sequences prediction. The TIMIT dataset taken for experiments for speech recognition and results are compared with QLSTMS and LSTMs, OVLSTM with less learning parameters reaches better results.

Keywords: Recurrent neural network · Octonion valued long short term memory (OVLSTM) · Speech recognition

1 Introduction

During the most recent decade, deep neural network (DNN) has experienced a wide achievement in various domain applications. Specifically, automatic speech recognition systems (ASR) achievements have been amazingly improved with the usage of DNNs. Among them, recurrent neural networks (RNN) [1] have been appeared to adequately encode input sequences, expanding the exactness of neural network-based (ASR) systems [2]. However, vanilla RNNs endures from evaporating/detonating issues [3], or the absence of a memory system to remember patterns in exceptionally long or short sequences. These issues have been addressed by the introduction of long-short term memory (LSTM) RNN [4] with a gates

A. M. A. Musleh Al-Sartawi et al. (Eds.): EAMMIS 2021, LNNS 239, pp. 57–64, 2021.
https://doi.org/10.1007/978-3-030-77246-8_6

mechanism that permits the model to refresh or overlook data in memory cells, and to choose the content cell state to uncover in a network hidden state.

Speech recognition existing around us last few decades onwards in our everyday lives, like Siri in our phones, smartwatches, Amazon (Alexa), Google (google assistant), Cortana, etc. Speech recognition became has one of the mainstream in research due to the emergence of DNNs, and efficiency also improved drastically. Two main problems present in speech recognition are feature vector and variable-length present in the conventional model addressed by the hidden Markov model (HMM) with the help of dynamic time wraping [5].

Abilities of computers have been enhanced by the usage of deep learning algorithms so it gets what humans can do in all aspects, which includes speech processing. Especially from the introduction of artificial intelligence in speech recognition, received more attention for the last few decades, being the principle strategy for correspondence among individuals [2–4]. In recent ASR models, with unchanged input acoustic features, LSTM [12] has reached state-of-the-art performance. In speech processing, three major approaches are there, first one acoustic-phonetic approach –where phonetic words are broadly classified based on a set of features such as pitch, format frequency, and cepstral coefficient. In the second approach-pattern recognition does not require knowledge of speech but requires the Hidden Markov Model (HMM), template matching, and Artificial Neural Networks (ANN). The third approach is called artificial intelligence [9, 10]; the recognition process depends on the person's intelligence in analyzing, visualizing, and decision on the acquired acoustic features. Nowadays second and third approaches are used in ASR because of the high accuracy rate.

Composing of time-frame based signal sequence in speech with multidimensional acoustic features like Mel filter along with first and second-order time derivates are used in traditional ASR systems. By considering this, neural network architecture partially addresses the internal or external dependencies between the time-frames by increasing learning capability. To address this issue, high-dimensional algebra-based new octonion–valued neural networks are proposed in this paper.

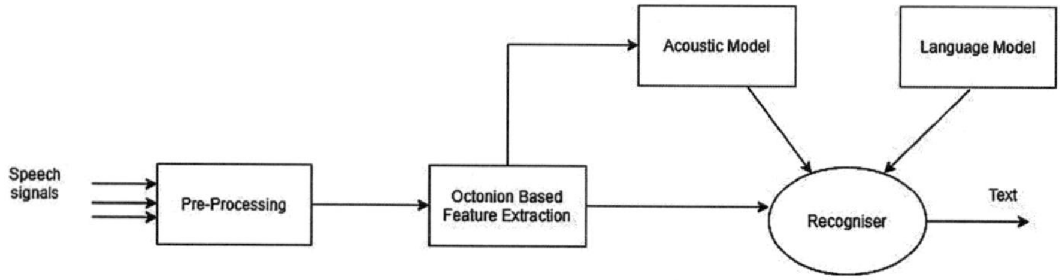

Fig. 1. Octonion based speech recognition.

Octonions with 8-dimensions are an extension of quaternion and hypercomplex numbers, which doesn't fall under the Clifford algebra category shown in Fig. 1. Octonions are applied in image processing, speech processing, and application fields with more dimensional objects. With one real and seven imaginary parts of octonion numbers, the pattern recognition task performed efficiently. LSTMs with octonion numbers eliminate interdependencies that exist among input features group and are trained with less learning parameters during the learning process than quaternion LSTMs and real-valued LSTMs. Deep octonion networks [16] are non-associative division algebra, whereas Clifford algebras [18] are associative but not division algebra, a multivalued bidirectional associative memory operating on a complex domain [17] and Octonion-valued neural networks [14, 25]. Considering all these facts into account, we introduced octonion-valued LSTMs for speech recognition.

The rest of the paper is organized as follows; Sect. 2 introducing octonion numbers as input features of eight dimensions with their domain-specific characteristics are discussed, in Sect. 3 more accuracy is achieved in speech recognition with OVLSTM when compare to real-valued LSTMs and QLSTMs with the ability to handle interdependencies among input features, experiments are discussed in Sect. 4 and finally, the conclusion is discussed in Sect. 5.

2 Octonion Algebra

The octonion algebra $\rtimes$ defines operations between octonion numbers. A quaternion number $\mathbb{Q}$ is an extension of a composite number defined in a four-dimensional space whereas an octonion number is hypercomplex with 8-dimensional, which is an extension of complex and quaternion numbers which has one real element and seven imaginary elements with basis $1, e_1, e_2, e_3, e_4, e_5, e_6, e_7$.

The octonion $\rtimes$ algebra is defined as.

$$\rtimes := \left\{ x = \sum_{0}^{7} [x]_q e_q | [x]_0, [x]_1 \ldots.., [x]_7 \in R \right\} \tag{1}$$

where $e_0 = 1$, e_q represent the octonion units, $0 \leq q \leq 7$ $e_1, e_2, \ldots, e_7$ are hyper imaginary units that satisfy $e_i^2 = -1$ and multiplication rules [14] given in Table 1.

Table 1. Octonion multiplication table, multiplying the elements in the *i*th row and *j*th column.

×	e0	e1	e_2	e_3	e_4	e_5	e_6	e_7
e0	e_0	e_1	e_2	e_3	e_4	e_5	e_6	e_7
e1	e_1	$-e_0$	e_3	$-e_2$	e_5	$-e_4$	$-e_7$	e_6
e_2	e_2	$-e_3$	$-e_0$	e_1	e_6	e_7	$-e_4$	$-e_5$
e_3	e_3	e_2	$-e_1$	$-e_0$	e_7	$-e_6$	e_5	$-e_4$
e_4	e_4	$-e_5$	$-e_6$	$-e_7$	$-e_0$	e_1	e_2	e_3
e_5	e_5	e_4	$-e_7$	e_6	$-e_1$	$-e_0$	$-e_3$	e_2
e_6	e_6	e_7	e_4	$-e_5$	$-e_2$	e_3	$-e_0$	$-e_1$
e_7	e_7	$-e_6$	e_5	e_4	$-e_3$	$-e_2$	e_1	$-e_0$

An octonion x can be represented as $x = x_0 + \vec{x}$, where $\vec{x} = x_1e_1 + x_2e_2 + x_3e_3 + x_4e_4 + x_5e_5 + x_6e_6 + x_7e_7$ and x_0 are called the vector part and the real part of x.

The addition of octonions is defined by a + b $=\sum_{q=0}^{7}([a]_q + [b]_q)$ e_q, from the multiplication table we can infer that ⋊ is not commutative nor associative because.

$$e_ie_j = -e_je_i \neq e_je_i \; \forall i \neq j,\ 0 < i, j \leq 7, \text{ and} \tag{2}$$

$$\left(e_ie_{j)}\right)e_k \neq e_i\left(e_je_k\right) = -e_i(e_je_k)\, for\ \ i \neq j \neq k \tag{3}$$

The norm of an octonion defined by $|x| = \sqrt{x_0^2 + x_1^2 + x_2^2 + x_3^2 + x_4^2 + x_5^2 + x_6^2 + x_7^2}$ and the conjugate of an octonion is $\overline{x{=}x_0{+}\vec{x}}$[13].

According to the Cayley-Dickson process of octonion ⋊ can be represented in quaternion algebra as:

$$p = p' + p''e, p', p'' \in \mathbb{Q}; \quad \mathbb{Q} = \{p_0 + p_1i + p_2j + p_3k | p_0, p_1, p_3 \in \mathbb{R}\} \tag{4}$$

Here, $\mathbb{Q}$ is the quaternion algebra with {1, i, j, k}, $i^2 = j^2 = k^2\ {-1}$, $ijk = -1$ and e is imaginary unit [7].

3 Octonion Long-Short Term Memory Neural Networks

Long short-term memory (LSTM) is a special kind of recurrent neural network (RNN) in artificial intelligence of deep learning field, introduced by Hochreiter and schmidhuber in 1997. Long term dependencies problem present in RNN can be avoided by LSTM. Bidirectional LSTM [11] processes the input sequences in both directions.i.e. forward and backward to get a good accuracy rate. With the octonion algebra properties and octonion valued long short-term memory (OVLSTM) speech recognition achieved more accuracy than RNN, on the grounds of octonion representation of the input, output, weights, and bias. The core element in the LSTM architecture is the cell [8] shown in Fig. 2, which controls dependencies between the features of input sequences and also stores the data for a longer length of time.

Mathematical representations of LSTM cells can describe by the following equations iteratively at time t are as follows:

$$\mathrm{i_t} = \sigma(w_{ix}x_t + w_{ix}x_{t-1} + b_i) \tag{5}$$

$$\mathrm{f}_t = \sigma(w_{fx}x_t + w_{cf}x_{t-1} + b_f) \tag{6}$$

$$c_t = \mathrm{f_t}\Theta c_{t-1} + i_t \cdot tanh(w_{xc}x_t + w_{ch}h_{t-1} + b_c) \tag{7}$$

$$o_t = \sigma(w_{ox}x_t + w_{oh}h_{t-1} + b_c) \tag{8}$$

$$h_t = o_t \cdot tanh(c_t) \tag{9}$$

where ***i, f, o, c*** and ***h*** are input, forget, output gates, cell states and the hidden vector of a LSTM with the same dimensionality with logistic sigmoid function $\boldsymbol{\sigma}$ and ***tanh*** activation functions, ***b'*** s are the bias vector and ***w***'s are the weight matrices for connecting different gates [6].

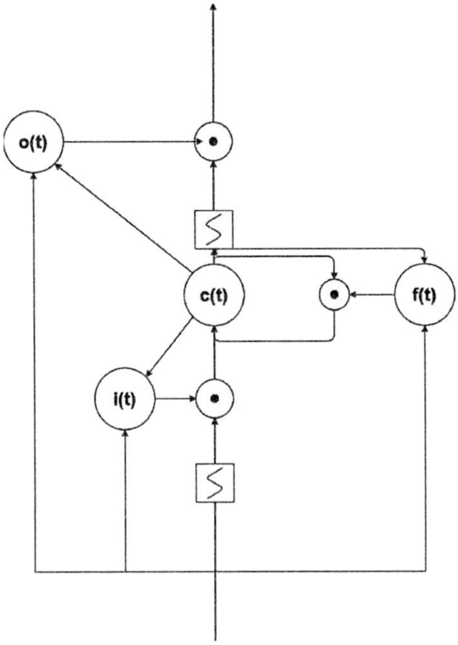

Fig. 2. LSTM Cell structure

4 Experiments

The proposed OVLSTM performance and effectiveness are evaluated on a phoneme recognition task. This division provides information about octonion acoustic features, the dataset, and finally observed results discussion. In this proposed method we used the TIMIT dataset [7] with a 462-speaker training dataset, a separate 50 speakers development set used, and a core test dataset with 192 sentences. An acoustic feature of octonion is formed as follows:

$$\rtimes = \mathbb{Q} + \mathbb{Q}'e \tag{10}$$

Here, $\mathbb{Q}$ a quaternion-valued frequency f at time t and energy $e\ (f,\ t)$ filter band frequency f defined as follows:

$$Q(f,t) = 1 + e(f,t)i + \frac{\partial e(f,t)}{\partial t}j + \frac{\partial^2 e(f,t)}{\partial^2 t}k, \tag{11}$$

$$Q(f,t) = 0 + \frac{\partial e(f,t)}{\partial t}i + \frac{\partial^2 e(f,t)}{\partial^2 t}j + \frac{\partial^3 e(f,t)}{\partial^3 t}k, \tag{12}$$

Slope view is described in the first derivative in Eq. 11, the concavity view described in the second-time derivative in 12, and third is describing the rate of change in the second derivative as in$Q(f,t)$. External relations between frequency bands and time interval are represented using quaternion's with four dimensions, and then input vector length with three energy function terms. The raw audio is converted into 40 coefficients of log Mel-filter-bank which results in a one-dimensional input vector of size 123. The proposed method uses N number of neurons from 256 to 1024 in L hidden layers with sizes (2–5), whereas the real-valued LSTM has 1024 input and hidden units. In QLSTM four-dimensional quaternion neurons 256(1024/4) are used and in OVLSTM eight-dimensional 128 octonion neurons are used with less learning parameters by using $128^2 \times 8 \approx 0.13M$ parameters but in QLSTM we are using $256^2 \times 4 \approx 0.26M$. This research used an open-source machine learning library of Python programming with Tensorflow and Keras models as architectures and activation function ReLu is used, as shown in Table 2 WER percentages of all three models are, LSTM with 87.3%, QLSTM 87.2%, and OVLSTM 86.8%. The least WER reported in OVLSTM with fewer l parameters(learning) of 0.13M by using 128 input features and 10 epochs. To compare loss values of LSTM, QLSTM and OVLSTM are present in column 2 of Table 2 with small differences, 160.69, 158.7, and 157.58.

Table 2. The phoneme error rate (PER) of LSTM, QLSTM, and OVLSTM

Models	WER (%)	Loss	parameters
sLSTM	87.3	160.69	4.0M
QLSTM	87.2	158.23	0.26M
OVLSTM	86.8	157.58	0.13M

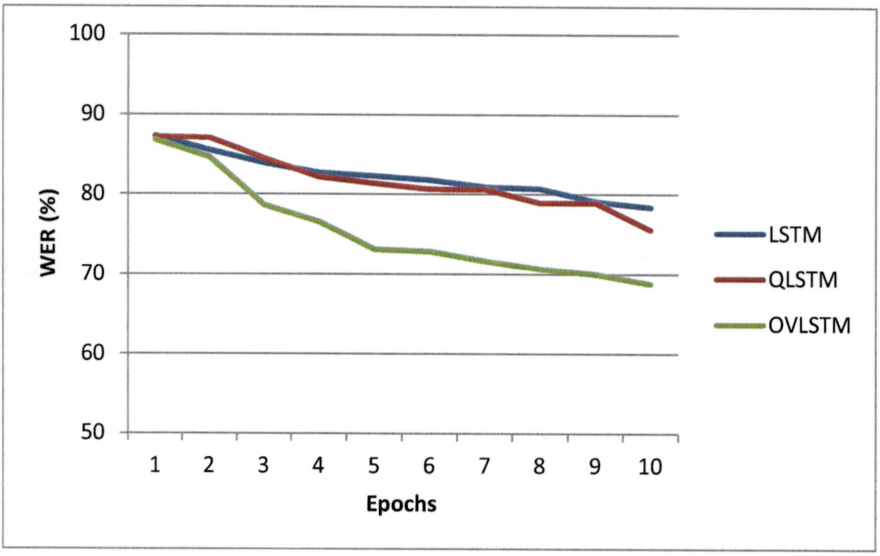

Fig. 3. WER % per epochs for all models

The reduction of number parameters in OVLSTM is not effecting performance in terms of WER percentage, from Fig. 3 we can see all three models WER values for 10 epochs. After each epoch WER values are documented along with test data, however from Fig. 3 we can observe that the best WER values of the three models are enhanced by increasing the number of epochs.

5 Conclusion

This paper proposes a novel octonion valued long-short term memory (OVLSTM) neural network for speech recognition based on acoustic features of words and more liable to handle series dependencies among the speech signals when compared to traditional speech recognition methods. This paper first introduces an octonion algebra and its properties, also the general architecture of octonion features input based automatic speech recognition system, and second an LSTM of recurrent neural networks with octonions to process multiple acoustic features. The projected model evaluated on the TIMIT dataset, OVLSTM obtain better results when compared to quaternion LSTM and real-valued LSTM with less learning parameters.

References

1. Rabiner, L.R., Juang, B.-H.: Fundamentals of Speech Recognition. PTR Prentice Hall, Englewood Cliffs (1993)
2. Singh, H., Bathla, A.K.: A survey on speech recognition. Int. J. Adv. Res. Comput. Eng. Technol. **2**(6), 2186–2189 (2013)
3. Anusuya, M.A., Katti, S.K.: Speech recognition by machine: a review. Int. J. Comput. Sci. Inf. Secur. **6**(3), 181–205 (2009)
4. Zhang, Y.: Speech recognition using deep learning algorithms. Technical report, Stanford Univ., Stanford, CA, USA, pp. 1-5 (2013)
5. Shahin, I.A., Nassif, B., Hamsa, S.: Novel cascaded Gaussian mixture model-deep neural network classi_er for speaker identi_cation in emotional talking environments. Neural Comput. Appl. (to be published)
6. Rama Murthy, G., Praveen, D.: Complex-valued neural associative memory on the complex hypercube. IEEE Conf. Cybern. Intell. Syst. Singapore, 1–3 December 2004
7. Parcollet, T., Zhang, Y., Morchid, M., Trabelsi, C., Linarès, G., de Mori, R., Bengio, Y.: Quaternion convolutional neural networks for end-to-end automatic speech recognition. In: Interspeech 2018, 19th Annual Conference of the International Speech Communication Association. Hyderabad, India **2–6**, 22–26 (2018)
8. Parcollet, T., Morchid, M., Linarès, G., De Mori, R.: Bidirectional Quaternion Long-Short Term Memory Recurrent Neural Networks For Speech Recognition, in November 2018
9. Hinton, G., Deng, L., Yu, D., Dahl, G.E., Mohamed, A.R., Jaitly, N., et al.: Deep neural networks for acoustic modeling in speech recognition: the shared views of four research groups. IEEE Signal Proces. Mag. **29**(6), 82–97 (2012)
10. Gales, M., Young, S.: The application of hidden Markov models in speech recognition. Found. Trends® Signal Proces. 1(3), 195–304 (2008)
11. Graves, A., Jaitly, N., Mohamed, A.R.: Hybrid speech recognition with deep bidirectional LSTM. In: 2013 IEEE Workshop on Automatic Speech Recognition and Understanding (ASRU). IEEE (2013)
12. Hochreiter, S., Schmidhuber, J.: Long short-term memory. Neural Comput. **9**(8), 1735–1780 (1997)

13. Xu, D., Zhang, L., Zhang, H.: Learning alogrithms in quaternion neural networks using GHR calculus. Neural Netw. World **27**(3), 271 (2017)
14. Popa, C.-A.: Octonion-valued neural networks. In: Artificial Neural Networks and Machine Learning-ICANN (2016)
15. Popa, C..-A.: Global exponential stability of neutral-type octonion-valued neural networks with time-varying delays. Neurocomputing **309**(2), 177–133 (2018)
16. Popa, C.-A.: Global exponential stability of neutral-type octonion-valued neural networks with time-varying delays. Neurocomputing 309(2) 177–133 (2018). Gaudet, C., Maida, A.: Deep quaternion networks. In: Proceedings of the IJCNN (2018)
17. Lee, D., Wang, W.: A multivalued bidirectional associative memory operating on a complex domain. Neural Netw. **11**(9), 1623–1635 (1998)
18. Vallejo, J., Bayro-Corrochano, E.: Clifford Hopfield neural networks. In: International Joint Conference on Neural Networks (IJCNN),pp. 3609–3612. IEEE, June 2008.

E-Learning Acceptance During the Covid-19 Outbreak: A Cross-sectional Study

Mohammed Habes[1](✉), Sana Ali[2], Atiqa Khalid[3], Hiba Abou Haykal[4], Mokhtar Elareshi[5], Tabinda Khan[3], and Abdulkairm Ziani[6]

[1] Yarmouk University, Irbid, Jordan
[2] Allama Iqbal Open University, Islamabad, Pakistan
[3] University of Health Sciences, Lahore, Pakistan
atiqakhalid@slmc.edu.pk
[4] Saint Joseph University, Beirut, Lebanon
[5] Al Ain University, Al Ain, UAE
mokhtar.elareshi@aau.ac.ae
[6] Ahlia University, Manama, Bahrain
aziani@ahlia.edu.bh

Abstract. The rise of Covid-19 raised several concerns for the educational policymakers and stakeholders all over the world, including Pakistan. During the pandemic, institutional closure was the only solution to hamper the virus spread. However, the adoption of E-learning largely sustained educational activities, raises several questions about its effectiveness. This study examines E-learning acceptance during the Covid-19 outbreak in Pakistan. It employed cross-sectional design and a random sample of $n = 314$ students from Sahiwal Medical College, Sahiwal, Pakistan. The researcher further applied Media Dependency Theory to provide support to the conceptual model Theory. Findings indicated a strong significant relationship between students' positive perceptions towards E-learning during Covid-19. Respondents considered E-learning as an effective substitute for the traditional learning environment, leading to improved academic performance. The implications of the study lean on the fact that using E-learning during in unforeseen times has not only highlighted the role of internet-based learning, but it has also demonstrated a greater dependence on new media technology. We suggest that, suggesting that E-learning is a potential solution to the problem of sustaining educational activities during institutional closure.

Keywords: Media system dependency theory · Covid-19 · E-learning · Education · Academic performance · Educational activities

1 Introduction

Closing educational institutions nationwide during the Covid-19 to hamper the outbreak is of great importance. Investigators assessed the impact on mortality of closure and found to be associated with a significant decline in Covid-19's incidence and mortality rate. Countries that closed educational institutions earlier, when the incidence was low,

A. M. A. Musleh Al-Sartawi et al. (Eds.): EAMMIS 2021, LNNS 239, pp. 65–77, 2021.
https://doi.org/10.1007/978-3-030-77246-8_7

saw a considerable impact on curbing Covid-19 transmission [1]. The closure of educational institutions strongly affected the functioning of society. The UNESCO discovered on March 18, 2020 that 107 countries had imposed nationwide institutional closure strategies, which affected roughly half of the global student population. A week earlier, only 29 countries had imposed the relevant strategy [2]. Covid-19 affected approximately one billion students, including more than 16 million learners in Central Asia. However, although some students could opt for online education, not all institutions were prepared to tackle this unprecedented educational disruption [3, 4].

The educational closure and the practice of E-learning as an alternative was adapted to keep students safe from the infectious virus [5]. E-learning has its advantages due to its feasibility, its flexibility, and its better control over the environment [6]. Although E-learning has its advantages, different studies have shown its limitations e.g., social isolation, lack of student-teacher interaction, and communication issues [7].

While most countries have adopted this system, many underdeveloped and developing countries do not practice the E-learning approach including Pakistan [8]. However, during the Covid-19 outbreak, all schools, colleges, and even medical and dental universities switched to E-learning systems. Recognizing the academic loss caused by the cessation of learning activities, the WHO [9] has issued guidance on adjusting public health and social measures (PHSMs) while managing the risk of a resurgence of cases. This annex provides considerations for decision makers and educators on how or when to reopen or close schools in the context of Covid-19 [9]. Moreover, Pakistan's Higher Education Commission has also proposed continuing education through online courses and lectures [1].

Hence, medical institutes should also provide effective E-learning environments to their students to balance the cessation of learning activities. Some medical institutes are effectively using the online teaching system, while others are irritated by the complexity of E-learning. This study explored the effectiveness of E-learning during Covid-19 among the medical students of a small division.

This study is divided into FIVE sections, after its introduction, a literature review discusses students' perceptions of E-learning, its substitute for traditional educational methods and impacts on academic performance. This section also offers a theoretical framework discussing the MSDT. The third section provides the methods used here, followed by the results section. The fourth section discussing the E-learning acceptance in relation to academic performance in higher education, and finally the study's implications, limitations and further research were provided.

2 Literature Review and the Theoretical Framework

2.1 Students' Perceptions of E-Learning

With the evolution in internet technology, E-learning has become an encouraging solution for educational concerns. E-learning chiefly facilitated students in improving their learning skills and obtaining extra curriculum material through different online platforms [10, 11]. Popovici and Mironov [12] conducted a preliminary pilot study to examine students' E-learning dependency at the University of Bucharest, Romania. The results revealed positive experiences regarding the use of e-learning technologies. This study also found the medium and advanced level of students' competencies based on E-learning tools. Gaffar and Septyandi [13] also investigated students' perceptions of social-media usage for learning purposes in Australia using mixed method. Findings showed that adaptability and better learning results were the most striking perceived advantages of e-learning and instruction. Students rely highly on E-learning, which is also benefiting them in multiple ways.

To further examine students' perceptions and dependency on E-learning, Hadullo et al. [14] chose Taiwanese university students who indicated they widely preferred E-learning, although they faced several challenges in this area, such as time, course uphold material provision, motivation to be involved in the class, interaction between educators and students and others. Similarly, Ladyshewsky [6] showed that students relied heavily on E-learning due to its perceived benefits and perceived usefulness. Participants also expressed their satisfaction with E-learning and considered it a substitute for the formal learning environment [15–17]. Ali et al. [15] indicated that respondents were satisfied with E-learning for their educational tasks and also considered E-learning as a source which could be used to search for informational data for study purposes. Keller and Cernerud [18] indicated that teaching and learning are both very challenging today, especially during the Covid-19 outbreak. Generally, both students and teachers rely on E-learning but, due to the current crisis, its importance has increased highly.

Furthermore, Elango et al. [19] utilized a cross-sectional design of undergraduate students from Jordan and the UAE. Respondents revealed that the potential benefits of E-learning were positioned in a diminishing way as they did not have to resort to malpractice such as plagiarism and cheating. They also gained from informal yet influential learning opportunities. They could attend their classes on time, attain supporting study material, and benefit from ease of interaction with their teachers.

Moreover et al. [20] also investigated E-learning dependency among Saudi students, with a sample of n = 72 students from King Abdul Aziz University. The findings revealed that perceived usefulness and perceived beneficial outcomes were the two primary reasons behind E-learning acceptance and dependency. Coldwell et al. [21] indicated that E-learning mainly facilitates the on-time delivery of educational content in several scenarios, such as distance learning, network learning, and online learning. These benefits make online learning even more acceptable for the students. They further examined the impacts of demographical variables on E-learning acceptance and usage among university students. The findings showed that dependency on online learning was due to its ease of access, its perceived usefulness and its perceived ease of use. Students considered E-learning as a great source of acquiring new skills and improving educational experiences.

Furthermore, Kamsin [22] investigated whether E-learning could be a substitute for conventional learning and whether it could improve the quality of education and revealed that 57% of students and instructors considered e-learning a substitute for conventional learning and a means of solving problems that never existed in the traditional education system.

H1: There is a significant relationship between E-learning and students' perceptions of acceptance.

2.2 E-Learning as a Substitute for Traditional Educational Methods

Technology facilitates almost every field of life, including education; technology usage has become an integral part of our daily lives. Today, the global-village concept facilitates communication, collaboration, education, and information with ease of access and more significant benefits [23]. Fischer et al. [24] examined students' perceptions regarding E-learning as a substitute for traditional learning methods. Data revealed a strong, significant interest in E-learning as an extension of learning. Respondents considered E-learning an adequate substitute for traditional learning environments. Goyal [25] considered that E-learning is the future of education, offering brighter prospects to students. With efficient bandwidth network, internet-based learning is a need of almost every student. E-learning is an effective alternative to traditional learning environments as it is growing briskly as an inevitable part of formal and informal education procedures. Due to its distinct features, students also prefer E-learning, indicating users' appreciation arising from E-learning's unique characteristics and benefits.

Owens and Price [26] examined the impact of modern E-learning and information technologies (IT) on the UK higher education. The results showed that E-learning and other technologies were valuable means of providing modern education to students and of potentially developing student interest in exciting ways. Rana et al. [27] conducted a study to investigate the benefits and challenges of E-learning. They found that respondents considered E-learning an adequate substitute for traditional learning patterns and policies. However, participants also expressed some challenges regarding E-learning's implication and usage, such as the unavailability of sufficient financial, social, cultural, local, national, and regional resources to meet modern study requirements.

Mohajan [28] found that students and teachers are still grappling with its implementation. Although education policymakers are also concerned with E-learning and implementation, there is still a long way to go. The E-learning policymakers and stakeholders also need to focus on some other areas, such as social, environmental, personal and cultural factors, in this regard.

Furthermore, Arrey-ndip et al. [29] examined Cameroonian secondary students' perceptions regarding E-learning as a substitute for conventional learning. The results revealed that school administration largely relied on E-learning to improve students' learning skills. Obi et al. [30] stated that E-learning helps students to stay connected to national institutions and to avail themselves of what international institutions can offer. Due to increased opportunities, students can today expand their exposure and attain better opportunities in different study disciplines. De Giusti [4] scrutinized the effectiveness of E-learning for US postgraduate programs in medicine and showed that asynchronous E-learning had been widely adapted and had replaced conventional learning as it provided ease of access and ease of schedule to residents with busy schedules.

H2: Students significantly consider E-learning as a substitute for formal learning.

2.3 Impacts on Academic Performance

E-learning has widely facilitated the use of information and communication technology to guide students and instructors. E-learning has upgraded the existing education system and improved students' academic performance [31]. Ali [32] examined social-media usage patterns among Pakistan young generation and found a strong positive correlation between social-media usage and students improved academic performance. Likewise, Alghizzawi et al. [33] investigated the impacts of internet-based learning on university students' academic performance in Jordan. The findings showed that YouTube was an essential source of E-learning for the students. Respondents also revealed that E-learning was improving students' academic grades as it provided them with extra study material and also kept them connected to their classmates and teachers. Zare et al. [34] stated that ITs give many prospects to students. Students can access educational material and information from all over the world due to ease of access and perceived usefulness. Students prefer to rely on E-learning which further results in improved academic performance.

Mothibi [35] investigated the proposed correlation between E-learning and students' academic achievement using a literature-review approach. Cited literature also witnessed a strong, positive correlation between E-learning and students' academic achievements. The study concluded that students' instructors were also resorting to E-learning to facilitate their students, further validating the significance of E-learning as a practical study resource. Stevens et al. [36] examined students' tend to get involved in E-learning and its impacts on their overall academic grades. The results showed that students were more likely to adopt E-learning regarding the designed course.

Moreover, De Araujo et al. [37] analyzed E-learning's role and its effects on managing cognitive challenges during the online courses. The results of the evaluation survey indicated that E-learning largely facilitated the students and positively impacted their academic grades. Franklin and Nahari [38] noted that one of the significant reasons behind switching to E-learning is the decreased cost and the ease of access to educational institutions worldwide. Universities also acknowledge the importance of E-learning in improving students' academic performance and motivating them to adopt E-learning. Many universities introduced E-learning and replaced the traditional face-to-face learning environment, which also led to improved teaching standards in general. In this context, Alhumaid et al. [3] investigated E-learning implementation during the Covid-19 outbreak in Pakistan. The results indicated that teachers considered E-learning an essential substitute for traditional learning, especially during a crisis such as Covid-19. Students were capable of availing themselves of supportive educational material through the internet, which also improved their overall academic grades. Therefore, in unforeseen times, E-learning has chiefly facilitated the resumption of the educational process. Students and institutions rely on E-learning to sustain their education, leading to improved academic performance [39].

H3: There is a positive impact of E-learning on students' academic performance (Fig. 1).

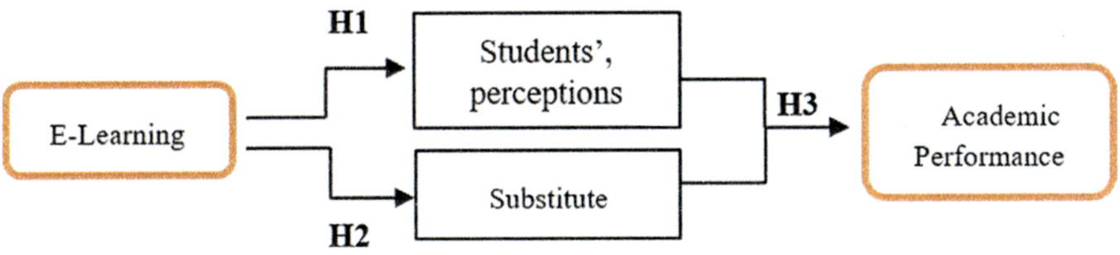

Fig. 1. The study's conceptual framework

2.4 Media System Dependency Theory (MSDT)

MSDT supports the conceptual background of the current investigation. The relevant approach was first proposed by Sandra Ball-Rokeach. This study represented MSDT as describing the role and necessity of media in everyday life [40]. However, the current investigation uses Brough and Li's [41] argument that MSDT provides a strong framework for micro-level studies where individual dependency is of primary concern.

Another benefit of MSDT is that is provides the dynamic of the cause-and-effect relationship as it highlights the reason for dependence and its potential effects. Moreover, in the current healthcare crisis, the rise of web-based technology has largely facilitated the education and learning process. Young students use social media to maintain their educational activities. During the Covid-19, web-based dependency is the only way for them to resume educational activities.

Despite E-learning being much appreciated by formal learning practitioners, youngsters consider it a vital learning source [42]. Today, students are also getting enrolled through social-media platforms so as to continue their educational activities. Institutions are continuing education through their learning-management systems and students are even resorting to E-learning to mitigate the impacts of Covid-19. Thus, the knowledge management during the current healthcare emergency also contributes to its significance in general [43].

3 Methodology

3.1 Population and Sampling

This study was inspired by several previous studies [3, 11, 21] and used cross-sectional design and close-ended, structured questionnaires. It involved students from both private- and public-sector medical colleges in Sahiwal, Pakistan. However, the study randomly selected a sample of $n = 314$ students from Sahiwal Medical College, using a convenient sampling technique as it is very feasible to manage with few rules governing how the sample should be collected [44]. The relative time and cost required to carry out a convenience sample are small in comparison with probability sampling techniques. The questionnaire was distributed using Google forms due to lockdown educational closure.

4 Data Analysis

This section discusses the data analysis starting with the demographic characteristics which is summarized in Table 1.

Table 1. Demographic characteristics

Variable	Factor	No.	%
Gender	Male	106	33.7%
	Female	208	66.0%
Study year	1st year	57	18.1%
	2nd year	90	28.6%
	3rd year	51	16.2%
	4th year	96	30.5%
	Final year	20	6.3%
Residence	Urban	81	25.7%
	Rural	233	74.2%

4.1 Reliability and Validity

Reliability deals with a measuring tool's stability and consistency, while validity tells us about its accuracy [45]. In this regard, the current investigation also involves validity and reliability analysis of research instruments to ensure the results' generalizability.

4.2 Intercoder Reliability Analysis and Univariate Pearson Correlation

As shown in Table 2 below, as part of the analysis it was important to conduct the IRA. This usually is undertaken to check the reliability of the research instrument [46], with the Cronbach Alpha value (0.886). The UPC was run to determine whether there is a linear component of association between two continuous variables [28]. The outcomes respectively affirmed that the study instrument was reliable with high validity.

Table 2. The research instrument of IRA and UPC

	Hypothesis	Cronbach Alpha	IRA status	Pearson correlation	Sign	Status
H1	EL > SP	.721	Reliable	.515	.000	Valid
H2	EL > SFL	.729	Reliable	.609	.000	Valid
H3	EL > AP	.753	Reliable	.457	.004	Valid

Note: EL = E-learning, S = Students' perceptions, SFL = Substitute of formal learning, AP = Academic performance.

5 Results

When the respondents were asked about their perceptions regarding Zoom as the best educational software, 80.0% of them strongly agreed that the relevant software was the best compared with others. Similarly, 24.4% strongly disagreed that E-learning helped them to utilize the time more effectively, although 20.0% agreed that it helped them to spend the time in a positive and effective manner as the majority of individuals (29.3%) slightly agreed that they frequently used the internet for learning purposes. Moreover, most respondents agreed that (31.7%) was helpful for smart study as they (35.6%) considered E-learning a convenient way of learning during the lockdown. Some (38.5%) perceived E-learning as positively impacting daily study routine. These results are consistent with the study conducted other studies [3] (Table 3).

Table 3. Relationship between E-learning and Students' perceptions (*n* = *314*)

Variable	Mean*	Variance	SD
Predictor variables: E-learning			
- Zoom is the best educational software practiced so far	1.32	.629	.793
- E-learning helps to utilize the time more effectively	3.16	1.766	1.329
- I often use the net for learning purposes	3.21	1.623	1.274
Explanatory variables: Perception			
- Online learning is helpful for smart study	2.85	1.338	1.157
- E-learning is a convenient way of learning during the lockdown	2.65	1.398	1.182
- E-learning created a positive impact on daily study routine	2.99	1.556	1.247

*Mean (min = 1 and max = 5)

Sujit et al. [47] stated that IT is an essential source of information and education during the current century. Our respondents (47.4%) agreed that E-learning was a substitute for formal education as they were (28.9%) fully satisfied with the online learning tools. Likewise, most (55.0%) considered it the best medium for pursuing educational activities if lockdown was ever imposed again in the future. Nonetheless, for most respondents (89.8%), E-learning has positively affected their educational performance, helping them utilizing their time during vacations thoroughly (80.2%) and exams preparation (85.6%). These results are compatible with Alghizzawi et al. [11] who found positive impacts of E-learning on students' academic performance (Table 4).

5.1 Analysis of Variance (ANOVA)

To examine any potential differences among the group means of variables, one-way analysis of variance (ANOVA) was employed. It is one of the most preferred tests in experimental designs to test the variations in factor and levels [48]. The results indicated that gender and locality were controlled and did not differ based on these two variables (Table 5). However, the responses varied based on the year of study, which shows that its effects are not controlled.

Table 4. Summary of Relationship between E-learning and substitute and impacts (*n = 314)*

Variable	Mean	Variance	SD
Explanatory variables: Substitute			
- E-learning as best study substitute in lockdown	1.58	.768	.877
- Satisfaction with online tools of learning	1.85	1.086	1.042
- E-learning can be the best continuing solution for education if lockdown is ever imposed again in future	1.72	.669	.818
Explanatory variables: Impact			
- E-learning positively impacts on studies	2.84	1.864	1.365
- E-learning helps students to utilize their time during vacations positively	3.18	1.423	1.193
- E-learning is positively affecting students' exam preparation	2.70	1.787	1.337
*Mean (min = 1 and max = 5)			

Table 5. One-way anova regarding demographic characteristics

Variable		*N*	Mean difference	Levene statistics	*F*	Sign
Gender	Male	106	.089	3.703	1.410	.000
	Female	208				
Year of study	1st year	57	.037	4.392	.717	.630
	2nd year	90				
	3rd year	51				
	4th year	96				
	Final year	20				
Locality	Rural	81	.438	.810	.808	.000
	Urban	233				

5.2 Linear Regression Analysis

The study used the regression-analysis technique to test the proposed relationship between the study variables. It is one of the most preferred hypothesis-testing tools, enabling us to validate our research propositions [47]. Regression analysis indeed provides in-depth details about the research assumptions and their proposed associations (Table 6). We found a strong, significant relationship between E-learning and students' perceptions (β = .597, p $\leq$ 0000), substitute (β = .965, p $\leq$ 0.000), and impact (β = 2.047, p $\leq$ 0.005). Thus, this validates the research hypothesis and confirms other research [49].

Table 6. Linear regression analysis

Hypothesis		*t*	*Adjusted R-Square*	*f*	*β*	Sign	Relationship
H1	EL > SP	5.196	.263	112.171	.597	.000***	Positive
H2	EL > SFL	6.468	.369	184.219	.965	.000***	Positive
H3	EL > AP	16.706	.022	7.890	2.047	.005**	Positive

6 Discussion and Conclusion

This study explored the effectiveness of E-learning during Covid-19 among the medical students of a small division. The current results are strongly consistent with the study conducted by Habes et al. [50], who found a strong, significant relationship between students and E-learning acceptance. With the significance value of $p \leq 0.000$, the respondents indicated a positive opinion regarding E-learning acceptance. Likewise, students also considered E-learning as a substitute for traditional learning, which also highlights the broader acceptance of E-learning among the young generation confirming similar studies [3].

Finally, with the significance value of $p \leq 0.005$, the results validated the third hypothesis assuming a strong, positive relationship between students' academic performance and E-learning. Kamsin [22] indicated that both students and teachers consider remote learning as the only way to maintain their education. Although the closure of educational institutions was an essential part of mitigating the impacts of Covid-19, it challenged the educational system in a developing country such as Pakistan. In such a situation, students' positive attitude towards E-learning is an optimistic approach for taking alternatives under consideration [11].

The higher reliance on E-learning during Covid-19 also ensures continuing education if the outbreak continues for long. Even in the future, E-learning will be a highly considered approach for coping with crises. Moreover, most learners are aware of the modern trends in education and learning. They resort greatly to virtual systems to learn new things and to modify their educational capabilities. The convergence of classical learning with modern, technology-based learning also shows that E-learning is more than merely a substitute for conventional education. It has introduced new ways to cope with challenges and has become an integral part of the modern educational system [51].

Thus, during the second wave of the Covid-19 outbreak, social-media dependency is a prominent advantage of web-based technology. Due to educational institutions' closure, students cannot formally attend classes to continue their academic activities. To continue their learning process, students at the medical college depend mainly on social media to sustain their educational activities, enhance their learning abilities and improve their academic performance [52]. This prominent dependency on digital media shows how social media is an essential component of risk management. Further, it expands the propositions of media-tendency theory in terms of social-media usage and its benefits. To sum up, due to social-media platforms' utility for educational purposes, students also consider E-learning a substitute for the informal learning process. Students no longer have to rely on a formal classroom to get their education and to continue their learning process. Similarly, using social media for educational purposes also improves their critical ability and enables them to cope effectively during a crisis [53]. Significantly, in a developing country such as Pakistan, where E-learning was merely a concept, the rise of Covid-19 has also affirmed its significance as an essential part of education.

7 The Limitations and Further Research

Mainly from Sahiwal Medical College and its relevance will be only in the domain of medical students. Second, the data were gathered using a convenience sampling method which further limits its scope. Yet the study recommended more investigations highlighting E-learning's role, its importance and its impacts on students' academic performance. Particularly from the viewpoint of instructors and parents, analyzing the importance of E-learning will bring out even more significant outcomes.

References

1. Khalid, A., Ali, S.: COVID-19 and its challenges for the healthcare system in Pakistan. Asian Bioeth. Rev. **12**(4), 551–564 (2020)
2. UNESCO. "COVID-19 and higher education: today and tomorrow. UNESCO International Institute for Higher Education in Latin America and the Caribbean, pp. 1–46 (2020)
3. Alhumaid, K., Ali, S., Waheed, A., Zahid, E., Habes, M.: COVID-19 and e-learning: perceptions and attitudes of teachers towards e-learning acceptance in the developing countries. Multicult. Educ. **6**(2), 100–115 (2020)
4. De Giusti, A.: Policy brief: education during COVID-19 and beyond. Rev. Iberoam. Tecnol. en Educ. y Educ. en Tecnol. **20**(26), e12 (2020)
5. Habes, M., Alghizzawi, M., Ali, S., SalihAlnaser, A., Salloum, S.A.: The relation among marketing ads, via digital media and mitigate (COVID-19) pandemic in Jordan. Int. J. Adv. Sci. **29**(7), 2326–12348 (2020)
6. Ladyshewsky, R.K.: E-learning compared with face to face: differences in the academic achievement of postgraduate business students. Australas. J. Educ. Technol. **20**(3), 316–336 (2004)
7. Arkorful, V., Abaidoo, N.: The role of e-learning, advantages and disadvantages of its adoption in higher education. Int. J. Instr. Technol. Distance Learn. **12**(1), 29–42 (2015)
8. Qureshi, I.A., Ilyas, K., Yasmin, R., Whitty, M.: Challenges of implementing e-learning in a Pakistani university. Knowl. Manag. E-Learning **4**(3), 310–324 (2012)
9. WHO. What we know about COVID-19 transmission in schools. Coronavirus Updates 39-World Health Organization (2020)
10. Elbasir, M., Elareshi, M., Habes, M., Aissani, R.: The influence of trust, security and reliability of multimedia payment on the adoption of EPS in Libya. Multicult. Educ. **6**(5), 53–68 (2020)
11. Alnaser, A.S., Al-Shibly, M.S., Alghizzawi, M., Habes, M.: Impacts of social media and demographical characteristics on university admissions: case of Jordanian private universities. PalArch's J. Archaeol. Egypt/Egyptol. **17**(11), 6433 (2020)
12. Popovici, A., Mironov, C.: Students' perception on using eLearning technologies. Procedia - Soc. Behav. Sci. **180**, 1514–1519 (2015)
13. Gaffar, M.R., Septyandi, C.B.: Using e-learning media during the Corona Faktor Yang Memengaruhi Kepuasan Mahasiswa Dalam Menggunakan Media E-Learning Selama periode Pandemik Corona, vol. 8, no. 1 (2020)

14. Hadullo, K., Oboko, R., Omwenga, E.: Factors affecting asynchronous e-learning quality in developing countries. A qualitative pre-study of JKUAT University. Int. J. Educ. Dev. Inf. Commun. Technol. **14**(1), 152–163 (2018)
15. Ali, S., Habes, M., Qamar, A.: Accessing the political information through new media: a review study. J. Res. Rev. Soc. Sci. Pakistan (2020)
16. Alnawafleh, H., Alghizzawi, M., Habes, M.: The impact of introducing international brands on the development of Jordanian tourism. Int. J. Inf. Technol. Lang. Stud. **3**(2) (2019)
17. Mehmood, S., Taswir, T.: The effects of social networking sites on the academic performance of students in college of applied sciences Nizwa, Oman. Int. J. Arts Commer. **2**(1), 111–125 (2013)
18. Keller, C., Cernerud, L.: Students' perceptions of e-learning in university education. J. Educ. Media **27**(1–2), 55–67 (2002)
19. Elango, R., Gudep, V.K., Selvam, M.: Quality of e-Learning: an analysis based on e-Learners' perception of e-Learning. Electron. J. E-Learn. **6**(1), 31–43 (2008)
20. Linjawi, A.I., Alfadda, L.S.: Students' perception, attitudes, and readiness toward online learning in dental education in Saudi Arabia: a cohort study. Adv. Med. Educ. Pract. **9**, 855–863 (2018)
21. Coldwell, J., Craig, A., Paterson, T., Mustard, J.: Online students: relationships between participation, demographics and academic performance. Electron. J. e-Learn. **6**(1), 19–28 (2008)
22. Kamsin, A.: Is E-learning the solution and substitute for conventional learning? Int. J. Comput. Internet Manag. **13**(3), 79–89 (2005)
23. Shahzad, A., Hassan, R., Aremu, A.Y., Hussain, A., Lodhi, R.N.: Effects of COVID-19 in E-learning on higher education institution students: the group comparison between male and female. Qual. Quant. **4**, 1–22 (2020)
24. Fischer, H., Heise, L., Heinz, M., Moebius, K., Koehler, T.: E-learning trends and hypes in academic teaching. Methodology and findings of a trend study. In: Proceedings of the International Conference e-Learning 2014 - Part of the Multi Conference on Computer Science and Information Systems, MCCSIS 2014, pp. 63–69 (2014)
25. Goyal, S.: e-Learning: future of education. J. Educ. Learn. **6**(4), 239 (2012)
26. Owens, J.D., Price, L.: Is e-learning replacing the traditional lecture? Educ. Train. **52**(2), 128–139 (2010)
27. Rana, H., Rajiv, R., Lal, M.: E-learning: issues and challenges. Int. J. Comput. Appl. **97**(5), 20–24 (2014)
28. Mohajan, H.K.: Two criteria for good measurements in research: validity and reliability. Ann. Spiru Haret Univ. Econ. Ser. **17**(4), 59–82 (2017)
29. Arrey-ndip, N.A., Carole, M.T.M., Carole, N.K., Njikam, A.: The current state and future of elearning in educational institutions in Cameroon : a case study of the City of Yaounde education in Cameroon and government's action towards ICTs in schools, pp. 141–153 (2006)
30. Obi, I.E., Charles-Okoli, A.N., Agunwa, C.C., Omotowo, B.I., Ndu, A.C., Agwu-Umahi, O.R.: e-learning readiness from perspectives of medical students: a survey in Nigeria. Niger. J. Clin. Pract. **21**(3), 293–300 (2018)
31. Nikdel Teymori, A., Fardin, M.A.: COVID-19 and educational challenges: a review of the benefits of online education. Ann. Mil. Heal. Sci. Res. **18**(3), 19–22 (2020)
32. Ali, S.: Social media usage among teenage girls in Rawalpindi and Islamabad. Glob. Media J. **16**(31:146), 1–9 (2018)
33. Alghizzawi, M., Habes, M., Salloum, S.A., Ghani, M.A., Mhamdi, C., Shaalan, K.: The effect of social media usage on students' e-learning acceptance in higher education: a case study from the United Arab Emirates. Int. J. Inf. Technol. Lang. Stud. **3**(3), 13–26 (2019)
34. Zare, M., Sarikhani, R., Salari, M., Mansouri, V.: The impact of e-learning on university students' academic achievement and creativity. J. Tech. Educ. Train. **8**(1), 25–33 (2016)

35. Mothibi, G.: A meta-analysis of the relationship between E e-learning and students' academic achievement in higher education. J. Educ. Pract. **6**(9), 6–10 (2015)
36. Stevens, N.T., et al.: Can e-learning improve the performance of undergraduate medical students in Clinical Microbiology examinations? BMC Med. Educ. **19**(1), 1–8 (2019)
37. De Araujo, T., et al.: Cognitive load and self-determination theories applied to e-learning: impact on students' participation and academic performance. PLoS ONE **11**(3), 1–21 (2016)
38. Franklin, U.E., Nahari, A.A.: The impact of e-learning on academic performance: Preliminary examination of King Khalid university. Int. J. Acad. Res. Progress. Educ. Dev. **7**(1) (2018)
39. Subedi, S., Nayaju, S., Subedi, S., Shah, S.K., Shah, J.M.: Impact of e-learning during COVID-19 pandemic among nursing students and teachers of Nepal. Int. J. Sci. Healthc. Res. **5**(3), 68–76 (2020)
40. Mehrad, J., Yousefi, Z.: introducing the theory of 'media system dependency' with emphasis on its potential application in theoretical framework of researches in the field of information science and knowledge Jafar. Int. J. Inf. Sci. Manag. **16**(1), 1–14 (2018)
41. Brough, M., Li, Z.: Media systems dependency, symbolic power, and human rights online video: learning from Burma's 'saffron revolution' and WITNESS's hub. Int. J. Commun. **7**(1), 281–304 (2013)
42. Alnaser, A.S., Habes, M., Alghizzawi, M., Ali, S.: The relation among marketing ads, via digital media and mitigate (COVID-19) pandemic in Jordan. Int. J. Adv. Sci. Technol. **29**(7), 12326–12348 (2020)
43. Almarzooq, Z.I., Lopes, M., Kochar, A.: Virtual learning during the Covid-19. J. Am. Cardiol. **75**(20), 2635–2638 (2020)
44. Abraham, J., Islamia, J.M., Saini, C., Islamia, J.M.: Using social media for educational purposes : approaches and challenges. February 2015 (2018)
45. Baruah, T.D.: Effectiveness of social media as a tool of communication and its potential for technology enabled connections: a micro-level study. Int. J. Sci. Res. Publ. **2**(5), 1–10 (2012)
46. Nili, A., Tate, M., Barros, A.: A critical analysis of inter-coder reliability methods in information systems research. In: Proceedings of the 28th Australasian Conference on Information Systems. ACIS 2017 (2017)
47. Kumar, S., Wotto, M., Bélanger, P.: E-learning, M-learning and D-learning: conceptual definition and comparative analysis. e-Learning Digit. Media **15**(4), 191–216 (2018)
48. Ostertagová, E., Ostertag, O.: Methodology and application of oneway ANOVA. Am. J. Mech. Eng. **1**(7), 256–261 (2013)
49. Babu, G.S., Srivedi, K.: Importance of e-learning in higher education: a study. Int. Res. Cult. Soc. **2**(5), 84–88 (2018)
50. Habes, M., Ali, S., Salloum, S.A., Elareshi, M., Ziani,A.-K.: Digital media and students' AP improvement: an empirical investigation of social TV. In: 2020 International Conference on Innovation and Intelligence for Informatics, Computing and Technologies (3ICT), pp. 1–6 (2020)
51. Edelhauser, E., Lupu-Dima, L.: Is Romania prepared for e-learning during the COVID-19 pandemic? Sustain **12**(13), 1–30 (2020)
52. Saeed Al-Maroof, R., Alhumaid, K., Salloum, S.: The continuous intention to use e-learning, from two different perspectives. Educ. Sci. **11**(1), 1–20 (2020)
53. Ali, S.: Combatting against Covid-19 & misinformation: a systematic review. Hum. Arenas, no. 0123456789 (2020)

Digital Social Contracts with AI Robots: Some Implications for Amazon.Com

Frank T. Lorne(✉) and Mahita Reddy Gogireddy

New York Institute of Technology, Vancouver, Canada
{florne,mgogired}@nyit.edu

Abstract. The use of AI robots on platforms in the digital world is ubiquitous. Amazon.com has AI features that significantly enhance users' productivity. There is no question that Amazon is subject to competition. Their competitors have their own standards. If standards can differentiate platforms, they could coexist. This study's evaluation of Amazon along public policy guidelines suggests that there is no reason for imposing additional regulations for Amazon based on its current trend of operations. This study proposes a new thinking, passing beyond the analogy between platform operations and city operations, notably along the dimension of *private* vs. *public spaces*, since people are familiar with thinking of issues in that paradigm as the pre-internet days.

Keywords: Artificial intelligence · Robots · Amazon · Digital social contracts

1 Introduction

The use of AI robots on platforms in the digital world is ubiquitous. Data management is likely to be an important question that any software programmer would need to address. Amazon.com, in the use of its powerful AWS, has AI features that would significantly enhance productivity of the users (https://aws.amazon.com/machine-learning/ai-services/). The increase in efficiency brought about by these features are likely to be of the 'productivity' type, rather than 'utility' focused AI aiming for consumption. Yet, there are AI usages and data management that go beyond managing productivity. This paper explores the broader issues. Data management of cities and digital platforms in relation to AI share similarities that have not been explored, particularly on how data can be used [15]. The similarities can help sort out some key issues dealing with why and how the emerging technology of Artificial Intelligence (AI) can properly assist humans in maximizing welfare. AI has potential usages in the management of cities and digital platforms, all requiring data as inputs for analysis. The presumption is that data is not dead objects. Data as Labor is a concept that has been brought up in recent academic writings. The proposition is that, currently, data suppliers are not adequately rewarded for their digital contributions to various internet platforms they participate in. Thus, 'high-quality data' is less likely to be contributed [25]. If a digital lifestyle is what will come as a complementary platform to enhance rather than displace our current lifestyle, it may be essential to examine the rules and regulations that make sense in the reality

A. M. A. Musleh Al-Sartawi et al. (Eds.): EAMMIS 2021, LNNS 239, pp. 78–89, 2021.
https://doi.org/10.1007/978-3-030-77246-8_8

space and apply them in the digital world which will lead to the creation of the digital economy and future sustainability [2]. [12] made the first attempt to open the discussion of different types of social contract regarding data management. This paper suggests a way to think through this issue by first going beyond the analogy between platform operations and city operations, notably along the dimension of private vs. public spaces, as people are more familiar with thinking about issues in that paradigm since the pre-internet days. While sometimes there cannot be a clear definition of what is private and public space, civilized citizens are expected to behave in certain ways in the public. At the same time, governments respect individual liberties for activities in private space provided that they are not breaking the laws of the land. Though this thinking and coordination of social activities are not perfectly structured, they have been ingrained into the culture of a civilized world along the spectrum of private vs public space for a long time and have continued without major misgivings. Therefore, there is no reason why that culture cannot be nourished in the internet platform world as well.

2 Analogy of Free Use as Public Space

The internet highway today is yesterday's railroad. Cities have been built along railroads and highways, just like internet platforms are being built today on the world wide web. The organization structures and cultures are different in every city. It is usually understood that people are free to move in and out of a city except in communist regimes of the world in the past. The same can be said about different platforms on the internet; each of them can have their own organizational structure and culture, but there is nothing that prevents individual users from joining or leaving a platform. The AI technology enters the internet highway in manners that have not been examined enough in the real world in the framework of private/public city lives. Robots, some employed privately, and some by the government, can roam the internet highway as well as public and private platforms in search of data to record and analyze. A privately operating platform need not grant access to other robots not designed by them. But if a privately operating platform provides free use of its software, it has a public space accessible like a city street, any robots can roam in that space. It is a matter of the company's own vision and mission (some will call a business model) to make their platform paid or unpaid, i.e., private or public. Google, for example, in allowing people to use Google Drive to store unlimited photos, also warns the users that those photos can be used by the company. Many internet/real world travelers choose to do so voluntarily, as it provides immense convenience.

3 The Publicness in Private and the Privateness in Public

Data is information, and thus, an economic public good in the sense that it can have multiple simultaneous usages. Economists have argued that the efficient pricing of a public good should be zero, implying that AI robots can freely utilize data [3]. In the framework of information as a public good and under zero transaction costs, the right to privacy is a red herring, even though it has been raised by many in the free world as considerations that need to be addressed. What could be a sustainable model for AI adoption in light of such concerns? Viewing the world as a binary choice of either complete or no privacy is too simplistic. That has never been the case for both the real and the digital world, as the real-world characterization of public or private is never binary, but rather entailing different degrees of publicness and privateness. For instance, Amazon took the opportunity to create an open-source software, 'Distance Assistant' which applies artificial intelligence and machine learning to the camera footage in 'public' workspaces to help site leaders identify high traffic areas and promote social distancing behavior in real-time during this Covid-19 pandemic. Amazon also has the brand affinity it needs to gain access to people's 'private' home space with its smart home initiatives, which are intended to become a big part of users' day-to-day routines. Thus, data sharing may ultimately be driven by users' needs on the one hand and their aversion to a compromise of their privacy on the other. This type of benefits/costs calculation certainly applies to internet users, as [20] has persuasively explained. Would the benefits/costs calculation be the same for both public and private spaces? In the real world of a 'free' society, it is commonly recognized that private data submitted to an organization is for the purpose of addressing specific information that the organization needs to obtain, and that the organization will indeed keep that information private and not let other entities use the data unless they are permitted to do so. An international law firm filed a federal lawsuit against Amazon.com Inc., alleging that the company's Alexa virtual assistant product violated privacy laws in Massachusetts and seven other states by making permanent recordings of children's voices without their consent [9], even though consumers need not place an Alexa in the private rooms of their children. We do not see this sentiment of approval-seeking to be widely adopted in public spaces for families who want to tour a city street with their kids. Although there are privacy groups that would push for the banning of closed-circuit TV monitoring in the public space of a city (e.g. Vancouver), such sentiment is not universal. Surely, individuals have a free choice to visit a CCTV enabled city or not, just like consumers have a free choice to install an Amazon virtual assistant or not. Public vs. private distinction aside, most people joining a digital platform believe that a user's yielding of personal information to an AI robot can enhance their customers' user experience. The discussion of the nature of a social contract with AI robots is a potpourri, a mixed bag of various concerns and an on-going process. New concepts had been proposed for a distinction between Data as Capital (DaC) vs. Data as Labor (DaL), suggesting the nature of 'social contract' being different for the two perspectives [12]. In formulating an economic thinking framework to categorize various types of social contracts, whether they are of a type of DaC, DaL, or a social contract with a government, we come to realize that it will be some particular economic rationales that can justify particular forms of social contract. For Amazon, we believe that a social contract has been used and practiced by the company that does not require additional government regulation as long as there is competition in the two-sided market model. Before we elaborate on features of the model, let us first evaluate social contract for AI in general.

4 Formulating a Framework for Social Contract Dialogue with AI Robots

Privacy issue usually can be addressed as an information problem. Information is a public good, per Economic Nobel Prize winner Paul Samuelson formulation. The aggregate demand for public good is a vertical summation of the individual demands. The treatment of privacy issues as an inter-individual problem of externalities can be a good direction of addressing the nature of social contracts using this formulation. For a community of a small number, inter-individual effects are contractual understanding that can be established relatively easily. This is a small extension of a Coasean approach [8] applied to the case of public goods. However, as the community size grows, transaction costs of inter-individual problems can rise rapidly, not precluding regulation being the most efficient solution dealing with the shaping of a 'general will', which is the notion of a social contract. A seeming anarchy over data of the internet and the robots roaming is not something we generally want to discuss. As Buchanan's long time writing partner, [26], expressed his detest in the idea of the conversation in a book titled, Explorations in the Theory of Anarchy (1972). We, too, may find unintuitive surprises in this discussion between the rights of data owners and data collectors.

The public-goodness of privacy remains at the heart of a debate concerning how this social contract should be structured. Undoubtedly, data is information and thus has its public goodness. Privacy is information and thus also has its public goodness. However, when a person is completely in his/her private space, even though another individual can gain value by having that information, the individual wants that potentially public information to be private. This is a matter of respecting an individual's basic rights. What is considered by an individual to be private is subjective, and if such information is produced in the physical environment of private property, the person has the right to turn something technically public into private. It is important to note that such rights to turn information into private is only considered fair if the information is produced in private space. If such information is produced in public space (including private digital companies that adopt free access), allowing individuals to keep complete privacy may not be socially efficient, as we will further explain in the following paragraphs.

To simplify the matter, as a public space problem, let's look at the multi-inter-individual effect problem as a vertical summation of only TWO demands, the private demand curve and the public demand curve. The social contract problem is thus reduced to a 'two-demand-curve' problem with the following behavior stipulation: A private individual treats a public demand 'spillover' as given and maximizes his own action to the spillover technology constraint. Likewise, the public entity treats the individual's demand 'spillover' as given and maximizes its 'public actions.' One can formulate this problem as a game-theoretical problem resulting in some sort of a prisoner dilemma. However, it is probably more fruitful for social contracting purposes to formulate it as a joint maximization problem. Whether a game-theoretical approach or a joint maximization approach is used, we need to make several additional assumptions so that discussions can be formulated into variations of simplistic diagrams. The following assumptions have been made: (1) The individual has preference over data, quantifiable as x, to be used for the individual's private consumption. The x has an ordinal meaning as well as a numerical meaning in terms of the quantity label on the x-axis. Therefore, $x = 1$ means the piece of data that the individual places the highest use on it; x = 2 has less use than x = 1. A certain quantity of data, n, denotes both the quantity of data as well as the nth ranking of importance. With this assumption, everyone will have a downward sloping demand for their own personal data, the demand curve that economists are familiar to work with. (2) The marginal value of x is non-negative. (3) The public preference over the use of individual data may or may not have the same ranking as an individual. We assume three categories of ranking: 1. Public does not have a ranking of preference over specific data; the public 'consumes' all data information equally, i.e., perfect anonymization. 2. Public has the same ordinal ranking as the individual, i.e., what the individual considers to be most important, the public considers to be most important too, e.g., various aspects of the look of a person. 3. Public has the opposite ordinal ranking to the individual, i.e., what the private individual cares most about is what the public cares the least about. Aside from various aspects of the look of a person, which can also be an example, it can be that the public cares most about the trivial things of a person, e.g., how his garbage can is placed in the alley.

The three assumptions allow several generic configurations for aggregate demand for data that has implications on what 'optimal privacy' may mean. The above three cases can be graphically illustrated below. For all cases, the blue line denotes private demand for data, the green line denotes public demand for individual data, and the red line denotes aggregate demand for data. It is the aggregate demand relative to private demand that defines optimal privacy.

Case I: Public has use of private data only in terms of quantity, not specific to the individual, the case of **No privacy** on data as shown in Fig. 1 – left chart.

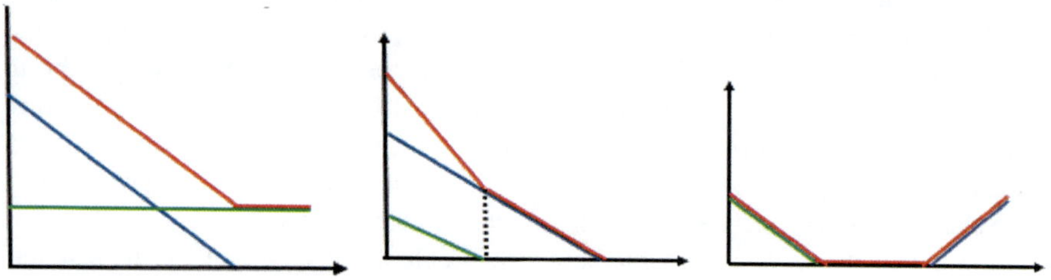

Fig. 1. Case I, Case II and Case III (Developed by authors)

Case II: Public ordinal ranking is the same as private, and private demand out weighted public demand, the case of **Optimal Privacy** as shown in Fig. 1 – center chart.

Case III: Public ordinal ranking in the opposite direction as private, the case of **Total Privacy** (Fig. 1 – right chart). The point of illustrating the three cases is not exploring which case is more prevalent for policy consideration but asserting the fact that a uniform, rigid, standardized privacy policy cannot be justified on efficiency grounds. The social contract needed to achieve social efficiency varies depending on the configuration of the private demand relative to the public demand for particular data sets. The analysis can also be mathematically formulated as a joint maximization of the two functions, U(x, y), the private utility function, and the public utility function Y(x, y). x is the private data generated by an individual, y is public space infrastructure that, by investing in it, it can enhance both the private and the public values. The joint maximization problem is simple and straightforward, with marginal equalities easily identifiable for both the x and y dimensions. The condition thus serves to define the optimal social welfare of an individual's publicness, the essence of a social contract between the individual and the public.

5 Social Contract with AI Added

How would robots complicate this simplified TWO PERSON demand inter-individual problem? The theoretical extension of that is simple, as the joint maximization problem would then become the sum of three functions, adding a Digital Platform function D(..) with an additional symbol, say a, added to the arguments of the function in addition to x and y to denote investment made to enable the robots, which can be interpreted as the algorithm of the robots designed by a platform. An increase in algorithm efforts will affect the utility that a private individual can get out of his private data, such as Amazon using AI to augment the quality of search results from customers' search queries to enhance overall shopping experience on Amazon.com platform. The social contract between AI robots and the data generators (i.e., individuals) with utility function U(x, y, a)) can be simplistically analyzed as the vertical summation of private demand and a digital platform aggregate demand, i.e., Ux + Dx, like the social contract of Ux + Yx as described by the assumptions and the three cases above. There will be a "no privacy" case, an "optimal privacy" case, and a "complete privacy" case for the social contracts. The preference of Dx is determined by the owner and the programmers of the platform. The crux of the matter is the interindividual effects assumed: What if Dx is in conflict with Ux?, i.e., what if the owners of Dx, in pursuit of their private benefits, structure their algorithm, a, that would reduce and not increase the utility of the users, Ux? In the TWO-PERSON case of interindividual effects of Ux and Yx, though the negative effect, i.e. conflict, is possible, the effect of that is very micro, and most likely not scalable as a broad community problem. That will not be the case in consideration of Dx. The effect of Dx on Ux is likely to be a business operation and scalable, i.e. affecting many individuals, as it is likely to be conducted as an organized program involving AI. When the integrity of the owner (or the enterprise) of Dx is in doubt, the need for a social contract is most acute. Broadly speaking, if certain publicness of an individual is expected as a matter of social norm, whether in public space in the city or in the cyberspace, a question can be asked for AI robots or the designers of them-to what extent that they will not misuse the publicness of individuals? Various computer and privacy rights experts have often addressed this question. For example, [11] documented stories of failed anonymization in public data sets, suggesting notions of 'differential privacy.' Recent Netflix documentary, Social Dilemma, also warned the public on possible AI misuse [18]. [29] asked whether profiling methods based on linked records will conflict with privacy and indeed cause discriminatory treatment, drawing attention to anonymization techniques used in EU General Data Protection Regulation. [10], in an interesting book and in a Youtube presentation at Google, articulated some guidelines for designers of AI to contemplate. He also suggested many new ideas on how designs can structure digital privacy. On the other hand, there are also writings teaching how ordinary citizens can be smart digital citizens, e.g., [25]. It will be a futile effort to second-guess the real intent of the owners of Dx. What may be more fruitful as an exercise of logical reasonings is to identify a business model that a business either explicitly or implicitly adopts. If we can identify the situation where within the context of a particular business model, social contract of the type that will automatically maximize social welfare in the aggregate, it will not be necessary to impose an explicit social contract, either by negotiation, by consent decree, or by regulations. We now turn to the consideration of such a model.

6 Digital Platforms as 'Two-Sided Markets'

The literature on two-sided markets is characterized by its emphasis on the behavior of the intermediary in the market, [22]. Generally speaking, a two-sided market analysis examines the choices by market intermediaries when there is some sort of interdependence or externality between agents that the intermediary serves. Amazon, for example, displays both one-sided and two-sided selling strategies. For some products, like books, Amazon buys at a wholesale price and sells for a retail price, which is a one-sided model. But for many other products, Amazon provides a web portal for producers to sell their goods to consumers at a retail price. As this decision of openness often depends on the platform rather than on purely technological features of the market, it may be better to use the term 'two-sided strategies' rather than 'two-sided markets [22]. The inquiry of how Dx can increase Ux comes to the frontier of discussion during the process where the intermediary adopts new methods to strengthen these 'two-sided strategies,' maximizing benefits for the buyers/sellers as well as itself. Thus, a two-sided strategy would require the platform to design an algorithm that can increase both the individuals on the seller side, and the individuals on the buyers side, i.e., Dx, Usx, and Ubx, with a focused intent of matching (as opposed to gathering general information for the sake of information gathering); the necessary but not sufficient requirement being that interaction between Dx, and the two types of individuals cannot be negative. There are profit-incentive reasons to ensure that the interactions on both sides of the market is not negative, because a business model of that type will be self-destructive, and is bound to fail in the business world. Thus, the assumption has to be that the interactions on both sides are positive, in an attempt to maximize the total size of the pie. We are aware of the fact that Amazon, which serves its end users and retailers on both sides, takes a margin on each product sold on its platform. However, Amazon's importance lies in how it makes the two sides feed each other - a two-sided network effect [7].

With focused objectives of gathering retail and wholesale data aiming for commerce, Amazon uses 'the force' (machine learning) to handle formulaic processes, which result in easy integration of suppliers. Under its automation campaign called 'Hands off the Wheel', AI & algorithms work to optimize the delays and efficiencies of product movement using the combination of Robots, Pods, and Associates towards the goal of speedier product delivery [14]. Many SMEs are flourishing due to the assistance provided by Amazon via its technology-supported seller services (FBA/SFP/Educational events). Amazon also proved its worth with key innovations that created direct network effects beyond what the 2-sided users can do themselves. The role of AI in Amazon's recommendation engine is enormous, as it generates 35 percent of the company's revenue

[1]. The business's recommendation engine tends to personalize the selection of items that consumers need to buy by gathering data from individual consumer interests and transactions. Using a data-driven '360-degree view' of an individual, Amazon identifies and makes recommendations to other clients, based on hundreds of touchpoints (data sources), that meet the same standards.

Amazon's AWS platform empowers companies, markets, and verticals of different kinds. Recently Amazon announced the general availability of AWS Glue DataBrew, a new visual data preparation tool that enables customers to clean and normalize data without writing code – a major time saver for many businesses [5]. Simultaneously, Amazon is also building a valuable ecosystem around its Alexa voice computing platform. Although Amazon wants to be everywhere, it has recognized that it cannot win the battle for the smart home alone. By investing in and open-sourcing Alexa, the company is converting potential competitors into valuable partners. Developments also suggest that Amazon views AI in healthcare as a frontier worth pursuing—and perhaps its next major revenue driver [7]. Amazon's entrance legitimizes the online pharmacy space by encouraging more consumers to consider getting their prescription medications delivered by an online pharmacy. A by-product of this effort is consumers empowering and educating themselves to make informed decisions when purchasing their prescription medication. Thus, we see, Amazon has designed everything for its 'two-sides' (sellers/buyers) around data input and output. Indeed, its role in this data market is pivotal. There are other platforms that also excel in formulating their 'two-sided' strategies. This process is characteristic of a Schumpeterian process that is the hallmark of Creative Destruction [23].

7 Further Formulations of the Two-Sided Markets

Two-sided markets as a matter of a business managing vendors and customers have been around since the early days of commerce, but two-sided markets with network effects are a more recent phenomenon in the digital world with AI now being augmented to the same. Let us try understanding the potential of these markets with the help of below scenarios. The analysis below is based on [28].

Scenario I: An industry which is characterized by a standard scale economy called Natural Monopoly as shown in Fig. 2 - left chart. D is the marginal valuation of the demand curve. AD is the average valuation of the demand curve. Shaded yellow area represents the total social welfare. The dotted yellow area represents the total fee net of costs awarded to the natural monopoly under fixed fee regulation. We have taken the case of a natural monopoly as the base of our explanation to indicate how similar economies of scale can be achieved by digital platforms by leveraging on network effects via a combination of linear and platform approach. Amazon's linear-platform hybrid model has forced many traditional brick-and-mortar retailers to shutter their windows but those who have managed to hold onto market share and grow despite the threat of Amazon managed to do so by adapting a linear-platform hybrid business of their own [13].

Scenario II: A 2-sided market as shown in Fig. 2 – center chart. While the curve AD, shaded yellow box of total welfare and dotted yellow box of total fee net of costs hold true for this diagram, they have been omitted for clarity purpose. The red arrows

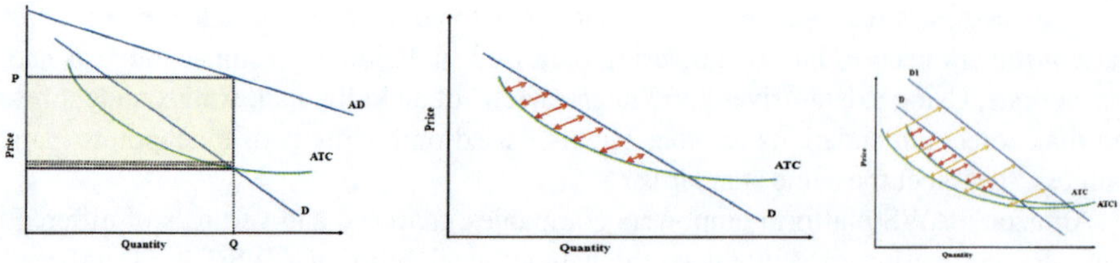

Fig. 2. Scenario I, II, and III (Developed by the authors)

signify the interactive process of a 2-sided market which is the characteristic of many digital platforms. Direct and indirect network effects generate value in a way which makes Amazon's role in this type of a market critical and is primarily the reason why many other players have not been able to disintermediate Amazon [16]. While the total fee net of costs is subject to competition, the total welfare remains to be the maximum [4].

Scenario III: A 2-sided market with network effects implemented by AI as shown in Fig. 2 – right chart. The yellow arrows denote the network effects. By employing AI robots, Amazon can increase its demand from D to D1 while simultaneously lowering its average total costs curve from ATC to ATC1. Total welfare in this situation will still be the maximum if the fee is subject to competition. This figure basically gives us a picture of how an efficient social contract can be formulated by a digital platform using AI in the absence of regulations. The formulation of AI effects in this figure goes beyond the original concept of the two-sided market suggested in [21].

8 Public Policy

Two-sided markets with AI are likely to tip toward a single dominant platform. As a result, it is not surprising that these markets are of interest to antitrust authorities. Economies of scale have given Amazon the flexibility to lower prices and build a series of increasingly powerful computing tools. Amazon's presence among this group of inspectable companies seems to be more based on its size than any verifiable anti-competitive behavior [17]. Undoubtedly the company is large, and the Covid-19-related shutdown of the economy has made its services even larger. Still, the answer to whether they will continue to serve maximizing total welfare will be a crucial question to answer. The analysis of the two-sided market in this paper and the discussion of privacies suggest several directions for future consideration:

(A) It is crucial to inquire whether the AI robots of a platform are true to the intent of creating network effects (yellow arrows in Fig. 2 – right chart and increasing total economic welfare under scale economies.

(B) Does the AI tools usage extend beyond bringing buyers and sellers together to create a monopoly of databases more than required? In the age of big data competition, it could be tempting for AI robot programmers to collect data for the sake of collecting it randomly, but that may be contrary to the spirit of Guideline A mentioned above.

And (C) In light of ever-expanding markets, are the fees that a platform charge for bringing buyers and sellers together subject to competition? Have competitors' businesses been maliciously affected? Although the study in this paper is specific to Amazon, these three directions of inquiry can also be applied to any other country that attempts to improve its economic welfare in their formulation of policies under Industrialization 4.0.

9 Conclusion

There is no question that Amazon is subject to competition. Their competitors each have their own standards. If standards can differentiate platforms from each other, they may be able to coexist successfully. Our evaluation of Amazon along the public policy guidelines suggests that there are no reasons for imposing additional regulations for Amazon based on its current trend of operations. The crucial question of whether Amazon is maximizing total welfare can be answered as overwhelmingly affirmative. Although Amazon engages in massive data collection, it focuses on productivity type of usages, subject itself to the force of competition. Amazon, like many enterprises in a capitalistic system, is exposed to Schumpeterian competition [23]. Even though it is now an evident winner in particular markets, there is no assurance that it will be the winner in all markets forever. The powerful force of the 4th Industrial Revolution [24] is an on-going process. Public policy, in its true intent to enable market competition, may not want to over-regulate a process that is just at the dawn of the new digital age. Most free world societies have a fair number of 'first-class citizens' who find comfort in city lives' existing public/private space. The same cannot be said about the internet digital world. Although it is a noble vision to strive to design a system that can provide 'first-class digital citizens' [12], it may also depend on whether the digital giants have a 'first-class robots of AI.' Robots are smart, but the question is whether they can be ethical, too, as they operate using mechanical algorithms only. The presence of good/bad cop in the city space is well known. Violation of real space 'social contracts' has hitherto been resolved by individuals withdrawing certain actions that they fear will intrude too much into their privacies. Some would confront a situation by resolving conflicts via the courts of laws if improper behaviors are discovered ex-post. A similar mechanism can be adopted for digital space with the support of government laws. However, there may be a case where a monopsony of government data is formed. The government AI robots may not be guided or constrained by ethical considerations. Under those circumstances, with government intervention of various sorts, the meaning of a broader type of social contract would have to be put in place. All these can be further examined. This paper lay out a thinking framework to take the first step in this direction.

Acknowledgement. The authors wish to thank Michael Hammond for comments. Support from NYIT is also gratefully acknowledged.

References

1. Arek, S.: Artificial Intelligence in retail innovation, by Amazon (2020). https://arekskuza.com/the-innovation-blog/amazon-and-artificial-intelligence-in-retail/

2. Al-Sartawi, A. (ed.): The big data-driven digital economy: artificial and computational intelligence In: Kacprzyk, J. (ed.) Studies in Computational Intelligence, vol. 974. Springer International Publishing, New York (2021)
3. Baye, M.R.: Microeconomics and business strategy. McGraw-Hill Irwin, New York (2010)
4. James, M.B.: Freedom in Constitutional Contract: Perspectives of a Political Economist. Texas A&M University Press, College Station and London (1977)
5. Businesswire: AWS Announces AWS Glue DataBrew (2020). https://www.businesswire.com/news/home/20201111005879/en/AWS-Announces-AWS-Glue-DataBrew
6. Carrascal, J.P., Riederer, C., Erramilli, V., Cherubini, M., de Oliveira, R.: Your Browsing Behavior for a Big Mac: Economics of Personal Information Online. In: Proceedings of the 22nd International Conference on World Wide Web, WWW'13, pp. 189–200. ACM, New York, USA (2013)
7. CBINSIGHTS: Amazon Strategy Teardown: Amazon's Barreling Into Physical Retail, Financial Services, Healthcare, And AI-Led Computing (n.d.). https://www.cbinsights.com/research/report/amazon-strategy-teardown/
8. Coase, R.H.: The Problem of Social Cost" Journal of Law and Economics, vol. 3, pp. 1–44. The University of Chicago Press (1960)
9. C.O. v. Amazon.com, Inc. et al.: Case 2:19-cv-00910 (2019)
10. Hartzog, W.: Privacy's Blueprint: The Battle to Control the Design of New Technologies (2018)
11. Heffetz, O., Ligett, K.: Privacy and data-based research. J. Econ. Perspect. Am. Econ. Assoc. **28**(2), 75–98 (2014)
12. Ibarra, I.A., Goff, L., Hernandez D.J., Lanier J., Weyl, E.G.: Should We Treat Data as Labor? Moving Beyond 'Free' AEA Papers and Proceedings, vol. 1, no. 1 (2018)
13. Johnson, N.: Platform vs. Linear: Business Models 101 (n.d.). https://www.applicoinc.com/blog/platform-vs-linear-business-models-101/
14. Kantrowitz, A.: Same network effects responsible for Amazon's rise might lead to its undoing with How Amazon automated work and put its people to better use (2020). https://hbr.org/2020/09/how-amazon-automated-work-and-put-its-people-to-better-use
15. Lorne, F.T.: Data in Cities and Digital Platforms: Some Implications for Digital Social Contracts with AI Robots. Mimeo. 14 March 2019
16. Malankar, J.: Same network effects responsible for Amazon's rise might lead to its undoing. Harvard Business Digital Initiatives (2015). https://digital.hbs.edu/platform-digit/submission/same-network-effects-responsible-for-amazons-rise-might-lead-to-its-undoing/
17. Newman, D.: Leave Amazon out of Big Tech regulation—it doesn't have a monopoly. Marketwatch (2020). https://www.marketwatch.com/story/leave-amazon-out-of-big-tech-regulation-it-doesnt-have-a-monopoly-2020-10-10
18. Orlowski, J.: The Social Dilemma [TV series] (2020). https://www.netflix.com/
19. Razzaque, A., Eldabi, T., Jalal-Karim, A.: An integrated framework to classify healthcare virtual communities. In: European, Mediterranean & Middle Eastern Conference on Information Systems 2012. Munich, Germany (2012)
20. Rinehart, W.: The Law & Economics of 'Owning Your Data'. Insight, April 10. American Action Forum (2018)
21. Rochet, J.-C., Tirole, J.: Platform competition in two-sided markets. J. Euro. Econ. Assoc. **1**(4), 990–1029 (2003)
22. Rysman, M.: The Economics of Two-Sided Markets. J. Econ. Perspect. **23**(3), 125–143 (2009). https://pubs.aeaweb.org/doi/pdf/10.1257/jep.23.3.125
23. Schumpeter, J.A.: Capitalism, Socialism and Democracy. Harper Colophon Books (1950)
24. Schwab, K.: The Fourth Industrial Revolution: what it means, how to respond (2016). https://www.weforum.org/agenda/2016/01/the-fourth-industrial-revolution-what-it-means-and-how-to-respond/

25. Szymielewicz, K.: Your digital identity has three layers, and you can only protect one of them. Quartz, 25 Jan 2019. https://qz.com/1525661/your-digital-identity-has-three-layers-and-you-can-only-protect-one-of-them/
26. Gordon, T.: Explorations in The Theory of Anarchy, Monograph published by The Center for the Study of Public Choice. Virginia Polytechnic Institute and State University (1972)
27. Memdani, L.: Demonetisation: a move towards cashless economy in India. Int. J. Electron. Bank. **2**(3), 205–211 (2020)
28. Yu, B.T., Ulukok, E.: Google's use of artificial intelligence (AI) in a two-sided advertising market, Loss of trust, and the Monopolization of Data. Mimeo (2021)
29. Wachter, S.: Normative challenges of identification in the internet of things: privacy, profiling, discrimination, and the GDPR. Comput. Law Secur. Rev. **34**(3), 436–449 (2017)
30. Wiggers, K.: Amazon claims its AI-powered packaging selector reduced product damage rates by 24%. Venture Beat (2020). https://venturebeat.com/2020/06/09/amazon-claims-its-ai-powered-packaging-selector-reduced-product-damage-rates-by-24/

Adaptations During COVID-19 and Smart Disease Detection Through 'Germica'

Rajkumar Rajasekaran[1], Jolly Masih[2], and Damandeep Kaur[3](✉)

[1] School of Computer Science and Engineering, Vellore Institute of Technology, Vellore, Tamilnadu, India
[2] Prestige Institute of Engineering Management and Research, Indore, India
[3] Dada Analtix Pvt Ltd., Vellore, India

Abstract. COVID-19 pandemic has shaken the world and has taught a lesson to maintain personal cleanliness and hygiene. In these tough times, only people with higher degree of cleanliness and hygiene could beat this deadly virus and survive the pandemic. Everyday millions of people are losing their lives due to COVID-19, which is majorly spreading due to lack of sanitization and unhygienic surroundings. Simple remedies to fight corona virus including washing hands with soaps, regular use of hand and air sanitizers, use of face mask etc. In public places such as railway stations, airports, restaurants, hospitals, cafes, shopping malls etc., the maintenance and monitoring of hygiene conditions of visiting people conditions becomes a big challenge. To help the government authorities and local administration in better surveillance and monitoring of Corona positive cases, or people contaminated with any other kind of germs and bacteria, we came up with the prototype design of 'Germica-A germ detection machine'. It is savvy medical and clinic framework, which has multiple germ detection features installed like UV lights to detect the germ intensity on hands or any other targeted body parts. Another feature of 'Germica' is germ detection cameras; which can take the image of different body parts, example nails and based to its colour, it can tell if a person is sick or healthy. In preparing the prototype of 'Germica', scientist have vigorously tested and trained the medical data to see its actual application and degree of accuracy. Small portable device like 'Germica' can help in maintaining hygiene conditions in hospitals, clinics and other places. 'Germica' in long run can help to combat with COVID-19 pandemic and other disease that spread through human carrier.

Keywords: COVID-19 · Coronavirus · Pandemic · Colour analysis · UV light · Multilayer perceptron

1 Introduction

As per WHO (World Health Organisation) report, by 11th June 2020, about 6,057,853 COVID-19 cases have been confirmed worldwide including 371,166 deaths [1] (see Fig. 1). On December 31, 2019, China for the first time reported WHO, about a few cases of pneumonia, who were not been able to get treated in Wuhan City in Hubei Province.

A. M. A. Musleh Al-Sartawi et al. (Eds.): EAMMIS 2021, LNNS 239, pp. 90–96, 2021.
https://doi.org/10.1007/978-3-030-77246-8_9

Within a span of few days, the disease crossed the borders of China and reached to other countries of the world through human carriers. The virus has been named SARS-CoV-2 and the disease is now called COVID-19. In February, 2020, WHO declared the COVID-19 as global pandemic and issued warming for other countries of the world to prepare themselves in order to fight the pandemic [2]. The COVID-19 pandemic has changed the world in last few days. It has caused detrimental loss of lives, livelihoods and has put the well-bring of mankind at a higher risk. The COVID-19 cases have taught the world to get better prepared about upcoming global heath challenges. It has unveiled the fact that how unprepared most health systems and global economies were to fight any pandemic. COVID-19 has posed major negative impact towards achieving the Sustainable Development Goal (SDGs) [2].

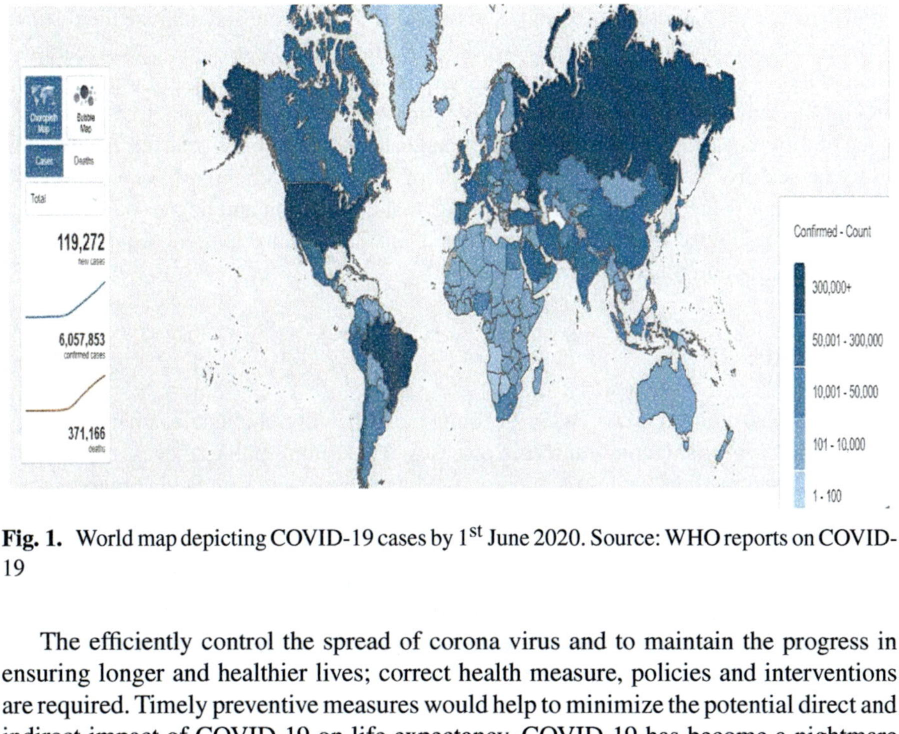

Fig. 1. World map depicting COVID-19 cases by 1st June 2020. Source: WHO reports on COVID-19

The efficiently control the spread of corona virus and to maintain the progress in ensuring longer and healthier lives; correct health measure, policies and interventions are required. Timely preventive measures would help to minimize the potential direct and indirect impact of COVID-19 on life expectancy. COVID-19 has become a nightmare due to excess mortality, especially among older adults, patients with pre-existing NCD conditions including diabetes, asthma, and hypertension. The COVID-19 pandemic has drawn the focus of WHO and other health organizations to rebuild strong health systems with increased facilities to quality health services, affordable cost of treatment and well-trained health workforce. COVID-19 pandemic also calls for the facilities which includes provision routine vaccinations, daily and basic hygiene and sanitation methods [3]. Looking at present scenario, USA is most vulnerable to the effect of Coronavirus, followed by Europe (see Fig. 2). This necessitates the urgency to invest in health systems, services and workforce [3]. Maintaining social distancing, restricting the public movement and physical contact are proven measures to reduce transmission of corona virus and to protect public health. However, in many countries like USA, Brazil, Rome etc. people are enjoying on beaches, making rallies on roads and public places against the government and not following the social distancing rules, thus worsening the situation further.

To help the government authorities and local administration in better surveillance and monitoring of Corona positive cases or people contaminated with any other kind of germs and bacteria, we came up with the prototype design of 'Germica-A germ detection machine. With the help of IoT sensors we can detect and predict some diseases by using smart devices, as it helps doctor to get more effective information in their clinical health network [4].

Use of AI technologies like drones, sensor, robots are important part as they highlights the strategies and frame the data collected in computing and are helpful in assisting the health care system to predict infections and provide medical care and analyze the results [5]. It is savvy medical and clinic framework, which has multiple germ detection features installed like UV lights to detect the germ intensity on hands or any other targeted body parts [6–8]. Another feature of 'Germica' is germ detection cameras; which can take the image of different body parts, example nails and based to its colour, it can tell if a person is sick or healthy. In preparing the prototype of 'Germica', scientist have vigorously tested and trained the medical data to see its actual application and degree of accuracy. Small portable device like 'Germica' can help in maintaining hygiene conditions in hospitals, clinics and other places [9–11].

2 Methods

From the spread story of COVID-19, it is quite clear that dirty and unclean hands are an open invitation to many communicable diseases. Thus, in all public places, especially in hospitals and medical care unit, keeping the hands clean and sanitized is mandatory. In the study, we aim to propose the prototype of germ detecting machine 'Germica'.

Study Area: Christian Medical College (CMC), Vellore, India

Sample Size: 1000 people visiting CMC hospital from 1st May 2020, to 31st May 2020 Sampling: Judgmental sampling is used. All the people entering the COVID-19 testing ward were required to undergo a quick scan under 'Germica' prototype.

Lab work: The physical prototype of 'Germica' is prepared at Vellore Institute of Technology, India under the supervision of Dr. Rajkumar Rajasekaram. The team also included researchers from Erasmus Medical College, the Netherlands. For initial testing of 'Germica', researchers selected India, since it has one of the highest number of COVID-19 patients. This small, portable gadget is equipped with UV rays camera. This camera would emit the UV rays on any targeted physical surface, example hands, to identify presence and intensity of germs on targeted surface. The high-resolution camera of 'Germica' could also capture high resolution skin images; example some white spots identified by HD camera of 'Germica' gave clues about a fungal infection on skin. This prototype also enables image capturing of nails, and performing colour analysis to get an idea about an overall health of a person. Pale nails indicate liver infection, while grey nails indicate fungal infection. Thus 'Germica' is a wholesome germ detection portable device which tells user about level of germ infestation of the surface of hands, takes a note of general conditions of the user by analysing the condition of skin and nails of hands. This portable device is very useful in prevention of communicable as well as fungal and nutrition deficiency-based diseases. This prototype is believed to produce upto 70% accurate results. However, the level of precision would rise, when medical trial of this prototype would be conducted on large scale. To create this type of prototype, researchers in past have used several algorithms such as CNN (Convolution Neural Networks), SVM (Support Vector Machine), Feature extraction, Quality analysis methods etc.

3 Results

In result section, we will have discussion about the steps which germ detector prototype 'Germica' would follow, in order to give a brief analysis about an individual's health.

3.1 Percentage Calculation

First, 'Germica' scans the image of the targetted body part of a person (for example hands). The image will be then further analysed under the UV light. Bacteria and some other germs (including corona virus) has property to glow under UV lights. Thus 'Germica' could play a major role in germ identification and calculation of germ percentage on targetted site of analysis (example germs present on the surface of hands). For calculation of germ percentage, we have to decide certain thresh-hold value. For instance, for the trial purpose, we have kept the *treshold value* as '100 bacteria/pixel of image captured' (this average value is decided after analysing average number of germs on hands of 1000 sample respondents). This value is selected by analysing the average number of bacteria which shown up under UV light in each pixel of the image captured of targetted body part. After deciding the thresh-hold value we calculate the *actual number* of bacteria found under UV light/pixel of image captured. For example, we found 130 bacteria under UV light/pixel of image captured. Finally, we calculate the percentage of germs present by using this formula *Germs percentage = (A/B) * 100,* where A is *actual number* of bacteria under UV light/pixel of image captured, which is 130 and B is *thresh-hold value,* average number of bacteria, which is 100.

Thus by using 'Germica', we found that germ percentage was {(130/100)*100} 30% on hands of a targeted person. If the germ percentage is above 50% then person is suggested to wash his hands before entering to ensure hygiene and sanitization of the surroundings (see Fig. 2)

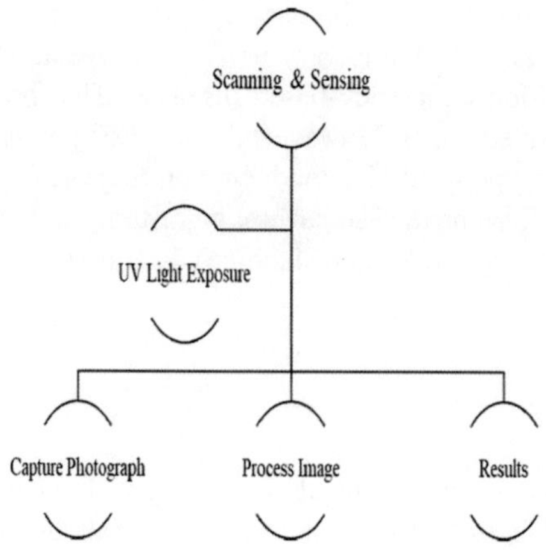

Fig. 2. The model diagram of the 'Germica'

3.2 Training and Testing the Dataset

In order to check the accuracy and reliability of proposed project, we run our project prototype on various data sets or the test case inputs. After getting the RGB percentage values to analyze nail colour, as well as the nail texture related values, we trained some of the medical data using Weka tool. At last through Multi-Perceptron algorithm we are going to detect the amount of accuracy present in each analysis (color and texture analysis). With the algorithm used for 'Germica' we are going to get about 70% germ detection and nail health detection accuracy. After running the prototype on 1000 people at Christian Medical College, Vellore, we found our prototype provided 77.77% accuracy for nail colour and 73.91% accuracy for nail texture cases. However higher accuracy was achieved by the prototype when we further split the test data and trained it again. Thus, data training can help in getting high accurate results in the medical prototypes likes 'Germica'. The prototype model of 'Germica' is able to detect germ percentage of all the people entering the campus of CMC, Vellore. It helped the local authorities to find potential COVID-19 patients, based on COVID-19 testing of people who were found to have germ percentage more than 30%. Other than this 'Germica' have also helped in identification of many other health ailments in the tested people based on their nail colour and texture. To support the results of 'Germica', CMC authorities have also used thermal sensors to measure the body temperature of the people who had higher germ percentage, as they were suspected to have COVID-19 infection. Use of 'Germica' provided about 70% accurate results in identification of COVID-19 positive patients (see Fig. 3). Since this prototype is still in testing phase, in further advancements, researchers are planning to incorporate thermal sensors on 'Germica' to measure the contact less body temperature of people entering in the hospital or any other place. Upcoming versions of 'Germica' might also include portable disinfectant spray carrying chamber, to quickly sanitize the hands of any person who has higher germ percentage, UV disinfectant scanner to disinfect other physical items which might have high germ contamination on their surface example-medical equipment, documents, stationary items etc. Artificial

Intelligence team is working on training and testing analysis of 'Germica' to launch this product in form of mobile app in coming future, where by using the mobile camera, the germ percentage and nail scan would bedone.

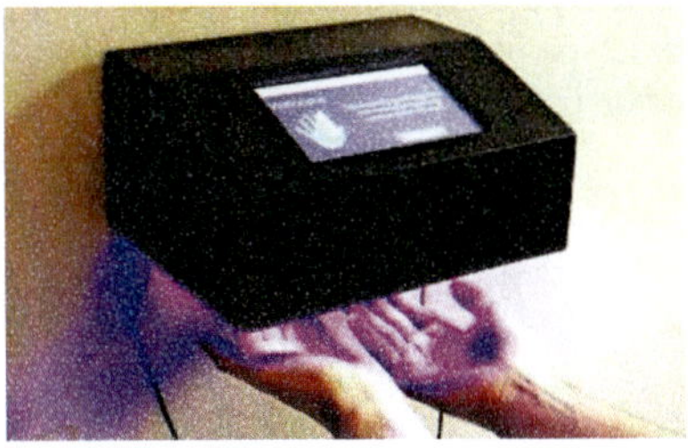

Fig. 3. First prototype of 'Germica' designed at Vellore Institute of Technology, Vellore, India

4 Conclusion

'Germica' is a game changing technology which in real-time provides health analysis related to germ percentage and other diseases by detecting the hands, skin, nail colour and texture. This prototype is in its initial stages of testing, and researchers are making continuous improvements and modifications in its features, design, weight and portability measures. Current prototype of 'Germica' provides about 70% accurate results after training and testing of the dataset. 'Germica' could be installed in any public place, work place, hospitals, offices, malls, theaters etc. to keep a check on health and hygiene of people. 'Germica' provides a hope of light to the world to fight COVID-19 pandemic and other communicable diseases in an efficient manner by providing health results in real-time.

References

1. Giannini, M.: Performance and quality improvement in healthcare organizations. Int. J. Healthcare Manage. **8**(3), 173–179 (2015)
2. Indi, T.S., Gunge, Y.A.: Early stage disease diagnosis system using human nail image processing. IJ Inform. Technol. Comput. Sci. **7**, 30–35 (2016)
3. Kavitha, M., Jayasankar, T., Venkatesh, P.M., Mani, G., Bharatiraja, C., Twala, B.: COVID-19 disease diagnosis using smart deep learning techniques. J. Appl. Sci. Eng. **24**(3), 271–277 (2021)
4. Khan, H., Kushwah, K.K., Singh, S., Urkude, H., Maurya, M.R., Sadasivuni, K.K.: Smart technologies driven approaches to tackle COVID-19 pandemic: a review. 3 Biotech **11**(2), 1–22 (2021)
5. Mente, R., Marulkar, S.V.: A review: fingernail images for disease detection. Int. J. Eng. Comput. Sci. **6**(11), 22830–22835 (2017)
6. Meraj, M., Singh, S.P., Johri, P., Quasim, M.T.: Detection and Prediction of Infectious Diseases Using IoT Sensors: A Review (2021). arXiv:2101.02029

7. Thakare, A., Meshram, S., Baradkar, H.M.: An automated medical support system for detecting human health conditions based on novel bicluster method. I J C T A, **10**(8), 223–228 (2017)
8. www.news18.com. Accessed on 06 Jan 2020
9. www.serverscheck.com. Accessed on 06 Jan 2020
10. www.thehindu.com. Accessed on 06 Jan 2020
11. www.theprint.in. Accessed on 06 Jan 2020

Facebook Advertising in the Era of Digital Economy

Sharma Megha[1(✉)], Dangwal Chandra Ramesh[1], and Masih Jolly[2]

[1] HNB Garhwal University, Srinagar, India
[2] Prestige College of Engineering and Management, Indore, India

Abstract. The present study explores the various drivers of Facebook advertising that influence the purchase intention of young consumers in digital economy. Further, the effect of demographic and usage variables on purchase intention is also ascertained. Both online and offline questionnaire survey of 241 Facebook account holder students were conducted. Data were analyzed using GLM ANOVA and Structural equation modeling. The study results suggest that Informativeness and Credibility of Facebook advertising have a significant impact on purchase intention young consumers whereas Entertainment showed no significant influence on purchase intention. No significant influence of usage characteristics, login frequency and mode of access as well as the demographic influence of gender, age and education on the perception of Facebook advertising driven purchase intention were found in the study. The study adds to the growing literature on Indian youth consumers by examining the effect of Facebook advertising on purchase intention. It also provides practitioners with some guidance on the types of advertising strategy they should employ when posting on Facebook in the digital era.

Keywords: Internet usage · Facebook advertising · Purchase intention · Digital economy · Young consumers

1 Introduction

The advent of the 21st century witnessed a boom in the advancement of technology which revolutionized our perception of the world. Over the last few years, the ingenuity of human knowledge has manifested itself in the form of cutting edge development in digital technology. Due to its distinctive strengths, emerging innovations have significantly expanded the information field of people and enterprises, reducing the cost of searching and processing information. The new evolutionary wave of digitalization has demanded changes in the economy, which has led to the emergence of the term "digital economy". Based on digital economy theory, the concept of "digital marketing" has become an integral part. Today, this technology has entrenched every perceivable facet of our lives and has altered the way we interact with our environment (Bolton et al. 2013). The widespread usage of the internet along with the advancement of technology has significantly influenced consumers' behaviour in general (Di Capua 2012). Consumers now have access to multiple online platforms and digital spaces that are adequate with

A. M. A. Musleh Al-Sartawi et al. (Eds.): EAMMIS 2021, LNNS 239, pp. 97–108, 2021.
https://doi.org/10.1007/978-3-030-77246-8_10

almost all the sufficient features which make them more user-friendly in terms of usage. These digital platforms become of particular interest to companies and organizations since they implicitly provide an understanding of the likes, preferences and the perceptions of the consumers towards various goods and services (Duffett 2015). Social media – Facebook, Twitter, YouTube, etc. - which remolded the face of communication, has become an ideal landscape for advertisers who are now appropriating a larger share of advertisement budget to social media than ever before. Small firms and businesses are also using social media for promoting their business activities as they find it a less expensive alternative than taking the risk of paying large sums to advertise in non –virtual media (Dehghani & Tumer 2015). Another drawback of advertising through traditional media is that it provides one-way communication as highlighted by Armstrong and Kotlar (2000). Subsequently, the advance features of Web 2.0 overcome this drawback of traditional mass media (Mulhern 2009).

According to Global Web Index 2018, the number of social media users worldwide stood at 3.196 billion in 2018. Saudi Arabia reported the largest growth rate of 32% closely followed by India at 31%. Therefore advertisements in social media which are now the lifeline of marketing communication has become an effective tool to get an insight into and influence consumers' tastes and preferences (Saxena and Khanna 2013). Current statistics reveal that amongst the various social media platforms Facebook is the largest online platform in the world (Statista, 2018). This necessitates a study to understand the influence of Facebook advertising on consumers' purchase intentions, especially the youth which has been impacted significantly.

Moreover, in response to global pandemic coronavirus (COVID-19) outbreak, Facebook has been continuously working to support global health officials and help people stay informed, safe and connected. The various features of Facebook advertising played a significant role to aware the public about pandemic and to stop the situation of panic buying.

The purpose of this study is to assess the influence of Facebook advertising on the purchase intention of young consumers in this digital era. Further, it will ascertain the influence of usage and demographic features on Facebook advertising driven purchase intentions. Studies examining the influence of Facebook advertising on purchase intention of young consumers appear to be rare if they exist. Therefore, this study is a modest yet significant attempt to add to the growing literature on Facebook advertising through empirical evidence on this subject. Second, the need for research in this field arises, especially in India, where studies on social media advertising, especially Facebook are in the embryonic stage. It is presumed that consumers' attitude towards social media advertising differs from one country to the other (Durvasula et al. 2001; Ferle et al. 2008). Besides, this study is pragmatic in nature as it deals with a special group of consumers who have access to the internet and frequently use Facebook in their lives. Moreover, previous studies (Duffett 2015; Roblyer et al. 2010) have focused more on factors that create consumers 'attitude towards Facebook advertising in general.

The rest of this paper is structured as follows. In the study, Sect. 1 presents a literature review of the constructs undertaken in the study with hypotheses and conceptual framework, while Sect. 2 focuses on research methodology, Sect. 3 results or findings of

SEM in the study. Section 4 presents discussion and implication while the last section presents future research.

2 Literature Review

2.1 Facebook Advertising and Purchase Intention

The rapid growth of social media platforms has notably changed the face of global marketing communication. This has increased the availability of choice offered to consumers while reducing the asymmetry in information which has resulted in well-informed customers. Earlier marketers used to target their potential consumer at a physical marketplace where buying and selling took place. This has now been replaced by virtual spaces. These digital spaces provide all the possible touch points to organizations for engaging consumers with products or campaigns. Yang & Lin (2014) have investigated the factors affecting Facebook stickiness amongst the young generation. The hedonic value was seen as an important value concern for Facebook users. The study revealed that the content of Facebook advertising enhances consumers' attitude towards brand and purchase intentions. Rohm, D. Kaltcheva & R. Milne (2013) disclosed consumer motive via social media (Facebook, Twitter) marketing that included product information (cognitive), entertainment (affective), incentives and promotion to encourage purchases (behavioural). The Facebook page of an organization positively impacts the stakeholders' attitude and purchase intention (Haigh et al. 2013). It was also found that companies which used social media increased the possibility of selling more (Edwards 2011). In a study conducted by Powers, Advincula et al. (2012) it was found that 20% of the consumers believed that social media was the reason of their final purchase; while another 20% agreed that it helped them in decision making regarding the purchase.

2.2 Credibility

The credibility of advertising is characterized as a consumers' perceptions of the reliability, trustworthiness, and believability of advertising in general. It has been examined that the credibility of ads creates a positive attitude towards a particular brand (MacKenzie and Lutz 1989). Advertising credibility has been validated to be a crucial factor that affects advertising value and attitude towards advertising in particular (Brackett and Carr 2001). DeGruttola (2017) stated that consumers consider the credibility of advertisement by their web site. It has also been suggested that the essential factor in advertising is the credibility of endorse (Goldsmith et al. 2000). As keep in mind the above discussion, the following hypothesis is generated:

H1: Credibility of Facebook advertising has a positively associated with Purchase Intention of young consumers.

2.3 Informativeness

Informativeness refers to the proficiency of the advertising to effectively inform consumers about product and service in order to satisfy their need (Ducoffe 1995). Rotzoll and Christians (1989) marked that the main constituting function of advertisement is its informative nature. It is considered as valuable source information for various products and services. Consumers increasingly resort to social media to collect information about various brands (Barreda et al. 2015; Heller Baird and Parasnis 2011; Naylor et al. 2012). It was found that there is a positive association between the informativeness level and the attitude of consumers towards advertising (Barreda et al. 2015; Bauer et al. 2005; Ducoffe 1995). The quality of information provided in Facebook advertising affects the users' perception towards companies and their product and services (Yuanxin and Pittana 2011). Taking into account the previous studies and findings in this context. Thus, the following hypothesis is formulated:

H2: Informativeness of Facebook advertising is positively associated with Purchase Intention of young consumers.

2.4 Entertainment

Ducoffe (1995) stated that entertainment is an essential feature of advertising not only in traditional media but in social media as well. The competency of advertising is measured by its entertainment, which can enhance the advertising experience of consumers (Alwitt and Prabhakar 1992). People´s feeling of enjoyment with a particular advertisement shaped the overall attitude of consumers toward them (Kamboj and Rahman 2017; Roblyer et al. 2010; Shavitt et al. 1998). With reference to mobile marketing, a multi-channel online marketing technique which aids the maximum amount of reach, it was found that the mobile marketing messages should be precise, creatively designed and entertaining in nature. High informative value with entertaining content develops a positive effect on consumers' attitude towards mobile marketing, which in turn leads to positive behavioural intention to use mobile marketing services (Bauer et al. 2005). As per the above discussed previous studies and findings, the following hypothesis is prepared:

H3: Entertainment of Facebook advertising is positively associated with Purchase Intention of young consumers.

2.5 Facebook Advertising and Usage and Demographic Factors

The attitude of consumers may or may not be consistent across demographic groups in all the population (Shavitt et al. 1998). The demographic factors are always considered to be important while determining the attitude towards advertising (Yuanxin and Pittana 2011). Men showed a more favourable attitude towards Facebook advertising, whereas women are not interested in Facebook activities (Global Web Index, 2018). It is found that internet usage influences the behavioural pattern of a user (Punj 2011). Chandra (2012) revealed that the frequency of using social media significantly affect the attitude of consumers towards Social media advertising. Facebook advertising was found more effective when millennial member spent two or more hours on per login session and frequently update their profile (Duffett 2015). On the basis of the above mentioned studies, the following hypothesis is developed:

H4: There is a significant impact of demographic and usage characteristics on young consumer purchase intention.

3 Conceptual Framework

On the basis of the above discussion, hypothesized relationships are shown through the conceptual model in Fig. 1 which represents the linkages between Facebook advertising, Purchase Intentions and Demographic and Usage Factors.

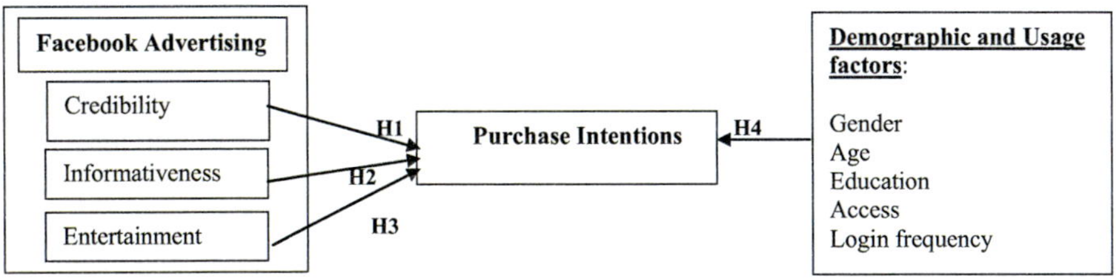

Fig. 1. The proposed conceptual model

4 Research Methodology

4.1 Instrument Design

The survey questionnaire was developed in two-section. The first section comprised of 20 questions relating to the key constructs which included Facebook advertising features and purchase intentions. The second section incorporated 5 questions related to the respondents' internet usage and demographic information. All the measurement scales were taken from existing literature and were improved to fit the context. The improvement was done in part of language which easily understood to the respondents. The items in the survey instruments were adapted from the studies of Ducoffe (1995), Khasawneh (2018) and Brackett and Carr (2001). The questionnaire was also verified by experts, which was then followed by a pre-testing survey on 50 students. The constructs were weighed using multi-item 5-point Likert scales with responses varying from 1 for strongly disagree to 5 for strongly agree.

4.2 Sample Design and Data Collection

The study was conducted in India using questionnaire survey method. A judgemental sample of 241 students was collected from HNB Garhwal University. Students aged between 18–30 years were chosen as a sample because this young segment of consumers has been showing distinct characteristics like easy access to credit card, rising purchasing power, and their acceptance to technology (Chao and Schor, 1998). Advertising displayed on Facebook was selected as the study context because it has the highest reach among young consumers (Kamboj and Rahman 2017; Roblyer et al. 2010). Many organizations are engaging consumers and building consumer relationships by creating Facebook pages (Islam and Rahman 2016). All the students were Indian nationals. A structured questionnaire was used in the study. Before distributing the questionnaire, students were given a 10 min presentation on the purpose of and a basic understanding of research constructs undertaken in the study. The data was retrieved from respondents through both offline and online survey. Out of the 241 distributed questionnaires, 234 responses were collected back. After rejecting 25 incomplete responses, a set of 209 completed questionnaires were undertaken for the study.

4.3 Respondents' Profile

The demographic and internet usage profile of the respondents summarizes the details information about respondents. The sample (n = 209) comprised 51.7% (108) male and 48.3% (101) female respondents. Approximately 48% were aged between 18–21 years, 29% between 22–26 years, and 26% between 27–30 years. 51.6% of the students were enrolled in the bachelor's programme, 33% in the Masters and 15.3% in other courses offered by the University. The majority of respondents accessed Facebook by using a mobile device (52.6%). Sixty- four per cent of respondents checked Facebook daily and 20.1% checked 2–4 times a week. They used Facebook for socializing, entertainment and information. They also participated in discussions, quizzes, games, and surveys and gave feedback on products and services.

5 Data Analysis and Interpretation

5.1 Reliability and Validity

The data were analyzed using structured equation modeling via AMOS 21 software. To check the dimensionality of the constructs as well as removing items which do not relate to its underlying constructs Exploratory Factor analysis (EFA) was used. EFA was run with all the indicators variables of Facebook advertising and purchase intention. It was observed that all the items of the scale showed community (>0.5) and factor loading (>0.6) and therefore retained. EFA results suggest all the constructs be one-dimensional and explained variance of more than 50% in the indicator scores. The value of Cronbach's α for the constructs was greater than denoted accepted reliability (α > 0.70). Conformity factor analysis (CFA) was run to test the discriminant and convergent validity of constructs. Greater than recommended cut off values of average variance extracted (AVE > 0.5) and composite reliability (CR > 0.70) were obtained (Fornell and Larcker 1981). All the loadings of CFA are exhibited in Table 1 at significant t values (p < .01) and significant convergent validity (>.05) (Hair 2006).

Discriminant validity of the constructs is also demonstrated in this study. Discriminant validity "assess the degree to which measures of different concepts are distinct" (Bagozzi and Yi 2012). It is verified by having the square root value of the AVE of each variable is greater than its correlation values and all the other variable correlation values (Fornell and Larcker 1981).

Table 2 shows the descriptive statistics and correlation values of the study, and presents that the square root of average variance explained (AVE) for each construct, is higher than the correlation of the construct with other constructs This indicates a significant level of validity.

Table 1. Results of exploratory factor analysis and reliability analysis

Construct	Items	Factor loadings	Cronbach's α	Composite reliability	AVE
Facebook advertising			0.886		
Credibility				0.85	0.54
	FA11	0.78			
	FA10	0.78			
	FA14	0.70			
	FA17	0.71			
	FA12	0.69			
Informativeness				0.83	0.55
	FA1	0.68			
	FA2	0.71			
	FA3	0.79			
	FA4	0.77			
Entertainment				0.80	0.58
	FA9	0.74			
	FA7	0.75			
	FA8	0.79			
Purchase intention			.849	0.86	0.56
	PI2	0.83			
	PI1	0.79			
	PI5	0.74			
	PI4	0.72			
	PI3	0.64			

Source: Authors' own findings

Table 2. Descriptive statistics and correlations

Variable	Mean	SD	CRE	INFO	ENT	PI	SqR AVE
Credibility	2.67	0.61					0.735
Informativeness	2.70	0.57	0.678*				0.752
Entertainment	2.93	0.65	0.629*	0.746*			0.762
Purchase Intention	2.94	0.74	0.656*	0.701*	0.632*		0.749

Source: Authors' own finding

In order to assess the effect of usage characteristics and demographic factors (predictors) on Purchase intention (dependent variable), GLM ANOVA was used to quantify the relationship between several independent variables and dependent variable, which can be measured by the higher standard errors (an example of this is the low number of respondents that logged in to Facebook at least once a month). GLM considers this problem and accordingly normalizes the outcomes.

Table 3. Effect of internet usage and demographic factors on purchase intention

Independent variables	Type III Df	P-Value
Access*Login frequency	7	0.6790
Access	2	0.515
Login frequency	4	0.795
Gender*Age*Education	11	0.211
Gender	1	0.026
Age	2	0.736
Education	4	0.816

Source: Authors' own finding
Note: Significant difference at $p < 0.001$.

Table 4 shows that the usage characteristics reveal that access and login frequency have no significant influence on Facebook advertising driven purchase intention. Whereas demographic factors age and education also show a similar trend and there is very little impact of gender ($p = 0.026$) on purchase intention, but not a significant level. Moreover, the relation is also checked by the interaction of variables. First, the usage characteristics, access and login frequency ($p = 0.6790$) interacted and they showed the not significant level. Similarly, demographic factors ($p = 0.211$) were interacted together to know their significant level for Facebook advertising driven purchase intention. No significant relation was found. Hence H4 is rejected.

5.2 Results of Structural Model

In order to check the relationship of Facebook advertising features credibility, informativeness, and entertainment with the purchase intention, a model was run and a good fit with the data was obtained by the model (CMIN$\chi^2 = 2.181$, $p < 0.05$, df = 113, TLI = 0.90, CFI = 0.925, and RMSEA = 0.075). The results of the structural analysis, the path coefficients, and the t values of the relationships are presented in Table 3 and Fig. 2. Therefore, the fit of the model was adequate as all indices were within the threshold limits as CFI > 0.9 and RMSEA < 0.08 (Hu and Bentler 1999). This significant impact of Facebook advertising on purchase intention is logically supported.

The hypotheses H_1 and H_2 were accepted whereas H_3 was rejected. H_1 asserts that Credibility of Facebook advertising positively influences the purchase intention. This hypothesis is fully supported by the data (path coefficient = 0.30). Informativeness of Facebook advertising (H_2) is highly assisted by the data with path coefficient 0.44. In addition to this, no significant impact of entertainment feature of Facebook advertising on purchase intention of young consumers was found. Hence, H3 is rejected.

Table 4. Hypotheses-testing results

Hypotheses	Path estimates	T values	P value	Test results
H1:CredibilityPurchase intention	0.30	4.671	0.02	Accepted
H2:InformativenessPurchase intention	0.44	5.391	0.03	Accepted
H3:EntertainmentPurchase intention	0.13	1.737	0.15	Rejected

Source: Authors' own finding

5.3 Discussion and Implication

Facebook advertising in the broadest sense is interpreted as a marketing aimed at increasing competitiveness and ensuring economic growth. So the digital economy has a direct link in web marketing development. By the concept of digital marketing, we mean modern marketing using digital technology to achieve more global and broader results in less time and cost. The findings of the study suggest that Facebook advertising has a positive influence on the purchase intentions of young consumers. Credibility and Informativeness of Facebook advertising positively influence the purchase intention of young consumers. However, no significant impact of entertainment of Facebook advertising on the young consumers purchase intention was found which is inconsistent with the findings of previous studies (Ducoffe 1995; Grant and O'Donohoe 2007). Moreover, demographic and usage factors also do not significantly influence the Facebook advertising driven purchase intention, which is supported by a study (Duffett 2015). The findings of the research will shed light on the important factors that any marketers should take into consideration while planning to position advertising through Facebook or their Facebook pages. The most compelling variable of Facebook advertising is the informativeness. As previous studies suggest the main purpose of any advertisement is to impart awareness or knowledge about the product and service (Tsang et al. 2004; Rotzoll et al. 1989). Therefore, it is very essential for marketers to make advertisement which carries valuable content about the products and services. Further, advertisers should give value to the credibility aspect of advertisement. Furthermore, the Covid-19 outbreak has significantly impacted the daily business owners and consumer behaviour, and as a result, these are the times when small business owners and advertisers need to find the most appropriate channel that help them to build personalized and lasting relationships with their audiences from safe distance. Thus, Facebook is the right place to build brand awareness with online audience.

5.4 Research Limitation

Despite its contribution, this study accepts certain limits. The selection of respondents is confined to young consumers. To generalize the findings, future research can be replicated with other age group population, e.g. middle age. It is also considered that the use of probability sampling is one the best methods to choose for data collection however the present study has employed judgemental sampling technique. A different sampling technique could be used in future research that may reflect the population design better.

The sample size was relatively small. Future could be carried out with a bigger sample size. The study was based on Indian origin. A comparative study could also be conducted to assess if any cultural bias exists. The study is limited to Facebook. Future research can consider other social networking sites. This study has taken three features of Facebook advertising. Future research can evaluate more features of Facebook advertising.

References

Alwitt, L.F., Prabhakar, P.R.: Functional and belief dimensions of attitudes to television advertising. J. Ad. Res. **32**, 30–42 (1992)

Armstorg, G., Kotlar, P.: Marketing: An Introduction. Prentice Hall, Upper Saddle River, NJ (2000)

Aydin, G., Karamehmet, B.: A comparative study on attitudes towards SMS advertising and mobile application advertising. Int. J. Mob. Commun. **15**(5), 514 (2017). https://doi.org/10.1504/IJMC.2017.10005357

Bannister, A., Kiefer, J., Nellums, J.: College students' perceptions of and behaviors regarding Facebook© advertising: an exploratory study. Catal. Multidiscip. Rev. Undergrad. Sch. Univ. S. Mississippi **3**(1) (2013). https://doi.org/10.18785/cat.0301.02

Bagozzi, R.P., Yi, Y.: Specification, evaluation and interpretation of structural equation models. J. Acad. Mark. Sci. **40**, 8–34 (2012)

Barreda, A.A., Bilgihan, A., Nusair, K., Okumus, F.: Generating brand awareness in online social networks. Comput. Hum. Behav. **50**, 600–609 (2015). https://doi.org/10.1016/j.chb.2015.03.023

Bauer, H.H., Reichardt, T., Barnes, S.J., Neumann, M.M.: Driving Consumer Acceptance of Mobile Marketing: A Theoretical Framework and Empirical Study **6**, 12 (2005)

Brackett, L.K., Carr, B.N.: Cyberspace advertising vs other media: consumer vs mature student attitudes. J. Ad Res. **41**(5), 23–32 (2001). https://doi.org/10.2501/JAR-41-5-23-32

Bolton, R.N., Parasuraman, A., Hoefnagels, A., Migchels, N., Kabadayi, S., Gruber, T., Loureiro, Y., KandSolnet, D.: Understanding generation Y and their use of social media: a review and research agenda. J. Serv. Manage. **24**(3), 245–267 (2013)

Chandra, B., Goswami, S., Chouhan, V.: Investigating attitude towards online advertising on social media-nan empirical study. Manage. Insight **8**(1), 1–44 (2012)

Chao, A., Schor, J.B.: Empirical tests of status consumption: evidence from women's cosmetics. J. Econ. Psychol. **19**(1), 107–131 (1998). https://doi.org/10.1016/S0167-4870(97)00038-X

DeGruttola, M.: Why Credibility leads to Conversions: Improving Facebook Ad Performance available at: https://stackla.com/resources/blog/credibility-leads-conversions-improving-facebook-ad-performance/. Accessed 26 Oct 2018

Digital in 2018: World's internet users pass the 4 billion mark. Accessed on 30 Jan 2018. https://wearesocial.com/uk/blog/2018/01/global-digital-report-2018

Ducoffe, R.H.: How consumers assess the value of advertising. J. Current Issues Res. Ad. **17**(1), 1–18 (1995). https://doi.org/10.1080/10641734.1995.10505022

Duffett, R.G.: Facebook advertising's influence on intention-to-purchase and purchase amongst Millennials. Int. Res. **25**(4), 498–526 (2015). https://doi.org/10.1108/IntR-01-2014-0020

Durvasula, S., Lysonski, S., Watson, J.: Does vanity describe other cultures? A cross-cultural examination of the vanity scale. J. Consum. Aff. **35**(1), 180–199 (2001). https://doi.org/10.1111/j.1745-6606.2001.tb00108.x

Edwards, S.M.: A social media mindset. J. Interact. Ad. **12**(1), 1–3 (2011). https://doi.org/10.1080/15252019.2011.10722186

Ferle, C.L., Edwards, S.M., Lee, W.-N.: Culture, attitudes, and media patterns in China, Taiwan, and the U.S.: balancing standardization and localization decisions. J. Global Mark. **21**(3), 191–205 (2008). https://doi.org/10.1080/08911760802152017

Fornell, C., Larcker, D.F.: Evaluating structural equation models with unobservable variables and measurement error. J. Mark. Res. **18**(1), 39 (1981). https://doi.org/10.2307/3151312

Gao, T., Sultan, F., Rohm, A.J.: Factors influencing Chinese youth consumers acceptance of mobile marketing. J. Consum. Mark. **27**(7), 574–583 (2010). https://doi.org/10.1108/07363761011086326

Goldsmith, R.E., Lafferty, B.A., Newell, S.J.: The influence of corporate credibility on consumer attitudes and purchase intent. Corp. Reputation Rev. **3**(4), 304–318 (2000). https://doi.org/10.1057/palgrave.crr.1540122

Grant, I., O'Donohoe, S.: Why young consumers are not open to mobile marketing communication. Int. J. Ad. **26**(2), 223–246 (2007). https://doi.org/10.1080/10803548.2007.11073008

Hair, J.F.: Multivariate data analysis, 6th edn. Pearson Prentice Hall, Upper Saddle River, N.J. (2006)

Haigh, M.M., Brubaker, P., Whiteside, E.: Facebook: examining the information presented and its impact on stakeholders. Corp. Commun. Int. j. **18**(1), 52–69 (2013). https://doi.org/10.1108/13563281311294128

Heller Baird, C., Parasnis, G.: From social media to social customer relationship management. Strat. Leadersh. **39**(5), 30–37 (2011). https://doi.org/10.1108/10878571111161507

Hu, L., Bentler, P.M.: Cutoff criteria for fit indexes in covariance structure analysis: conventional criteria versus new alternatives. Struct. Eqn. Model. Multidiscip. j. **6**(1), 1–55 (1999). https://doi.org/10.1080/10705519909540118

Islam, J.U., Rahman, Z.: The transpiring journey of customer engagement research in marketing: a systematic review of the past decade. Manage. Decis. **54**(8), 2008–2034 (2016). https://doi.org/10.1108/MD-01-2016-0028

Kamboj, S., Rahman, Z.: Understanding customer participation in online brand communities: literature review and future research agenda. Qual. Mark. Res. Int. j. **20**(3), 306–334 (2017). https://doi.org/10.1108/QMR-08-2016-0069

Karson, E.J., McCloy, S.D., Bonner, P.G.: An examination of consumers' attitudes and beliefs towards web site advertising. J. Current Issues Res. Adv. **28**(2), 77–91 (2006). https://doi.org/10.1080/10641734.2006.10505200

Khasawneh, M.H.A.: Developing and validating a comprehensive model of factors influencing consumer acceptance of SMS advertising: empirical evidence using SEM-PLS. 33 (2018)

MacKenzie, S.B., Lutz, R.J.: An empirical examination of the structural antecedents of attitude toward the ad in an advertising pretesting context. J. Mark. **53**(2), 48 (1989). https://doi.org/10.2307/1251413

Mulhern, F.: Integrated marketing communications: from media channels to digital connectivity. J. Mark. Commun. **15**(2–3), 85–101 (2009). https://doi.org/10.1080/13527260902757506

Naylor, R.W., Lamberton, C.P., West, P.M.: Beyond the "Like" button: the impact of mere virtual presence on brand evaluations and purchase intentions in social media settings. J. Mark. **76**(6), 105–120 (2012). https://doi.org/10.1509/jm.11.0105

Powers, T., Advincula, D., Austin, M.S., Graiko, S., Snyder, J.: Digital and social media in the purchase decision process: a special report from the advertising research foundation. J. Ad. Res. **52**(4), 479–489 (2012). https://doi.org/10.2501/JAR-52-4-479-489

Punj, G.: Effect of consumer beliefs on online purchase behaviour the influence of demographic characteristics and consumption values. J. Interact. Mark. **25**(3), 134–144 (2011)

Roblyer, M.D., McDaniel, M., Webb, M., Herman, J., Witty, J.V.: Findings on facebook in higher education: a comparison of college faculty and student uses and perceptions of social networking sites. Internet Higher Educ. **13**(3), 134–140 (2010). https://doi.org/10.1016/j.iheduc.2010.03.002

Rotzoll, K.B., Christians, C.G.: Advertising agency practitioners perceptions of ethical decisions. J. Q. **57**(3), 425–430 (1980). https://doi.org/10.1177/107769908005700307

Rohm, A., Kaltcheva, D.V., Milne, R.G.: A mixed-method approach to examining brand-consumer interactions driven by social media. J. Res. Interact. Mark. **7**(4), 295–311 (2013). https://doi.org/10.1108/JRIM-01-2013-0009

Saxena, A., Khanna, U.: Advertising on social network Sites: a structural equation modelling approach. Vision J. Bus. Persp. **17**(1), 17–25 (2013). https://doi.org/10.1177/0972262912469560

Shavitt, S., Lowrey, P.M., Haefner, J.E.: Public attitudes towards advertising: more favourable than you might think. J. Ad. Res. **38**, 7–22 (1998)

Tsang, M.M., Ho, S.-C., Liang, T.-P.: Consumer attitudes toward mobile advertising: an empirical study. Int. J. Electron. Commer. **8**(3), 65–78 (2004). https://doi.org/10.1080/10864415.2004.11044301

Yang, H.-L., Lin, C.-L.: Why do people stick to Facebook website? A value theory-based view. Inf. Technol. People **27**(1), 21–37 (2014). https://doi.org/10.1108/ITP-11-2012-0130

Yuanxin, M., Pittana, N.: A Study Investigating Attitude towards Social Media Advertising (Master Thesis). Umea School of Business, Sweden (2011)

Zhou, Z., Bao, Y.: Users attitudes towards web advertising: effects of internet motivation and internet ability. Adv. Consum. Res. **29**, 71–78 (2002)

Life-Saving APP: Snake Classification 'Venomous and Non-venomous' Using fast.ai Based on Indian Species

D. Lakshmi[1], Ramesh Chandra Panda[2(✉)], Amrita[3], and Abhinav Prakash[4]

[1] School of Computer Science and Engineering, VIT Bhopal University, Bhopal, Madhya Pradesh, India
lakshmi.d@vitbhopal.ac.in

[2] Research and Development Cell, Mechanical Engineering, Synergy Institute of Engineering and Technology, Dhenkenal, Orissa, India

[3] Computer Science and Engineering, SET, Sharda University, Greater Noida, UP, India

[4] Department of Computer Science and Engineering, Gandhi Institute for Technological Advancement, Bhubaneswar, India

Abstract. In tropical India, farming and agriculture are the major sources of livelihood. In the rural area, snake bite is a common medical emergency as well. Due to snakebite in most of the rural places human survival is a question mark. In India, the mass media declared that numbers of human deaths are increasing due to snake bite. This causality is equal to other disasters like floods, cyclones, and lightning [1]. The identification of snake species is greatly important to administer the appropriate drug to the affected person. Journal of Association of Physicians of India reveals that standard anti-snake venom injections are given to most of the victims [2] in all the circumstances. The reason for this kind of treatment procedure is due to an inadequate mechanism to distinguish the species of snake. In view of this issue, human life saving mobile apps are created using deep learning algorithms Convolutional Neural Network (CNN) using transfer learning techniques are used in this research study. The pre-trained models such as ResNet34 and SqueezNet are used to identify the snake specie

Keywords: Transfer learning · Snake classification · Convolution neural network · Deep neural network · Multi-class image classification

1 Introduction

In recent days smart phones are used widely. Hence, the implementation of the handy technology for societal issues becomes feasible. In India, especially in the state of Odisha each year the death rate is high due to snakebite. Based on the mass media publication, each year around 80,000 Indians died due to snakebite alone. In order to support the public, deep machine learning models are used in this research study. Using the fast.ai library and Goggle Colab the Convolution Neural Network (CNN) model is used for training the dataset. Fast.ai is a Machine Learning library used for transfer learning

A. M. A. Musleh Al-Sartawi et al. (Eds.): EAMMIS 2021, LNNS 239, pp. 109–115, 2021.
https://doi.org/10.1007/978-3-030-77246-8_11

models. The flexibility of its architecture enables it to deploy the application across the different platforms like CPU, GPU, and TPU processor, mobile devices and edge devices.

Deep neural networks are data-hungry systems [3], hence the more data you feed them, the more powerful and accurate they become. The fast.ai library simplifies training and produces accurate neural networks with improved execution time.

In the field of biology, dangerous snakes are termed as venomous. The term venomous is applied to any species that bite to inject their poison. Only twenty percent of snakes are venomous, among seven percent have the venomous ability to kill a human. This research is carried out to classify Indian snake species, whether it is venomous or not using transfer learning models Resnet34 and Squeeznet1.1 [4] which are trained on ImageNet datasets [5]. We freeze all the layers of the model so that it will not get trained while training our model and added a new loss output layer which is the only layer trainable.

The current paper presents a fine-grained image recognition problem. It is one of multi-class classification, namely determining the breed of Indian snakes. Finegrained recognition is the task of distinguishing between visually very similar objects such as identifying the species of a bird, the breed of a dog or the model of an aircraft. The apparent differences between such categories are very subtle. It's nearly impossible to distinguish the breed for the human eye.

There is no research yet on the classification of snakes based on its type, especially based on Indian snake species. This research is aimed to classify venomous or nonvenomous. Based on this study a mobile app is launched at the Google Play store for the public to recognize its type. This mobile App would help anyone to undergo a treatment based on the type of snake. Each year in India more than 15 thousand people are dying due to snake bite. Snakes also play a vital role in the ecosystem, especially killing the rat and preserving the crops until the crop harvest. This App would help the farmers and the public to take a decision to dial to the rescue department to rescue the snake.

2 Literature Review

In general, there are two categories of snake namely, venomous and non-venomous. Both types have differences in the structure of the teeth. [6–8] research works are carried out based on the taxonomic part of the snake's body using feature extraction and image processing techniques. Whereas, [9] research work is based on the snake bite image of the victim in predicting venomous snake bite or non-venomous snake bite using Chain Code and K Nearest Neighbor.

2.1 Overview of Transfer Learning

The transfer learning technique supports using an existing CNN feature extraction and the associated trained network weights, and transferring the learned parameters from one problem to another one. As it was mentioned earlier CNN exploits the hierarchical distribution of the parameter representations. The lower layers of a CNN, contain more generic features that are useful for many tasks, but higher layers of the CNN become more specific to the details of the domain classes contained in the original dataset.

Since modern CNN consumes more execution time and computational power to train on ImageNet, it is common to see people release their final CNN trained weight for the benefit of others who can apply the pre-trained networks in a wide range of vision-related problems [10]. A basic strategy for transfer-learning is to take a CNN pre-trained model on a larger dataset. We need to remove the fully-connected layer, and then treat the rest of the CNN as a fixed feature extractor for the new problem-specific dataset. Once the features are extracted, a linear classifier can be trained for the Snake dataset.

Deep networks are data-hungry systems, the more the data we feed, the more powerful and accurate results are obtained. Whereas, transfer learning is used to improve a learner from one domain to another related domain by transferring knowledge of previously trained parameters. Especially, transfer learning works well when the training dataset is smaller in size.

For better accuracy in the field of image classification, the deep neural models must train on at least 1000 images per class. But in our research, we tried Google images to collect the snake images; we have obtained only near about 200 images per class. In our dataset, there are 13 classes of snakes and in each class, there are nearly 200 images. That's why we took the advantage of transfer learning to overcome data insufficiency.

2.2 Overview of fast.ai

Fast.ai includes several pretrained models of torch vision such as, variants of ResNet architecture, variants of Squeezenet model, variants of VGG model, Alexnet and Densenet. Fast.ai also has several other models too, such as YOLO V3's Darknet architecture, WideResNet and Unet architecture too. Fast.ai is the recent advancement in the Python deep learning neural net library. It helps in transfer learning Convolutional Neural Network (CNN) models to achieve a great result in a lesser inference time. The Fast.ai library works based on research into deep learning best practices undertaken at Fast.ai, including "out of the box" support for the vision, text, tabular, and collaborative filtering models [11]. It also helps in building a model with the pretrained architecture and weights to obtain the best accuracy in the field of computer vision. The weights of pre-trained models need to be altered greatly in the last layer of the model, whereas. Fast.ai is a deep learning library for Python. It is built on top of PyTorch, and provides a high level API to various common deep learning applications and data types.

3 Research Methodology

3.1 Data Collection

The total number of snake images considered for training is 2388 and testing 596 respectively. All the images are taken from the 'Google Images' of all the Indian snake species. There are 13 classes, among each class having 200 snake images are considered in training the learning models. The name of the snake classes/categories are 1. 'Banded_Racer' (Non-Venomous), 2. 'Checkered_Keelback'

(Non-Venomous), 3. 'Common_Krait' Venomous, 4. 'Common_Rat_Snake' (Non-Venomous), 5. 'Common_Sand_Boa' (Non-Venomous), 6. 'Common_Trinket' (Non-Venomous), 7. 'Green_Tree_Vine' (Non-Venomous), 8. 'Indian_Rock_Python' (Non-Venomous), 9. 'King_Cobra' (Venomous), 10. 'Monocle_Cobra' (Venomous), 11. 'Russells_Viper' (venomous), 12. 'Saw-scaled_Viper' (Venomous), 13. 'Spectacled_Cobra' (Venomous). Out of 13 classes, 6 are 'Venoms' category and the remaining 7 are 'Non-Venomous' category. The transfer learning models used are Resnet34 and Squeeznet1.1. There are 13 classes of images considered in this experimental model.

3.2 Predictions with ResNet 34

The total numbers of images considered for training are 2388 and testing 596 respectively. All the images are taken from 'Google Images'. There are two classes considered based on whether they are poisonous or not. The names of the classes are 'Venomous' and 'Non-Venomous'. The learning models used are Resnet34 and Squeeznet1_1. The successful prediction or accuracy of the deep neural networks is possible because of additional layers. The intuitive phenomenon behind these networks functionality is a progressive learning of the layers.

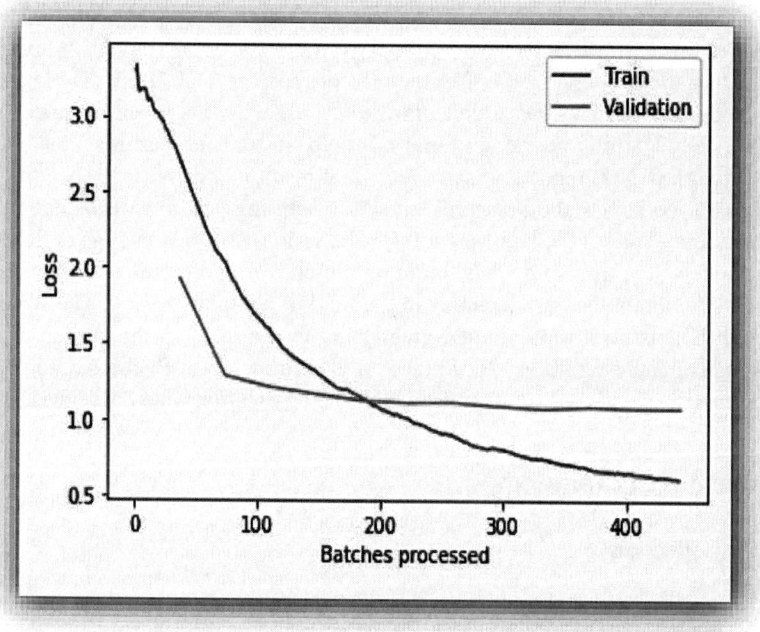

Fig. 1. ResNet34 training vs validation loss

When we train a neural network using gradient based optimization techniques, vanishing gradient problems need to be taken care of. The Residual Network or ResNet model solves this problem automatically. It is a block of two convolution layers with a small filter size, as well the same number of filters. ResNet adds the output of the second layer as the input of the first convolutional layer [12]. When the network is deep, after the application of the chain rule, the gradients naturally vanish towards zero. ResNet34 consists of a residual network with 34 parameter layers and it performs 3.6 billion FLOPS and 21.282 million parameters [13]. The problem of training deep networks has been overcome with the introduction of the special learning block 'The Residual Block'.

By training snake dataset using Resnet34 and Fast.ai for 12 epochs we have obtained an average training loss of 1.18735, average validation loss of 1.190062, average error rate of 0.363531 and average time per epoch is 18:17. In the 12th epoch, the training loss is 0.6611784, the validation loss is 1.075198, the error rate is 0.320470 and the time taken in the 12th epoch is 17:15. Accuracy of the ResNet34 model during training and testing is illustrated in Fig. 1. Among 12 epochs train loss reduces gradually, whereas, validation loss decreases gradually until 6 epochs; afterward the rate of changes becomes smaller as shown in Fig. 1.

3.3 Predictions with SqueezeNet

The building block of SqueezeNet is named as fire module, which contains two layers: a squeeze layer and an expand layer. The SqueezeNet model only contains 1 × 1 filter, which means it works like a fully-connected layer working on feature points in the same position. In other words, it doesn't have the ability of spatial abstract. One of its benefits is to reduce the depth of the feature map. Reducing depth means the 3 × 3 filters in

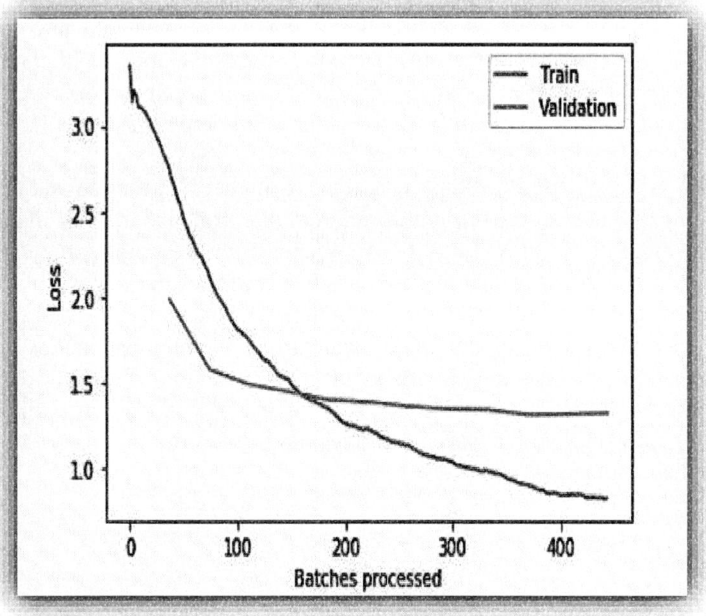

Fig. 2. SqueezeNet training vs validation loss

the expanded layer have fewer computations to do. It boosts the speed as it uses 1 × 1 filter, instead of 3 × 3 filter as it requires 9 times computation. By intuition, too much squeezing limits information flow; too few 3 × 3 filters limit space resolution [14].

By training snake dataset using SqueezeNet and Fast.ai for 12 epochs we have obtained an average training loss of 1.367807, average validation loss of 1.430801, average error rate of 0.441706 and average time per epoch is 04:06. In the 12th epoch the training loss is 0.727643, the validation loss is 1.221001, the error rate is 0.377651 and time elapsed in the 12th epoch is 03:19. Among 12 epochs train loss reduces gradually, whereas, validation loss decreases gradually until 4 epochs; afterwards the rate of changes becomes smaller as shown in Fig. 2.

4 Results and Discussion

Training a model is the prime most stages of deep learning approach. Training is to make CNN work with more accuracy and high performance. There are 13 classes of images carried out with the training models Restnet34 and Squeeznet1_1. In our experimentation, we found that Squeeznet1.1 produces higher accuracy with lesser inference time, whereas Resnet34 produces better accuracy with more inference time.

5 Optimization and Smartphone (Android) Application

Adam optimizer is used to optimize the trained model. The optimization technique is used to optimize the model and make it as a suitable file to deploy in the smartphone. Optimizing a trained TensorFlow model makes it Lite to implement into the Smartphone application. It also provides useful tips regarding snakes, symptoms after snake bite, curing mechanism and preventive steps. This is an offline app that works well without internet connectivity only for identifying the snakes. This app is made available for free download in the Google play store.

TensorFlow Lite is used for deploying the ML model into mobile and embedded devices. For deploying our model into the android platform also to work in the offline mode, we've used TensorFlow Lite library. Whereas, for the training the dataset, we used TensorFlow library. The same dataset has been used for training in Fast.ai of this paper. Subsequently, we used Mobilenet architecture with the transfer learning approach. The Mobilenet architecture model works well for the mobile applications due to its lightweight nature. After training, we saved the model into TensorFlow saved extension (.pb). Using TensorFlow lite converter, we've converted our model into (.tflite) format for deploying into Android [15]. We used 'Java' and 'Android Studio' for building the android application.

6 Conclusion

In this paper, we provide a smart way to classify the snake image either 'Venomous' or non-venomous' using deep machine learning. By using the ResNet34 and SqeezeNet the image is predicted. This app name 'Project AntiVenom' can work well with low-end smart phones. It's an offline app, works well without the internet. Delay is the major cause of death due to snake bite. If the affected person is admitted to the hospital within 7 h after the snake bite the survival rate is high. This app helps in all the spheres of safeguarding the affected persons. Across India, state-wise death rates due to snakebite are statistically discussed. This research study would become one of the prime uses of the life-saving model of the rural population.

References

1. https://www.downtoearth.org.in/news/wildlife-biodiversity/serpentine-problem-58396
2. Pandey, P.C., Bajaj, S., Srivastava, A.: A clinico-epidemiological profile of neuroparalytic snake bite: using low dose ASV in a tertiary care centre from North India. J. Assoc. Phys. India 64 (2016)
3. LeCun, Y., Bengio, Y., Hinton, G.: Deep learning. Nature **521**(7553), 436–444 (2015)
4. Bianco, S., Cadene, R., Celona, L., Napoletano, P.: Benchmark analysis of representative deep neural network architectures. IEEE Access **6**, 64270–64277 (2018)
5. Krizhevsky, A., Sutskever, I., Hinton, G.: ImageNet classification with deep convolutional neural networks. Neural Inf. Process. Syst. 25 (2012). https://doi.org/10.1145/3065386
6. James, A.: Snake classification from images. In: PeerJ Preprints 5, v1 (2017)
7. Slagboom, J., Kool, J., Harrison, R., Casewell, N.: Haemotoxic snake venoms: their functional activity, impact on snake-bite victims and pharmaceutical promise. Br. J. Haematol. **177**(6), 947–959 (2017). https://doi.org/10.1111/bjh.14591
8. James, A.P., Mathews, B., Sugathan, S., Raveendran, D.K.: Discriminative histogram taxonomy features for snake species identification. Hum. Cent. Comput. Inf. Sci. **4**(1), 1–11 (2014). https://doi.org/10.1186/s13673-014-0003-0
9. Adiwijaya, R.M.P., Utama, D.Q.: Snake bite classification using Chain code and k nearest neighbor. In: The 2ndInternational Conference on Data and Information Science (2019)
10. Simonyan, K., Zisserman, A.: Very deep convolutional networks for large-scale image recognition. In: Conference Paper at ICLR (2015)
11. Howard, J., Gugger, S.: fastai: a layered API for deep learning. Inf. **11**(2) (2020)
12. He, K., Zhang, X., Ren, S., Sun, J.: Identity mappings in deep residual networks. **9908**, 630–645 (2016)
13. He, K., Zhang, X., Ren, S., Sun, J.: Deep residual learning for image recognition, pp. 770–778 (2016)
14. Landola, F., Moskewicz, M., Ashraf, K., Han, S., Dally, W., Keutzer, K.: SqueezeNet: AlexNet-level accuracy with 50x fewer parameters and textless 1MB model size (2016)
15. Dr, E., Kamalanaban, M., Gopinath, S.: Medicine box: doctor's prescription recognition using deep machine learning. Int. J. Eng. Technol. **7**, 114–117 (2018)

Digital Media Usage Among Arab Journalists During Covid-19 Outbreak

Abdul-Karim Ziani[1], Mokhtar Elareshi[2](✉), Mohammed Habes[3], Khalaf Mohammed Tahat[3], and Sana Ali[4]

[1] Ahlia University, Manama, Bahrain
aziani@ahlia.edu.bh
[2] Al Ain University, Al Ain, UAE
mokhtar.elareshi@aau.ac.ae
[3] Yarmouk University, Irbid, Jordan
{mohammad.habes,khalaf.tahat}@yu.edu.jo
[4] Allama Iqbal Open University, Islamabad, Pakistan

Abstract. The digitalization of media and its convergence in web-based platforms greatly influenced the journalists' perceptions. Social media is a contemporary source of information especially during the crisis situations when direct social connectivity is difficult. The current article scrutinized the potential relationship between Arab Journalists, their opinion, usage and dependency on social media. The researchers selected a sample of $n = 1{,}443$ journalists from different Arab countries, and employed uses and gratification theories as the primary theoretical support. Data gathered through online surveys indicated that social media well serves the journalists to gather information they need during the Covid-19 outbreak. Respondents expressed their strong dependency on online platforms due to its usefulness. Overall, the study found positive opinion and attitudes regarding social media reliance among Arab journalists. Thus, the researchers recommend more studies to highlight the journalists' dependency on social media even in the ordinary situations to further elaborate its importance.

Keywords: Arab journalists · Media dependency · Digital media · Online journalism · Online news · Covid-19

1 Introduction

The emergence of social media technology has greatly revolution our daily lives. From doctors to lawyers, teachers, and journalists, all depend on social media for information, education, and other purposes. This increased dependency is positive in nature as online information is comparatively more accessible and available [1]. Especially, journalists consider social media as a primary component of their professional practices. For this purpose, they take interest in online information to enhancer their knowledge and awareness [1–4]. Due to greater availability of social media, now journalists are able to resume their activated with less obstacles. Today, journalists cannot pursue their

A. M. A. Musleh Al-Sartawi et al. (Eds.): EAMMIS 2021, LNNS 239, pp. 116–129, 2021.
https://doi.org/10.1007/978-3-030-77246-8_12

activities without social media. Studies also have highlighted the importance of social media for journalists and their professional practices [2, 5–8]. [9] considers social media integration in journalists as a progress and development of journalistic practices.

As today, researchers are more efficient, they can access different information gathering platforms and even they can also communicate with the public to know about their opinions. As result, new organizations also motivate journalists to prefer social media for information gathering and expect them to be familiar with online platforms' usage [10]. Moreover, today people are also connected through social media. Consequently, the adoption of digital platforms has also transformed the ways journalists communicate regarding their professional practices [2, 11].

The power of social media can be estimated by the fact that today more than 30 percent of world population is using social networking platforms on daily basis [12]. However, still a little is known about social media adoption, integration, implementation, and dependency by the Arab journalists [2, 8, 11, 13]. This study focused to fill the gap regarding social media adoption among the Arab journalists particularly during the Covid-19 outbreak. It examines the perceptions of Arab journalists regarding social media usage and dependency during the healthcare crisis. This research will favorable dig out the relevant facts and highlight the digital media dependency also as a source of risk management system.

This study is divided into five sections, after its introduction, a literature review discusses digital media usage by journalists, with a theoretical framework discussing the UGDT and its implications on journalism. The third section reviews the research methodology, followed by the results section. The fifth section discussing the adoption of social media by Arab journalists and its implications on the field, and finally the study's limitations and further research were provided.

2 Literature Review

2.1 Social Media and Journalists

Media digitalization and convergence into social networking platforms have been an interesting topic for academics for the past two decades [6]. The widespread use of mobile Internet connections, smart and digital devices resulted in more journalists relying on these advanced technologies for professional practices [2, 11, 14]. As noted by Lewis and Molyneux [6], journalists attain much benefits from social media to create more news content to engage the wider audience. Others critics also argue that the Internet and digital technologies have contributed much to improve journalism and added more value to journalistic practices [9, 15].

Today, social media provides much more accessibility and availability of information to the journalists, which is playing a significant role in improving their reporting and writing skills [13, 16–18]. Indeed integrating social media in journalism is a greater investment which is giving much more fruitful outcomes [1, 19]. In this context, Cottrill [4] noted that more than 75% of journalists in the UK, Germany, Sweden, the US, Australia and Finland daily login to their social media profiles for news reporting. However, this usage varies from nation to nation, yet the ratio is much higher. Similarly, Twitter and Facebook are the most preferred social networking platforms both by the news organizations and individual journalists [1]. A survey conducted by Rodrigues [20] also affirmed the impotence of social media for journalism in Iran that found social media usage is highly common among Iranian journalists. They use it for reporting on different topic to increase awareness among the masses.

Moreover, the young journalists also consider social media as less strict and comparatively having lower traditional norms regarding reporting practices. For the young journalists, digital environment supports freedom of expression and freedom to access information, which is another positive aspect of integration digital media in journalism [6]. For instance, American journalists prefer social media to share their personally covered stories, and opinion-based articles. While, German journalists use social media more to share news from their organization, that are filtered by editors and other gatekeepers [1, 2, 4]. Yet, the adoption and dependency on social media does not mean that journalists have quit using traditional patterns of reporting. As traditional media still contains more value and authentication, they consider conventional media platforms as much more important part of their profession [13, 17]. For journalists, social media is sometimes also loaded with gigantic amount of disinformation and misinformation, that can harm audiences, their personal reputation and professional career as well [21]. A study conducted by Ali [22] also validated the existing misinformation on social media as she found misinformation during the crisis situation of Covid-19. However, still the social media weighs much importance for the journalists worldwide [20].

2.2 Uses, Gratifications and Dependency Theories

Theoretically, this study is built on the psychological and cognitive orientations aspect by adopting the uses, gratifications and dependency models. As proposed by Sandra Ball-Rokeach and Melvin DeFleur, the main premise of the dependency theory arises from the assumption, linked to interdependent relationships between the media system, that individuals (journalists/users/audiences) rely on media to satisfy and gratify their goals through use of the medium content [23–26]. This theory highlights how people (e.g., Arab journalists) actively choose and use medium (e.g., social media) and how they become dependent on it. The more they rely on a medium as an important sources of information, the more their influence and role increase and becomes more important and more central to them [14].

Thus, the relationship between the intensity of dependence and the degree of influence on people of the method is also established. Also, the societies also become more complex due to their dependency on a medium [27, 28]. As a result, this can lead, in turn, to different patterns of media exposure and use. Eventually, lit results in cognitive, affective and behavioral effects arising from media usage [29]. The process of reliance on media can be examined from either a macro or micro level. While the former looks at the role of media in terms of how Arab journalists use and depend on media to meet their specific goals. The latter focuses on how they will become more dependent on a specific social media that will satisfy a variety of their needs rather than a few needs [30, 31].

The macro level also looks at the interdependence between users, the media system and the larger social system. In other words, people who have a high need to be concerned with what is going on tend to use and expose themselves to more media [32]. For example, people need the media to know what is going on around them, journalists need people for their information, and society needs media for entertainment and information [33].

2.3 The Implications of Journalists' Dependence on Social Media

The main question regarding media dependency is the explanation of when and why individuals are exposed to the media content and what the effects of such exposure are on their beliefs and behaviors (e.g., Arab journalists). However, news media, e.g., online newspapers, hold the power to set the momentous apart from the trivial, as they seem to give special attention to some stories, while leaving aside others that would impact on consumers' views and perceptions of these stories [34]. For example, Saleh [35] indicated that Egyptian journalists rely on social media to get ideas and public stories, to search for information and to read readers' comments and reflections.

Users' dependence on the media also increases when there are few sources of information and little availability of alternative sources of information. For example, Pole and Gulyas [36] found that UK journalists prefer to use social media as a professional tool in their work and perceive it as a vital part of their daily journalism work. This is also an indication of how journalistic work has changed to a large extent since the emergence of social media, particularly Facebook, Twitter and Instagram [1, 6]. Furthermore, some also indicated that journalists are also reliant on social media to become more engaged with their audience and peers [6, 37, 38]. Therefore, social media has an impact on journalists use and production of information; however, there are few studies on the adoption of social media for main news sources and stories in the Arab world. Hence, there is a need to understand such adoption in unforeseen times and how such platforms are perceived by Arab journalists.

This paper, therefore, sought to fill such gap and answer a four basic research questions, while providing a foundation for future research dealing with the relationship between social media and Arab journalists' perceptions: How do Arab journalists expose and depend on social media in unforeseen times? What are they most preferred social media apps? What topics are they interested in such time? Are there any significant differences in their attitudes towards the adoption of social media and their demographic characteristics?

3 Research Methodology

3.1 Participants

The main data was collected from a random sample of n = 1443 Arab journalists from most Arab countries, who reported regularly accessing social media for information related to unforeseen times to use in their news stories. Arab journalists (aged 20 years old and over) who belong to different Facebook groups were contacted with the online survey link. These groups are: Association of Libyan Media (Abroad) (46.8K members); Syrian Journalists Association (13.6K members); Open Media Library Group (13.4K members); Electronic Media Syndicate (12.6K members); The National Council of Algerian Journalists (6.7K members); Forum of Originality and Renewal in Arab Media

Research (6.5K members); National Syndicate of Tunisian Journalists (3.8K members); Palestinian Journalists Syndicate (2.7K members); Young Journalists Group in the Iraqi Journalists Syndicate (2.4K members); Arab Journalists Network (2.3K members); and Bahrain Journalists Association (951 members). The data-collection period was from April to September 2020. Regarding the participants' nationality, the results indicated that the responses came from the following countries: Mauritania (n = 128), Jordan (n = 123), Algeria (n = 104), Tunisia (n = 100), Iraq (99), Egypt (n = 93), Morocco (n = 92), Kuwait (n = 87), Sudan (n = 83), Saudi Arabia (n = 71), Lebanon (69), Libya (63), Yemen and Palestine (both 59), Syria (n = 58), Bahrain (54), UAE (n = 42), Oman (n = 36), and Qatar (n = 23). Table 1 below summarizes the demographic sample.

Table 1. The sample characteristics ($N = 1443$)

Items		N	%
Gender	Male	1114	77.2%
	Female	329	22.8%
Age	20–30	466	32.3%
	31–40	720	49.9%
	41–50	167	11.6%
	51 +	90	6.2%
Education level	High school	11	0.8%
	University	1142	79.1%
	Postgraduate	290	20.1%
Work type	Independent journalist	285	19.8%
	Online newspaper	360	24.9%
	Traditional newspaper	273	18.9%
	TV/Radio	375	26.0%
	News agency	150	10.4%

3.2 The Questionnaire and Measurement

A pre-structured questionnaire was written in Arabic – the official language of Arabs. The included questions were about social-media exposure, general social-media news-topic interests and personal details. It consisted of different sections of close-ended responses and statements. All the respondents took part voluntarily and were free to withdraw from the survey any time. This paper will report findings regarding social-media exposure, news topics of interest and personal details only.

Furthermore, the questions were designed as follows. Q1: "In general, how much do you use social media in journalism tasks?" ("always", "sometimes", "rare"). Q2: "With especially regard to Covid-19 news, to what extent do you depend on social media?" ("high", "average", "low", "don't use"). Q3: "What are social media news topics that you interested / follow / depend on?", with n = 20 items measured on a three-point scale ("very important", "less important", "not important at all") (Cronbach's alpha = .924). Q4: "What is your most preference social media used to follow the Covid-19 news?" (Twitter, Facebook, Instagram, Snapchat, YouTube, WhatsApp, Viber), measured on a four-point rating scale ("daily", "sometime a week", "rare", "don't use"). Finally, respondents were asked to report their demographic characteristics (gender, age, education level, work type, nationality). The reliability of questionnaire was examined by using Interceder Reliability Assessment. With the Cronbach Alpha Value of .851, the research tool is validated. Likewise, the validity of the questionnaire was tested by reviewing it and also by using the Bivariate Pearson Correlation Analysis ensuring that the words were linguistically clear. Table 2 below summarizes the Univariate Pearson Correlation analysis. In order to examine the extent of statistically significant differences between demographic variables, a different statistical coefficient test was adopted via SPSS software 26.0 version [39]. The data are available upon request.

Table 2. Validity assessment of research tool

	.662**	.551**	.556**	.553**	.349**
.662**		.643**	.593**	.552**	.148**
.551	.643***		.684**	.699**	.250
.556**	.593**	.684**		.747**	.309**
.553***	.552**	.699**	.747		.306**
.349***	.148**	.250**	.309**	.306**	

4 Results

4.1 Social-Media Exposure

When asked about social-media exposure, respondents reported different use and interest in social media in general. 84% of respondents indicated using it "always", while 10.4% used it "sometimes" and 5.6% used it "rarely". These results confirmed the emergence of social media in the Arab world in general [40, 41].

4.1.1 Demographic Differences in Using Social-Media Sites

To examine the demographic profiles of different social media among Arab journalists, two statistical tests were used: The Mann-Whitney U-test to compare gender differences and the Kruskal-Wallis test to compare significant differences between other demographic variables where there were three or more groups.

In terms of exposure, social media were reportedly used more by male (85%) and female journalists (80.5%) (U (1443) = 176513.5; exact p = .113, 2-tailed). Those aged (31–40) (89.3%) were more likely than those aged (20–30) (85.4%), (41–50) (79%) or (51 +) (43.3%) to be exposed to social media regularly (X2 (3, N = 1443) = 174.8, p < .000). Interestingly, those journalists with university-level education (87.45%) were reportedly exposed more to social media than were those with postgraduate-level education (71.7%) or with high-school education (54.4%) (X2 (2, N = 1443) = 51.5, p < .000). Those journalists who work in traditional newspapers (89.7%) were more slightly to do so than those working for online newspapers (87.5%), TV/radio (85.6%), independent journalists (77.2%) and news-agency journalists (74%) (X2 (4, N = 1443) = 28.5, p < .000).

Regarding Covid-19 news, the majority (75%) indicated that they relied on social media for such news, while 17% responded with "average", 5.5% with "low" and 2.6% with "don't use". Female journalists (76.6%) were slightly more likely than males (74.4%) using social media for Covid-19 news information (U (1443) = 177147.0; exact p = .225, 2-tailed). Those aged (20–30) (82.4%) were more likely to depend on social media than those aged (31–40) (77.1%), (41–50) (57.5%), (51 +) (51.1%) (X^2 (3, N = 1443) = 75.6, p < .000). Those with only high-school education (81.8%) were more likely to depend on social media than those at university level (77.6%) or postgraduate level (64.1%) (X^2 (2, N = 1443) = 19.5, p < .000). Those journalists who worked for online newspapers (86.7%) were more likely to do so than those in news-agency journalists (75.3%), traditional newspapers (74%), or TV/radio (72.8%) or independent journalists (63.5%) (X^2 (4, N = 1443) = 28.5, p = .000).

4.2 Social-Media Preference

When asked to state how often they used social-media apps as an information source (listed in Table 3, with higher percentage meaning a higher level of preference and dependence). Interestingly to note that some apps became the main-preference social-media platforms for our respondents. Our results confirmed the emergence of WhatsApp and its owner, Facebook. These two platforms were clearly named as the most frequently preferred apps and as the apps that were most relied on. Twitter and YouTube closely followed, while Instagram, Viber and Snapchat were also mentioned, with some variation evident between these online platforms. These results are consistent with several examples of past research indicating the importance of preferred social media by journalists and other users [1, 6, 11, 17, 37, 42].

4.2.1 Demographic Differences in Social-Media Preferences.

In terms of gender differences, female Arab journalists (62.3%) were more likely than males (50.5%) to prefer Facebook as a main source for their news gathering and for producing news and information (U (1443) = 157794.5; exact p = .000, 2-tailed); this was followed by Instagram (48.9% and 36.8% respectively) (U (1443) = 157777.5; exact p = .000, 2-tailed) and WhatsApp (80.5% and 64.7% respectively) (U (1443) = 151481.5; exact p = .000, 2-tailed). Regarding age, those ranging between (20–30) (55.6%) were more likely than other age groups to prefer Twitter (compared with those

Table 3. Social-media preferences (N = 1443)

	Daily%	Average%	Rare%	Don't use%
WhatsApp	68.3	16.0	13.2	2.4
Facebook	53.2	23.6	19.5	3.6
Twitter	49.3	20.8	25.2	1.7
YouTube	42.4	37.3	17.0	3.3
Instagram	39.6	22.1	36.0	2.3
Viber	38.2	12.1	46.4	2.4
Snapchat	30.8	27.2	38.3	3.7

aged 31–40 (51.8%), 41–50 (35.9%) and 51 + (21.1%) (X^2 (3, N = 1443) = 38.1, p = .000). Other age-group results were as follows: Viber (48.9%, 34.7%, 31.7%, 22.2% respectively) (X^2 (3, N = 1443) = 62.6, p = .000); Instagram (53.4%, 39.6%, 16.8%, 10% respectively) (X^2 (3, N = 1443) = 127.3, p = .000); Snapchat (49.4%, 24%, 23.3%, 6.7% respectively) (X^2 (3, N = 1443) = 106.6, p = .000); YouTube (48.3%, 46.1%, 41.8%, 10% respectively) (X^2 (3, N = 1443) = 51.2, p = .000); WhatsApp (73.8%, 70.4%, 62.2%, 47.3% respectively) (X^2 (3, N = 1443) = 26.3, p = .000). Regarding education level, those with high-school level education (81.8%) were more likely than those with university-level (56.5%) or postgraduate-level education (19.7%) to prefer Twitter for their news and information (X2 (2, N = 1443) = 102.4, p = .000). The education-level results for WhatsApp were 81.8%, 70.7%, 58.6% respectively (X2 (2, N = 1443) = 9.9, p = .000). Facebook was preferred by those with postgraduate-level education (59.7%) rather than those with university-level education (52.1%) or high-school education only (0.0%) (X2 (2, N = 1443) = 10.0, p = .007). For Viber, the figures were 16.2%, 11.1% and 0.0% respectively) (X2 (2, N = 1443) = 66.7, p = .000). Instagram was preferred by those with university-level education (42.5%) rather than those with high-school level education (36.4%) or those with postgraduate-level education (28.3%) (X2 (2, N = 1443) = 29.4, p = .000). Snapchat was also preferred by those with university-level education (35%) than those with postgraduate-level education (15.2%) or high-school level (0.0%) (X2 (2, N = 1443) = 64.4, p = .000). For YouTube, the figures were 45%, 33.8% and 0.0% respectively (X2 (2, N = 1443) = 34.4, p = .000).

Regarding job type, those who worked in news agencies (77.3%) were more likely to prefer Twitter as the main source for their news than those who worked in traditional newspapers (75.8%), online newspapers (57.2%), or TV/radio (38.1%), or independent journalists (13.7%) (X^2 (4, N = 1443) = 312.9, p < .000). The rest of the results were as follows: Facebook (independent journalists 58.2%, online newspapers 58.1%, news agencies 57.3%, TV/radio 49.9%, traditional newspapers 44%) (X^2 (4, N = 1443) = 19.1, p < .001); Instagram (traditional newspapers 73.6%, news agencies 41.3%, online newspapers 34.4%, TV/radio 29.9%, independent journalists 25.3%) (X^2 (4, N = 1443) = 181.9, p < .000); Snapchat (traditional newspapers 59.7%, news agency 32%, online newspapers 29.4%, TV/radio 19.3%, independent journalists 19.2%) (X^2 (4, N = 1443)

= 145.3, p < .000); YouTube (traditional newspapers 63%, online newspapers 46.4%, news agencies 44.7%, TV/radio 32%, independent journalists 30.2%) (X^2 (4, N = 1443) = 109.9, p < .000); WhatsApp (traditional newspapers 88.3%, news agencies 86.7%, online newspapers 76.9%, TV/radio 55.2%, independent journalists 46%) (X^2 (4, N = 1443) = 183.6, p < .000); Viber (traditional newspapers 69.6%, news agencies 54%, TV/radio 32.8%, online newspapers 31.7%, independent journalists 15.1%) (X^2 (4, N = 1443) = 249.0, p < .000).

4.3 Social-Media News Topics

One of the study's interests was in exploring what Arab journalists identify as social-media news topics of importance and what they are personally interested in. Respondents were asked to evaluate a list of social-media news topics in terms of their perceived importance. These topics were selected to cover a wide range of issues in order to see how much the respondents depended on social-media news topics for their coverage. The results are summarized in Table 3.

Table 4. Ranking of news topics by their importance (N = 1443)

News topics	Mean*	S. D	% "saying Very Important"
- Methods of preventing Covid-19 infection	2.63	.573	67.2
- Domestic-violence cases as a result of quarantine	2.59	.567	63.4
- Peoples' news and jokes about the quarantine	2.52	.625	59.1
- News about the standards and safety issues to avoid Covid-19	2.51	.634	59.0
- Topics for future expectations of Covid-19	2.48	.648	56.9
- Covid-19 infections around the world	2.48	.651	56.8
- Causes of the spread of Covid-19	2.44	.665	53.5
- News about the number of Covid-19 victims around the world	2.43	.721	57.2
- Press conferences at the international level	2.41	.713	54.7
- News of Covid-19 symptoms	2.40	.659	50.1
- Patients' news and their suffering	2.38	.694	50.0
- General and local social news	2.37	.745	53.5
- Press conferences at the national level	2.35	.742	51.2
- Cases of accusations between countries over their responsibility for spreading Covid-19	2.34	.623	42.5

(continued)

Table 4. (*continued*)

News topics	Mean*	S. D	% "saying Very Important"
- Information about vaccinations and available vaccinations	2.33	.729	55.5
- Social-solidarity issues	2.31	.743	48.0
- Work and health-insurance issues and monthly salary	2.25	.735	43.0
- Preparations to cope with Covid-19	2.20	.774	41.6
- Food-supply issues during the quarantine	2.16	.807	42.0
- News about the numbers of Covid-19 victims at the national level	2.12	.723	32.7

*Higher scores = greater importance. Scale ranges from 3 = very important to 1 = not important at all

According to Table 4, the majority of respondents unsurprisingly indicated that the most important types of social-media topic was those related to Covid-19 infection. The highest rated news items were those also related to methods of preventing Covid-19, domestic-violence cases as a result of quarantine, peoples' news and jokes about the quarantine, standards and safety of Covid-19 measures, expectations of Covid-19, Covid-19 infections around the world, causes of the spread of Covid-19 and the numbers of Covid-19 victims around the world. There are few studies on the adoption of social media for main news sources and stories in the Arab world.

4.4 Analysis of Variance

To examine any potential difference based on participants' demographical characteristics, One-Way Analysis of Variance was also conducted. Table 5 summarizes the results of One-Way Analysis of Variance. Thus, the Analysis of Variance indicated that there are not any differences among respondents concerning the participants' demographical data.

Table 5. Test of homogeneity of variances

	Gender	Age	Education	Work	Nationality
Mean	1.23	1.92	2.19	2.62	10.65
Standard Deviation	.420	.826	.414	1.298	6.039
Levene Statistics	10.882	14.341	16.779	9.754	8.730
F Value	2.194	20.827	7.777	5.442	6.461
Sig	.020***	.000***	.000***	.000***	.000***

5 Discussion and Conclusion

It is argued that the adoption of social media has changed the way in which media professionals, including journalists, communicate and write their reports and stories [2, 11]. An online survey examined the adoption of social media for main news sources and stories by Arab journalists in Covid-19 pandemic [2, 5, 43, 44]. It examined their exposure and social media used, as well as the topics of interest that they preferred, the better to understand the pattern of how they depend on social media as their main sources of news and information [13, 16–18]. These journalists (1443) reported regular accessing and use of social media for news and information related to the pandemic in reporting their news stories.

In response to RQ1, the findings reported above indicate that our respondents viewed and used different social media to a large extent to gather information for, and to promote, news stories [2, 17], approving that new technologies can be adopted by news organizations to provide news and information and even to "frame a situation" especially during such unforeseen times, e.g., Covid-19 [11, 23]. Furthermore, focusing on social-media exposure, the study found differences between our respondents with the most usage by younger and better educated male journalists and by those working in traditional newspapers. The exception to this, in terms of most usage, is the example of younger females working for online newspapers. When it comes to Covid-19 news, it seems that younger females with high-school level education, who work for online newspapers, were slightly more likely to depend on social media for their coverage of such news. Perhaps the use of mobile devices has enhanced such use so that Arab journalists are more likely to use social media as a habit and also in their work. Previous research has indicated the dominance of social media as a main source of news in the Arab world. Mansour [2] reported that most professionals and Egyptian journalists accepted social media as a source of information and news that could be used in their own reporting. Others indicate that social media are used as a news-reporting mechanism and a news-sharing system especially during such uncertain times [4, 45, 46].

In response to RQ2, it is clear that social media have become powerful tools for journalists and users to capture information flow and for news organizations to reach more people than ever before [14, 34]. Interestingly, the newer app (WhatsApp) has superseded the older app (Facebook) in terms of the most preferred platforms, although the former is already owned by the latter. These apps are followed by Twitter, YouTube and Instagram [6, 37]. These findings are in agreement with and consistent with those indicating that for journalists being active in social media is an "obvious and necessary step in journalism's digital-first transformation" [6]. This could be with the intention of seeking news sources, or news ideas, as well as exposing themselves to their unruly public and to public concerns.

Indeed, social media are used by journalists to the extent that they see it as a "gold mine of source materials", especially in circulating breaking news (e.g., Covid-19) [34]. This is an indication that, with the world around us full of social-media content generated by the public and by journalists, social media for this reason (Covid-19 news) has made an impact on journalistic perceptions and, of course, on practices [37]. Although the online data methodology used here was not equipped to recruit statistically representative samples of the national populations of the countries involved, the robustness of its findings is reinforced by their similarity to earlier research evidence, especially in such an unforeseen time (RQ3). For example, social media are often the place where news is released first and where journalists place news to reach out to wider audiences. Such an assumption could lead to use of, gratification from, and dependence on social media in journalism, as it undoubtedly invites multiple daily use of such platforms [47, 48]. Research in the Arab and non-Arab world has confirmed that journalists are increasingly viewing, and depending on, social media for news and information [2, 4, 17, 49, 50], creating their own social-media pages [14] and promoting their own news production [2, 4, 37].

Finally, for its limitations, it is important to bear in mind that, because the data were gathered through an online environment, the generalizability and the findings of this research are limited to our respondents. Nevertheless, given the time spent on carrying out this research (Covid-19), and the cost of it, our findings have perhaps provided a reasonable cross-national section with 1443 responses. Therefore, the growing research into social media and journalism can be described as a young and immature area of research in which more studies are needed to examine the assumption about social media's impact on the future of journalism.

References

1. Heravi, B., Harrower, N.: Introduction to citizen journalism and social media archiving minitrack. In: 48th Hawaii International Conference on System Sciences (2015)
2. Mansour, E.: The adoption and use of social media as a source of information by Egyptian government journalists. J. Librariansh. Inf. Sci. **50**(1), 48–67 (2018). https://doi.org/10.1177/0961000616669977
3. Hanna, M., Sanders, K.: Journalism education in Britain: who are the students and what do they want? J. Pract. **1**(3), 404–420 (2007)
4. Cottrill, E.: Journalists' use of social media evolves. In: AXIA (2015). https://www.axiapr.com/blog/journalists-use-of-social-media-evolves. Accessed 14 Jan 2021
5. Khamis, S., Vaughn, K.: Cyberactivism in the Egyptian revolution: how civic engagement and citizen journalism tilted the balance. Arab Media Soc. 13 (2011)
6. Lewis, S.C., Molyneux, L.: A decade of research on social media and journalism: assumptions, blind spots, and a way forward. Media Commun. 6(4) (2018) doi: https://doi.org/10.17645/mac.v6i4.1562. (News and Participation through and beyond Proprietary)
7. Bossetta, M.: The digital architectures of social media: comparing political campaigning on Facebook, Twitter, Instagram, and Snapchat in the 2016 U.S. election. J. Mass Commun. Q. **95**(2), 471–496 (2018). https://doi.org/10.1177/1077699018763307
8. Harshini, P., Madhumidha, V.: Explosion of digital journalism through social media. Int. J. Linguist. Lit. Transl. **2**(2), 55–58 (2019). https://doi.org/10.32996/ijllt.2019.2.2.10
9. Thomas, C.: The development of journalism in the face of social media relationship to the audience: a study on social media's impact on a journalist's role, method and relationship to the audience, pp. 1–72 (2013)
10. Safori, A.O.: Social media's impact on a journalist's role. J. Sci. Educ. **19**(1), 148–162 (2018)
11. Fatima, S.S.: Understanding the construction of journalistic frames during crisis communication. Södertörn University|School of Social Sciences (2020)
12. Schwab, K.: The fourth industrial revolution: what it means and how to respond. In: World Economic Forum, 14 Jan 2016. https://www.weforum.org/agenda/2016/01/the-fourth-industrial-revolution-what-it-means-and-how-to-respond/. Accessed 04 Feb 2021

13. Allan, S.: Online News: Journalism and the Internet. Open University Press, Maidenhead (2006)
14. Lee, J.: The double-edged sword: The effects of journalists' social media activities on audience perceptions of journalists and their news products. J. Comput. Commun. **20**(3), 312–329 (2015). https://doi.org/10.1111/jcc4.12113
15. Alhumaid, K., Ali, S., Waheed, A., Zahid, E., Habes, M.: COVID-19 and e-learning: perceptions and attitudes of teachers towards e-learning acceptance in the developing countries. Multicult. Educ. **6**(2), 100–115 (2020). https://doi.org/10.5281/zenodo.4060121
16. Sundar, S.S.: Effect of source attribution on perception of online news stories. J. Mass Commun. Q. **75**(1), 55–68 (1998)
17. Pavlik, J.: The impact of technology on journalism. J. Stud. **1**(2), 229–237 (2000)
18. Finberg, H., Stone, M.L., Lynch, D.: Digital journalism credibility study. In: Online News Association, vol. 3, Washington, DC, pp. 2003-Finberg, H., Stone, M. (2002). Digital J. (2002)
19. Jardine, C.G., Boerner, F.U., Boyd, A.D., Driedger, S.M.: The more the better? A comparison of the information sources used by the public during two infectious disease outbreaks. PLoS One **10**(10), e0140028 (2015). https://doi.org/10.1371/journal.pone.0140028
20. Rodrigues, U.: Social media's impact on journalism: a study of media's coverage of anti-corruption protests in India. Glob. Media J. Aust. Ed. **8**(1), 1–10 (2014)
21. Ireton, C., Posetti, J.: Journalism, "fake news" and Disinformation. UNESCO, Paris, France (2018)
22. Ali, S.: Combatting against Covid-19 & misinformation: a systematic review. Hum. Arenas, 0123456789 (2020). doi: https://doi.org/10.1007/s42087-020-00139-1
23. Habes, M., Alghizzawi, M., Salloum, S.A., Mhamdi, C.: Effects of Facebook personal news sharing on building social capital in Jordanian universities. In: Al-Emran, M., Shaalan, K., Hassanien, A.E. (eds.) Recent Advances in Intelligent Systems and Smart Applications, pp. 653–670. Springer, Cham (2021)
24. Elbasir, M., Elareshi, M., Habes, M., Aissani, R.: The influence of trust, security and reliability of multimedia payment on the adoption of EPS in Libya. Multicult. Educ. **6**(5), 53–68 (2020). https://doi.org/10.5281/zenodo.4312133
25. Lin, Y.: Media dependency theory. In: Kaid, L.L., Bacha, H., Encyclopedia of Political Communication (eds.) Sage Publications Inc. (2008)
26. Ball-Rokeach, S., DeFleur, M.: A dependency model of mass-media effects. Commun. Res. **3**(1), 3–21 (1976). https://doi.org/10.1177/009365027600300101
27. Rosengren, K.: Uses and gratifications a paradigm outlined. In: Blumler, J., Katz, E. (eds.) The Uses of Mass Communications: Current Perspectives on Gratifications Research, pp. 269–286. SAGE, London (1974)
28. Katz, E.: Communications research since Lazarsfeld. Public Opin. Q. **51**, S25–S45 (1987)
29. Tahat, K., Habes, M., Tahat, D.: Towards employing new media in journalism major: a cross sectional study at Yarmouk University, Jordan. Palarch's J. Archaeol. Egypt/Egyptol. **17**(7), 15404–15422 (2020)
30. Dimmick, J., Chen, Y., Li, Z.: Competition between the Internet and traditional news media: the gratification-opportunities niche dimension. J. Media Econ. **17**(1), 19–33 (2004)
31. Ziani, A., Elareshi, M., Alrashid, M.: Social impact of digital media: growth pattern of facebook in the Arab world. J. Commun. Media Watch VIII(II), 177–191 (2017). doi: https://doi.org/10.15655/mw/2017/v8i2/49011
32. Ali, S., Habes, M., Qamar, A.: Accessing the political information through new media: a review study. J. Res. Rev. Soc. Sci. Pakistan, December 2020
33. Katz, E., Blumler, J.G., Gurevitch, M.: Uses and gratifications research. Public Opin. Q. **37**(4), 509–523 (1973)
34. Jallow, A.Y.: The emerging of global journalism and social media. Glob. Media J. **13**(25), 1–10 (2015)

35. Saleh, H.: Social Media in Egyptian Newspapers: New Opportunity. LAMBERT Academic (2016)
36. Pole, K., Gulyas, A.: Social journalism study 2013 United Kingdom. Polit. Sci. (2015)
37. Djerf-Pierre, M., Lindgren, M., Budinski, M.: The role of journalism on YouTube: audience engagement with 'superbug' reporting. Media Commun. **7**(1), 235–247 (2019). https://doi.org/10.17645/mac.v7i1.1758
38. Swart, J., Peters, C., Broersma, M.: Shedding light on the dark social: the connective role of news and journalism in social media communities 772063N. New Media Soceity **20**(11), 4329–4345 (2018). https://doi.org/10.1177/1461444818772063
39. Arkkelin, D.: Using SPSS to understand research and data analysis. Psychol. Curr. Mater. **2014**, 194 (2014)
40. Dennis, E., Martin, J., Wood, R.: Media use in the Middle East: a seven-nation survey. North Western University, Qatar (2017). http://mideastmedia.org/survey/2017/. Accessed 23 Jan 2018
41. Elareshi, M., Ziani, A.: Digital and interactive social media among Middle East women: empirical TAM study. Media Watch **10**(2), 235–250 (2019). https://doi.org/10.15655/mw/2019/v10i2/49642
42. Geçer, E., Yıldırım, M., Akgül, Ö.: Sources of information in times of health crisis: evidence from Turkey during COVID-19. J. Public Heal. Theory Pract. 1(7) (2020). doi: https://doi.org/10.1007/s10389-020-01393-x
43. Al Nashmi, E., Alkazemi, M.F., Wanta, W.: Journalism and mass communication education in the Arab world: towards a typology. Int. Commun. Gaz. **80**(5), 403–425 (2018). https://doi.org/10.1177/1748048517747492
44. WHO: Coronavirus disease (COVID-19) pandemic (2020). https://www.who.int/. Accessed 31 Aug 2020
45. Bruns, A., Highfield, T., Burgess, J.: The Arab spring and social media audiences: English and Arabic Twitter users and their networks. Am. Behav. Sci. **57**(7), 871–898 (2013). https://doi.org/10.1177/0002764213479374s
46. Habes, M., Alghizzawi, M., Khalaf, R., Salloum, S.A., Ghani, M.A.: The relationship between social media and academic performance: Facebook perspective. Int. J. Inf. Technol. Lang. Stud. 2(1) (2018)
47. Powers, M., Zambrano, S.V.: How journalists use social media in France and the United States: analyzing technology use across journalistic fields. New Media Soc. **20**(8), 2728–2744 (2017). https://doi.org/10.1177/1461444817731566
48. Salloum, S.A., Al-Emra, M., Habes, M., Alghizzawi, M.: Understanding the impact of social media practices on e-learning systems acceptance. In: International Conference on Advanced Intelligent Systems and Informatics, pp. 360–369. Springer, Cham (2019)
49. Ziani, A., Elareshi, M., Alrashid, M., Al-Jaber, K.: Journalism education in the GCC region: university students' and professionalism perspectives. Media Watch **9**(1), 52–68 (2017). https://doi.org/10.15655/mw/2018/v9i1/49275
50. El Issawi, F.: Women and media: libyan female journalists from Gaddafi media to post-revolution: case study. CyberOrient, 8(1) (2014)

Industry 4.0 Management for Sustainable Technologies

Impact of E-procurement on Supply Chain Performance

Ali M. Albinkalil(✉)

Bahrain National Gas Company (Banagas), Sakhir, Kingdom of Bahrain

Abstract. The advent of the internet has revolutionized the procurement's sourcing processes in firms among multiple sectors and industries. It has stimulated cost-effective ways besides various innovative applications that enterprises could leverage from it. In this paper, we are going to spot the light on the main factors which are impacting the supply chain performance. These factors are E-procurement, Supplier relationship and supplier integration. The literature review will discuss the benefit of automating procurement processes and how it will add value to the organization, followed by the need for strengthening the relationship between the organization and their key suppliers, especially for complex and bottleneck products. Finally, supplier integration to which extent it will assist the organization to be adaptive to the unforeseen situation in the external environment. This paper will examine the impact of the E-procurement on the performance of the supply chain and discusses the main benefits and challenges that the organization may face during its implementation. Moreover, to demonstrate that supplier relationship contributes to influencing supply chain performance in a positive way, furthermore, to analyze the perception of business users.

Keywords: Electronic procurement · Supply chain management · Supplier relationship · Supplier integration · Supply chain performance · Cost savings

1 Introduction

Business is all about the commercial transaction. These transactions result in the essential revenue that businesses need to survive in the market. Procuring services and product is a vital mission of any organization; it can directly influence an organization's profitability [79]. Purchasing the right goods and services at the right quality at the right time with the right price and deliver it to the right place is the main target of procurement department [74]. The evolution of the Internet has revolutionized the sourcing and procurement processes in organizations in every sector. It has stimulated cost-effective ways besides various innovative applications that enterprises could take advantage of. Nowadays, many businesses application is using information technology (IT) as a way of procurement for B2B, B2C, which is also known as E-procurement represents the flow of information among internet base networks, at all stakeholders in the "supply chain," whether within the enterprise, among businesses or between (B2C). Supply chain operation technologies can be described as using ICT to enhance the output, operational routine, and logistics activities in the supply chain channel. Productivity involves

A. M. A. Musleh Al-Sartawi et al. (Eds.): EAMMIS 2021, LNNS 239, pp. 133–143, 2021.
https://doi.org/10.1007/978-3-030-77246-8_13

the improvement of normal working processes responsible for the management of the capacity and inventory.

E-procurement can be described as the automation of the firm's procurement functions using web-based applications. Procurement considers as the largest expense goods in an organization's cost structure [31, 48, 72]. Many firms use e-procurement to obtain contracts to attain advantages such as increasing their employees' efficiency and saving costs by getting quicker and cheaper goods and services. Besides, it minimizes corruption and improves transparency in procurement services through the firms' employees and manager. In earlier days, procurement transactions were processed manually. Thus, it takes a long time to accomplish a single task [32], as the buyer drafts request for quotations (RFQ), prepare purchase requests, obtain management endorsement, and issue purchase orders in manual ways. While commercial transactions and communications between parties have occurred manually in earlier days, these transactions' mechanisms have evolved rapidly, especially with the advent of the internet, computers, and other communication technologies. The procurement process has had re-engineered and restructured with the advent of the internet; thus, business worldwide tends to implement "e-procurement" to improve their supply chain management and streamline the procurement functions [26]. Information technology (IT) has an essential role in evolving supply chain management (SCM). Hence, an ideal system that helps aggregate scattered information and share information through applications in a secure and flexible manner is highly desirable [13, 75].

In 1970, the available technologies used for supply chain were finite such as customized forms, innovative warehouse management techniques, and straightforward information processing procedures. In the 1980s, the most significant technological innovation has arrived, such as electronic equipment able to read the bar code such as, optical scanner, which is customized for each item. Enters the bar code for each item into the store server database and establishes the current cost of that product. In the 1990s, improved inventory cost-saving, and customer service attained through cooperation among supply chain partners. For instance, invest in supply chain technologies, for example, EDI "Electronic data interchange", "E-commerce", "Vendor Managed Inventory" VMI. In the 2000s, cooperation among supply chain partners increased by adapting the "Enterprise Planning system" (ERP). Which in turn lead to additional developments in the organization such as "warehouse management system" WMS, "Transportation management system" TMS, "Customer relationship management" CRM, and "Vendor Relationship Management" VRM; these systems result in improving the accuracy of forecasting on consumer-demand and has changed the way of interaction between supply chain partners. E-procurement has been attracted more attention from industries and individuals. In Nowadays global business scenario, the use of recent technology in organizations has gained immense importance. As a result, organizations are obliged to restructure their business process from traditional manners to electronic platforms [47, 73].

2 Literature Review

2.1 E-procurement

The procurement function is the "process of acquiring goods, works, and services, covering both acquisitions from third parties and in-house providers" [28]. It has an influential role in attaining cost competitiveness. In most manufacturing firms, the cost of materials varies between 40–70% of the cost of goods sold. The cost of holding and handling material could be substantial [4]; they have stated that the estimation price of preparing a single purchase order could vary between 50$ to 200$. Furthermore, the transaction cost could be higher if the products procured through the tendering process. These extra costs could appear in inviting suppliers to participate in the specific tender, evaluating suppliers' technical capabilities, preparing items or contract specifications, technical and commercial evaluation of the received offers and selecting the winner bidder, and other related activities could incur very high costs.

In earlier days, procurement transactions processed manually, resulting in slow transaction accomplishment and even slow handling of the whole procurement processes. However, nowadays, firms become more conscious of how they react to different influences and pressure in the procurement department. Before the launch of "e-procurement", Usually, strategic procurement interact with routine administrative work, for example, turning purchase requisition to purchase order, guarantee the proper distribution of the received invoices for each transaction. Simultaneously, the strategic sides are usually ignored in the process, where the buying organization has a slight impact on suppliers' selection and required items. Multiple and frequent changes in business advancements and technological shifts have increased the level of competition in the industry. Hence, it has increased the need for innovation as an inevitable condition for survival and growth in an intensively competitive environment. These advancements are realized in cost improvements, process improvements, new markets, customer satisfaction, and technological competence [30]. Thus, web-based procurement technologies aim to boost procurement processes and be more efficient by allowing buyers to focus on strategic tasks [27, 71]. E-procurement is business-to-business "B2B" purchasing process through electronic platforms. It has been used to facilitate, manage, and report on the enterprises purchasing functions. E-Procurement system assists managers in the decision-making stage by providing all related information up to date and organized. Nowadays, many firms worldwide strive for e-procurement implementation to digitize their operations, which will allow the organization to improve their efficiency and effectiveness [15, 60, 80].

E-procurement can be defined as a comprehensive procedure in which organizations utilize information technology systems to establish a contract to procure items or services or to purchase products or services for instant payment "Purchasing" [76]. Likewise, [54] described e-procurement as "the use of information technology to facilitate business-to-business purchase transactions for materials and services." From both diffusions, we realize that: E-procurement is not just an online purchasing process system. However, it considers as a communication tool between the organization and its suppliers. [9, 19] defined that e-procurement is "the application of e-commerce in procurement, It involves the use of various forms of Information Technology (IT), such as electronic mail (e-mail), Electronic Data Interchange (EDI), and electronic marketplace (e-marketplace), to automate and streamline the procurement process in business organizations". Simultaneously, the [1] describes e-procurement system as "creating private web-based that automate communications, transactions, and collaboration between supply chain partners".

2.2 E-Procurement Functions

E-procurement involves activities such as "Enterprise Resources Planning" ERP; "E-maintenance, Repair, and Operations" E-MRO; E-tendering, E-reverse auction; E-marketplaces; E-sourcing, and E-informing [18]. E-procurement consists of e-catalogue, e-dossiers, e-submission, e-signature, e-auction, and e-notice [54]. [35] emphasized that e-tendering, e-evaluation, e-negotiation, e-informing e-design, and e-sourcing are the prime functions through which e-procurement participates in supply chain performance. The automation of these purchasing processes by leveraging information technologies such as EDI, Internet, and ERP will help the organization's linking their material management system with suppliers. E-Procurement system consists of six essential processes: E-design [34], E-Informing [79], E-negotiation [77], E-evaluation [12, 41, 66], E-tendering [37, 40, 53], and E-sourcing [17, 52, 59].

2.3 E-Procurement Features

The application of e-procurement functions has several advantages to the organization. E-procurement increase geographical outreach [7]; it allows building a new relationship with new vendors and provides better control over supplier relationships [4, 5]. The benefits of e-procurement have been proved in a vast number of research [1, 3, 23]. According to these researches, e-procurement allows organizations to be operational decentralize procurement processes and functions and centralize strategic processes due to the higher transparency offered by the e-procurement system. For example, in Africa, they require eliminating corruption; enhance the responsibility for procurement activity in the public sector, and openness enforced the e-procurement system's adaption. Most African countries decided on legal reform to facilitate the implementation of the procurement framework and public procurement. In Tanzania, for example, the government supports implementing an e-procurement system to enable proper communication, invoicing, valuation, allotment, oversight, and openness in the advertisement to ensure that the public sector does all purchasing activities are integrated and processed through the internet [10, 59]. Organizations who have implemented e-procurement have attained significant cost saving by 8–15% [81], resulting in improved coordination and collaboration within the organization, saving on transactional costs and increasing competitive sourcing [69].

2.4 Drawback of E-procurement

Few authors have mentioned the obstacles to the application of an e-procurement system. According to [6, 37, 38, 69, 74] have mentioned that employees are unwilling to change employees, lack adequate knowledge of the technology, lack reliance on the technology. At the same time, [2, 22] indicated that the system's high initial investment cost, less flexibility, and the inability to access organizations that do not have internet services, lack of confidentiality, and information insecurity in an electronic transaction. Members of the management team face challenges on convincing top management to invest in technology; they must show that these technologies are in line with the enterprise's strategic objectives [39, 64]. Inadequacies in government policies and legislation are area to be highlighted in the system. The government tendering procedures obliged the buying companies interested in participating in the tender to purchase the printed tender documents in person from the physical offices. This procedure prevents using the e-tendering system and presents a massive relapse for the government trying to initiate the electronic government system. The use of different e-procurement system standards prevents users from communicating with each other due to that difference in systems [58]. The most common reasons indicated by firms regarding implementing e-procurement are that they do not intend to adopt e-procurement due to their high costs. Some firms believe that their size is too small to benefit e-procurement—resistance of the management to support the idea of implementing e-procurement in their organization [78].

3 Supplier Relationship

One of the primary functions of procurement; they are responsible for searching and evaluating new vendors, supporting strategic objectives of the organization by improving sourcing portfolio, strengthening the long-term relationship with key partner, selecting the suitable ICT to streamline procurement activities and related processes. Thus, it will create a fit environment for effective cross-functional teams [67, 70]. Due to the intense demand for complex products, buyers and sellers increase their collaboration, resulting in reliable, trustworthy, and partner key vendors. In this scenario, the vendors must be flexible and can take the risk to co-develop customized items [30]. [33] described SRM as a "process involved in managing preferred suppliers and finding new ones whilst reducing costs, making procurement predictable and repeatable, pooling buyer experience and extracting the benefits of supplier partnerships."

The most crucial areas for vendors and key partner selection are their performance history, production capabilities, business structure, technical and engineering competencies, financial and commitment position, reputation, reliability, integrity, honesty, warranties, price, quality, services, and delivery [21, 49, 61, 76]. At the same time, [20] have proposed the main two types of vendors' selection. The first type, single source, is the only source that can satisfy the buying organization's needs most of the time; it is considered the original equipment manufacturer (OEM). The second type is buying from several suppliers "Multiple Sourcing"; in this type, the buying organization decides to divide the ordering quantities between selling companies as one vendor cannot supply all the buying organization's requirements and create a persistent competitive environment. A study conducted by [30] showed that the relationship between the organization and its suppliers could last for more than ten years due to their collaboration in providing complex and unique items. In contrast, the relationship with suppliers who offer simple and not complicated products cold lasts less due to e-marketplace. To meet today's market needs, managers must look for new sources of value creation and opportunities and directions of ensuring the continuous improvement of processes executed in buying organization and vendor relationship. [43] point out that adaptability and resiliently are the two essential components of the resistant supply chains, building adaptive and flexible relationships with vendors is an opportunity for sustainable flow functions in case of serious inconvenience situation coming from the external environment. Adaptive vendor relationship can be considered as the restructure of vendors' base and changes in purchasing activities and resources response to significant uncertainty coming from the macro-environment of the supply chain [11]. The capability to adapt efficiently to supply chain uncertainty called purchasing flexibility [55]. This strategy recommends managing and maintains a relationship with vendors for bottleneck items [42, 50]. [8] refers to supplier relationship management as the development and design of an organization's vendors portfolio and the relationships that the organization has with its vendors. At the same time, [66] denotes supplier relationship management as a process that promotes various manners on how organizations should deal with various vendors to work with each other for shared benefits. Recent supported debate by [57] distinguishing among CRM and SRM, state that: "*supplier relationship management is the process of engaging in activities of setting up, developing, stabilizing and dissolving relationships with in-suppliers as well as the observation of out-suppliers to create and enhance value within relationships.*"

3.1 Supplier Integration and Supply Chain Performance

Supplier integration defines as "the unified control of several successive or similar economic or especially industrial processes formerly carried on independently" [20]. Also, it refers to how the organization collaborates with its vendors to manage their inter-organizational processes that lead to customer demand fulfilment [36]. Supplier integration is considered the long-term relationship between the firm and its vendors [78]. Several firms worldwide are building a collaborative, joint benefit partnership with supply chain members because of the intense rivalry environment globally [75]. However, by integration between buying organization and supplier, buying organizations could share related information with the suppliers to ensure their obligation to avoid such issues. Supplier integration benefits are that it enhances responsiveness, flexibility, and timesaving [13, 24]. Also, it plays a vital role in reducing transaction costs by reducing uncertainties and minimizing the cost of production [25]. Hence, supplier integration positively influences operational performance [14, 78]. When the organization integrates with its supplier, it will significantly reduce opportunistic behaviors under shared visions and collective goals [65].

Successful supply chain performance relies on the robust commitment and mutual trust between supply chain members. Effective supply chain planning represents trust and shares information between the partners, as it is a primary requirement to attain successful supply chain management [45]. [62] Identified supply chain performance as "on-time delivery, reduced lead time, and responsiveness, and cost reduction, conformance to specifications, process improvements, and time-to-markets as constituents of supply chain performance." [16] debate that organizations that do not measure their supply chain performances are in the dark about their supply chain efficiency as they estimate their target inventory should be. These organizations are probably bear a high price for their goods and services; thus, they become less competitive in the market. The relationships between integration and SME performance have been extensively studied [40, 44, 46]. These findings confirm that integration can be transformed into competitive capabilities, thus contributing to positive supply chain performance. A study conducted by [63] showed that e-procurement and supply chain integration is responsible for enhancing supply chain performance. E-procurement is considered a vital resource, based on the "Resource-Based View" (RBV) theory; RBV theory is an organization.

4 Conclusion and Limitation

In summary, this paper spots light on the main factors that affect the supply chain's performance. These factors are e-procurement, supplier relationship, and supplier integration. This study provided evidence through the literature review that these factors significantly influence the supply chain performance in a positive way. E-procurement contributes to eliminating paperwork, significantly reducing costs, improving transparency, minimizing the purchasing cycle, and streamlining the procurement process. E-procurement improves the effectiveness and efficiency of procurement functions, its assist the organization to geographical outreach to build a new collaborative relationship with multiple suppliers worldwide. Supplier relationship management contributes to strengthening the relationship with key suppliers to ensure a sustainable flow of materials and services. There are two main types of supplier relationships: relationship with the OEM; this relationship mainly focuses on building a long-term relationship with a key supplier to ensure sustainable flow for complicated items. The other type is multiple sourcing where the organizations endeavor to attain the best price available at the market. Furthermore, supplier relationship management provides better control over supplier relationships. Finally, the literature revealed the fundamental role supplier integration contributes to minimizing the risk of receiving wrong items or late in delivering the required material by working closely with the supplier and proper sharing and communicating the related information to its vendors to ensure adequately performing of their obligation and to avoid such issues [51, 78].

4.1 Recommendations

The implementation of e-procurement system will add value and bring immense advantages to all sectors and industries. Since the e-procurement system strongly relies on ICT, the organization needs to build a suitable infrastructure based on recent technologies to enable the procurement department to effectively and efficiently do the procurement function. Additionally, managers in future can leverage this study and design policies based on this model. The organization should also move toward a collaborative partnership relationship with its supplier rather than an arms-length relationship to ensure the sustainability of their processes. One of the top challenges to e-procurement implementation in any organization are the high investment costs, made worse by existing tight budgetary measures [76]. The costs to implement e-procurement could be quite significant, and the returns of investment could take more than three years [68]. This study constitutes a couple of theoretical implications and practical implications. For instance: [29, 56] studies have been conducted regarding the impact of e-procurement on supplier relationship. Accordingly, the current study contributor to the vast body of global knowledge in the e-procurement area. From the practical perspective managers can empirically examine the e-procurement model and reflected in policies and strategies that will aid the respective paganization in streamlining their procurement strategies.

References

1. Aberdeen: e-Procurement: Don't Believe the Anti-Hype. Aberdeen Group, 29 August, Boston (MA) (2001)
2. Aduwo, E.B., Ibem, E.O., Tunji-Olayeni, P., Uwakonye, O.U., Ayo-Vaughan, E.K.: Barriers to the uptake of e-Procurement in the Nigerian building industry. Int. J. Appl. Theor. Appl. Inf. Technol. **89**(1), 133–147 (2016)
3. Al-Sartawi, A.: Social media disclosure of intellectual capital and firm value. Int. J. Learn. Intell. Capital **17**(4), 312–323 (2020)
4. Attaran, M., Attaran, S.: Catch the wave of e-procurement. Ind. Manage. **11**(3), 16–21 (2002)
5. Baily, P., Farmer, D., Crocker, B., Jessop, D., Jones, D.: Procurement Principles and Management, 10th edn. Prentice Hall, Harlow (2008)
6. Bello, W.A., Iyagba, R.O.: Comparative analysis of barriers to E-procurement among quantity surveyors in UK and Nigeria. Scott. J. Arts Soc. Sci. Sci. Stud. **14**(2), 175–187 (2013)
7. Bendoly, E., Schoenherr, T.: ERP system and implementation-process benefits: implications for B2B e-procurement. Int. J. Oper. Prod. Manage. **25**(4), 304–319 (2005)
8. Boddy, D., Macbeth, D., Wagner, B.: Implementing collaboration between organizations: an empirical study of supply chain partnering. J. Manage. Stud. **37**(7), 1003–1018 (2000)
9. Cagliano, R., Caniato, F., Spina, G.: E-business strategy: how companies are shaping their supply chain through the Internet. Int. J. Oper. Prod. Manage. **23**(10), 1142–1162 (2003)
10. Cameron, S.: Making a difference. J. Supply Chain Manage. **12**(3), 50–52 (2007)
11. Chan, H.K., Wang, Y.C., Luong Lee, H.S., Chan Felix, T.S.: Flexibility and adaptability in supply chains: a lesson learnt from a practitioner. Supply Chain Manage. Int. J. **14**(6), 407–410 (2009)
12. Chang, H.H., Wong, K.H.: Adoption of e-procurement and participation of e-marketplace on firm performance: trust as a moderator. Inf. Manage. **47**(5–6), 262–270 (2012)
13. Chen, Y.J., Deng, M.: Information sharing in a manufacturer–supplier relationship: suppliers' incentive and production efficiency. Prod. Oper. Manage. **24**(4), 619–633 (2015)
14. Cohen, M.A., Agrawal, N.: An empirical investigation of supplier management practices. T.W.S. Operations and Information Management Department, University of Pennsylvania (Ed.), Philadelphia (1996)
15. Concha, G., Astudillo, H., Porrúa, M., Pimenta, C.: E-Government procurement observatory, maturity model and early measurements. Gov. Inf. Q. **29**, S43–S50 (2012). https://doi.org/10.1016/j.giq.2011.08.005
16. Cook, M., Hagey, R.: Why companies flunk supply-chain 101. J. Bus. Strat. **24**(4), 35–42 (2003)
17. Croom, S.: The impact of web-based procurement on the management of operating resources supply. J. Supply Chain Manage. **36**(1), 4–13 (2000)
18. Davila, A., Gupta, M., Palmer, R.: Moving procurement systems to the Internet: adoption and use of e-procurement technology models. Eur. Manage. J. **21**(1), 11–23 (2003)
19. De Boer, L., Harink, J., Heijboer, G.: A conceptual model for assessing the impact of electronic procurement. Eur. J. Purch. Supply Manage. **8**(1), 25–33 (2002)

20. Demirtas, E.A., Ustun, O.: An integrated multi objective decision making process for supplier selection and order allocation. Omega **36**(1), 76–90 (2008)
21. Dickson, G.W.: An analysis of vendor selection systems and decisions. J. Purch. **2**(1), 5–17 (1966)
22. Eadie, R., Perera, S., Heaney, G.: A cross-discipline comparison of rankings for e-procurement drivers and barriers within UK Construction Organizations. J. Inf. Technol. Constr. **15**, 217–223 (2010)
23. Eyholzer, K., Hunziker, D.: The used of the internet in procurement: an empirical analysis. In: ECIS 2000 Proceedings, vol. 105 (2000)
24. Fawcett, S.E., Osterhaus, P., Magnan, G.M., Brau, J.C., McCarter, M.W.: Information sharing and supply chain performance: the role of connectivity and willingness. Supply Chain Manage. **12**(5), 358–368 (2007)
25. Flynn, B.B., Huo, B., Zhao, X.: The impact of supply chain integration on performance: a contingency and configuration approach. J. Oper. Manage. **28**(1), 58–71 (2010)
26. Foulds, L.R., West, M., Jin, Y.: Application of project management in e-procurement. Int. J. Logist. Syst. Manage. **4**(2), 135–154 (2008)
27. Giunipero, L.C., Sawchuk, C.: E-procurement Plus: Changing the Way Corporation Buy. JGC Enterprises, New York, NY, USA (2000)
28. Gordon Murray, J.: Public procurement strategy for accelerating the economic recovery. Supply Chain Manage. **14**(6), 429–434 (2009)
29. Gunasekaran, A., McGaughey, R.E., Ngai, E.W., Rai, B.K.: E-procurement adoption in the Southcoast SMEs. Int. J. Prod. Econ. **122**(1), 161–175 (2009)
30. Gupta, M., Sikarwar, T.S.: Modelling credit risk management and bank's profitability. Int. J. Electron. Bank. **2**(2), 170–183 (2020)
31. Harelimana, J.B.: The impact of e-procurement on the performance of public institutions in Rwanda. Glob. J. Manage. Bus. Res. **18**(2) (2018)
32. Hawking, P., Stein, A., Wyld, D., Forster, S.: E-procurement: is the ugly duckling actually a swan down under. Asia Pac. J. Mark. Logist. **16**(1), 1–26 (2004)
33. Herrmann, J.W., Hodgson, B.: SRM: leveraging the supply base for competitive advantage. In: Proceedings of the SMTA International Conference, vol. 1, pp. 1–10 (2001)
34. Hsin Chang, H., Tsai, Y., Hsu, C.: E-procurement and supply chain performance. Supply Chain Manage. **18**(1), 34–51 (2013)
35. Hugos, M.H.: Essentials of Supply Chain Management, 3rd edn. Wiley, Hoboken, NJ (2011)
36. Huo, B.: The impact of supply chain integration on company performance: an organizational capability perspective. Supply Chain Manage. **17**(6), 596–610 (2012)
37. Ibem, E.O., Laryea, S.: E-tendering in the South African construction industry. Int. J. Constr. Manage. **17**(4), 310–328 (2017)
38. Ibem, E.,Electronic (e-) procurement adoption and users' experience in the Nigerian construction sector.Int. J. Constr. Educ. Res., 1–19 (2020)
39. Kengpol, A., Tuominen, M.: A framework for group decision support systems: an application in the evaluation of information technology for logistics firm. Int. J. Prod. Econ. **101**(1), 159–171 (2005)
40. Kim, J.I., Shunk, D.L.: Matching indirect procurement process with different B2B e-procurement systems. Comput. Ind. **53**(2), 153–164 (2004)
41. Kim, S.W.: An investigation on the direct and indirect effect of supply chain integration on firm performance. Int. J. Prod. Econ. **119**(2), 328–346 (2009)
42. Kraljic, P.: Purchasing must become supply management. Harvard Bus. Rev. **61**(5), 109–117 (1983)
43. Kramarz, M., Kramarz, W.: The flexibility and specialization of resources-competitive strategies of materials decoupling points of a network supply chain of metallurgic products. Proc. Soc. Behav. Sci. **111**, 741–750 (2014)

44. Kristal, M.M., Huang, X., Roth, A.V.: The effect of an ambidextrous supply chain strategy on combinative competitive capabilities and business performance. J. Oper. Manage. **28**(5), 415–429 (2010)
45. Kwon, I.W.G., Suh, T.: Factors affecting the level of trust and commitment in supply chain relationships. J. Supply Chain Manage. **40**(1), 4–14 (2004)
46. Lau, A.K.W., Yam, R.C.M., Tang, E.P.Y.: Supply chain product co-development, product modularity and product performance: empirical evidence from Hong Kong manufacturers. Ind. Manage. Data Syst. **107**(7), 1036–1065 (2007)
47. Lee, H.L., Whang, S.: E-business and supply chain integration. Stanford University, Stanford, CA, Stanford Global Supply Chain Management Forum (2001)
48. Lennon, C.: Achieving bottom line results in a flat economy: leveraging procurement business services. Computer Sciences Corporation, El Segundo, CA (2002)
49. Levary, R.R.: Ranking foreign suppliers based on supply risk. Supply Chain Manage. Int. J. **12**(6), 392–394 (2007)
50. Li, S., Ragu Nathan, B., Ragu-Nathan, T.S., Subba Rao, S.: The impact of supply chain management practices on competitive advantage and organizational performance. Omega **34**(2), 107–124 (2006)
51. Liljestrand, K., Christopher, M., Andersson, D.: Using a transport portfolio framework to reduce carbon footprint. Int. J. Logist. Manage. **26**(2), 296–312 (2015)
52. Lysons, K., Farrington, B.: Purchasing and Supply Chain Management, 8th edn. Pearson, Harlow, UK (2012)
53. Matano, F., Musau, E., Nyaboga, Y.B.: Effects of e-procurement implementation practices on procurement of goods, works and services in the national youth service, Nairobi City County. Int. Acad. J. Procurement Supply Chain Manage. **3**(2), 63–82 (2020)
54. McCue, C., Roman, A.V.: E-procurement: myth or reality. J. Pub. Procurement **12**(2), 221–248 (2012)
55. Mendonça Tachizawa, E., Thomsen, G.: Drivers and sources of supply flexibility: an exploratory study. Int. J. Oper. Prod. Manage. **213**(27), 1115–1136 (2007)
56. Min, H., Galle, W.: E-purchasing: profiles of adopters and nonadopters. Ind. Market. Manage. **32**(3), 227–233 (2003)
57. Moeller, S., Fassnacht, M., Klose, S.: A framework for supplier relationship management (SRM). J. Bus. Bus. Market. **13**(4), 69–94 (2008)
58. Mohd Nawi, M.N., Roslan, S., Salleh, N.A., Zulhumadi, F., Harun, A.N.: The benefits and challenges of E-procurement implementation: a case study of Malaysian Company. Int. J. Econ. Financ. Issues **6**(S7), 329–332 (2016)
59. Mwangi, G.K., Mwangangi, P.: Influence of Procurement Management on Performance of Retail Chain Storesin Nairobi City County, Kenya. J. Int. Bus. Innov. Strateg. Manage. **1**(7), 235–245 (2018)
60. Ombat, K.O.: The relationship between procurement systems and performance of procurement function in commercial banks in Kenya. Int. J. Econ. Commer. Manage. **3**(12), 697–723 (2015)
61. Pal, P., Kumar, B.: 16T: toward a dynamic vendor evaluation model in integrated SCM processes. Supply Chain Manage. Int. J. **13**(6), 391–397 (2008)
62. Panayides, P.M., Lun, Y.V.: The impact of trust on innovativeness and supply chain performance. Int. J. Prod. Econ. **122**(1), 35–46 (2009)
63. Pattanayak, D., Punyatoya, P.: Effect of supply chain technology internalization and e-procurement on supply chain performance. Bus. Process Manage. J. **26**(6), 1425–1442 (2019)
64. Poirier, C.C.: Supplier relationship management: an advanced supply chain management technique (2010). An Advanced-SCM-Technique. https://www.scribd.com/doc/39098530/SRM

65. Prajogo, D.I., Oke, A., Olhager, J.: Supply chain processes: linking supply logistics integration, supply performance, lean processes and competitive performance. Int. J. Oper. Prod. Manage. **30**(3), 312–335 (2015)
66. Presutti, W.D.: Supply management and e-procurement: creating value added in the supply chain. Ind. Market. Manage. **33**(2), 219–26 (2003)
67. Pujawan, I.N., Goyal, S.K.: Electronic procurement and manufacturing strategic objectives. Int. J. Logist. Syst. Manage. **1**(2–3), 227–243 (2005)
68. Rajkumar, T.M.: E-procurement: business and technical issues. Inf. Syst. Manage. **18**(4), 52–60 (2001)
69. Rankin, J.H., Chen, Y., Christian, A.J.: E-procurement in the Atlantic Canadian AEC industry. J. Inf. Technol. Constr. (ITcon) **11**(6), 75–87 (2006)
70. Razzaque, A.: M-learning improves knowledge sharing over e-Learning platforms to build higher education students' social capital. SAGE Open **10**(2), 1–9 (2020)
71. Razzaque, A., Jalal-Karim, A.: Conceptual healthcare knowledge management model for adaptability and interoperability of HER. In: European, Mediterranean & Middle Eastern Conference on Information Systems, EMCIS 2010, Abu-Dhabi, UAE (2010)
72. Razzaque, A., Karolak, M.: Building a knowledge management system for the e-health knowledge society. J. Econ. Develop. Manage. IT Financ. Market. **2**(2), 23–40 (2011)
73. Razzaque, A., Eldabi, T., Jalal-Karim, A.: Physician virtual community and medical decision-making: mediating role of knowledge sharing. J. Enterp. Inf. Manage. **26**(5), 500–515 (2013)
74. Rinderle-Ma, S.: Towards the automation of e-negotiation process based on web services – a modeling approach. J. Procurement Manage. **6**, 1–12 (2005)
75. Sharif, A.M., Alshawi, S., Kamal, M.M., Eldabi, T., Mazhar, A.: Exploring the role of supplier relationship management for sustainable operations: an OR perspective. J. Oper. Res. Soc. **65**(6), 963–978 (2014)
76. Sharma, S.K., Bhat, A.: Identification and assessment of supply chain risk: development of AHP model for supply chain risk prioritization. Int. J. Agile Syst. Manage. **5**(4), 350–369 (2012)
77. Smart, A.: Exploring the business case for e-procurement. Int. J. Phys. Distrib. Logist. Manage. **40**(3), 181–201 (2010)
78. Smuts, H.: Determinants of supplier-retailer collaboration: evidence from an international study. Int. J. Oper. Prod. Manage. **26**(1), 24–49 (2008)
79. Srivastava, S.C., Teo, T., Devaraj, S.: You can't bribe a computer: dealing with the societal challenge of corruption through ICT. MIS Q. **40**(2), 511–526 (2016)
80. Tai, Y.M., Ho, C.F., Wu, W.H.: The performance impact of implementing web-based e-procurement systems. Int. J. Prod. Res. **48**(18), 5397–5414 (2010)
81. Zhao, F.: Impact of national culture on e-government development: a global study. Internet. Res. **21**(3), 362–380 (2011)

Predicting Pyramid Geometric Solar Water Still Plant Efficiency Using RBF Based Multi-layer Perceptron

Rajeev Raghuvanshi[1(✉)], MD Ilyas[2], and Surbhi Sharma[2]

[1] Dr. APJ Abdul Kalam University, Indore, MP, India
[2] Computer Science and Engineering, Prestige Institute of Engineering Management and Research, Indore, MP, India
{mdilyas,surbhi}@piemr.edu.in

Abstract. Solar still plant helps to produce fresh water. Development of an accurate predictive model is required on order to monitor and maintain. The machine learning (ML) based model deploys time-series data analysis for the prediction of performance of the still water plant. A review of pyramid geometry water still plant is first prepared. That provides an understanding of components of the still. Recent efforts for improving solar water plant performance are reviewed. The model applies neural network, and is based on the improved weight initialized multi-layer perceptron (). The performance is measured using the experiments on water-production and predictive values of production. The results indicate the superiority of proposed IWMLP. RMSE is 0.0056. A comparison with MLP is also carried out which results in RMSE = 0.0153. The experimental results show that the use of (i) good quality of datasets for learning, (ii) normalization, and (iii) weight initialization strategies improve the prediction accuracy of the model algorithm.

Keywords: Pyramid geometry solar still plant · Predictive data modeling · Time series prediction · MLP · Performance analysis

1 Introduction

Solar energy is a one of unlimited source of energy which can be utilized in various applications, such as water harvesting, power generation, transportation systems, sensor applications [1]. The solar water distillation is also an essential application. But for obtaining the better production with these distillation plants, their monitor is need for maintenance [2]. Recent researches prove that the predictive analytics of time-series help and enhances the productivity. The machine learning techniques are used in designing the predictive models [3]. Recent research study on predicting model based on multi-layer perceptron model show the applicability to the water-production of a solar seawater greenhouse desalination unit [4].

The proposed work focuses on analyzing the solar radiation data and solar still plant yield for predicting the one step a-head water still plant performance.

The proposed task includes the following key objectives to be accomplished.

A. M. A. Musleh Al-Sartawi et al. (Eds.): EAMMIS 2021, LNNS 239, pp. 144–153, 2021.
https://doi.org/10.1007/978-3-030-77246-8_14

1. Collecting dataset for solar radiation
2. Study of pyramid geometry solar water still plant production yield
3. Establishing relationship among solar radiation and water production
4. Preparing an enhanced multilayer Perceptron model for predicting solar water plant production yield
5. Comparing the proposed neural network model with existing model

Section 2 gives the literature review of the researches related to the presented work. Section 3 describes the architecture of the pyramid geometry solar water plant and experiments for measuring performance. Section 4 describes the different techniques which help to modify the ML model prediction-accuracy. The section gives an enhanced model IWMLP proposed in this paper. Section 5 gives the result of the performance of the proposed data model. Section 6 gives the conclusion and discusses the future extension.

2 Literature Review

This section discussed various efforts and studies that are similar in our direction to get support and providing understanding about the proposed data modeling (Table 1).

3 Research Methodology

The proposed work is intended to design and developed enhance predictive model which predict the performance of water production by pyramid geometry solar still plants. The solar water still plants purify the raw water into drinkable water. Figure 1 shows a pyramid-shaped still plant. The different components of the model are also given. The model contains four slops to create the pyramid- like shape. It also contains a box like water tank which is filled with water. The box dimension of the plant is 120 * 120 * 25 cm. The body is designed with the help of a GI sheet of 3.5 MM thickness. The key advantage of the pyramid solar still plant is that it can observe solar energy from any one side of cover. Therefore pyramid geometric solar still gain higher solar radiation than other side type solar still plants. Additionally, in this plant only one cover wall is getting solar radiation and other side remains at lower temperature. This idea enhances the property of condensation by maintaining temperature difference between water surface and cover. The slopes of water still plant are covered with a glass (see Fig. 1). The glass cover is made with 5 MM thickness [5].

In the pyramid still, all the four slops are having 45° inclined. The sunlight is passed through the glass cover and heated th e water, for the evaporation process. The generated steam condensed over the glass cover and walls. Thus it is collected using a drainage system. Additionally, all the water is poured into a basin outside the plant. The basin to collect the freshwater is developed with a metal sheet of GI (Galvanized Iron). The basin also consists of a measuring jar which is used to find the water production [5].

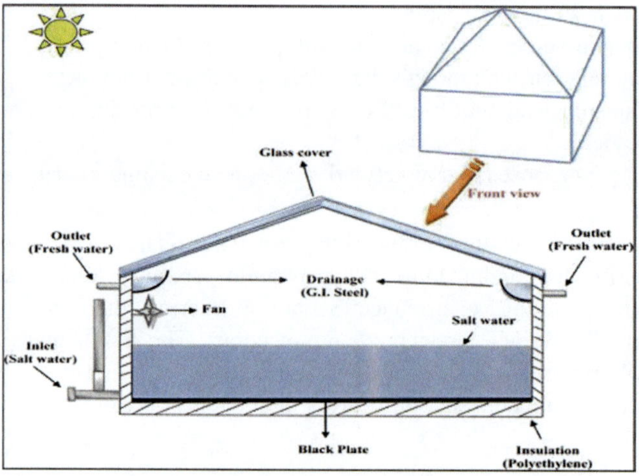

Fig. 1. Pyramid geometry solar still (Developed by authors)

Proposed Work:
The machine learning and the communication technology when combined together it becomes evolutionary. In this context the IoT (Internet of Things) enabled devices are playing an essential role. That is type of wireless sensor network which is used to collect data and send the data to a nearest base station for making essential decisions. Therefore IoT devices are very useful in various different applications such as precise farming, wild life protection and various industrial processes. In order to support the large sc ale water harvesting using solar still plants, the IoT may also becomes very useful for monitoring and maintenance task.

Figure 2 demonstrate an IoT based network for supporting the proposed application. In this context the two clusters of devices are shown in this figure, the nodes with the blue color shows the IoT enabled devices connected with the solar still plants. These devices are collecting the different parameters from the solar still plant and communicate it to the relevant gateway node. The gateway nodes in this figure are demonstrated using red color nodes. These nodes (i.e. gateway nodes) are connected with base station which is a server to collect and process the data.

In our proposed system the base station is a server which consists of a machine learning model. That can collect and process the data for prediction of the solar water still plant's performance or yield. Therefore an experimental model based on neural network is proposed. The proposed model is demonstrated in Fig. 3.

The proposed predictive model used in this experiment is demonstrated in Fig. 3. The model contains a provision to accept the dataset which is used here in a vector format. The dataset contains different attributes which is demonstrated in Fig. 4. These attributes are first preprocessed for removing the missing data and noisy information. The preprocessed data is stored in a database for further use with training process.

On the other hand for training with these time series data the two versions of MLP is implemented. Both the models are described in [reference of previous paper]. In addition of that the previously introduced IWMLP is modified with RBF (radial basis function).

Table 1. Research Work with Author's Name

Ref. no.	Authors	Contributions
[4]	Taleb Zarei, Reza Behyad	Predicting the water production of a solar seawater greenhouse desalination unit using multi-layer perceptron model
[5]	R. Raghuvanshi, D. Verma, M. K. Deshpande	Performance Monitoring And Prediction System For Double Slop And Pyramid Geometry Solar Still Plant
[6]	F. Cavallaro	Presents Takagi-Sugeno Fuzzy Inference System to builds a synthetic index to assess sustainability for energy
[7]	Q. A. Nezhad et al.	Investigate fuzzy controllers based on Takagi -Sugeno & Mamdani model and comparing with similar controllers
[8]	T. Takagi et al.	Fuzzy Identification Systems and Applications. Demonstrate two industrial processes for water cleaning and steel-making process using IF-THEN rules
[9]	P. Angelov et al.	Design of Takagi-Sugeno Models. It evolving structure and learn with real-time data using on-line recursive clustering and estimation of the consequent parameters
[10]	Shanmugan	Fuzzy modeling of floating cum tilted – wick solar still using a set of rules based on solar radiation and distillate output
[11]	Panchal	Study double slope solar still productivity to find performance parameters and conclude black die will increase the output of solar still, sprinkler increase the condensation rate
[12]	Shanmugam et al.	Fuzzy modeling of single slope single basin solar still and interpretation of fuzzy rules using Mamdani model
[13]	H. Kulkarni et al.	Evaluate different designs of solar still for phase change material. Phase change and energy storage materials can enhance energy of system. Top cover cooling is also improving condensation
[14]	J. I. Orisaleye et al.	Developed and evaluate solar water still with characterization of water quality before and after distillation
[15]	M. Chandrashekara et al.	Indirect methods are suitable for medium and large scale desalination, and direct methods for small scale
[16]	H. N. Panchal et al.	Attempt to evaluate the impact of design, climate and operational parameters for a still
[17]	M. T. Chaichan et al.	Investigates use of energy extracted by solar heater where Paraffin wax is used as a PCM. The system concentrating, heating, and productivity, has increased
[18]	K. Rabhi et al.	A modified solar still with pin fins absorber and external condenser. And find thermal behavior and water production performance
[19]	D.B. Singh et al.	Analyze partially covered hybrid photovoltaic thermal flat plate collector solar still. And show relationship between experimental and theoretical values

(continued)

Table 1. (*continued*)

[20]	W. M. El-Maghlany	Presents the optimum inclination angles of the glass cover of the double slope solar still, for maximum energy collection
[21]	A. M. Manokar et al.	Study the performance of a photoVolatic panel integrated solar still and show that maximum distillate output with sidewall and bottom insulation
[22]	F. Suarez et al.	The performance of a direct contact membrane distillation system driven by salt- gradient solar ponds was investigated
[23]	A. E. Kabeel et al.	Analyze effect of heat exchange mechanisms to augment water production
[24]	R. S. Hansen et al.	The performance of inclined type solar still was investigated using different wick materials. New materials are suggested for absorption, capillary rise, porosity, water repellence, and heat transfer co-efficient
[25]	R. Sathyamurthy et al.	Review geometrical shapes of solar still and conclude that geometry in solar still influences the yield of water
[26]	B. Gupta et al.	Offer the design of modified single slope solar still which includes walls white painting, and water sprinkler
[27]	K. H. Nayi et al.	Review development in pyramid solar still and show pyramid still is more efficient and economical
[28]	H. Al-Madhhachi et al.	Investigates the factors that affect water production using experiments to find influence of evaporation temperature, vapor volume, Peltier current and input power
[29]	R. Ullah et al.	Introduce energy parameters and impacts of membrane properties, mass and heat transfer, feed water
[30]	Z. M. Omara et al.	Concerned with double layer wick material, reflectors and influence of water depth on performance

In place of last hidden layer the RBF kernel function for distance computation.

$$RBF = \exp\left(\frac{\|x - x'\|^2}{2\sigma^2}\right)$$

After learning with both the models the model can be used for prediction. Thus some samples from the preprocessed data are used for prediction and validation of learning algorithm's performance. After training the model accept last recent obtained patterns of attributes and predict the value for next time or one step ahead. This section explains the working of the proposed methodology; additionally the next section provides the results analysis of the experiments.

Results Analysis

The proposed modified variants of the predictive systems are described in this section. Based on performance of prediction the implemented models are compared. During the experiments the following format of data is obtained as given in Table 2.

Table 2 provides the actual observational values of the water production yield, additionally their predictive values are also obtained. But due to bulk amount of data we

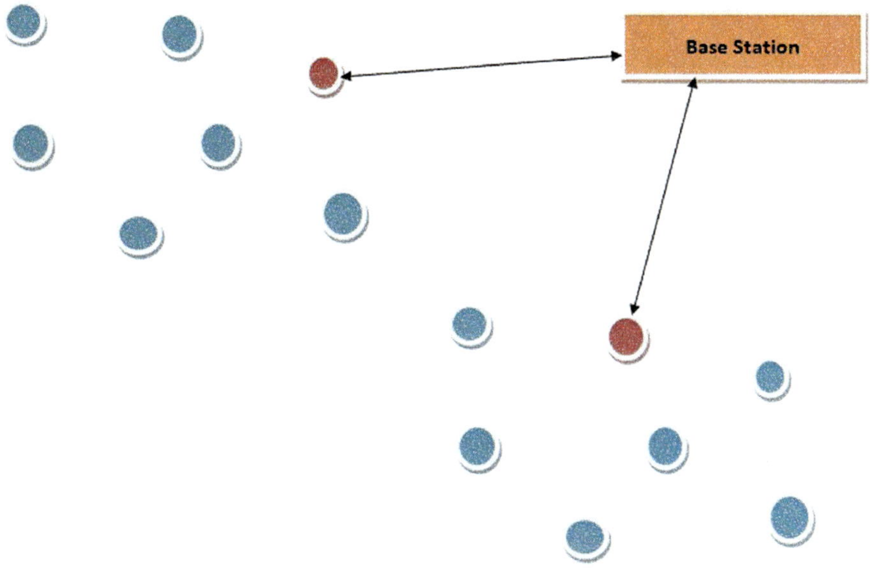

Fig. 2. Example of Required IoT Network (Developed by authors)

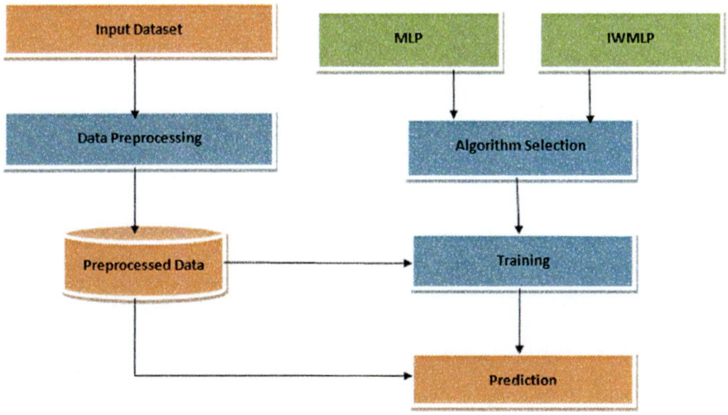

Fig. 3. Proposed Predictive Model (Developed by authors)

are reporting here the mean value of predicted production yield and actual production yield. Their graphical representation is demonstrated in Fig. 4. The figure contains the line graph for prediction using both the data models and the actual observation values. These values are taken for 30 days as sample size and per day mean values are reported using the X axis of line graph. On the other hand, the Y axis contains the production yield of solar water still plant. According to the obtain ed experimental results we can clearly see that the green and blue lines are much closer than each other additionally red line deviate most of the time with the actual pattern. Thus, first the prediction deviation is measured and reported in Fig. 5.

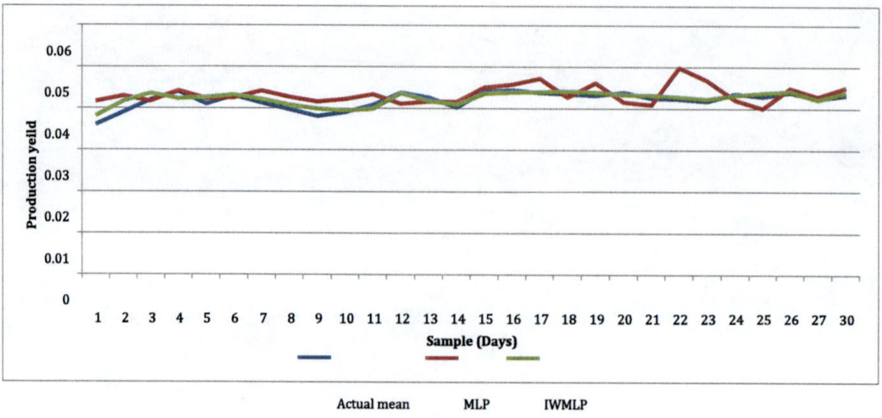

Fig. 4. Experimental observations for actual Vs predicted data (Developed by authors)

Table 2 Experiment details (Developed by authors)

S.No.	basin Temperature (°C)	Water Temp (°C)	Glass Temperature (°C)	Solar Radiation (W/m²)	Distilled water yield	Instantaneous Efficiency (%)
					Experimental	Experimental
1	43	48.9	37	800	0.010	12.47
2	45	54.4	41	850	0.040	11.33

Table 3. RMSE details for MLP and IWMLP (Developed by authors)

	MLP	IWMLP
RMSE	0.0153	0.00565

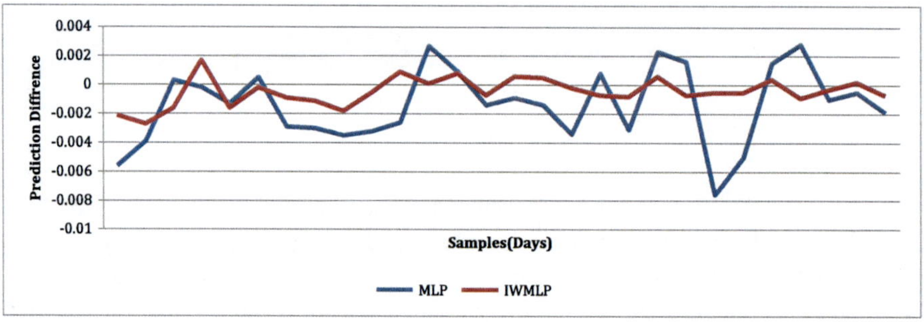

Fig. 5. Prediction difference (Developed by authors)

The prediction difference is calculated based on the following formula:

$$Diff = Actual\ Value - Predicted\ Value$$

Therefore, the prediction different can be found in both the directions positive as well as negative. In this diagram the X axis contains the sample size of 30 days and Y axis demonstrates the deviation in predicted value. According to the diagram the prediction closer to 0 is much accurate, thus the red line provides much close prediction as compared to blue line.

Thus IWMLP is much accurate as compared to normal MLP. Further to compute the RMSE the following formula is used.

$$RMSE = \sqrt{\frac{\sum_{i=1}^{N}(x_i - px_i)^2}{N}}$$

Where the is actual observation and is predicted value, additionally the indicates the individual sample, and N is the total samples. The measured RMSE is calculated and reported using Table 3 and Fig. 6. The figure is a bar graph representation of measured RMSE. It indicates less RMSE and higher accurate results.

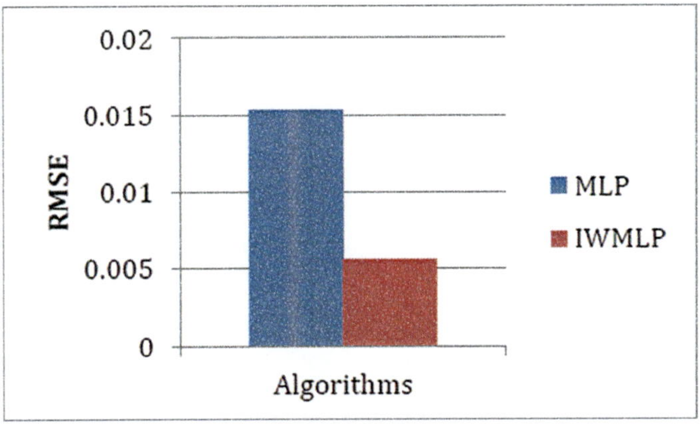

Fig. 6. RMSE comparison for MLP and IWMLP (Developed by authors)

According to the experimental results the IWMLP demonstrate low error rate and higher precise predictive values.

4 Results and Conclusion

The main aim of the proposed work to enhance the accuracy of the model is accomplished, this section provides the summary of entire study and their feasible future extension is also discussed.

A. Conclusion

Sustainable source of energy and living resource are become essential, in this context a significant amount of research work is observed as research area. In this presented work for fresh water harvesting pyramid geometric solar still water plant were studied.

Additionally two different datasets namely solar radiation and still plant performance dataset is combined to establish relationship among them. Further a new variant of neural network is proposed for predicting the accurate performance of solar water plant. Thus a detailed design overview of the proposed system is included in this paper. Finally the implementation of the model was done and their performance is evaluated and compared with the traditional multi-layer Perceptron model. Based on comparative performance the proposed technique found efficient and accurate than the traditional MLP model. The proposed MLP model's RMSE found very fewer as 0.00565. Thus the proposed model improves the predictive performance of the system.

B. Main Results are as Follows

1. The pyramid geometry based solar water still plants can improve the water production up to 30–45%.
2. The outlier removal from training data improves the learning performance of the ML algorithm
3. Normalization and weight initialization techniques help in improving the performance of the predictive algorithm.

References

1. Shaikh, M.R.S., Waghmare, S.B., Labade, S.S., Fuke, P.V., Tekale, A.: A review paper on electricity generation from solar energy. Int. J. Res. Appl. Sci. Eng. Technol. **5**(IX) (2017)
2. https://energypedia.info/wiki/Solar_Water_Purification_in_India
3. Çınar, Z.M., Nuhu, A.A., Zeeshan, Q., Korhan, O., Asmael, M., Safaei, B.: Machine learning in predictive maintenance towards sustainable smart manufacturing in industry 4.0. Sustainability **12**, 8211 (2020). https://doi.org/10.3390/su12198211
4. Zarei, T., Behyad, R.: Predicting the water production of a solar seawater greenhouse desalination unit using multi-layer perceptron model. Solar Energy **177**, 595–603 (2019)
5. Raghuvanshi, R., Verma, D., Deshpande, M.K.: Performance monitoring and prediction system for double slop and pyramid geometry solar still plant. Int. J. Sci. Technol. Res. **9**(2) (2020)
6. Cavallaro, F.: A Takagi – Sugeno fuzzy inference system for developing a sustainability index of biomass. Sustainability **7**,12359–12371 (2015)
7. Nezhad, Q.A., Zand, J.P., Hoseini, S.S.: An investigation on fuzzy logic controllers (Takagi-Sugeno & Mamdani). Inverse Pendulum Syst. (IJFLS) **3**(3) (2013)
8. Takagi, T., Sugeno, M.: Fuzzy identification of systems and its applications to modeling and control. IEEE, vol. SMC-15, Jan/Feb (1985)
9. Angelov, D.F.: On-line design of Takagi-Sugeno models. In: IFSA 2003. LNAI, vol. 2715, pp. 576–584 (2003)
10. Shanmugan, S.: Fuzzy logic modeling of floating cum titled–wick solar still. Int. J. Recent Sci. Res. **4**(5), 579–582, May 2013
11. Panchal, N.H.: Effect of different parameter on double slope solar still productivity. Int. J. Adv. Eng. Sci. **1**(2) (2011)
12. Shanmugan, S., Krishnamoorthi, G.: Fuzzy logic modeling of single slope single basin solar still. Int. J. Fuzzy Math. Syst. **3**(2), 125–134 (2013)

13. Kulkarni, H., Kute, C., Patel, C., Tavse, A., Dhumne, L.R.: Experimental investigation and performance evaluation of solar still using phase change material (IRJET) **05**(05), May 2018. e-ISSN: 2395-0056
14. Orisaleye, J.I., Ismail, S.O., Ogbonnaya, M., Ogundare, A.A.: Development and performance evaluation of solar water still, Tome XI (2018)|Fascicule 1 [January–March] (2018)
15. Chandrashekara, M., Yadav, A.: Water desalination system using solar heat: a review. Renew. Sustain. Energy Rev. **67**, 1308–1330 (2017)
16. Panchal, H.N., Patel, S.: Effect of various parameters on augmentation of distillate output of solar still: a review. Technol. Econ. Smart Grids Sustain Energy **1**, 4 (2016). https://doi.org/10.1007/s40866-016-0005-2
17. Chaichan, M.T., Kazem, H.A.: Water solar distiller productivity enhancement using concentrating solar water heater and phase change material (PCM). Case Stud. Thermal Eng. **5**, 151–159 (2015)
18. Rabhi, K., Nciri, R., Nasri, F., Ali, C., Bach, H.B.: Experimental performance analysis of a modified single-basin single-slope solar still with pin fins absorber and condenser. Desalination **416**, 86–93 (2017)
19. Singh, D.B., Yadav, J.K., Dwivedi, V.K., Kumar, S., Tiwari, G.N., Al-Helal, I.M.: Experimental studies of active solar still integrated with two hybrid PVT collectors. Sol. Energy **130**, 207–223 (2016)
20. El-Maghlany, W.M.: An approach to optimization of double slope solar still geometry for maximum collected solar energy. Alexandria Eng. J. **54**, 823–828 (2015)
21. Manokar, A.M., Winston, D.P., Kabeel, A.E., Sathyamurthy, R.: Sustainable fresh water and power production by integrating PV panel in inclined solar still. J. Clean. Prod. **172**, 2711e2719 (2018)
22. Suarez, F., Urtubia, R.: Tackling the water-energy nexus: an assessment of membrane distillation driven by salt-gradient solar ponds. Clean Tech. Environ. Policy. https://doi.org/10.1007/s10098-016-1210-3
23. Kabeel, A.E., Kumar, T.A., Denkenberger, D.C., Sathyamurthy, R.: Performance enhancement of solar still through efficient heat exchange mechanism–a review. Appl. Thermal Eng. **114,** 815–836 (2017)
24. Hansen, R.S., Narayanan, C.S., Murugavel, K.K.: Performance analysis on inclined solar still with different new wick materials and wire mesh. Desalination **358**, 1–8 (2015)
25. Sathyamurthy, R., Harris Samuel, D.G., Nagarajan, P.K., Arunkumar, T.: Geometrical variations in solar stills for improving th e fresh water yield—a review. Desalin. Water Treat. 1–15 (2016). https://doi.org/10.1080/19443994.2015.1136241
26. Gupta, B., Sharma, R., Shankar, P., Baredar, P.: Performance enhancement of modified solar still using water sprinkler: an experimental approach. Perspect. Sci. **8**, 191–194 (2016)
27. Nayi, K.H., Kalpesh, V.M.: Pyramid solar still: a comprehensive review. Renew. Sustain. Energy Rev. **81**, 136–148 (2018)
28. Madhhachi, H.A., Min, G.: Key factors affecting the water production in a thermoelectric distillation system. Energy Conv. Manage. **165,** 459–464 (2018)
29. Ullah, R., Khraisheh, M., Esteves, R.J., McLeskey Jr., J.T., AlGhouti, M., Gad-el-Hak, M., Tafreshi, H.V.: Energy efficiency of direct contact membrane distillation. Desalination **433**, 56–67 (2018)
30. Omara, Z.M., Kabeel, A.E., Abdullah, A.S., Essa, F.A.: Experimental investigation of corrugated absorber solar still with wick and reflectors. Desalination **381**, 111–116 (2016)

Comparative Analysis of Machine Learning Algorithms for Stock Market Prediction During COVID-19 Outbreak

Jolly Masih[1,2(✉)], Rajkumar Rajasekaran[3], Neha Saini[4], and Damandeep Kaur[5]

[1] Prestige Institute of Engineering Management, Indore, India
[2] Erasmus School of Economics, Erasmus University, Rotterdam, The Netherlands
[3] Vellore Institute of Technology, Vellore, India
[4] Birla Institute of Technology, Mesra, Ranchi, India
[5] DaDa Analtix Pvt Ltd, Vellore, India

Abstract. In the current period of time, when there is a havoc across the world due to COVID-19 virus outbreak, it becomes very important to foresee the impact of this pandemic on the world economy. This has attracted us to analyze and predict the stock market prices of some international IT international companies which provide employment to thousands of people and create revenue for many countries namely Google, Microsoft, Apple and Amazon. In this study, we have implemented algorithms such as SVM and LSTM on stock market data to see if major IT companies see a rise or fall during the COVID-19 pandemic. We have also used ARIMA forecasting method to predict the stocks of above mentioned 4 companies . This paper provides a simple but original statistical analysis of the impact of the COVID-19 pandemic on stock market risk for 4 major IT companies of the world. Results revealed that while some businesses like personal computers from Microsoft, I phone handsets, sale of luxury and fashion goods at Amazon has declined during the pandemic, thus leading to fall of stocks. However, some prominent other segments like online shopping, cloud computing and streaming video from Amazon, oversees Office, Dynamics, Skype, LinkedIn Intelligent Cloud from Microsoft, Google's ad sales during the crisis and issue of cheap bonds by Apple came out to be the winning corporate strategies to fight the negative economic effect of COVID-19 and to stabilize the situation of stocks in coming months. This study may help investors and companies to sustain the tide of economic fall.

Keywords: COVID-19 · Stock prices · Stock market volatility · Google · Microsoft · Apple · Amazon

1 Introduction

It is believed that the impact of COVID-19 pandemic is going to be huge on the global economic growth. Due to corona virus, the Organization for Economic Co-operation and Development (OECD) has already halved the global gross domestic product (GDP) growth projection for 2020. This pandemic has also impacted the Indian economy

A. M. A. Musleh Al-Sartawi et al. (Eds.): EAMMIS 2021, LNNS 239, pp. 154–161, 2021.
https://doi.org/10.1007/978-3-030-77246-8_15

adversely [1]. Owing to adverse effects of a pandemic, some economists have done research using statistical methods like ANN (Artificial Neural Network) algorithm for stock market prediction. Stock valuation was made with respect to some socio-economic factors [2]. Capital market across the world suffered miserably due to this pandemic [3]. For prediction of stock market, most researchers have focused on Artificial Neural Networks (ANN) [1]. While using ANN, the machine trains itself and by detecting various patterns it assigns the weights again in every epoch until the desired results are obtained [2]. The spread of the COVID-19 has severely affected the financial markets all over the world. It has created an exceptional level of risk, causing investors to bear considerable losses in a very short span of time [3, 4]. With the Corona pandemic prolonged to haunt the global equity market, and most Asian and European markets registered big losses over the economic fallout. The Indian stock market had been witnessing a roller coaster ride where Sensex ended 1,627.73 points higher at 29,915.96, while the Nifty was up 482 points at 8,745.45 [5, 7]. All the investors big and small were keeping an eye on the global indices count on the latest cases, and the government involvement to understand their role in the current share of market volatility [7]. Another study analyzed the potential effects, that the corona virus will have on the stock markets [8–10]. The most popular five technology companies in the world – Apple, Facebook, Microsoft and Google and Amazon had lost collectively $623.23 billion in the market cap in the month of March after the WHO declared the Corona virus outbreak a public health emergency. Apple which had a market cap of $1.417 lost the most [11–13]. Apple, which has reported its second quarter earnings, showed elastic growth for the company in spite the impact of COVID-19 pandemic on its retail business and supply chain. It was seen that Apple was facing manufacturing delays due to corona virus related lockdowns [14]. Apple had got a strong hike in the start of 2020, after reporting solid December quarter earnings and a record high for the stock. However, due to the outbreak of COVID-19 and its spread across the world, Apple claimed that it won't be able to generate revenue guidance for the March quarter. Apple's stock had fallen nearly 19%. It is expected that Apple would be able to reach back its highest record of shares in the next 12 months [15]. With Stock market prediction we aim to determine the future values of several companies' stocks. A detailed comparative study will be conducted between the outcomes of these techniques to state which algorithm is most suitable for achieving higher accuracy [16–19]. ARIMA forecasting is another famous technique to predict the prices or stocks [20]. This may help investors and companies to sustain the tide of economic fall. In this study, we have implemented algorithms such as SVM and LSTM on stock market data to see if major IT companies see a rise or fall. Thus, the paper tries to study the economic impact of the pandemic by providing evidence that COVID-19 has affected the stock market of the country using latest technologies like ARIMA, SVM and LSTM. Our paper contributes to the literature by documenting the impact of COVID-19 on the stock market of various five major IT companies. Moreover, it also provides a reference for assessing stock markets after the pandemic subsides.

2 Methodology

Data Used: There are generally two sources that publish the stock prices of companies in India, namely, BSE (Bombay Stock Market) and NSE (National Stock Market). Here, we have focus on NSE stock prices instead of BSE. The reason behind using NSE data is that BSE stocks are more volatile in nature as compared to those of NSE stocks. Less volatile prices are preferred by analysts as it helps and enables to estimate the stock prices soon. It thus aids, to increase the accuracy of the machine learning model used.

We have considered the daily data from four major IT companies, namely Apple, Google, Microsoft and Amazon from 20th March 2019 to 20th March 2020. Since, all these 4 IT giants are USA based companies, thus to support our results, we also considered secondary data in results and discussion about NYSE (New York Stock Exchange).

There was an option of directly downloading the data from YAHOO FINANCE using pandas_datareader in python. We have opted international companies such as Apple, Google, Microsoft and Amazon as they fall in the category of giant companies. Their data were merged to form a single CSV for further pre-processing, analysis and prediction using machine learning algorithms.

It was noticed that there was a sudden and huge fall in stock prices of most of the companies. This change was not predictable. These changes attracted us to try and predict the stock market prices of some companies for next 30 days. So, we can see the difference between the predicted and actual prices. Analysis of the mentioned companies was carried out using seaborn and matplotlib libraries in python. We implemented two algorithms, namely SVM and LSTM models on all the companies mentioned before. In the final stage, we have implemented ARIMA best fit model to predict the stock prices of major IT companies, owing to the COVID-19 pandemic scenario.

3 Results and Discussion

As the coronavirus continued to spread across the world, the world's financial markets began to crash. The NYSE started plummeting in February and March, as more countries and companies across the world had to deal with mandated stay-at-home orders and social distancing guidelines (see Fig. 1). It was a similar story for **Apple's** stock price. The price reached its lowest point of $224.37 on March 23, causing Apple to dip below the $1 trillion market capitalization. China began recovering in March, with improved Apple's supply chain in the region. Elsewhere, however, Apple shuttered all of its retail locations outside of China, and they remained closed throughout April and early May. As the world transitioned to remote learning and work-from-home policies, Apple's stock began recovering again. The company eventually regained its $1 trillion valuation, and throughout April, Apple's share price has largely recovered to pre-coronavirus levels. On April 30, Apple reported earnings that exceeded Wall Street expectations. Although revenue was still down from the pre-pandemic levels, the company's results proved to investors that it could maintain a relatively strong position during COVID-19 [21–24].

Talking about **Google,** Alphabet (parent company of Google) stock fell more than 13% in March, which marked its worst month in last five years. Investors sold Alphabet shares due to the COVID-19 pandemic. So far, the coronavirus has weakened the demand of advertising services, which is an Alphabet's main revenue source. Since advertising sales contribute most of Alphabet's revenue, the stock tends to be sensitive to the marketing industry's health [25].

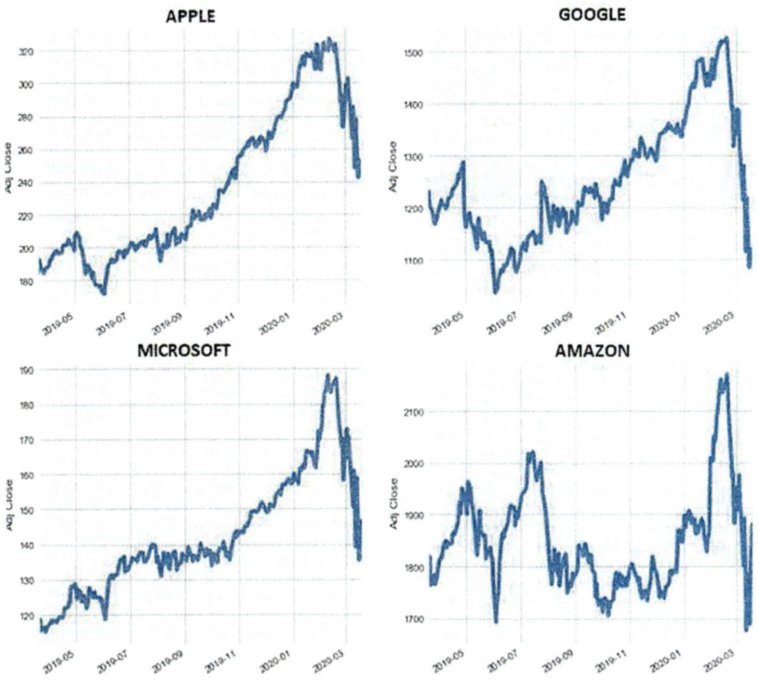

Fig. 1. Historical view of the closing price. The figure illustrates the drastic drop in the stock market of the listed companies during 2020 for months February and March due to COVID-19 pandemic.

Microsoft, stated that COVID-19 had minimal impact on revenue in its latest quarter. Its Stocks continued to rise with a minor downfall even in COVID-19 economic shut down. Microsoft's fiscal third-quarter went better than expected. For the period ended March 31, the software giant reported revenue of $35 billion, up 15%. For the final weeks of the quarter, Microsoft noted a slowdown in transaction licensing, particularly in small and medium businesses, and a reduction in advertising spend in LinkedIn due to the impact of COVID-19 [26–31].

Amazon.com Inc on May 2, 2020 reported it could post its first quarterly loss in five years, even as revenue surges because it is spending at least $4 billion in response to the coronavirus pandemic, including plans to test its workforce for COVID-19.Shares of Amazon, the world's largest online retailer, fell 5% in after-hours trade [32]

It is visible in the graph that certain histogram (representing daily return of stock) represented higher fluctuation as compared to average histogram curve.

Amongst all 4 companies, the maximum fluctuation was observed in case of Amazon, since due to lockdown the online delivery of goods was adversely affected. However, this fluctuation in the Amazon was considered temporary since Amazon own multiple ventures like cloud computing and online streaming serves which helped the company to regain the strength of its stocks. Apple, Google and Microsoft also witnessed some odd days when their stock prices fluctuated more than expected due to the impact of COVID-19 (see Fig. 2) [25–32].

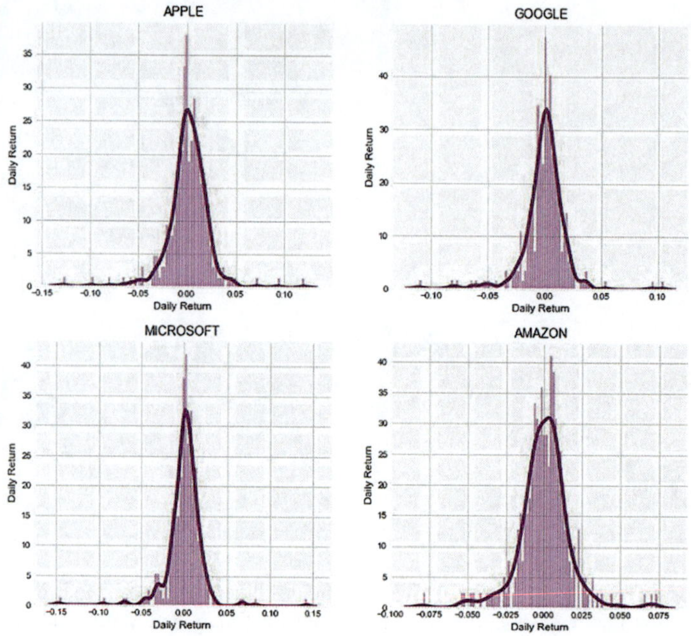

Fig. 2. Daily return of the stock on average histogram. The figure reflects the spikes which are moving out of normality curve, due to COVID-19 fluctuations in stock prizes.

Apple is the most valuable brand in the world. According to Forbes, Apple was worth $205.5 Billion in 2019. The company has a strong following across the world. This is because people consider owning Apple products to be a great luxury. Due to this, market analysts usually give Apple's stock a high rating. They consider it to be a blue-chip stock. But during the COVID-19 pandemic, even blue-chip companies are suffering. The stock market is very unstable currently. When the COVID-19 pandemic started spreading in February, Apple's stock was at a price of 308.66 USD as of 3 February 2020. The stock plummeted initially, falling as low as 224.37 USD as of 23 March 2020. But the company has recovered since and now stands at a price of 318.89 USD. While the demand for Apple's iPhones and Wearables, the company has consolidated with a rise in subscription to its premium services. Overall, the company is holding up well during the pandemic [21, 23].

Google said on its first-quarter earnings call that search revenues were down 15% in March and April, while YouTube revenue growth decelerated to high single digits by end March [26]. GOOGL stock on Jan. 16 closed at 1,450.16, reaching a $1 trillion market cap. It hit an all-time high of 1,530.74 in intraday trading on Feb. 19. Amid the coronavirus-driven market sell-off, Google stock hit a low of 1,054.13 on March 23. Google stock has formed a *cup-with-handle chart pattern*, and trades about 10% below its proper entry point of 1,530.84 by March end [27].

Microsoft said its gaming segment "benefited from increased engagement following stay-at-home guidelines." The company reported that its Search business "was negatively impacted by reductions in advertising spend, particularly in the industries most impacted by COVID-19." The company said it returned $9.9 billion to holders in the quarter in dividends and stock repurchases, up 33% from the December quarter. The stock repurchases totalled $6 billion. The performance of Microsoft should not necessarily be a surprise.

Amazon has unsettled investors in the past with heavy spending on cloud data centres, streaming video and voice-controlled gadgets, which often paid off in the form of new businesses. Since mid-February, Amazon's shares have risen by more than 10%, while the broader stock market has sunk. With online shopping booming as millions of people stay at home and avoid traditional stores, the Amazon's stock has climbed 9% in averages of stock prices over the period of last one year. However, the stock prices suddenly inflated during the months of February and March due to the impact of COVID-19 pandemic.

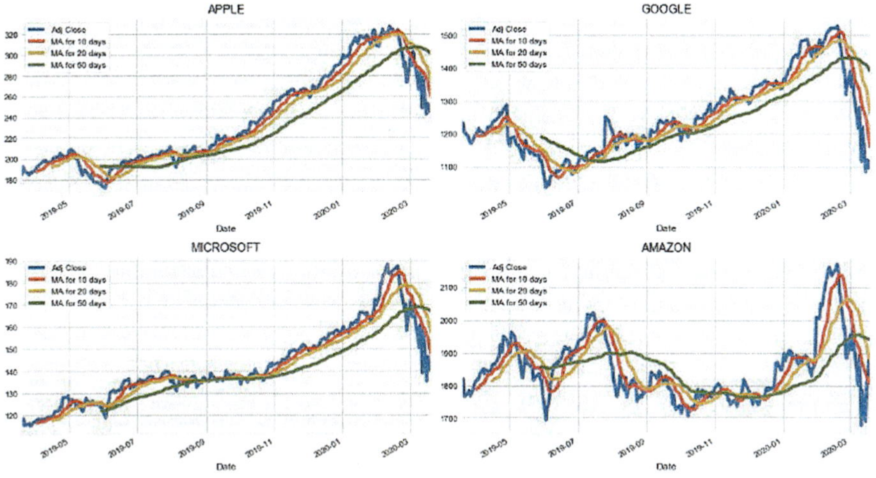

Fig. 3. Plot of all the moving averages. The figure compares the moving averages of stock prices of 4 major IT giant companies. An upward trend was found in moving.

Google and Apple are now roughly flat for the year, after each falling more than 30% from the February [28]. Google stock has dropped nearly 10% from its all-time high as the coronavirus impacted the economy worldwide. Google stocks had plunged 34% from on March 23 due to COVID-19 outbreak. In the March quarter, Google bought $8.5 billion of its own stock, topping the December quarter's $6.1 billion [27]. Amazon's strategy of landing in internet streaming service, came out as clear winner during the locked down. Microsoft, has also benefited from the video-calling services (see Fig. 3). Till February 27th, Apple's stock has been on a rollercoaster ride, with shares rising as high as $278.41 within a span of a few hours before quickly plummeting as low as the $260 s. The coronavirus outbreak prompted many of Apple's suppliers to suspend production at their factories in China [22]. Apple shares are virtually flat year to-date, compared with a 12% drop in the S&P 500 Index. Apple reported in a regulatory filing that funds will go toward general corporate purposes, including share repurchases and dividend payments. During the six months ended March 28, Apple spent $38.5 billion to repurchase its own stock [21].

4 Conclusion

On the basis of analysis, we can conclude that the stock prices of major IT companies would be adversely for sometimes, but after a few months the stock prices will get stable and might again start to rise. Considering prediction of stock prices as of now, we have got an accuracy of 75.95% using the SVM algorithm and ARIMA best fit forecast model. This drop in accuracy was intended, as described before due to increased volatility in the stock market due to COVID-19 outbreak. This has led to dynamic changes in the stock prices, making it very difficult to predict the future values given the current situation. This paper provides a simple but original statistical analysis of the impact of the COVID-19 pandemic on stock market risk for 4 major IT companies of the world. The financial markets have seen dramatic movement on an unprecedented scale due COVID-19 outbreak. Results revealed that while some businesses like personal computers from Microsoft, Iphone handsets, sale of luxury and fashion goods at Amazon has reduced thus leading to fall of stocks. But along with the time some prominent other segments like online shopping, cloud computing and streaming video from Amazon, oversees Office, Dynamics, Skype, LinkedIn Intelligent Cloud from Microsoft, Google's ads sales during the crisis and issue of cheap bonds by Apple came out to be the winning corporate strategies to fight the negative economic effect of COVID-19 and to stabilize the situation of stocks in coming months.

References

1. Ahmar, A.S., del Val, E.B.: SutteARIMA: short-term forecasting method, a case: COVID-19 and stock market in Spain. Sci. Total Environ. Article no. 138883 (2020)
2. BjoernKrollner, G.F., Vanstone, B.: Financial time series forecasting with machine learning techniques: a survey. In: European Symposium on Artificial Neural Networks: Computational and Machine Learning, Bruges, Belgium, April 2010
3. Kim, K.J.: Financial time series forecasting using support vector machines. Neuro-computing **55** (2003)
4. Mayank, S.R.Y., Patel, K.B.: Stock prediction using artificial neural network. Int. J. Innov. Res. Sci. Eng. Technol. (2014)
5. Shah, V.H.: Machine learning techniques for stock price prediction, pp. 1–18 (2007)
6. www.angelbroking.com. Accessed 28 May 2020
7. www.businesstoday.in. Accessed 28 May 2020
8. www.cnbc.com. Accessed 28 May 2020
9. www.doi.org. Accessed 28 May 2020
10. www.dx.doi.org. Accessed 28 May 2020
11. www.economictimes.indiatimes.com. Accessed 28 May 2020
12. www.financialexpress.com. Accessed 28 May 2020
13. www.livemint.com. Accessed 28 May 2020

14. www.macrumors.com. Accessed 28 May 2020
15. www.moneycontrol.com. Accessed 28 May 2020
16. www.moneycontrol.com. Accessed 28 May 2020
17. www.outlookindia.com. Accessed 28 May 2020
18. www.researchgate.net. Accessed 28 May 2020
19. www.theverge.com. Accessed 28 May 2020
20. www.voxeu.org. Accessed 28 May 2020

Bitcoin Prediction Using Ensemble Modelling

K. Govinda[1], R. Rajkumar[1], and Jolly Masih[2](✉)

[1] SCOPE, VIT University, Vellore, India
{kgovinda,rrajkumar}@vit.ac.in
[2] Prestige Institute of Engineering Management and Research, Indore, India
jmasih@piemr.edu.in

Abstract. Bitcoin is one of the recent phases for the cryptocurrency front. It uses peer to peer technology for operation with no authority or bank making note or controlling it any for or factor. Its transactions are managed and it is issued out collectively by the network in place. Bitcoin is open-source – Its layout is public i.e. nobody owns or controls bitcoin but everyone can participate. What is awe inspiring is the exciting use bitcoin provides unlike any of its predecessors are the Faster peer to peer Transactions, the Worldwide acceptance and the Lowest processing rate in the market, the main goal of the work is to find a better methodology on which the traders can depend on while investing in the bitcoins, so that they can be sure of conformed returns and in turn increase the overall investment in the cryptocurrency as a whole.

Keywords: Blockchain · Cryptocurrency · Bitcoin · Transactions

1 Introduction

The term BITCOIN has an unknown origin but it came into existence in 2008 by a person using the alias of Satoshi Nakamoto. The transactions are verified by its nodes via cryptography and logged into a publicly accessible ledger i.e. BLOCKCHAIN [1]. They are created as a reward system of mining (as in crypto currency mining) In 2017, it was seen that Bitcoin had about 3 to 6 million users. It's also been used as an investment forum instead of just as a currency because the rate of increase of this currency is independent from the economic factors which rule the normal currency of any nation as this currency is based on the transfer from individual to individual as it exists in limited quantity i.e. 21,000,000 Bitcoins exist in supply limit. The proposed work ideally would want to predict the value of Bitcoin with as high accuracy as possible using the tools. As Bitcoin is a very much volatile currency its prediction can be really helpful. Proposed work uses the concept of machine learning i.e. train multiple models of test data to find out which model works the best for the dataset, which means we are going to use Ensemble Machine Learning model along with LSTSM (RNN model) [2].

RNN model is the only model dealing with the data at all points with respect to time and making use of those checkpoints to give an understanding into the desired result that is the required result. We plan to use Random Forest under the Ensemble modelling as it creates multiple decision trees using the bagging method, the basic idea of bagging

A. M. A. Musleh Al-Sartawi et al. (Eds.): EAMMIS 2021, LNNS 239, pp. 162–167, 2021.
https://doi.org/10.1007/978-3-030-77246-8_16

is making a considerable combination of learning models to increase the overall result. Simply – RF builds multiple decision trees and merges them to get an accurate and stable result in prediction. It is defined as a growing list of records known as blocks; these blocks are connected via cryptography. Each block has a cryptographic hash of the preceding block, the transaction data and the timestamp. By the default engineering of its design it is impervious to any sort of modifications intended to the data [3].

Hash
It is similar to a unique fingerprint. It identifies a block along with the entire content of a block. Every block has a unique hash. For the addition of every new block there is the generation of a new hash value. The basic working of blockchain is based on the principle of hashing. A hash function in a blockchain is a mathematical function which is used in converting a variable, numbers and strings to hash. Bitcoin works on the principal of blockchain [4]. Here verification works on the basis of the Verify: 1 = ? v (Signature, messages, Public Key).

As we have been shown by the creators that each and every bitcoin has a digital signature. Market of the cryptocurrency is very small compared to the traditional currency and the currency is also very volatile which is not seen as the good indicator for long term investments. The bitcoin market has seen explosive growth during the time 2013–2019. The current market cap of the bitcoin at 88,259,447,888 USD. Around 80% of all the trading is done in USD [5]. Machine Learning is the study of mathematical and statistical models which is used by computers to perform a certain task provided by the user command. In ML we are able to build and train a model on a certain required dataset. After the training of the model it can be tested to certify the accuracy of the model, to check if the accuracy is greater than certain ratio so that it can be used to predict the required problem.

2 Proposed Method

The prediction of value of bitcoin, we are using Ensemble Learning. Let us provide an example of what ensemble learning model deals with. Suppose you are to buy a new vehicle, you probably wouldn't walk up to the dealership and buy the first car that you see, you will research online and contrast and compare various models to the point where your requirements are met. Ensemble Learning uses multiple ML algorithms and we select the particular result which satisfies our requirements. Here proposed method uses three models namely:

Linear Regression, Random Forest and Long Short Term Memory (LSTM) model [6].

2.1 Linear Regression

Linear Regression is established to be the most common model of machine learning known to the users. It is the statistical approach to modelling a relationship in between two dependent variables. The representation in a linear regression is a linear equation that combines some specific set of input value that is taken as (x) the solution to which is said to be the predicted output for that particular set of input values (y). And both the input values (x) and the output value are numeric. The linear equation assigns one scale factor to each of the input value and column, called as coefficient and represented by the capital letter Beta (B). One additional coefficient is required to be added/utilized, which provides the line with an additional degree of freedom (e.g. like moving up and down on a two- dimensional plot) and is called as the intercept or the bias coefficient. The case in which one explanatory variable is available is called as simple linear regression whereas the case in which multiple explanatory variables are available are called as the multiple linear regression [7].

2.2 Long Short Term Memory (LSTM)

LSTM is an artificial recurrent neural network architecture that is mainly used in the field of the deep learning. LSTM can process multiple sequence of data due this reason it is the most commercially available AI model in use. LSTM model is trained using the Recurring Neural networks Backpropagation Through Time methodology and overcomes the vanishing gradient problem. As such, LSTM can be used to create large recurrent networks that in turn can be used to address difficult sequence problems in machine learning and achieve state-of-the-art results [8].

2.3 Random Forest Regression

Decision algorithm makes decision trees that are vulnerable to high variance if they are not pruned or in simple words EDA that is proper data preprocessing is not done. This high variance can be harnessed and would be reduced by creating multiple trees with different samples of the training dataset that are just different views of the problems and combining all their predictions. This methodology is known bootstrap aggregation or also known as bagging for short. We can limit the features(rows) that the greedy algorithm evaluates at each instance while a tree is created to force the decision trees to be different. And this is known as the Random Forest Regression Algorithm. So, in simple words Random Forest Regression are an ensemble learning method for the classification and regression and prediction which are work by creating multiple decision tree at training an outputting the class that is that is the mode of the classification or mean prediction of the individual tree. The problem is that the previous approaches for price prediction can't be used in the long run so to solve this problem we have shifted our focus in the new emerging field of the machine learning. Here we took our data set with the help of the Kraken API [9]. After we have the data then we find the shape of the data (that is how many rows and how many columns we have) after that we set the date as the index of the data then we correlate the data using the correlation function then we take out the parameters for matching then we break the data into the train cases and the test cases. Then we load our model in the program and pass the train cases so that our program can train on these cases and then we check the formula that is made by passing the test cases in the model with the train cases we have stored [10].

The result we got after passing the train data in all the model were:

Table 1. Model vs accuracy

Linear regression	99.04%
Random Forest Regression	98.5%
LSTM	79.4%

After observation we can say that Linear regression was the most accurate followed by Random Forest and then LSTM. Here LSTM was not the most accurate due lesser training time. LSTM accuracy can be improved by training it on different test- train ratios of the dataset [11].

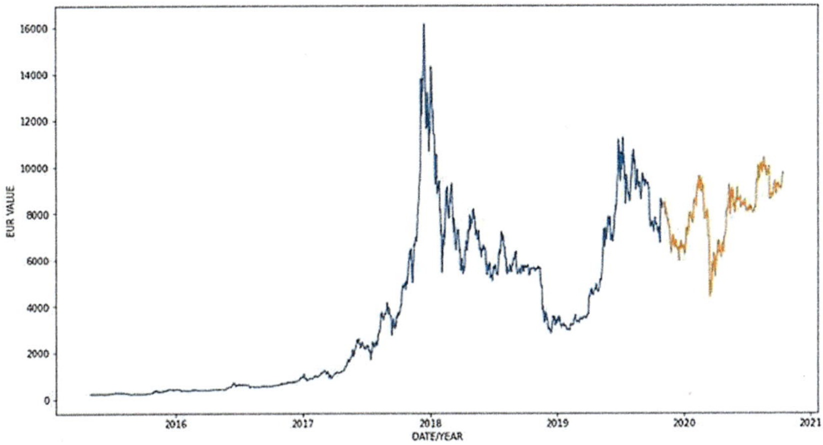

Fig. 1. Predicted result using Random Forest Regression

So here in Fig. 1 and 2 above we can see the relation between the predicted result and the real result. After this we can save our model and then we can call it in future and can pass the parameters based on which it can tell us the predicted price of the bitcoin of the certain date [12–14 15].

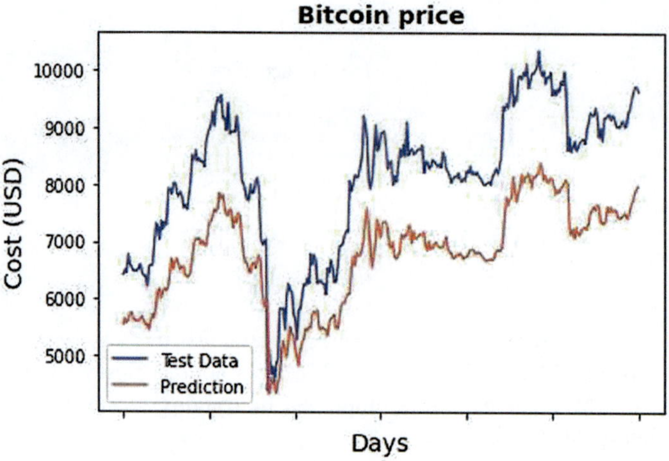

Fig. 2. Predicted result using LSTM

3 Conclusion

So, we were able to predict whether to buy or to sell bitcoin up to certain descent level. The problem that we are facing is that our sentimental analysis algorithm can predict till the three-day prior only as we only get three day old google trends data only. The problem with our second approach is that it is in its infancy and to predict for future more accurately, we need to modify our data preprocessing and get a larger data set to improve the model that we are using. Now the future work/output that needs to be done on this work or can be done for the sake of improvement is that we need to combine a sentimental analysis and machine learning and do learning on the basis of how human reacts on internet regarding certain news on the internet about the cryptocurrencies and based on their social media usage and discussion of bitcoin between people and introducing the Wikipedia search data and then passing all this data as the parameter to model and training the model on these large number of parameters and then improving the overall accuracy of the model for prediction of bitcoin price of future.

References

1. Brownlee, J.: Machine learning mastery with python. Mach. Learn. Mastery **527**, 100–120 (2016)
2. Brownlee, J.: Deep Learning for Time Series Forecasting: Predict the Future with MLPs, CNNs and LSTMs in Python. Machine Learning Mastery (2018)
3. Brownlee, J.: Generative Adversarial Networks with Python: Deep Learning Generative Models for Image Synthesis and Image Translation. Machine Learning Mastery (2019)
4. Gandal, N., Hamrick, J.T., Moore, T., Oberman, T.: Price manipulation in the Bitcoin ecosystem. J. Monetary Econ. **95**, 86–96 (2018)
5. Iglesias de Ussel, J.: Bitcoin: a new way to understand payment systems (Doctoral dissertation, Massachusetts Institute of Technology) (2015)

6. Kristoufek, L.: What are the main drivers of the Bitcoin price? Evidence from wavelet coherence analysis. PLoS ONE **10**(4), (2015)
7. Madan, I., Saluja, S., Zhao, A.: Automated bitcoin trading via machine learning algorithms (2015). http://cs229.stanford.edu/proj2014/Isaac%20Madan,20
8. Mallqui, D.C., Fernandes, R.A.: Predicting the direction, maximum, minimum and closing prices of daily Bitcoin exchange rate using machine learning techniques. Appl. Soft Comput. **75**, 596–606 (2019)
9. Matta, M., Lunesu, I., Marchesi, M.: Bitcoin spread prediction using social and web search media. In: UMAP Workshops, pp. 1–10, June 2015
10. Mirtaheri, M., Abu-El-Haija, S., Morstatter, F., Steeg, G. V., Galstyan, A.: Identifying and analyzing cryptocurrency manipulations in social media. *arXiv preprint* arXiv:1902.03110 (2019)
11. Nakamoto, S., Bitcoin, A.: A peer-to-peer electronic cash system. Bitcoin (2008). https://bitcoin.org/bitcoin.pdf,4
12. Popper, N.: Digital Gold: The Untold Story of Bitcoin. Penguin, New York (2015)
13. Gupta, M., Sikarwar, T.S.: Modelling credit risk management and bank's profitability. Int. J. Electron. Bank. **2**(2), 170–183 (2020)
14. Memdani, L.: Demonetisation: a move towards cashless economy in India. Int. J. Electron. Bank. **2**(3), 205–211 (2020)

Structural Modeling of the Lean Supply Chain Enablers

Hemant Sharma[1(✉)] and Nagendra Sohani[2]

[1] Prestige Institute of Engineering Management and Research, Indore, India
[2] Institute of Engineering and Technology, Devi Ahliya Vishwavidyalaya, Indore, India

Abstract. Today the role of industry 4.0 plays a very important role in enhancing any supply chain network, as the Industry 4.0 supply chain uses advanced analytics and Big Data for information and visibility of product. Latest data information is available to support real-time decision-making and bring visibility to the entire supply chain, both within and outside the organizations. The main focus of the paper is to identify various Lean supply chain management enablers (LSCMEs) in automobile industry sector with the help of literature review and experts opinion. Thereafter with the help of (AHP) Analytical Hierarchy Process to find the rankings of the enablers by calculating their weightage. This paper has identified 28 key LSCMEs and developed an integrated model, which is very helpful to identify and classify the important enablers as per their ranking obtained and will help understanding and exploring the direct and indirect effects of each enabler on the LSCM implementation.

Keywords: Lean supply chain management · Lean supply chain enablers · AHP

1 Introduction

The role of analytics in industry 4.0 has become very vital today as nearly all supply chain managers rely on analytics for information and optimization of production as well as for the enhancement of the whole supply chain network The Industry 4.0 supply chain uses advanced analytics and Big Data which not only results in the lean focused efficient supply chain but also a supply chain capable of greater responsiveness to disruption, because of which a manufacturing companies can gain greater insight into their overall supply situation.

Lean practices and implementation has become very popular globally as its aims are to eliminate the wastes for the improvement of performance. This is the result of increasing competition in the global business, because of which many organizations are looking for ways to gain competitive advantage among their competitors. It is observed that the market competition is shifting very fast from product manufacturing to service sector like supply chain management, therefore supply chain optimization has become a key factor for their existence [47]. Another study suggests that selecting any supply chain strategy primarily depend upon the nature of demand [17]. Lean supply chain strategy is one of the most suitable for the product or market having stable demand [9].

A. M. A. Musleh Al-Sartawi et al. (Eds.): EAMMIS 2021, LNNS 239, pp. 168–180, 2021.
https://doi.org/10.1007/978-3-030-77246-8_17

To gain the competitive advantage now many researchers are progressively working for the implementation of lean concept in the supply chain [11, 31, 49]. Some argued about the lean implementation as it is the very basis of LSCM [2]. There are several case studies done previously on how implementations of lean in the supply chain have resulted in important improvements [14, 32, 48]. Lean is an evolving concept and fast gaining its popularity as one of the important supply chain management strategy [23]. Lean had emerged as most crucial and effective improvement technique which is adopted by many manufacturing as well as service sectors across the world over the past few decades for enhancing the production line as well as the SC [27]. The integrated approach of lean helps in eliminating the non-value added activities and also in reducing the defects [6]. Today the organizations are experiencing an intensive pressure for building up its productive future consistently by capitalizing the opportunities in the market [35]. One of the main challenges is to increase the integration with key suppliers and customers within the SC network and spread the lean principles and practices in every SC's knot [29].

In order that the SC strategy could be well implemented, the identification of the best strategy should be done before that [34]. So, it is very essential when the company starts developing its SC strategy, the very key success enablers to the successful implementation of SC strategy has to be identified. Also, there exists a strong need for the hierarchical structuring of the enablers that contribute towards the effective implementation of improved LSCM. Today there are several methodologies developed and applied by researchers in various disciplines for modeling the enablers, but the selection of the best methodology relies on the type of problem, its usefulness, and the form of outcome needed by the practitioners and researchers. The best-known methods among many are the Analytical Hierarchy Process (AHP), Analytic Network Process (ANP) and the interpretive structural modeling (ISM) [13]. The comparison among the above three methods were done and their feasibility of usage as per the environment requirement and need was also proposed [19]. The present study originates from the very aim to explore the rankings and the interactions among LSCEs and how strongly they influence the implementation of lean in the entire SC system. Hence, there exist a strong need of the study that not only explores the appropriate set of enablers and but also finds the interactions among them. So this study is an attempt to accomplish the above purpose. There are 35 lean enablers identified after the rigorous analysis of collected articles from the literature review. In order the strategy to be well implemented, it has to be appropriate with the practices of the company. Therefore in-depth interview and brainstorming sessions were conducted with the group of experts like the Director and the Operations Managers of selected industries, educationists and consultants to gain any information about if the 35 variables identified from the literature relevant with the need of the company. After finishing this step, then 7 out of 35 variables were removed because the experts stated that those are either irrelevant or repetitive.

2 Literature Review

This section provides a review of literature about the lean supply chain enablers for the successful lean supply chain management implementation. A literature review is conducted to address the above mentioned research objective. The definition of literature review is "a systematic, explicit, and reproducible design for identifying, evaluating, and interpreting the existing body of recorded documents" [15, 16]. A lean supply chain must allow a flow of information, goods, services and technology from suppliers to customers without waste [48].

The primary focus of lean supply chain principles is to identify various types of waste in the supply chain process. Waste is an activity that does not create any value to the customer or product. In supply chain management, wastes are created by improper information flow, material flow and money flow in the system [25]. Information and inventory have close relationship in the system and dependent on each other [7]. One of the main objectives of lean principles is to control inventory in the system. It is understood that the need of avoiding information waste in the lean supply chain is to bring down inventory levels of the organization. Information is considered to a major cause of wastes in the lean supply chain management system.

A conceptual frame work for lean supply chain implementation based on the literature review was developed [3]. A lean supply chain should act as per the customers' needs and demands thus encouraging all stakeholders of the supply chain to produce under a pull system rather than push system [25]. Thus it is effective and efficient to flow the information from the end customer to all tiers of the lean supply chain [46]. Some expert discussed a study using fuzzy association rule approach to evaluate the leanness of an Indian modular switches manufacturing organization [45]. And some presented a framework based on some important SCM operational constructs, like customer service, inventory management, flexibility, time to market, finance, ordering cycle time, quality, and market [8]. The application of JIT, suppliers and customers relationship in the lean supply chain management was also elaborated [12, 18]. Two most important strategies in the supply chain were explored [21]. The strategies which include "lean and agility" are termed as "generic" supply chain strategies. "Lean" works best under the conditions of high-volume, low-variety and predictable environments, whereas "agility" is highly needed in a less predictable environment where the demand for variety is high.

The understanding of the company's current context is important for the implementation of lean supply chain and organizations should adopt the practices and processes that are effective and efficient in their context [1, 4]. There are some findings that the research related to the lean supply chain practices implementation usually neglects the supply chain contexts [42]. An integrated framework based on fuzzy quality function deployment and analytical hierarchy process (AHP) to enhance leanness by analyzing enablers of lean supply chain linked with lean attributes [50]. Eight practices acts as pillars of lean SCM implementation which are information technology management, supplier management, logistics management, top management commitment, elimination of waste, JIT production, customer relationship management and continuous improvement [25]. Leagile SCM factors were analyzed and performed empirical analysis to identify leagile factors suitable for Indian manufacturing industry [39]. An integrated fuzzy QFD and TOPSIS approach for the enhancement of leanness in supply chain was presented [22]. There exist the findings in their research that the industries should adopt the concept of lean manufacturing in totality, not through the island approach for overall productivity improvements [11].

The interrelationship between the pillars was analyzed and constructs of the framework of LSCM in Indian manufacturing industries [39]. Six pillars of lean supply chain namely strategic management, logistics management, supplier management, marketing management, manufacturing management and collaboration management. Empirically investigation about the effect of a set of contextual variables (i.e. plant size, supply chain level, level of onshore suppliers and age of the lean manufacturing initiative) on the implementation of lean supply chain practices is done [43]. Investigation about the relationships among the implementation of LSC practices was also done and examined 27 lean supply chain practices from different sectors located in the Southern Brazil [44].

In the beginning, literature review was conducted in which articles related to "lean supply chain" are collected from well know SCOPUS and Web of Science database as per review protocol. After the rigorous analysis of collected articles lean supply chain enablers are identified. A group of experts is formed to collect their opinion which consists of academicians, practitioners and consultants. A total of 35 lean supply chain enablers were identified. Brainstorming sessions were conducted to finalized lean supply chain enablers. The objective of research study was discussed with the experts in the first meeting. All the identified lean supply chain enablers were discussed and similar attribute lean supply chain enablers grouped in to five groups namely strategic, managerial, operational, technological and socio-cultural in the second meeting. In the next meeting, the experts finalized the relevant lean supply chain enablers, discard the same meaning lean supply chain enablers and merged the same meaning lean supply chain enablers. Twenty eight lean supply chain enablers were finalized for the further research as shown in Table 1.

This study proposed AHP methodology to weight the enablers of lean supply chain, and finally to rank them which will be helpful for further analyzing their structural interrelationship.

Table 1. LSCEs with description

Code	LSCEs	Description
LSCE1	Top management commitment	The top management should provide essential infrastructure and training to the employees to implement operation strategy of the organization
LSCE2	Total quality management	TQM focuses on market requirements and satisfying the customer needs & requires the involvement of every employee
LSCE3	Performance measurement	By identifying performance metrics and applying performance evaluation methods, performance of lean supply chain organization can be studied which helps in improving the efficiency of the organization

(*continued*)

Table 1. (*continued*)

Code	LSCEs	Description
LSCE4	Long term planning	Long term planning addresses decisions that helps to determine overall strategic directions for the implementation of lean supply chain
LSCE5	Quality improvement	Quality improvement at every stage leads to successful implementation of lean supply chain
LSCE6	Supplier management	The suppliers play an important role in the implementation of lean supply chain management in the organization
LSCE7	Logistic management	Logistics management is one of the pillars of supply chain management to implement successful lean practices in the lean supply chainmanagement for any organization.
LSCE8	Inventory management	Inventory management means the right stock, at the right levels, in the right place, at the right time, and at the right cost
LSCE9	Customer relationship management	It is beneficial to interact with the potential customer to enhance the performance of lean supply chain
LSCE10	Knowledge and information management	Knowledge and information management helps to promote all kinds of innovation to improve the productivity of organization
LSCE11	Value chain management	Value chain management is key to optimizing business activities, operations and maximizing profit
LSCE12	Marketing management	Marketing management strategy has a significant impact on the customer satisfaction in a LSCM
LSCE13	Continuous improvement (Kaizen)	Continuous improvement initiates the activities that are improving the success rate and reducing the failure rate in the organizations
LSCE14	JIT production	Just-in-time (JIT) production also known as toyota production system (TPS) aims to reduce the time within the production system as well as response time from suppliers to customers
LSCE15	Eliminate wastes	Waste activities is to reduce setup time and defects in the manufacturing operations. Waste can be group into seven wastes: over production inventory, motion, over-processing, defects, transportation and waiting.

(*continued*)

Table 1. *(continued)*

Code	LSCEs	Description
LSCE16	Kanban system or pull system	Kanban plays a crucial role in pull scheduling, it is an information system connecting production and delivery in which upstream supplier does not produce until the downstream customer signals a need
LSCE17	Housekeeping (5S)	Housekeeping (5S) involves the principle of waste elimination in the workplace organization. 5S in the Japanese words namely seiri, seiton seiso, seiketsu, and shitsuke
LSCE18	Total productive maintenance	TPM is the team based maintenance process designed to maximize machine availability and performance of product quality
LSCE19	Reduction in lead time	Reduction in lead time is important enabler to improve efficiency of organization
LSCE20	Information Technology	The perspectives of information technology is essential to control and access information flow across the supply chain activities
LSCE21	Standardization of work	Work standardization refers to operational procedures on the shop floor that ensure customer satisfaction
LSCE22	Production Automation	Production automation results in reduced down times, enhance the productivity, improved quality, safety and energy savings in the lean supply chain
LSCE23	Efficient replenishment	Replenishment is based upon consumer demand and point of sale information
LSCE24	Coordination and collaboration among SC	Coordination and collaborative relationship among the stakeholders of the lean supply chain is essential for achieving its full benefits
LSCE25	Human resource training and education	Training and education related to specific work is important to build necessary skills in the employee and make them competent to do work
LSCE26	Employee participation and empowerment	Employee participation and empowerment involves ability to participate in the decision making in the implementation of lean supply chain
LSCE27	Change management (Cultural change)	The creation of a supportive organizational culture is an important platform for the implementation of lean supply chain management

3 Research Methodology

In this research, interviews with experts of industry, consultants as well as educationists was conducted with brainstorming sessions along with the synthesis of literature review, the key variables were identified and thereafter AHP methodology have been employed to achieve research objective. The methodologies and the results respectively are discussed in sequence.

3.1 Analytic Hierarchy Process (AHP)

"The foundation of the Analytic Hierarchy Process (AHP) is a set of axioms that carefully delimits the scope of the problem environment" [36]. The very base of it depends on the welldefined mathematical structure of consistent matrices and their associated eigenvector's ability to generate true or approximate weights [37]. The AHP methodology compares various criteria's, or alternatives with respect to a criterion, in a natural and in a pair wise mode. For doing this, the AHP uses a fundamental scale of absolute numbers that has already been proven in practice and also validated by physical and decision problem experiments. This fundamental scale has been shown to be a scale that captures individual preferences with respect to quantitative or qualitative attributes just as well or better than other scales [36, 37]. Then it converts the individual preferences into ratio scale weights that can be combined into a linear additive weight for each alternative. The resultant can be finally used to compare and rank the alternatives and, hence, support the decision maker in making a choice among alternatives. In this study, all the criteria have been rated from 1 to 9 versus all other criteria [20, 39]. Based on the ratings obtained through the brainstorming sessions with the experts, matrices are formed and the priorities are synthesized using the methodology of AHP.

The AHP decomposes a complex problem into a hierarchy, in which each level has particular characteristics. These elements, in turn, are further deconstructed into sub elements, thereby developing a hierarchical representation of the problem until at the lowest level.

AHP process consists of the following seven steps.

1. Defining an unstructured problem and determining its goal.
2. Structuring the hierarchy from the top (objectives from a decision-makers viewpoint) through intermediate levels (criteria on which subsequent levels depend) to the lowest level, which contains a list of alternatives.
3. Employ a pair-wise comparison approach. The pair-wise comparison matrix A, in which the element aij of the matrix is the relative importance of the ith factor with respect to the jth factor, could be calculated as

$$A = [a_{ij}] = \begin{bmatrix} 1 & a12 & \cdots & a1n \\ 1/a12 & 1 & \cdots & a2n \\ \vdots & \vdots & \vdots & \vdots \\ 1/a1n & 1/a2n & \cdots & 1 \end{bmatrix}$$

There are n(n − 1)/2 judgments require developing the set of matrices in step 3. Reciprocals are then automatically assigned to each pair-wise comparison, where n is the matrix size.

4. Hierarchical synthesis is then utilized to weight the eigenvectors according to weights of criteria. The sum is for all the weighted eigenvectors corresponding to those in the next lower hierarchy level.
5. Having made all pair-wise comparisons, consistency is id entified by using the Eigen value λ_{max}, to calculate the consistency index. The largest Eigen value, λ_{max}, will be

$$\lambda_{max} = \sum\nolimits_{ij}^{n} aij\frac{wj}{wi}$$

Where λ_{max} is the principal or largest Eigen value of positive real values in a judgment matrix; Wj is the weight of jth factor Wi is the weight of ith factor.

6. The consistency test in which each pair-wise comparison contains numerous decision elements for the consistency index (CI), which measures the entire consistency judgment for each comparison matrix and the hierarchy structure. The CI and consistency ratio (CR) is utilized to assess the consistency of the comparison matrix. The judgment consistency can be checked by taking the CR of CI with the appropriate value. The CR should not exceed a value 0.10 for its acceptance. For the CR value >0.10, the judgment matrix is inconsistent and to acquire a consistent matrix, judgments should be reviewed and improved. Hierarchy construction is the first step in the problem-solving process. The goal of an AHP decision is to find the rankings of all the enablers as per their weightage.

3.1.1 AHP Application and Calculations

Twenty eight enablers which are identified are grouped in five different categories namely strategic, managerial, operational, technological and socio-cultural which are the main criteria's and pair wise comparison matrix is formed by applying AHP the results of which are as shown in the Table 2. and their weights are found. The strategic criterion is having maximum weight as per the results and the socio-cultural criterion is having the least weight.

Then the pair wise comparison of sub criteria's is formed by applying AHP. The result of the sub criteria of strategic group shows that the code LSCE1 is having the

Table 2. Pair wise comparison matrix of the main criteria

	Strategic	Managerial	Operational	Technological	Socio-cultural	Weight
Strategic	1.00	3.00	3.00	5.00	5.00	0.44
Managerial	0.33	1.00	3.00	5.00	7.00	0.29
Operational	0.33	0.33	1.00	3.00	7.00	0.16
Technological	0.20	0.20	0.33	1.00	3.00	0.07
Socio – cultural	0.20	0.14	0.14	0.33	1.00	0.04

maximum weight and the code LSCE5 is having the minimum weight. After that the globalized weight is calculated and the final priorities of LSCEs are found out as shown in the Table 3.

Table 3. Final priorities of LSCEs

Main LSCEs	Weight	Sub criteria	Weight	Globalized weight	Ranking
Strategic LSCEs	0.44	LSCE1	0.58	0.2552	1
		LSCE2	0.2	0.088	3
		LSCE3	0.12	0.0528	6
		LSCE4	0.066	0.02904	10
		LSCE5	0.035	0.0154	16
Managerial LSCEs	0.29	LSCE6	0.42	0.1218	2
		LSCE7	0.22	0.0638	5
		LSCE8	0.15	0.0435	7
		LSCE9	0.09	0.0261	12
		LSCE10	0.06	0.0174	14
		LSCE11	0.04	0.0116	17
		LSCE12	0.02	0.0058	23
Operational LSCEs	0.16	LSCE13	0.45	0.072	4
		LSCE14	0.26	0.0416	8
		LSCE15	0.17	0.0272	11
		LSCE16	0.06	0.0096	19
		LSCE17	0.04	0.0064	21
		LSCE18	0.02	0.0032	25
Technological LSCEs	0.07	LSCE19	0.5	0.035	9
		LSCE20	0.24	0.0168	15
		LSCE21	0.13	0.0091	20
		LSCE22	0.09	0.0063	22
		LSCE23	0.04	0.0028	26
Socio cultural LSCEs	0.04	LSCE24	0.49	0.0196	13
		LSCE25	0.28	0.0112	18
		LSCE26	0.12	0.0048	24
		LSCE27	0.07	0.0028	27
		LSCE28	0.04	0.0016	28

4 Results

The main objective of this research is to study, examine and find the weight of identified enablers for their ranking by applying AHP methodology. The study identified 35 LSCEs by reviewing a number of research articles and discussion with experts. The present study shows the utilization of an innovative approach to the LSCM implementation in different organization. From brainstorming sessions and analysis the number of LSCEs are reduced from 35 to 28 for further analysis. This model has been developed on the basis of literature review and input from experts. The results elaborates that the Top management commitment is the top ranking enabler, Supplier management, Total quality management and Continuous improvement are ranked after that. These are the top enablers on which the organization should focus the most for lean implementation in supply chain. The second level enablers according to the decreasing ranking order are Logistic Management, Performance Management, Inventory Management, JIT Production, Reduction in lead time, Long term Planning and Eliminate Wastes. These enablers are intermediately on which the focus to be given after first level enablers. The third level enablers consists Customer Relationship Management, Coordination and Collaboration among SC, of Knowledge and Information Management, Information Technology, Quality Improvement, Value Chain Management, Human Resource Training and Education, Kanban System or Pull System and Standardization of Work which are arranged as per their ranking obtained. These enablers should be least focused by the organization for lean implementation. The rest of eight enablers which are having lowest ranking could be eliminated as these are having very low weightage.

5 Conclusion

In present scenario, almost all the organizations are facing high competition globally. As observed the quality has now become the most critical issue in the era of globalization. Therefore, the organizations are forced and bounded for exploring various areas in supply chain enhancement for delivering products at minimum costs for having a cutting edge advantage over their competitors. So, for achieving and maintaining quality at marginal cost; implementation of LSCM has now emerged to be the most optimum solution. To achieve this objective, it becomes very important and essential to enumerate various LSCEs and examine them very minutely. Literature reports various LSCEs which are significant for implementing LSCM in any organization. The present study identifies 28 key LSCEs by reviewing several research articles in peer-reviewed journals along with the brainstorming sessions with the group of experts consisting of academicians, practitioners and consultants.

The present study proposes a hierarchical model of LSCEs by employing AHP methodology that may be helpful for practitioners of any organization to classify and key out the significant LSCEs and to interpret the results and the effects of each LSCE on LSCM implementation. This study proposed the AHP approach to weight the enablers of lean supply chain which will be helpful in future for analyzing the structural interrelationship among enablers of lean supply chain and classify these enablers of lean supply chain based on dependence power and driving power.. The organization should have a main focus on the enablers with the highest priority followed by the enables that have less priority. This process will help to ensure improvement in overall performance in terms of efficiency and will be a great contribution in operations management as per the optimization of resources is concern, whereas to reduce the inputs and increase the output which is also termed as enhancing the productivity.

References

1. Achanga, P., Shehab, E., Roy, R., Nelder, G.: Critical success factors for lean implementation within SMEs. J. Manufact. Technol. Manage. **17**(4), 460–471 (2006)
2. Agus, A., Hajinoor, M.S.: Lean production supply chain management as driver towards enhancing product quality and business performance: case study of manufacturing companies in Malaysia. Int. J. Qual. Reliab. Manage. **29**(1), 92–121 (2012)
3. Anand, G., Kodali, R.: A conceptual framework for lean supply chain and its implementation. Int. J. Value Chain Manage. **2**(3), 313–357 (2008)
4. Anvari, A., Zulkifli, N., Yusuff, R.M., Hojjati, S.M.H., Ismail, Y.: A proposed dynamic model for a lean roadmap. Afr. J. Bus. Manage. **5**(16), 6727–6737 (2011)
5. Arif-Uz-Zaman, K., Nazmul Ahsan, A.M.M.: Lean supply chain performance measurement. Int. J. Prod. Perform. Manage. **63**(5), 588–612 (2014)
6. Assarlind, M., Aaboen, L.: Forces affecting one Lean Six Sigma adoption process. Int. J. Lean Six Sigma **5**(3), 324–340 (2013)
7. Baum, D.: Creating the lean enterprise. Profit, 44 November. http://www.oracle.com/oramag/profit/04-nov/p44lean.html. Accessed 14 May 2014
8. Blos, M., Hoeflich, S., Miyagi, P.: A general supply chain continuity management framework. Procedia Comput. Sci. **55**, 1160–1164 (2015)
9. Christopher, M., Towill, R.D.: Developing market specific supply chain strategies. Int. J. Logist. Manage. **13**(1), 1–14 (2002)
10. Cudney, E., Elrod, C.: A comparative analysis of integrating lean concepts into supply chain management in manufacturing and service industries. Int. J. Lean Six Sigma **2**(1), 5–22 (2011)
11. Dave, Y., Sohani, N.: Improving productivity through Lean practices in central India-based manufacturing industries. Int. J. Lean Six Sigma **10**(2), 601–621 (2019)
12. Dong, Y., Carter, C., Dresner, M.: JIT purchasing and performance: an exploratory analysis of buyer and supplier perspectives. J. Oper. Manage. **19**(4), 471–483 (2001)
13. Drohomeretski, E., Gouvea da Costa, S.E., Pinheiro de Lima, E., Garbuio, P.A.D.R.: Lean, Six Sigma and Lean Six Sigma: an analysis based on operations strategy. Int. J. Prod. Res. **52**(3), 804–824 (2014)
14. Eriksson P.E.: Improving construction supply chain collaboration and performance: a lean construction pilot project. Int. J. Suppl. Chain Manag. **15**(5), 394–403 (2010)
15. Fink, D.: Guidelines for the successful adoption of information technology in small and medium enterprises. Int. J. Inf. Manage. **18**(4), 243–253 (1998)
16. Fink, A.: Conducting Research Literature Reviews: From Paper to the Internet. Sage, Thousand Oaks (1998)
17. Fisher, M.L.: What is the right supply chain for your product? Harv. Bus. Rev. **75**(2), 105–116 (1997)
18. Green, K., Inman, R., Birou, L., Whitten, D.: Total JIT (T-JIT) and its impact on supply chain competency and organizational performance. Int. J. Prod. Econ. **147**, 125–135 (2014)
19. Gorane, S.J., Kant, R.: Modelling the SCM implementation barriers: an integrated ISM-fuzzy MICMAC approach. J. Model. Manage. **10**(2), 158–178 (2015)

20. Hafeez, K., Zhang, Y., Malak, N.: Determining key capabilities of a firm using analytical hierarchy process. Int. J. Prod. Econ. **76**(1), 39–51 (2002)
21. Hallgren, M., Olhager, J.: Lean and agile supply chain: external and internal drivers and performance outcomes. Int. J. Oper. Prod. Manage. **29**(10), 976–999 (2009)
22. Haq, A.N., Boddu, V.: An integrated fuzzy QFD and TOPSIS approach to enhance leanness in supply chain. Int. J. Bus. Perform. Supply Chain Model. **7**(2), 171–188 (2015)
23. Hines, P., Holweg, M., Rich, N.: Learning to evolve: a review of contemporary lean thinking. Int. J. Oper. Prod. Manage. **24**(10), 994–1011 (2004)
24. Jasti, N., Kodali, R.: Lean production: literature review and trends. Int. J. Prod. Res. **53**(3), 1–19 (2014)
25. Jasti, N., Kodali, R.: A critical review of lean supply chain management frameworks: proposed framework. Prod. Plan. Control **26**(13), 1051–1068 (2015)
26. Kumar, A., Ozdamar, L., Ning Zhang, C.: Supply chain redesign in the healthcare industry of Singapore. Supply Chain Manage. Int. J. **13**(2), 95–103 (2008)
27. Maleyeff, J.: The continuing evolution of Lean Six Sigma. TQM J. **24**(6), 542–555 (2012)
28. Marodin, G., Frank, A., Tortorella, G., Saurin, T.: Contextual factors and lean production implementation in the Brazilian automotive supply chain. Supply Chain Manage. Int. J. **21**(4), 417–432 (2016)
29. Martínez-Jurado. P, Moyano-Fuentes. J.: Lean management and sustainability: a literature review. J. Clean. Prod. 1–17 (2013, In Press)
30. Netland, T., Schloetzer, J., Ferdows, K.: Implementing lean: the effect of takt time. In: Proceedings of Euroma Conference 2015, Neuchatel, 28–30 June (2015)
31. Oliver, N.: JIT: issues and items for research agenda. Int. J. Phys. Distrib. Logist. Manage. **20**(7), 3–11 (1993)
32. Perez, C., Castro, R., Simons, D., Gimenez, G.: Development of lean supply chains: a case study of the Catalan pork sector. Supply Chain Manage. Int. J. **15**(1), 55–68 (2010)
33. Piercy, N., Rich, N.: Lean transformation in the pure service environment: the case of the call service centre. Int. J. Oper. Prod. Manage. **29**(1), 54–76 (2009)
34. Qrunfleh, S., Tarafdar, M.: Lean and agile supply chain strategies and supply chain responsiveness: the role of strategic supplier partnership and postponement. Supply Chain Manage. Int. J. **6**, 571–582 (2013)
35. Psomas, E.L., Antony, J.: The effectiveness of the ISO 9001 quality management system in service companies. Int. J. Prod. Res. **53**(7), 2089–2099 (2015)
36. Saaty, T.L.: Axiomatic foundation of the analytic hierarchy process. Manage. Sci. **32**, 841–855 (1986)
37. Saaty, T.: How to make a decision: the analytic hierarchy process. Interfaces **24**, 19–43 (1994)
38. Saaty, T.L.: Fundamentals of Decision Making and Priority Theory with the Analytic Hierarchy Process, vol. 6. RWS Publications, Pittsburgh (2000)
39. Soni, G., Kodali, R.: Evaluating reliability and validity of lean, agile and leagile supply chain constructs in Indian manufacturing industry. Prod. Plan. Control **23**(10–11), 864–884 (2012)
40. Taylor, D.H.: Strategic considerations in the development of lean agri-food supply chains: a case study of the UK pork sector. Supply Chain Manage. Int. J. **11**(3), 271–280 (2006)
41. Taylor, D.: Parallel incremental transformation strategy: an approach to the development of lean supply chains. Int. J. Logist. Res. Appl. **2**(3), 305–323 (1999)
42. Theagarajan, S., Manohar, H.: Lean management practices to improve supply chain performance of leather footwear industry. In: International Conference on Industrial Engineering and Operations Management, Dubai, 3–5 March (2015)
43. Tortorella, G., Miorando, R., Tlapa, D.: Implementation of lean supply chain: an empirical research on the effect of context. TQM J. **29**(4), 610–623 (2017)
44. Tortorella, G., Giglio, R., Fettermmann, D.C., Tlapa, D.: Lean supply chain practices: an exploratory study on their relationship. Int. J. Logist. Manage. **29**(3), 1049–1076 (2018)

45. Vinodh, S., Joy, D.: Structural equation modelling of lean manufacturing practices. Int. J. Prod. Res. **50**(6), 1598–1607 (2011)
46. Vitasek, K., Manrodt, K.B., Abbott, J.: What makes a lean supply chain? Supply Chain Manage. Rev. **9**(7), 39–45 (2005)
47. Vonderembse, M.A., Uppal, M., Huang, S.H., Dismukes, J.P.: De-signing supply chains: towards theory development. Int. J. Prod. Econ. **100**(2), 223–238 (2006)
48. Wee, H., Wu, S.: Lean supply chain and its effect on product cost and quality: a case study on ford motor company. Supply Chain Manage. Int. J. **14**(5), 335–341 (2009)
49. Womack, J.P., Jones, D.T.: From lean production to the lean enterprise. Harv. Bus. Rev. **72**(2), 93–103 (1994)
50. Zarei, M., Fakhrzad, M.B., Paghaleh, M.J.: Food supply chain leanness using a developed QFD model. J. Food Eng. **102**(1), 25–33 (2011)

Increasing Productivity and Quality with IoT Technologies in Industrial Treatment Systems

Adem Kayar[1(✉)], Fatih Öztürk[2], and Hakan Ceyhan[1]

[1] Istanbul Commerce University, Istanbul, Turkey
{akayar,hakan.ceyhan}@mfd.com.tr
[2] Istanbul Medeniyet University, Istanbul, Turkey
fatih.ozturk@medeniyet.edu.tr

Abstract. Today, enterprises that produce in almost every sector are looking for ways to increase efficiency and quality in their production processes. In the globalizing world economy, the conditions of competition are increasing day by day, it is desired to keep under control by monitoring the cost of each product produced. We are witnessing that digital applications, which have entered our lives rapidly in recent years, are now included in the production processes. These new technologies, where digitalization in production is defined as Industry 4.0, can be seen in different application areas in all manufacturing enterprises . Industry 4.0, which has many components such as Internet of Things (IoT), big data, cyber security, augmented reality and autonomous robots, aims to increase efficiency and quality in production processes. IoT devices are one of these new technologies that are encountered in Industry 4.0 applications and used in product life cycle management, especially in mass production enterprises. IoT devices that instantly read production data and transfer them to the cloud platform have been used in many different industrial applications in recent years. At all stages of production, IoT devices that instantly read the data and transfer it to the cloud platform, can use different data such as malfunctions and warnings, number of products produced, pressure, temperature, waste and energy. While these data are important criteria for the traceability of production, they are used to increase productivity and quality. In this study, while analyzing the use of IoT technologies in Industry 4.0 applications, a real application in which industrial treatment systems are monitored with IoT technologies will be mentioned.

Keywords: Smart factory · Smart manufacturing · Industry 4.0 · IoT · Production systems

1 Introduction

Today, we witness that digital technologies are used in different applications in many parts of our lives. Digitalization is heading towards a completely different process with the rapid development in information technologies in recent years, especially with the introduction of smart phones into our lives. Day does not pass so that a new application we

A. M. A. Musleh Al-Sartawi et al. (Eds.): EAMMIS 2021, LNNS 239, pp. 181–188, 2021.
https://doi.org/10.1007/978-3-030-77246-8_18

use in our daily lives does not enter our lives. Especially during the Covid-19 pandemic process we are in, we can see how much digital applications make our lives easier.

The importance of digitalization and Industry 4.0 was once again clearly understood during the Covid-19 outbreak. It has been observed that performing operations at different stages of production with minimum human use is very important for the sustainability of production under such epidemic conditions [1]. Nowadays, when digitalization has entered our lives so intensely, digitalization of manufacturing or digital transformation in production are encountered with industrial applications at the same intensity.

Digitalization and Industry 4.0 are met with interest in almost every sector. When Industry 4.0 was first mentioned at the Hannover fair held in Germany in 2011, no one thought the process would develop so quickly. We see that rapidly developing digital transformation applications are used in production lines all over the world. It is seen that iot, big data, autonomous robots, simulation, system integration, cyber security and augmented reality (AR) applications, which are considered as the main components of Industry 4.0, are preferred for different purposes in different processes in production lines. Digital transformation applications play an important role, especially in production lines where continuous production is carried out [2].

Changes in social and economic systems that emerged with the rapid development of new technologies led to a global revolution known as the 4th Industrial Revolution. Industry 4.0 manifests itself with the rapid development of the Internet of Things (IoT), artificial intelligence and machine learning; It radically changes human lifestyles and contributes significantly to the development of new production methods [3].

In order to develop applications suitable for the smart production concept, it is necessary to read and save the production data in the database from the moment the production starts and interface designs that show meaningful values with this information. The electrical connections of the sensors and devices on the machine must be made correctly in order to collect and record machine data. Electrically wired sensors and devices are connected to the digital or analog input modules of the Programmable Logic Controller (PLC) system. In some applications, the transfer of sensors and devices on the machine to the PLC control system can be done via distributed I/O. In particular, sensor groups located in different locations on the machine are read by the PLC control system using industrial communication protocols after electrical connections are made to the distributed sensor modules [4].

The transfer of production data read by the PLC device to the cloud platform is done by industrial IoT devices. IoT devices, which instantly read the data realized in production and transfer them to the cloud platform, can write malfunctions and warning information, other detailed information about production (product numbers, pressure, temperature, waste and energy, etc.) data to the database on the cloud platform. These data are very important for the traceability of production. With the algorithms to be created using this production data, malfunctions can be tracked and the continuity of production can be ensured without allowing malfunctions with predictive maintenance software. Thanks to these smart algorithms, while productivity increases in manufacturing, the quality of the manufactured products also increases. In this study, the use of IoT devices in Industry 4.0 applications is explained. By talking about a real application where industrial treatment devices are monitored with IoT devices, the advantages of smart production are emphasized.

2 Digital Technologies

Many different technologies are used during digital transformation applications in manufacturing enterprises. These technologies consist of many different applications designed for the management and control of manufacturing processes.

2.1 PLC

PLCs are often the main part of automated systems in industry [5]. They are very efficient and reliable in applications involving sequential control and synchronization of processes and auxiliary elements in the manufacturing, chemical and process industries [6].

Currently, there are many PLC manufacturers in the world. Most companies produce a variety of models that vary in size, cost, and complexity to meet the needs of specific applications. These producers are listed in Table 1 [7].

Table 1. List of PLC manufacturer in the global market

No	Manufacturer	No	Manufacturer
1	Siemens	9	Panasonic
2	ABB	10	Idec
3	Schneider (Modicon)	11	Keyence
4	Rockwell (Allen-Bradley	12	Toshiba
5	Mitsubishi	13	Fuji
6	GE-Fanuc	14	Beckhoff
7	Omron	15	Bosch Rexroth
8	Koyo	16	Rockwell/Allen-Bradley

2.2 Supervisory Control and Data Collection (SCADA)

Historically, Supervisory Control and Data Acquisition (SCADA) systems are the core of industry automation. SCADA is the system responsible for monitoring, controlling and coordination of business procedures in all devices of the industry.

Upstream SCADA acts as an interface for human control and Manufacturing Execution Systems (MES), facilitating decision making and control of the industrial process. The general structure of the SCADA system includes multiple sensors and actuators controlled and transmitted via Remote Terminal Units (RTU). It also includes Human

Machine Interfaces (HMIs) so that human actions can be included in the system based on production and safety requirements, allowing control of the process [8].

People with many years of machine and process knowledge are specialized in solving the malfunctions in machinery and production lines. After the failure of the production lines, the data saved in the database can be used both for monitoring the process and for predictive maintenance. These data read from the PLC system and stored in the database give us the operating performance of the machine, malfunctions and current status of the devices on the machine [9].

2.3 IoT

With the emergence of the fourth industrial revolution (Industry 4.0), production systems have been transformed into Semiconductor Industry. Internet of Things (IoT) and other emerging technologies play an important role in this transformation. Replacing manufacturing companies in computer integrated manufacturing (CIM), the so-called smart parts-based manufacturing system addresses these concerns well. The smart part carries operating instructions for the manufacturer or machine [10].

Over the past decade, significant growth has been observed in sustainable supply chain management (SSCM) and industry 4.0 research. Big data analytics, Internet of Things (IoT), deep learning, simulation and prediction techniques; It is used to meet the needs of contemporary supply chains, such as flexibility, increased productivity, less waste, optimization of resources inside and outside the factory, and more [11].

2.4 Cloud Computing

Computing resources offered as a service over the Internet are called cloud computing. It follows the optional property and payment model. Today, the cloud offers all kinds of services needed for individuals and industries. Initially, the cloud provides three types of service models: Infrastructure level (IaaS), Development environment level or Platform as a Service (PaaS) and software level or Software as a Service (SaaS). Data center resources are mapped to cloud user role through hypervisor. Hypervisor is a virtual machine manager used to perform VM (Virtual Machine) creation and maintenance with high service availability [12].

In impersonation or masked attacks, an attacker illegally uses system resources after breaching a legitimate user's account. This class of attacks becomes a major challenge in cloud systems because the massive amount of system resources, alternative deployment models, and the distribution of user controls and activities across several virtual machines with different environments greatly increase the complexity of detection [13].

Cloud computing has been widely adopted over the past decade thanks to its many advantages such as reduced deployment cost, physical resource sharing between VMs (Virtual Machines) using virtualization, and flexible and flexible on-demand services. However, in addition to the traditional threats present in every computing environment, cloud computing has opened the door to many additional threats as customers share the same physical resources due to the physical colocation paradigm [14].

Step 1: Connect and Monitor: In the first step, a connection to the control system such as PLC in the field is provided in order to read the machine data. This type of connection can be in a different protocol depending on the features of the devices used. The data received from the control system are instantly recorded in the database. The performance of the machine or system is followed by monitoring these recorded data.

Step 2: Analyze and Predict: In the second step, analyzes are made about the read data. Faults that occur or sensor data out of limits are detected. Users are notified when a machine runs out of specification. This step allows users to search, filter and share linked asset data. It improves the customer experience by offering new digital services for products. It allows users to export data for further analysis than required. Analyzes real-time asset data to maximize performance and health. It visualizes workflows and data analysis for quick response. Predicts presence errors to prevent unplanned downtime. Defines and calculates key performance indicators (KPIs) of healthy and performing assets. It informs users if KPI thresholds are exceeded, triggering actions such as email notifications. It empowers all users to use the system with its ability to easily access, analyze and interpret large data sets. By sharing visualizations internally and externally, it speeds up the data-based decision-making process.

Predicts sequences of events that lead to field faults and fail before they occur. Plan preventive maintenance before the equipment fails. Predicts field performance to improve product quality. It increases customer satisfaction by suggesting fixes to customers before performance drops It increases operational efficiency and lowers costs.

Step 3: Digitalize and Transform: In the third step, it contributes to the monitoring and management of all processes of the enterprise by working integrated with ERP software. It means digitalization of all processes, from warehouse management to suppliers and even customer feedback. This step automates insights to drive product and process optimization from suppliers.

3 Application Example of IoT Technologies

Industry 4.0 aims to enable objects to communicate with each other and with people by monitoring physical processes with cyber-physical systems in modular smart factories, thus making decentralized decisions, i.e. smart production [15].

IoT devices are the most important of the most used new technological products to realize smart factory or smart production applications. In the example here, work has been done with the IoT2040 device of Siemens [16].

3.1 Current Status of the Industrial Water Treatment Device

Periodic maintenance of industrial purification devices produced in the factory and sold to customers is required. At the same time, the sand and activated carbon filters used in the treatment devices should change according to their usage time. It was not possible to monitor the industrial treatment devices sold by the manufacturer company to its customers and to make the necessary intervention in current situations. However, when the customer had a problem with the industrial treatment device, he contacted the manufacturer and requested technical support.

This situation prevented the manufacturer from planning and detecting problems with the use of their devices. On the other hand, the replacement of the sand and active carbon filters used in the device was made by the customer. The customer could change these filters by procuring them from other companies or sub-industry. This situation both affected the working performance of the purification device and meant that the manufacturer could not sell sand and active carbon filters, which are consumable materials. It was decided to continuously monitor the treatment system with the IoT device in order to eliminate these negative situations and to monitor the treatment device.

3.2 Topology Structure of the System

The control of the industrial water purification device is made by Siemens Simatic S7-1200 PLC CPU 1214 DC control system. The entire working scenario of the device is programmed in the PLC control system according to the algorithm requested by the manufacturer. In the PLC control system, the fault information about the treatment device, the occupancy rates of the sand and active carbon filters, the flow rate of the treated water, the temperature and pressure of the water are instantly read from the sensors and processed according to the designed algorithm. These data are read from the PLC with the IoT device and transmitted to the industrial cloud platform, Siemens MindSphere platform.

3.3 Result of the IoT Solution

As a result of instant monitoring of the industrial treatment system using IoT technologies, it is seen that both the manufacturer and the customer provide great advantages. When looked at by the customer, the advantages can be listed as follows.

1. Industrial water treatment system is monitored instantly by the manufacturer.
2. The continuity of the system is ensured by giving feedback before the system fails.
3. Malfunctions in the system are solved in a much shorter time by the technical team of the manufacturer.

When looked at by the manufacturer, the advantages can be listed as follows.

1. The performance of the industrial water treatment system produced is instantly monitored by the company's technical team.
2. Feedback is made to the product development department according to the frequency of malfunctions in the system and the malfunction type. Thus, the manufacturer company can develop products for the water treatment systems it will produce.
3. By providing feedback before the system fails, it can ensure continuity of the system without stopping, increasing customer satisfaction.
4. When the time comes for the replacement of sand and activated carbon filters, it generates income from both service fee and consumable sales by directing the technical support team.

4 Conclusion

We can clearly see that digitalization and Industry 4.0 applications provide enormous advantages to production enterprises. Thanks to IoT technologies, which are frequently used in Industry 4.0 applications, the traceability of production is possible. The data obtained with the traceability of the production can be analyzed and solutions that will increase the performance and quality of the processes can be created [17, 18].

The application of IoT technologies and cloud computing in production processes has started to gain more importance, especially with the emergence of 5G technologies. In IoT and cloud computing, the fact that the internet speed could not catch up with the process speed could create problems in monitoring production processes. Both the rapid development in IoT and Cloud computing technologies and the increase in internet connection speeds with new technologies will enable the IoT applications to be used more in production in the coming days.

Digitalization in production and Industry 4.0 applications appear almost every day with different applications in different sectors [15].

In recent years, we have witnessed that digitalization is developing very rapidly. This process will continue in full swing from now on in all sectors. Businesses that can apply this process to their own production processes as soon as possible will gain great advantage, and businesses that miss time may face the danger of extinction.

References

1. Kayar, A.: Pandemi Sonrası Yeni Dünya Düzeninde Teknoloji Yönetimi ve İnsani Dijitalizasyon, Pandemi Sonrası Endüstri 4.0 ve Dijitalizasyon: Covid-19 ve Otomotiv Sektör Analizi, pp. 537–552. Hiperyayın (2020)
2. Kayar, A., Öztürk, F., Kayacan, Ö.: Fast fault solving methods in smart manufacturing lines with augmented reality applications. In: Recent Advances in Data Science and Business Analytics (y-BIS 2019 Conference), Istanbul/Turkey, 25–28 September 2019, pp. 182–187 (2019)
3. Arfi, W.B., Nasr, I.B., Khvatova, T., Zaied, Y.B.: Understanding acceptance of eHealthcare by IoT natives and IoT immigrants: An integrated model of UTAUT, perceived risk, and financial cost. Technol. Forecast. Soc. Change **163**, (2021)
4. Kayar, A., Öztürk, F.: Machine data collection and analysis in industry 4.0 applications. In: 2th International Conference on Life and Engıneering Sciences (ICOLES 2019), Istanbul/Turkey, 27–29 June 2019, pp. 317–322 (2019)
5. Bayındır, R., Çetinceviz, Y.: A water pumping control system with a programmable logic controller (PLC) and industrial wireless modules for industrial plants - an experimental setup. ISA Trans. **50**(2), 321–328 (2011)
6. Saad, N., Arrofiq, M.: A PLC-based modified-fuzzy controller for PWM-driven induction motor drive with constant V/Hz ratio control. Robot. Comput. Integrat. Manufact. **28**(2), 95–112 (2012)
7. Alphonsus, E.R., Abdullah, M.O.: A review on the applications of programmable logic controllers (PLCs). Renew. Sustain. Energy Rev. **60**, 1185–1205 (2016)
8. Fernández, P., de Fernández, M.Á.F., Candás, J.L., Arboleya, P.A.: An IoT opensource platform for photovoltaic plants supervision. Int. J. Electr. Power Energy Syst. **125**, 106540 (2021)

9. Kayar, A., Öztürk, F.: Predictive maintenance in Industry 4.0 applications. In: 2th International Conference on Life and Engineering Sciences (ICOLES 2019), Istanbul/Turkey, 27–29 June 2019, p. 16 (2019)
10. Chen, Y.-Q., Zhou, B., Zhang, M., Chen, C.-M.: Using IoT technology for computer-integrated manufacturing systems in the semiconductor industry. Appl. Soft Comput. J. **89**, 106065 (2020)
11. Mastos, T.D., Nizamis, A., Vafeiadis, T., Alexopoulos, N., Ntinas, C., Gkortzis, D., Papadopoulos, A., Ioannidis, D., Tzovaras, D.: Industry 4.0 sustainable supply chains: an application of an IoT enabled scrap metal management solution. J. Clean. Prod. **269**, 122377 (2020)
12. Narayana, K.E., Jayashree, K.: Survey on cross virtual machine side channel attack detection and properties of cloud computing as sustainable material. In: Materials Today: Proceedings (2020). in press
13. Kholidy, H.A.: Detecting impersonation attacks in cloud computing environments using a centric user profiling approach. Future Gen. Comput. Syst. **117**, 299–320 (2021)
14. Demigha, O., Larguet, R.: Hardware-based solutions for trusted cloud computing. Comput. Secur. **103**, (2021)
15. Kayar, A., Öztürk, F. : Smart reactor production monitoring system-an industry 4.0 application in the chemical industry. J. Ind. Policy Technol. Manage. **3**(2), 157–173 (2020)
16. Siemens, A.G.: Siemens industry mall web site. https://mall.industry.siemens.com/mall/tr/tr/Catalog/Products/10321262?tree=CatlogTree#More%20information. Accessed 31 Jan 2021
17. Öztürk, F.: Qualität, Effizienzsteigerung und integrierte Managementsystemen im türkischen Eisenbahnsektor. Soc. Nat. Sci. J. **8**(2), 14–19 (2014). Print ISSN 1804-4158, Online ISSN 1804-9710
18. Celik, N., Öztürk, F.: The upcoming issues of industry 4.0 on occupational health and safety specialized on turkey example. Int. J. Econ. Bus. Manage. Res. **1**(5), 236–256 (2017). ISSN 2456-7760

Intelligent Community Detection: Comparative Study (COVID19 Dataset)

Jenan Moosa[1](✉), Wasan Awad[1], and Tatiana Kalganova[2]

[1] Ahlia University, Manama, Bahrain
{jhasan,wawad}@ahlia.edu.bh
[2] Brunel University, London, UK
tatiana.kalganova@brunel.ac.uk

Abstract. Community detection is an important tool for analyzing networks; it can help us understand the structures and functional characteristics. Network communities represent a principled way of organizing real-world networks into densely connected groups of nodes, whereas a community is a cluster of nodes that are strongly connected to each other in a subnetwork than to the rest of the network. This has remarkable results in various fields, e.g., social science, bibliometrics, marketing and recommendations, biology etc. This research proposes a dataset based on COVID19 distribution. The network dataset is formed by tracing the transmission of the virus among the world countries. This experiment demonstrates the spread of COVID19 and its mutated strain. Several algorithms such as Girvan Newman, Greedy, Louvain, Clustering, and Label Propagation are implemented on the proposed dataset, and the results are evaluated using the modularity measure. The proposed COVID 19 dataset demonstrated the properties and dynamic behavior similar to Zachry's Karate dataset.

Keywords: Community detection · Modularity · COVID19 · Contact tracing

1 Introduction

Living in an era of technology, a dramatic amount of data is being produced. Having all this data will not only help in performing the task it was created for, but it can also facilitate in executing new functions and outcomes. As raw data and its' relationships can be transformed into meaningful information and structures when applying the appropriate mining tools.

An epochal type of data is being excessively generated since December 2019, when a novel virus named COVID-19 emerged, and while the first case was discovered in Wuhan, China, it did not take long for the disease to travel across the globe and infect every continent (except Antarctica) [1], causing widespread infections and deaths due to its contagious characteristics and no medically proven treatment [2]. The COVID-19 pandemic has been termed as the most consequential global crisis since the World Wars. Because of the rapid prevalence of this virus, health organizations all over the world tend to track and store all data related to this pandemic. This includes the contact tracing,

A. M. A. Musleh Al-Sartawi et al. (Eds.): EAMMIS 2021, LNNS 239, pp. 189–196, 2021.
https://doi.org/10.1007/978-3-030-77246-8_19

number of cases, number of deaths, etc. The availability of rich textual data from various online sources can be used to understand the growth, nature and spread of COVID-19 [3].

According to World Health Organization (WHO) [4], contact tracing is the process of identifying, assessing, and managing people who have been exposed to a disease to prevent onward transmission. When systematically applied, contact tracing will break the chains of transmission of an infectious disease and is thus an essential public health tool for controlling infectious disease outbreaks. When contact tracing data is compiled it can be represented by a network, and hence structured into a graph, which can be analyzed using graph mining techniques.

In this paper, the main concern is to use the contact tracing data to create a new network dataset, which is motivated by the open-source efforts by the health organizations. The rest of this paper is organized as follows. Section 2 reviews the basic concepts of com-munity detection and its most common evaluation measure, in addition to a few algorithms applied to solve this problem. Section 3 presents the proposed dataset. Experimental result with discussion is given in Sect. 4 and conclusion in Sect. 5.

2 Literature Review

A network is consisted of connected communities that are created by the communicated individuals, and a community is a densely connected group of nodes that is also connected to the rest of the network [1]. In different research fields, a community is also referred to as group, cluster, cohesive subgroup, or module. The main goal of community detection is to detect and discover the strongly connected members as compared to the rest of the members, and thereby form the people of similar taste, choices and preferences in a virtual cluster or a community [2]. The extracted communities in the community detection procedure can considerably analyse building ontology for semantic web, detecting topics in tagging systems, analysing the behaviour of a community, and personalized search and recommendations. Bedi and Sharma (2016) stated that any network can be outlined in a graph [3] and hence detection of communities can be utilized. Community detection problem is getting more attention, as different algorithms and techniques have been proposed so far, which includes traditional algorithms, evolutionary algorithms etc. Various algorithms and approaches were tested; such as heuristic [4, 5] hierarchical clustering [6, 7], spectral clustering [8], label propagation [9]. In addition to evolutionary algorithms [10, 11], neural networks [12], etc.

To evaluate the performance of a certain algorithm, it is implemented on a predefined dataset. The datasets are basically networks represented by graphs. The social networks used were common among most research studies. Real-world datasets which are open-source (e.g. Zachry's Karate Club, Bottlenose Dolphins, American College Football and American Political Books) were used in a distinct number of researches [11, 13–15]. However, Synthetic networks such Lancichinetti–Fortunato–Radicchi [16] and Girvan-Newman [17] were also benchmarked in a number of researches [11, 13]. Several performance evaluation measures were also proposed and used, e.g. (Normalized Mutual Information NMI [18], modularity [19], time complexity and Omega index [20]).

2.1 Evaluation Measure

Modularity (normally denoted as Q) was proposed in [21]; they noted that in practical situations, an algorithm will be used to find communities within a network for which communities are not known ahead of time, and this raises a question: how do we know if the detected communities are good ones? And how good the structure found is? Moreover, the output of the algorithm is in the form of dendogram (which represents a hierarchy of possible community divisions), so we need to determine the best divisions of the network. Those questions led Newman & Girvan to designate Modularity which is a measure of the quality of a particular division of a network.

$$Q = \sum_{c=1}^{n} \left[\frac{L_c}{m} - \left(\frac{k_c}{2m} \right)^2 \right] [25] \tag{1}$$

Where the sum iterates over all communities c, m is the number of edges, L_c is the number of intra-community links for community c, k_c is the sum of degrees of the nodes in community c. If the number of within-community edges is no better than random, we will get $Q = 0$. Values approaching $Q = 1$, which is the maximum, indicate strong community structure. In practice, values for such networks typically fall in the range from about 0.3 to 0.7. In general, this unnormalized measure will not reach a value of 1, even on a perfectly mixed network [19].

2.2 Community Detection Algorithms

This section lists a number of the most used algorithms to solve the community detection problems, that will later be implemented and evaluated on two different datasets.

Louvain Best partition [6] a hierarchical clustering algorithm, that recursively merges communities into a single node and executes the modularity clustering on the condensed graphs.

Greedy [5] is a heuristic method that finds communities in graph using Clauset-Newman-Moore greedy modularity maximization. This method begins with each node in its own community and joins the pair of communities that most increases modularity until no such pair exists.

Newman 2008 [7] clusters the network into several modules using modularity maximization by spectral methods. Supports directed and undirected networks. Edge weights are ignored.

Spectral Clustering [8] is a type of unsupervised community discovery algorithm. The number of communities k should be given in advance.

Kernighan–Lin [4] is a heuristic method that partitions a network into two sets by iteratively swapping pairs of nodes to reduce the edge cut between the two sets.

Asynchronous label propagation [9] initializes each node with a unique label, it repeatedly sets the label of a node to be the label that appears most frequently among that nodes' neighbors. The algorithm is asynchronous because each node is updated without waiting for updates on the remaining nodes.

3 Proposed Dataset

As any network can be outlined in a graph, and the graph is composed of a set of nodes which can be individuals or entities, and edges that represent the connections and interactions between the nodes [22]. This research creates and studies COIVID19 datasets, to examine how the virus is spreading among nodes. According to CDC (Centers for Disease Control and Prevention, US), the use of digital contact tracing tools may help with certain case investigation and contact tracing activities but will not replace the need for a large public health workforce. As the complete clinical picture of COVID-19 is not fully known, the virus can be spread before symptoms occur or when no symptoms are present. Therefore, case investigation and contact tracing activities need to be studied. This type of analysis helps us understand the transmission of the virus, the distribution of countries based on their infection. It might also help predict where the virus might strike next, or how a mutated virus will behave, and therefore take the precautions needed.

The dataset proposed in this paper is based on the spread of virus between countries. An open-source contact tracing data was used to follow the spread of virus from January to March 2020, between the countries worldwide, which started in China and expanded to other countries. In this dataset, each country is represented by a node, and an edge is used when a country has a contact infected person from another country. (World Countries, Fig. 1) The dataset consists of 36 nodes (countries), and 75 edges.

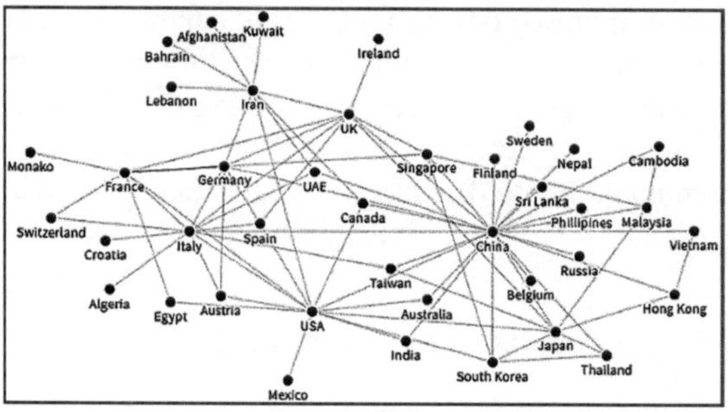

Fig. 1. World countries proposed network

4 Results and Discussion

Data science, defined broadly, will play a central role in the global response to the COVID-19 pandemic [23]. All the scientific efforts necessitate that the data brought to service for the analysis should be open source to promote the extension, validation, and collaboration of the work in the fight against the global pandemic. The open-source tracing data is utilized to create a novel network dataset, that is concerned with the spread of COVID19. And therefore, dissect this dataset using various community detection algorithms to discover the clusters within the network. Subsequently, the results are evaluated using Modularity value which is one of the most common evaluation measures in this field.

4.1 Experiments

The aim of this experiment is to compare the different community detection algorithms on two datasets. The first dataset is (Zachry's Karate Club) [24], it was used in the literature and its structure is somewhat close to the proposed one. It is a social network of friendships between 34 members of a karate club at a US university in the 1970s. And the second one is the dataset proposed in this paper (World Countries), it consists of 36 nodes and 75 edges. Upon detecting the communities within these two networks, the results will be tested in terms of number of communities detected (k) and modularity value; the latter is calculated to measure the quality of a disjoint partition of a network.

Experiment A: Investigation of Literature Dataset (Zachry's Karate Club Dataset)
This section presents the simulation results for the Zachry's Karate club dataset (as an example of commonly used datasets) tested on community detection algorithms Table 1.

No	Algorithm	k	Modularity value
1	Louvain	4	0.4155982905982906
2	Greedy	3	0.3806706114398416
3	GN 2008	4	0.4188034188034188
4	Spectral Clustering Pre-defined (k)	3	0.399079552925705
		4	0.41116042077580445
		5	0.3901216305062452
5	Kernighan–Lin	2	0.3717948717948704
6	ALP	5	0.39439513477975024

Table 1. Karate club dataset results

The algorithms were already tested in previous research papers, the main idea of this experiment is to compare these existing examined results with the results of the proposed dataset generates. It is remarkable that the highest modularity measure was achieved by Girvan Newman algorithm, followed by spectral clustering when the number of detected communities was set to 4. Furthermore, the lowest modularity measure was achieved by Kernighan–Lin algorithm, although most algorithms gave convergent values.

Experiment B: Investigation of Proposed Dataset (COVID 19 World Countries)
On the other hand, Table 2 sums up the studies regarding testing the algorithms explained in the literature on the World Countries designed dataset.

No	Algorithm	k	Modularity value
1	Louvain	4	0.34450328707085465
2	Greedy	3	0.3333637691745792
3	GN 2008	4	0.33245069393718046
4	Spectral Clustering -Pre-defined (k)	3	0.3328159240321386
		4	0.3149196493791074
		5	0.32286340394448354
5	Kernighan–Lin	2	0.2534696859021175
6	ALP	6	0.3048758217677133

Table 2. World countries dataset results

As can be seen in Table 2, the modularity values for the proposed dataset are likewise close, except for Kernighan–Lin which seems to give the lowest value, it also gave the lowest with karate dataset. While the highest modularity value was achieved by Louvain, which was not the case in experiment A. Louvain algorithm randomly order all nodes in the network, then, one by one, it will remove and insert each node in a different community until no significant increase in modularity is verified [25]. The modularity measure achieved by the proposed dataset, are comparatively close to the measure produced in the Zachry's Karate dataset. For spectral clustering algorithm, the number of communities should be set in advance, so the common numbers observed from other algorithms (3, 4 and 5) were selected and applied, and the modularity gave its best result when the number of communities was 3. This algorithm also achieved promising results on both datasets.

In general, it is intelligible that the countries are distributed geographically. However, as can be seen in the visual representation (Figs. 2, 3, 4, 5, 6 and 7), although some algorithms generate the same number of communities, the nodes formation in each community is distinct.

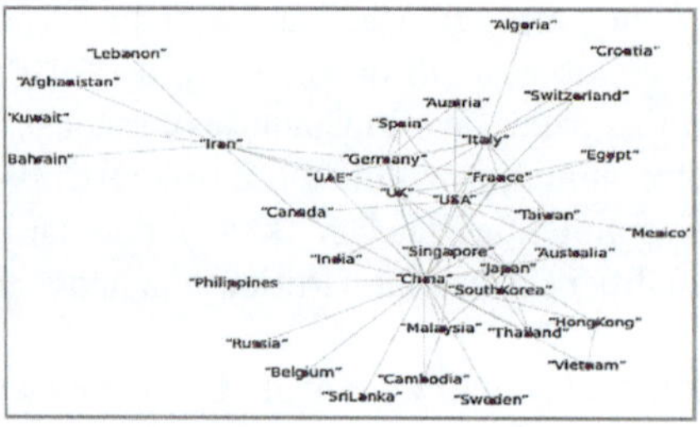

Fig. 2. Louvain-world countries Communities k = 4

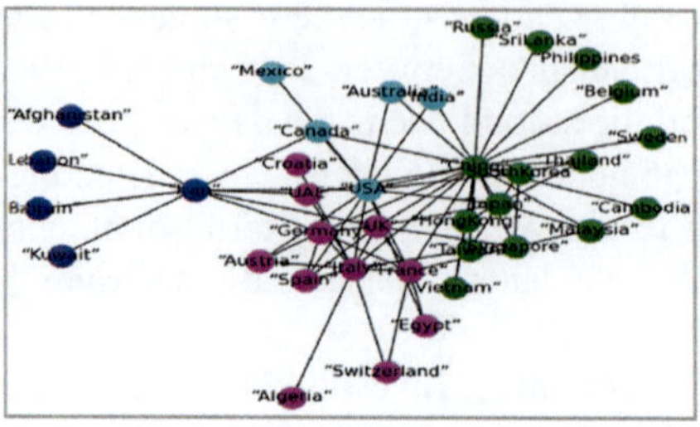

Fig. 3. GN-world countries Communities k = 4

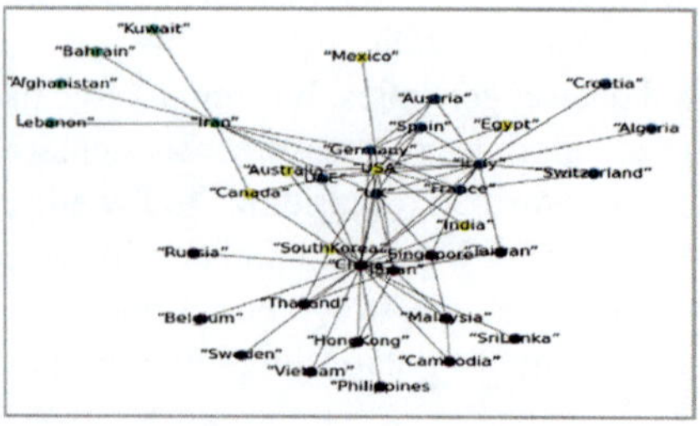

Fig. 4. SC world countries communities k = 4

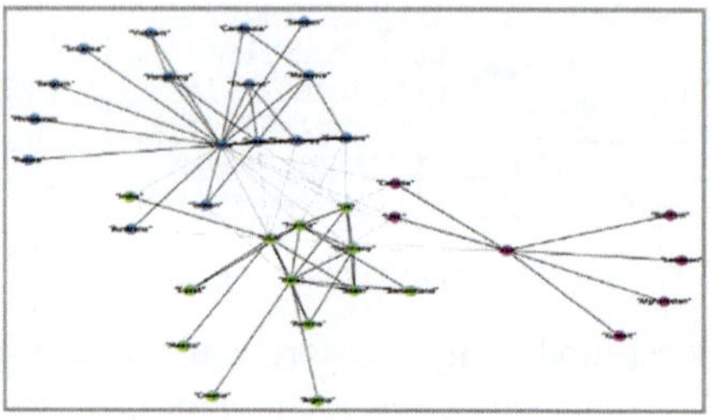

Fig. 5. Greedy world countries communities k = 3

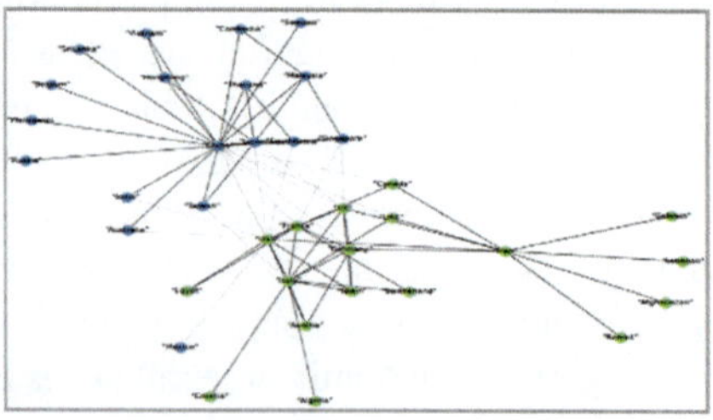

Fig. 6. Kernighan-Lin world countries communities k = 2

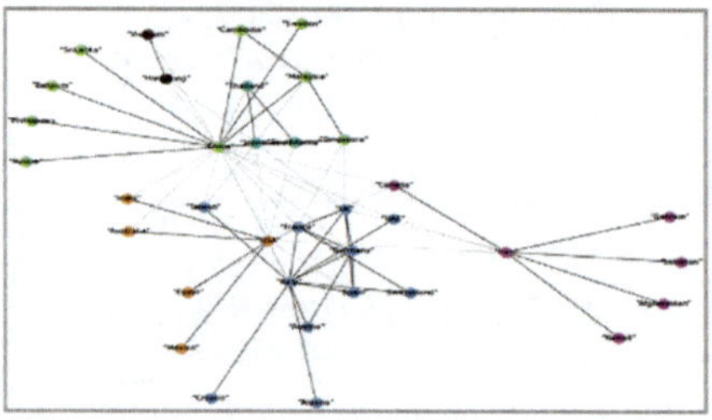

Fig. 7. ALP world countries communities k = 6

For example, Louvain in Fig. 2, classifies "Canada" in a community with ["Iran", "UAE", "Lebanon", "Kuwait", "Bahrain", "Afghanistan"] whereas in Girvan Newman (Fig. 3), "Canada" does not belong to this community even though the other six countries do. And in Spectral clustering when k = 4 (Fig. 4), the community that contains the majority of these countries did not include "Canada" neither "UAE".

Another interesting finding is that "China" and "Iran" are acting as nodes in the network, they both consistently have the focus in the subnetwork, except in Kernighan-Lin (Fig. 6), as there are only two communities. Additionally, countries Europian countries "Italy, Spain, Germany, France" are more difficult to be classified and separated, their community is consistent in all algorithms. It is also explicit that some countries belonged to the same clusters among all algorithms, such as [France, Italy, Germany, Spain, Austria, Algeria, Switzerland, Croatia, UK].

In fact, the exact community was generated by Louvain, GN, and Spectral Clustering, unless Louvain's also contained Egypt, and Spectral Clustering contained UAE. Another common community includes [China, Singapore, Malaysia, Cambodia, Sri-Lanka, Russia, Philippines, Sweden, and Belgium]. Furthermore, Label Propagation algorithm would have produced only 4 communities and its' results would have been very close to Louvain's if it had merged the East Asian countries in one community instead of three.

5 Conclusion and Future Work

This paper proposes a novel dataset based on the coronavirus contact tracing within the world countries. Since it has become a global concern, the records are utilized and transformed into a network, where each country is represented by a node, and their relationships modelled by edges. Analysis of diverse algorithm was carried out on the newly generated dataset, for comparison purpose in terms of numbers of detected communities and modularity performance measure. Louvain algorithm tend to score the highest modularity value. The average number of detected communities is 4, and the average modularity value is 0.317. Analysis of communities generated by algorithms have demonstrated the common features of communities identified in COVID 19 dataset that are not always generated by geographical locations.

The main strength among all algorithms was the consistency in countries' allocation, regardless the number of detected communities. The consistency of the results demonstrate that the proposed dataset was created with balanced number of communities, and the results were comparable with the literature dataset Zachry's Karate Club.

As future work, research need to be conducted on how to improve the accuracy and increase the modularity performance measure value of community detection in such networks (i.e., social networks), by designing a hybrid intelligent community detection technique.

References

1. Harenberg, S., et al.: Community detection in large-scale networks: a survey and empirical evaluation **6** (2014). https://doi.org/10.1002/wics.1319
2. Verma, A., Bharadwaj, K.K.: Identifying community structure in a multi-relational network employing non-negative tensor factorization and GA k-means clustering **7**, 1–22 (2017). https://doi.org/10.1002/widm.1196
3. Bedi, P., Sharma, C.: Community detection in social networks, vol. 6, pp. 115–135 (2016). https://doi.org/10.1002/widm.1178
4. Campari, R.: A general optimization technique for high quality community detection in complex networks, pp. 1–19 (2014)
5. Clauset, A., Newman, M.E.J., Moore, C.: Finding community structure in very large networks, pp. 1–6 (2004)

6. Lu, H., Halappanavar, M., Kalyanaraman, A.: Parallel heuristics for scalable community detection. Parallel Comput. **47**, 19–37 (2015). https://doi.org/10.1016/j.parco.2015.03.003
7. Leicht, E.A., Newman, M.E.J.: Community structure in directed networks, pp. 1–5 (2008)
8. Von Luxburg, U.: A tutorial on spectral clustering, pp. 1–32 (2007)
9. Raghavan, U.N., Kumara, S.: Near linear time algorithm to detect community structures in large-scale networks **1**, pp. 1–12 (2007)
10. Tian, Y., Yang, S., Zhang, X., Member, S.: An evolutionary multiobjective optimization based fuzzy. IEEE Trans. Fuzzy Syst. **PP**(c), 1 (2019). https://doi.org/10.1109/tfuzz.2019.2945241
11. Cheng, F., Cui, T., Su, Y., Niu, Y., Zhang, X.: A local information based multi-objective evolutionary algorithm for community detection in complex networks. Appl. Soft Comput. J. **69**, 357–367 (2018). https://doi.org/10.1016/j.asoc.2018.04.037
12. Bruna, J.: Community detection with graph neural networks community detection with graph neural networks, May 2017
13. Bilal, S., Abdelouahab, M.: Evolutionary algorithm and modularity for detecting communities in networks. Phys. A **473**, 89–96 (2017). https://doi.org/10.1016/j.physa.2017.01.018
14. Rahimi, S., Abdollahpouri, A., Moradi, P.: A multi-objective particle swarm optimization algorithm for community detection in complex networks. Swarm Evol. Comput. **39**, 297–309 (2018). https://doi.org/10.1016/j.swevo.2017.10.009
15. Yuanyuan, M., Xiyu, L.: Quantum inspired evolutionary algorithm for community detection in complex networks. Phys. Lett. A **382**(34), 2305–2312 (2018). https://doi.org/10.1016/j.physleta.2018.05.044
16. Lancichinetti, A., Fortunato, S.: Community detection algorithms: a comparative analysis, pp. 1–12 (2010)
17. Girvan, M., Newman, M.E.J.: Community structure in social and biological networks, vol. 99, no. 12 (2002)
18. Danon, L., Albert, D., Duch, J.: Comparing community structure identification, vol. 09008 (2005). https://doi.org/10.1088/1742-5468/2005/09/p09008
19. Newman, M.E.J.: Modularity and community structure in networks, vol. 103, no. 23, pp. 8577–8582 (2006)
20. Xie, J., Szymanski, B.K.: Detection in social networks (2012)
21. Newman, M.E.J., Girvan, M.: Finding and evaluating community structure in networks. Phys. Rev. E Stat. Nonlinear Soft Matter Phys. **69**(2), 1–16 (2004). https://doi.org/10.1103/physreve.69.026113
22. Bedi, P., Sharma, C.: Community detection in social networks. Wiley Interdiscip. Rev. Data Min. Knowl. Discov. **6**(3), 115–135 (2016). https://doi.org/10.1002/widm.1178
23. Usman, M., Iqbal, W., Mary, Q., Qadir, J.: Leveraging data science to combat COVID-19: a comprehensive review (2020). https://doi.org/10.13140/rg.2.2.12685.28644/4
24. Zachary, W.W., Zachary, W.W.: Flow Model for Conflict An Information Fission in Small Groups. J. Anthropol. Res. **33**(4), 452–473 (1977)
25. Blondel, V.D., Guillaume, J., Lefebvre, E.: Fast unfolding of communities in large networks, pp. 1–12 (2008)

An Exploration into the Posture of Robo-Advisor Globally

Ismail Canoz(✉)

Istanbul Medeniyet University, Istanbul, Turkey
ismail.canoz@medeniyet.edu.tr

Abstract. One of the critical outputs of financial technology is robo-advisor providing asset management services. The study aims to present the current global view of robo-advisors, based on their characteristics of being disruptive innovation. This study also discusses robo-advisor's future panorama. In more detail, the study will compare their effectiveness in asset management based on both countries and companies. The information reported in this study is based on a known data provider. However, it obtained some of that information from secondary data. The results of the research have been reported under four headings by evaluating these data: (1) global awareness, (2) country-based comparison, (3) company-based comparison, and (4) global market outlook. The results indicate that the number of robo-advisor users, assets under management amount, and the penetration rate increased even in 2020, as covered by the Covid-19 pandemic. Conversely, assets under management amount per person decreased.

Keywords: Fintech · Disruptive innovation · Robo-advisor · Wealth management

1 Introduction

Robots and Artificial Intelligence (AI) are already transforming all types of industries. They could replace nearly half of existing jobs in the next 20 years, as the penetration rate of automated technology is increasing at approximately 20% per year. This transformation brought about by the technological revolution threatens all established principles both in the economy and in the labor market [2].

The industries where technological transformation has shown its impact concretely are financial institutions. Recently, financial technology (fintech) is the most emphasized issue by banks, other financial institutions, and financial start-up companies. The term fintech is beyond e-banking and consumer digitalization. It focuses on improving innovative technology tools to meet current and potential customers' financial needs and demands [5]. However, while fintech offers benefits for customers, it also threatens some professions disruptively. One of fintech's disruptive innovations that can reduce the labor force is Robo-Advisor (RA), which provides asset and asset management services. Supportively, many fintech followers believe that Robo-Advisors (RAs) are fully automated machines that make investment decisions without any human interaction and

A. M. A. Musleh Al-Sartawi et al. (Eds.): EAMMIS 2021, LNNS 239, pp. 197–205, 2021.
https://doi.org/10.1007/978-3-030-77246-8_20

can fully substitute professionals. They underline the dangers and destructive effects of excessive automation and claim that unmanned automated services push specialized human advisors into a kind of Luddite [12].

RAs, designed as a low-cost alternative to specialists, automatically create and manage investors' portfolios online. As traditional financial institutions such as large brokerage firms, asset management companies, and banks begin to offer their own RA services, they have grown in popularity in recent years [1].

The purpose of this text is to provide a discussion of the financial services offered by RAs and to demonstrate the current positions of some RAs in asset management. In addition, it is discussed how countries adapt to RA. The text is structured as follows: first, I will discuss RA's conceptual and theoretical background. Then, I will present their global awareness, country-based asset under management (AuM) comparisons, company-based AuM, and yield comparisons and market view results. The last section is a conclusion.

2 Conceptual Background

The traditional investment managers are experts investing on behalf of a third party, and they intend to offer customers the maximum return. Asset management services are provided by specialist financial institutions and are based on the long-term management of private individuals' assets. Individual investors' assets transferred to managing asset management companies become an investment portfolio. Companies provide portfolio management services to their customers by following a specific investment strategy [8].

While AI is changing people's living standards in almost every field, such as travel, health, education, and communication, it has deeply affected wealth management in recent years. It has made it possible to evolve from individual financial advisors to unmanned robotic systems following specific investment strategies. Some asset management companies have adopted AI-based services for clients to receive investment advice whenever they want. These services are fast accessible, cheaper, transparent, and impartial. Since algorithm-driven software provides the consultancy services just like experts, they are called RA. Also, a personal financial advisor may or may not assist them while customers benefit from RA services. If so, this is a kind of hybrid consulting.

An automated digital program or online portal representing RA bases its portfolio management on algorithm trading. Thanks to RAs, investors can get online investment advice and efficiently and quickly manage their portfolios without much human intervention. "These are no longer just robots, but really comprehensive personal finance platforms that contain additional products and tools and have the ability to talk to a human consultant," says Ken Schapiro, founder of Backend Benchmarking, a digital advisory firm in the US [9].

The history of RAs has begun ever since the US-based global crisis. Asset management company "Betterment" in the US launched the first RA in 2008. These innovative programs were running on an online interface, and their purpose was to rebalance investor assets within target-date funds. The fund's rebalancing is to make the fund more conservative as the target date approaches. Rebalance is done only when there is a skew in a portfolio's allocations due to market changes. The strategy monitored by RA is essentially passive, "buy and hold" [11]. Modern Portfolio Theory continues to be the primary framework used in them, according to a study analyzing RAs worldwide. In fact, the current trend is to advance and expand this framework rather than implement and evolve entirely new methods [4].

RA's goal is to interact with customers digitally to collect their personal information and manage their investments cheaply. Personal information often includes their risk preferences, assets, revenues, liabilities, and investment goals. RA asks customers to respond to them to create an online account. RA offers reasonable investment options in asset allocation and diversification based on customer data with computer algorithms. These options typically include low-cost mutual funds and exchange-traded funds (ETF). RA invests in the client's portfolio based on the proposed asset allocation, which the client can often change. Also, some of them harvest tax losses on taxable portfolios [6].

3 Theoretical Background

In order to understand mainstream innovation theory, it is necessary to absorb what technology means for businesses. Technology refers to how an enterprise transforms its information and data, human labor, or economic capital into more valuable products or services. Thus, digital advice, automatic portfolio rebalancing, and Goal-Based Investment workflows can all be considered technology from a business's perspective. New technologies can lead to changes in both the production of companies and customers' access to products. In this respect, technology is a process that develops over time, both internally and externally [12].

On the other hand, innovation refers to any change in existing technology used by a company. The effect of this change can be either disruptive or growth-sustaining. Sustainable innovation relates to improvements in product performance and strives for businesses to reduce their production costs or increase their output prices. Thus, product quality increases, competition decreases, and trade margins rise. However, disruptive innovation creates revolutionary products. These are generally cheaper, simpler, or easier to use. Simultaneously, it is more appropriate for them to address new customers or respond to existing customers' unique needs. For this reason, disruptive innovation can cause to exhibit the poor performance of the existing substitute product over time. However, none should forget that a product's disruption does not happen immediately overnight.

RA is admitted to disruptive innovation for a variety of reasons. It has advantages such as being less expensive than traditional financial advice, easy accessibility, being attractive to new customers, and replying to existing customers' new needs. A study in the US examines whether RAs reduce investors' demands for financial service providers' human financial advice. The results point to a robust negative relationship between using a RA and getting advice from a financial expert [7]. This situation is evidence that RA is disruptive innovation. In addition, RAs are an alternative for investors concerned about potential conflicts that may result from human financial advice.

4 Global Awareness of the Robo-Advisor

Recently, wealth management companies have begun to meet a new generation of customers. Their differential features compared to traditional clients are that they are open to technology, educated, want to have active and continuous control over their investments, and care about information from multiple sources rather than individual financial advisors [4]. Statista, a world-renowned and influential data provider, asked the following question to participants aged 18–64 in various countries: *"Have you ever used a RA for finance issues and investments?"* I also tabulate it [13–23].

The survey output revealed striking findings for some countries. Especially in Russia, 90% of respondents have an opinion about RA, surprisingly. Moreover, 25% of the participants benefit from RA. This finding is an excellent sign that the participants follow fintech closely. Although the usage rate is lower than in Russia, there is a high awareness of RA in Canada and the USA. Of the selected countries, Finland has the lowest RA usage at 3%, and 22% of respondents do not even know what RA is. Besides, with a few exceptions, it is possible to say that European countries' results are close.

Table 1. Robo-advisor usage of some countries in 2020

Countries	Yes	No, but I could imagine consulting a RA	No, I do not consider consulting a RA	Do not know	Participants
Spain	7	33	45	15	2.093
Austria	3	22	60	16	2.087
Italy	5	30	48	16	2.100
Germany	6	27	57	13	2.094
France	8	21	58	13	2.090
UK	6	25	57	12	2.093
Finland	3	18	57	22	3.115
Russia	25	38	27	10	2.095
Australia	8	21	58	12	1.050
US	7	27	57	10	4.173
Canada	7	27	56	10	2.092

Source: Statista Global Consumer Survey

5 Countries-Based Comparison of the Robo-Advisor

No matter how global the world is and the borders disappear, technology development is different even in developed countries. In parallel, it is normal for some technologies to become widespread, primarily in producing countries. Similarly, I guess that advances on RAs are more innovative in the US, although it has not yet fully come of age. I made

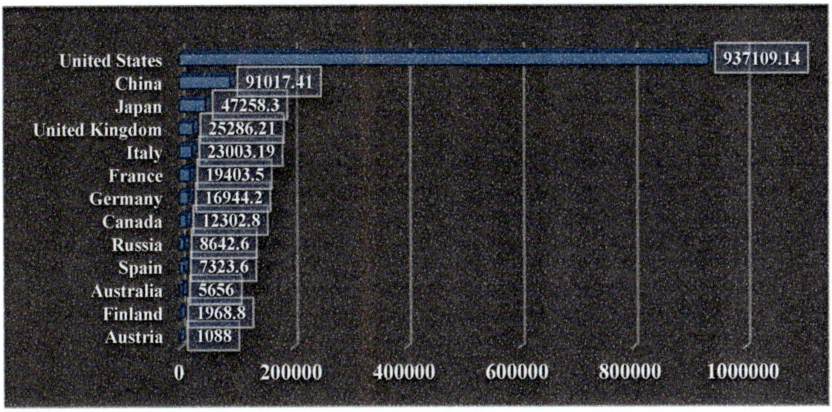

Fig. 1. **Asset under Management of Robo-Advisor (in a million US$). Source:** Statista (Forecast adjusted for the expected impact of COVID-19), January 2021. **Note:** It is graphed by the author.

this inference by comparing the AuM amount of RA, which belongs to some countries, in the database I used. Figure 1 below seems to prove my claim [10].

According to Fig. 1, the amount of AuM controlled by RAs in the USA is almost the entire world. This situation is an incredible statistic. Moreover, these are the most recent figures I could reach. I would like to state that the first five countries are not selected randomly; these are really the top five countries in terms of size. In Table 1, awareness of RA in Russia is high, but the amount of AuM is not that high. I think the main reason for this is that RA allows low investment accounts. Thus, there can be a large number of users even with minimal quantities. Perhaps the lack of trust in these systems yet pushes the investor to open accounts in amounts that may be at minimum risk. However, if I could find country-based user data, I would have reached a more detailed finding.

6 Company-Based Comparison of Robo-Advisor

Companies that provide RA services and strive to offer innovative products, especially in fintech, compete to deliver this service to more people and increase their AuM volumes. Figure 2 compares the value of assets some RAs manage [24].

On the other hand, Fig. 3 compares the returns of some RA [3]. Due to the Covid-19 pandemic, the turmoil in the markets in the first half of 2020, and severe volatility in some financial instruments undoubtedly forced portfolio managers. The situation is not much different in RAs. The returns of some selected RAs show that their investors have lost in this dire process. For example, in the first half of the year, there was a loss of approximately 3.8% in the S&P500 index.

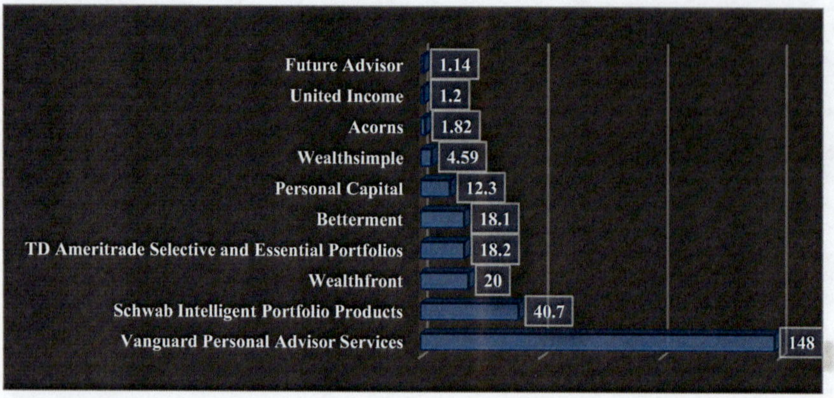

Fig. 2. AuM of Selected Robo-Advisors Globally in March 2020 (in billion US dollars). Source: Statista Research Department (Various sources (Backend Benchmarking); CrunchBase; PitchBook). **Note:** It is graphed by the author.

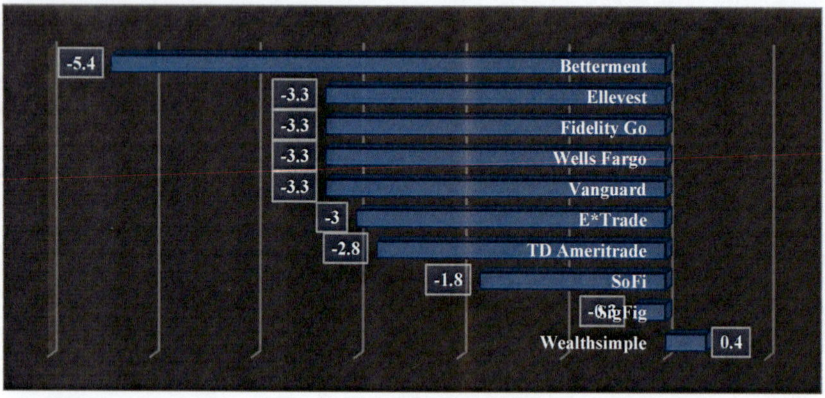

Fig. 3 Portfolio Returns of Selected Robo-Advisors Worldwide the First Half of 2020. Source: Statista (Various sources (Backend Benchmarking); Barron's). **Note:** It is graphed by the author.

7 Global Market Outlook of Robo-Advisors

RAs have only been around for more than ten years in our lives. The market outlook of RAs has been increasing over the years despite the virus that has affected the whole world since the beginning of 2020. This finding may indicate that it is expanding its destructiveness.

The same data provider I used before shows that RAs manage assets worth $987.494.140, 00 in 2020. Despite the pandemic, I am talking about an asset value that increased by approximately 20% compared to the previous year. Likewise, there is a severe increase in the number of global investors using it. However, as the number of users increases, the amount of AuM per person decreases. In addition, the increase in penetration rate is a concrete indicator of global developments in RAs [10] (Fig. 4).

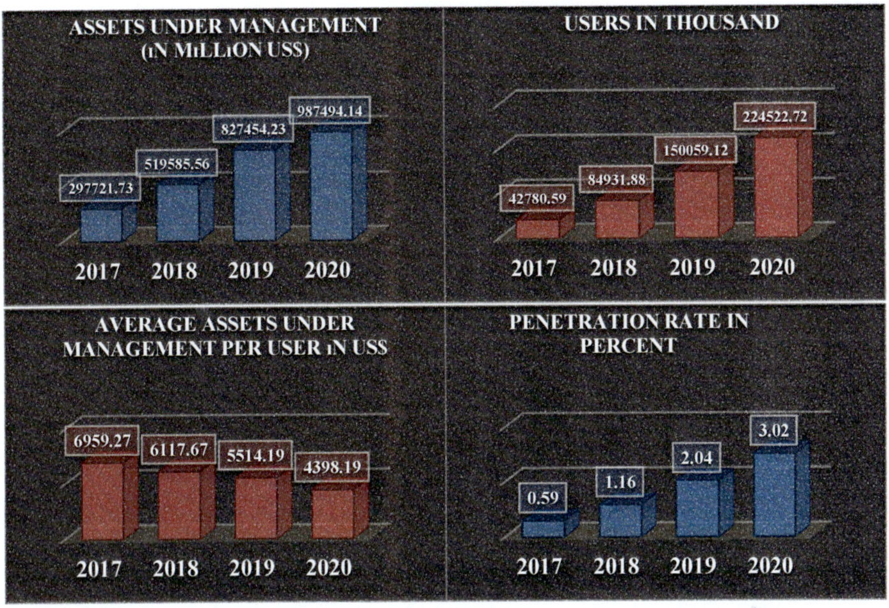

Fig. 4. The Market Outlook of Robo-Advisors Worldwide between 2017 and 2020. Source: Statista. **Note:** They are graphed by the author.

8 Conclusion

Despite being a disruptive innovation in the fintech space, the RA industry is still at an early stage. Hence, its possible effects on asset markets and financial systems are still waiting for recent studies. In the end, we will see whether RAs will be better for investors than human advisors in general. Due to their low commissions and low account balances, they can provide financial advice to young generations who cannot afford traditional financial advisors' investment advice for now. For this group, RAs seem clearly seen as a better option than human financial advisors. Furthermore, the older generation customers, who are increasingly open to technology and, therefore, expect and demand digital investment services offered by major fintech developers, contribute to the demographic change that has led to the increase in the use of fintech.

On the other hand, there is much less chance of a conflict of interest in a RA than a specialist. In the future, we can expect their advice to become more complex due to individual differences in risk preference and improvements in asset management strategies. The current trend towards hybrid consultants involving a RA working in partnership with a human consultant could possibly minimize their disruptive innovation. This type of hybrid consulting service can charge lower fees than conventional advisors and offer clients the opportunity to speak to a financial advisor.

References

1. Abraham, F., Schmukler, S.L., Tessada, J.: Robo-advisors: investing through machines. World Bank Research and Policy Briefs (2019)

2. Acemoglu, D., Restrepo, P.: Robots and jobs: evidence from US labor markets. J. Polit. Econ. **128**(6), 2188–2244 (2020)
3. Barron's: Portfolio returns of selected robo-advisors worldwide in the 1st half of 2020 [Graph]. In Statista, 31 July 2020. https://0-www-statista-com.seyhan.library.boun.edu.tr/statistics/755599/performance-of-selected-robo-advisors/. Accessed 27 Jan 2021
4. Beketov, M., Lehmann, K., Wittke, M.: Robo advisors: quantitative methods inside the robots. J. Asset Manage. **19**(6), 363–370 (2018)
5. Belanche, D., Casaló, L.V., Flavián, C.: Artificial Intelligence in FinTech: understanding robo-advisors adoption among customers. Ind. Manage. Data Syst. (2019)
6. Berger, R.: 7 Robo Advisers that Make Investing Effortless, Forbes (2015). http://www.forbes.com/sites/robertberger/2015/02/05/7-robo-advisors-that-makeinvesting-effortless/#75eaae1f7e48. Accessed 26 Jan 2021
7. Brenner, L., Meyll, T.: Robo-advisors: a substitute for human financial advice? J. Behav. Exp. Finance **25,** 100275 (2020)
8. Kordela, D.: Robo-advisors in asset management–the experience from Germany. Eur. J. Serv. Manage. **28**, 151–157 (2018)
9. Max, S.: Robo-Advisors Thrived Amid the Covid-19 Volatility. Here's Barron's Latest Robo Ranking, Barron's (2020). https://www.barrons.com/articles/the-best-robo-advisors-for-2020-according-to-barrons-exclusive-ranking-51596243960. Accessed 27 Jan 2021
10. Robo-Advisors - worldwide (n.d.): https://0-www-statista-com.seyhan.library.boun.edu.tr/outlook/337/100/robo-advisors/worldwide. Accessed 27 Jan 2021
11. Singh, I., Kaur, N.: Wealth management through robo advisory. Int. J. Res.-Granthaalayah **5**(6), 33–43 (2017)
12. Sironi, P.: FinTech Innovation: From Robo-Advisors to Goal Based Investing and Gamification. Wiley, Hoboken (2016)
13. Statista: Have you ever used a RoboAdvisor (algorithm-based digital program) for finance issues and investments? [Graph]. In Statista, 19 November 2020. https://0-www-statista-com.seyhan.library.boun.edu.tr/forecasts/1001481/roboadvisor-usage-in-spain. Accessed 27 Jan 2021
14. Statista: Have you ever used a RoboAdvisor (algorithm-based digital program) for finance issues and investments? [Graph]. In Statista, 19 November 2020. https://0-www-statista-com.seyhan.library.boun.edu.tr/forecasts/998833/roboadvisor-usage-in-germany. Accessed 27 Jan 2021
15. Statista: Have you ever used a RoboAdvisor (algorithm-based digital program) for finance issues and investments? [Graph]. In Statista, 19 November 2020. https://0-www-statista-com.seyhan.library.boun.edu.tr/forecasts/1188202/roboadvisor-usage-in-russia. Accessed 27 Jan 2021
16. Statista: Have you ever used a RoboAdvisor (algorithm-based digital program) for finance issues and investments? [Graph]. In Statista, 19 November 2020. https://0-www-statista-com.seyhan.library.boun.edu.tr/forecasts/1188128/roboadvisor-usage-in-finland. Accessed 27 Jan 2021
17. Statista: Have you ever used a RoboAdvisor (algorithm-based digital program) for finance issues and investments? [Graph]. In Statista, 19 November 2020. https://0-www-statista-com.seyhan.library.boun.edu.tr/forecasts/1187951/roboadvisor-usage-in-australia. Accessed 27 Jan 2021
18. Statista: Have you ever used a RoboAdvisor (algorithm-based digital program) for finance issues and investments? [Graph]. In Statista, 19 November 2020. https://0-www-statista-com.seyhan.library.boun.edu.tr/forecasts/1001255/roboadvisor-usage-in-austria. Accessed 27 Jan 2021

19. Statista: Have you ever used a RoboAdvisor (algorithm-based digital program) for finance issues and investments? [Graph]. In Statista, 19 November 2020. https://0-www-statista-com.seyhan.library.boun.edu.tr/forecasts/1000821/roboadvisor-usage-in-italy. Accessed 27 Jan 2021
20. Statista: Have you ever used a RoboAdvisor (algorithm-based digital program) for finance issues and investments? [Graph]. In Statista, 19 November 2020. https://0-www-statista-com.seyhan.library.boun.edu.tr/forecasts/998480/roboadvisor-usage-in-canada. Accessed 27 Jan 2021
21. Statista: Have you ever used a RoboAdvisor (algorithm-based digital program) for finance issues and investments? [Graph]. In Statista, 19 November 2020. https://0-www-statista-com.seyhan.library.boun.edu.tr/forecasts/998180/roboadvisor-usage-in-france. Accessed 27 Jan 2021
22. Statista: Have you ever used a RoboAdvisor (algorithm-based digital program) for finance issues and investments? [Graph]. In Statista, 19 November 2020. https://0-www-statista-com.seyhan.library.boun.edu.tr/forecasts/997865/roboadvisor-usage-in-the-uk. Accessed 27 Jan 2021
23. Statista: Have you ever used a RoboAdvisor (algorithm-based digital program) for finance issues and investments? [Graph]. In Statista, 19 November 2020. https://0-www-statista-com.seyhan.library.boun.edu.tr/forecasts/997189/roboadvisor-usage-in-the-us. Accessed 27 Jan 2021
24. Various sources (Backend Benchmarking): Value of assets under management of selected robo-advisors worldwide as of March 2020 (in billion U.S. dollars) [Graph]. In Statista, 4 May 2020. https://0-www-statista-com.seyhan.library.boun.edu.tr/statistics/573291/aum-of-selected-robo-advisors-globally/. Accessed 27 Jan 2021

A Review on Industry 4.0 Management for Sustainable Technologies

Mohammad Al-Shoqran[1](✉) and Sharaf Al Zub'i[2]

[1] Mathematical Sciences, Languages and General Studies Department, Ahlia University, Manama, Kingdom of Bahrain
malshoqran@ahlia.edu.bh

[2] Mobile Computing Department, Amman Arab University, Amman, Jordan
skalzubi@aau.edu.jo

Abstract. Many recent studies indicate that Industry 4.0 management plays an important role in sustainable development. This is because of the role of rapid communication technologies, and massive cloud information services in the acceleration of technological development. Therefore, many organizations around the world have adopted digital technologies, artificial intelligence, big data, and Industry 4.0 technologies, as strong enablers to achieve sustainable development goals.

This study presents a review of Industry 4.0 management in sustainable technologies via content analysis of some related articles with determining its strengths and advantages. In addition to examining the main components of Industry 4.0, which include IoT, and Smart Factory. Also, this paper explained the components and characteristics of Industry 4.0. Therefore, it is suggested that organizations around the world need to increase the use of I4.0 technologies in Its economic activities and to achieve sustainable development.

Keywords: Industry 4.0 · Sustainable development · Technologies · Management

1 Introduction

The German researchers Henning Kagermann, Wolf-Dieter Lukas, and Wolfgang Wahlster shaped the word Industry 4.0 as a paradigm change to sustain the potential competitiveness of the German economy [20]. Industry 4.0 aims to connect resources, services, and humans in real-time across Cyber Physical Systems and the Internet of Things (IoT) [16] growth. Sensors, actuators, and embedded software supplement physical components, allowing them the capacity to process and communicate data. The continuous networking in a production environment between all organizations and subsystems results in the Cyber Physical Systems. The digital and the real world are connected through this framework. Therefore, in periods of growing sophistication and modern versatility requirements [22], the linear automation principles of the 3rd industrial revolution can be solved.

A. M. A. Musleh Al-Sartawi et al. (Eds.): EAMMIS 2021, LNNS 239, pp. 206–217, 2021.
https://doi.org/10.1007/978-3-030-77246-8_21

Data was the basis of the 4th industrial revolution. A strategic consideration has been the way it can be collected, analysed, and used to take and improve decisions. Therefore, it is not only an organized or entirely different production that can have the competitive edge, but also the incorporation of product with digital services, i.e. how businesses filter information on the data produced by it to help decision-making. Manufacturing and distribution networks have steadily been augmented in recent decades by information technology supports, because the monitoring of increasingly sophisticated technologies, the requirements of multi-site production, and the support of logistics have become much more complicated activities. The unavoidable role of IT in industries has changed working and productivity conditions and its value is undeniable [6]. The new information technology has changed.

This paper examines the main components of Industry 4.0, which include Big Data, IoT, and Smart Plant. All these factors have an important interaction with the development and services of small and medium-sized businesses and improve their output [23]. Industry 4.0 industries, including Big Data, IoT, and Smart Factory, will overcome the technical problems confronting SMEs and eventually improve the competitive efficiency of companies. How Industry 4.0 encourages efficiency as seen in Fig. 1. This suggests that 5 Industry 4.0 variables have major e-effects on manufacturing and services. Production and services have essential impacts on the output of the manufacturing and service sectors. It also illustrates that Industry four boosts the efficiency of the manufacturing and service industry. The literature shows that Industry 4.0 have a positive effect on manufacturing and services and boosts efficiency [10].

2 Significance and Contribution of This Study

This comprehensive analysis aims to detect how sustainable Industry 4.0 research leads to the advancement of the Industry 4.0 agenda by mapping and synchronizing current research activities, defining research priorities, and discrepancies and possibilities for research development to better understand the connections between Industry 4.0 and Sustainable technology.

This research applied to current literature as well. This paper has helped to explain how academics can successfully handle the early stages of Sustainable Industry 4.0. The holistic review of this study applies a valuable aspect of Industry 4.0, which is to be called early in Sustainable Industry 4.0, to current studies, relative to a component of 8 significant characteristics.

3 Research Methodology

Due to its systematic and scientific approach to identifying and evaluating scientific works, a literature review has been selected as the right analytical approach for achieving the purpose of this paper. It promotes state-of-the-art characterisation and the recognition of the other important and contentious topics [9].

In addition, research methodology shows and explains the method of obtaining the scientific results in clear and regular steps by clarifying the research conducted method to achieve the objectives and to explain the methods of obtaining the results. The research approach for this research is a descriptive analysis that relies on content analysis for some research papers concerning Industry Management 4.0 for Sustainable Technologies, by determining their strengths, weaknesses, advantages, and disadvantages. in addition, this investigation presents some graphical presentation with their explanations.

This search consists of a study, from 2016, of industry 4.0 science papers in the Google Scholar data base (Industry 4.0) and reviews about industry 4.0. Over 15 papers were listed and the abstract and findings were examined to ascertain their significance in this study to provide an analysis of the various industrial developments with a focus on and dimensions of Industry 4.0. As seen in Fig. 1, the systematic examination technique took a 5-step approach

Fig. 1. Systematic examination technique

The following science papers were considered for this study after analysing the original articles to ascertain their validity and accuracy: from [1] to [21].

4 Industry 4.0 and the Industry Revolutions Momentum

The definition of "Advanced Manufacture, Smart Manufacture, SMART Factory," the Internet of Things, Internet of Everything" or the industrial internet is still not agreed, and some scholars could call it the same concept. However, we summarize that the 4th Industrial Revolution based on the literature review is identified by advanced digitalisation, convergence and usage of Internet and smart objects in industrial and logistical processes. The adoption of ICT combines the real and virtual worlds with cyber physical production processes, consisting of online social networks that connect the IT with mechanical and electronic components that interact with each other through a network. As seen in Fig. 2.

Industry 4.0 generates higher value chains and transforms conventional processes for human and manufacturing and organizational market patterns that influence society as a whole and the environment [20]. Students also pointed out the need to increase service orientation, which means that manufacturing firms are no longer concentrating only on their products, but that they need to change their value offerings for a mix of both products and services [13].

Digitalized production can be characterized as the system of linked and interdependent operations that specify the way that the organization conducts its business with clients, partners and supplier [10]. While there is no consensus among experts on the concept of the business model, most accept that the business model comprises a systematic systemic approach to understanding how companies conduct business [18].

Fig. 2. Industry 4.0 enabling technologies

According to stock (2016) a new business model, the company's quest for new means of generating and capturing value for its stakeholders is largely focused on identifying innovative methods of generating sales and defining value offers for its clients, vendors and partners. Even a good business model cannot be regarded as a lasting one due to the environmental dynamic, and the superior ability to redesign your business model before circumstances cause you to be used as a more and more vital source of comparative benefit in uncertain and nuanced circumstances [20]. Industry four provides new business markets that are also not completely used.

5 Significant Features of Industry 4.0

Industry 4.0 is the industrial production future. The fourth step of the industrial revolution, or Industry 4.0, is the era of robotics, digitized factories, and digitized services. Even so, the field of academia is the approach still cannot be described as the fundamental concept relating to the fourth industrial revolution is Industry 4.0. This creates difficulties in identifying its components. As seen in Table 1. Below, there are 8 industry characteristics.

Table 1. Industry 4.0 enabling technologies. Source: Ardito, L., Petruzzelli, A. M., Panniello, U., and Garavelli, A. C. (2019)

Enabling technology	Description
Advanced manufacturing	Advanced production refers to the latest technological advances which companies can make in the production of improved company products and/or processes. Advanced robotics [3], automation solutions are examples of these technologies
Additive manufacturing	Additional production reflects the technology set to develop computer controlled three-dimensional objects by layer. 3-D printing is the most representative technology in this field [8]
Augmented reality	Technological technologies that are still in their infancy are augmented-reality-based systems [21]. By bringing abstract objects in the physical world in order to improve one's present view of reality, they transform the environment around the staff in a digital interface
Simulation	Simulation is related to techniques that are used mainly for development simulation activities, so that operators can evaluate and refine machine settings before a physical change in direction for the next line of good [11]
Cloud computing	Cloud computing provides for an Information technology share of software and hardware to conveniently be stored and remotely accessible by a number of players [13]
Industrial IoT	The use of IoT technology in demand- and supply-focused processes is the reference of the industrial IoT [15]. It enables interoperability between computers and machinery with various protocols and architectures, allowing the availability of information in real time across the value chain
Cyber security	The need to secure sensitive information increases rapidly with the increased accessibility and the usage of pooled IT services. Thus, cyber security threats [5] technologies are necessary to ensure safe and efficient communications

(*continued*)

Table 1. (*continued*)

Enabling technology	Description
Big data analytics and customer profiling	In a sense of Industry 4.0, a vast volume of data comes from a number of resources, such as equipment and processes for output, supply chain actors and consumer systems [17]. Big data analytics and consumer profiling provide application tools that allow vast data sets to be evaluated and help decisions in real time

6 Industry 4.0 Components

In Industry 4.0 three components can be classified. Horizontal integration is the first one. It incorporates the idea of a new form of the global network value chain. The second is integration in the vertical. The idea is to excel in Hierarchical production line subsystems to create a high flexibility production line that is easy to configure. From the beginning to the end, the last part is engineering convergence across the whole supply chain to better customize products [2]. An organization should cooperate and compete with enterprises with identical characteristics for building an efficient horizontal integration structure. Items, financial reporting and knowledge can easily be linked across all these sectors. New management systems and models could also be available. Vertical integration offers a plant concept with several information and physical subsystems including, for example, supply management, actuator and sensor, value, and corporate planning. High versatility and ease of set up production lines must be ensured for sensor and actuator signal vertical integration across various phases of company resource planning levels [14].

From this integration, the extremely smart machines generate an automatically controllable operated computer. Reconfigured in various types of the product. In the production system, the large amounts of data collected and analyzed are clear [13]. The inclusion, during productive value creation, of End-To-End innovation into a series of operations calls for factors like customer needs representativeness, product design, installation, recycling, manufacturing engineering, industrial services, production planning and maintenance. The product architecture will influence the installations and production by using software platforms in the chain to ensure the products are customization.

The new technical development continued for many decades and has now taken place in the age of Industry 4.0 [14]. The modern German economic architecture was first established in 2011 under the Industry 4.0 Framework [20]. The first industrial revolution started in the last phase of the 18th century [17]. This turning point was characterized by both water-based and steam power manufacturing plants. The second industrial revolution started in the early 20th century and was characterized by mass manufacture of electricity technology. The third technical transition in the 1970s started with the characteristics of programmed output based on modern technology linked to the Internet. Finally, the fourth industrial revolution, Industry 4.0, continues to involve the characteristics of the development of the Cyber Physical System, focused on different data and information combinations [18].

The key features of the Cyber Physical System are based on the notion of meeting dynamic output requirements and advancing the information and skills of the entire manufacturing field. Industry 4.0 involves several technology and associated patterns, cloud-based development, numerous resource planning operations, IoT, and enhancement in social goods [19] and has an important role in digital English and education [22]. In addition, som of these features include cloud computing, virtual reality, customer engagement at various stages, sophisticated big data algorithms, smart sensors [4], handheld devices, IoT networks, position recognition, advanced human machinery, and 3D printing.

Academics in this community have defined Industry 4.0 from several perspectives. Industry 4.0, for example, is "the convergence of sophisticated physical machinery and networked sensor and software systems, used to predict, monitor and plan for improved market and social performance" (Industrial Internet Consortium 2017). Industry 4.0 is described by Henning (2013) as "a new level of organization and management of the value chain across the lifecycle of products". Industry 4.0 is defined by Hermann et al. (2016) as a collaborative term for value chain organization technology and principles.

7 Articles Review

Industry 4.0 Management and Sustainable Technologies has been studied widely during the last few years ago, this section includes the analysis for some articles related to Industry 4.0 Management for Sustainable Technologies. (see Table 2).

Based on the content analysis of the above-selected articles related to industry 4.0 management for in sustainable technologies, we concluded that most of these studies constrained on the important role of the technology of the fourth industrial revolution in various fields and for sustainable development especially. Where the studies presented and showed some perspectives, challenges, and suggestions related to the fourth industrial revolution. Also, it gave an identification of the prospect of industrial value creation in Industry 4.0. Also, it concluded that there is a need for policies for investment allocation for development and research to improve the manpower quality and to support the consumption of green energy. In addition, there is a shortage of quantitative studies of the ecological and social effects of Industry 4.0.

Table 2. Literature review of Industry 4.0 Management for Sustainable Technologies.

Articles	Strengths/advantages	Weaknesses/disadvantages
[1]	This study gave a review on technologies for manufacturing systems and illustrated the technology keys combined to their role in Industry 4.0 such as Cloud Computing, Big Data, Cybersecurity, and The Industrial Internet of Things. In addition to a discussion of the Smart Factory in I 4.0. Industry 4.0 environment is scanned, with a description of systems over the manufacturing environment and the enabling technologies The review concluded that I4.0 bases are the advanced technologies of automation and the main challenge of I4.0 is to how to make the systems of production more elastic and collaborative. Therefore, the use of enabling technologies is the main strategy of I4.0. So, in an industrial factor, the use of each performed technology separately will give a lower effect. While if it performed together the effect will be higher	The study did not address how to use the internet of services for medium or low-quality goods. So that we can arise the following question: is it possible to increase its competitiveness through the Internet of services?
[6]	This paper aimed to analyse the prospect of Industry 4.0 applications to create a circular economy and to achieve more sustainable industrialization. Also, the paper suggested an alternative solution that can be utilized to trace materials through the supply chain. The paper said that the economy is closed on systems prioritizing the linear models in consumption and production The paper concluded that there is a high prospect of Industry 4.0 to guarantee more sustainable manufacturing processes. This is evidenced via an analysis of the value drivers in Industry 4.0, the prospect of realignment value chains, and Novelty business models. Also, it concluded that there is a high prospect of Industry 4.0 components to guarantee more sustainable manufacturing processes, which is evidenced via analysing the value drivers of Industry 4.0	• The paper did show an example of the new nonlinear models to an alternative of the linear models in the economic components • The paper did show how to measure the impact of industry 4.0 applications
[7]	The paper explored a simulation of the reverse logistics supporting the virtual factory component of I4.0 through joint implementation of the Industry 4.0 Principles and Resolve model of circular economy (CE) approaches In conclusion, the paper proposed a suitable combination of based dispatching family rules and information-sharing And there is a need to focus on the cost of socially affected operations. Therefore, the integration of CE and I4.0 shows a decision of the real-time model of a sustainable reverse logistics system	There is a need to clarify how the implementation of the operational excellence technology in I4.0 gives worthy guidance to the pursuit of the sustainable execution of a circular SC

(*continued*)

Table 2. (*continued*)

Articles	Strengths/advantages	Weaknesses/disadvantages
[9]	This paper presented Industry 4.0 principles combined with a discussion and summary of its opportunities in terms of reliability engineering and for some of its challenges In addition, the paper proposed new research trends in many fields of this regard such as big data analysis, systems modelling, health management, the interaction of human-machine, and uncertainty The paper concluded that there is a need for attention to system modelling, production and maintenance and supply chain jointly optimization	The paper did not give a real-life example on how to embed the uncertainty in the modern research for example in climate uncertainty and many other fields of research
[10]	This investigation described the current relationship between sustainability and I4.0, some important matters, with an identification to research gaps regarding acting of it in the future Also, the investigation had analysed thematic network frame and the strategic diagram with an interpretation of the function of fourth technologies in this regard such as: • Thematic network structure of the cluster 'climate-change' • Thematic network structure of the cluster 'conceptual-framework • Thematic network structure of the cluster 'decision-making' This study assured that scientific efforts are constrained on how to improve environmental and economic aspects Also, it showed the perspectives, challenges, and suggestions for possible future work	• The study is based the English documents only • There is a need to suggest a solution to the expected challenges in the future • There is a difficulty to achieve a quantitative and qualitative mixture of TBL boundaries
[12]	This study aimed to discover the Systematic Impact of the technology development in the fourth Industry and its utilization in the sustainable energy transition in the developing countries using a conceptualization model. the conceptual model depends on causal loop diagram that is generated by system dynamics process In conclusion, key variables to support Industry 4.0 technologies is the share of the industry technology investment in Indonesia. Also, the steed concluded that there is a need for policies for investment allocation for development and research to improve the manpower quality and to support the consumption of green energy	There is no explanation of how to use the presented model in the future. in another word, how can the model be developed in the future to ensure its continued use?

(*continued*)

Table 2. (*continued*)

Articles	Strengths/advantages	Weaknesses/disadvantages
[22]	This paper identified the prospect of industrial value creation in Industry 4.0 regarding their contribution to sustainable development In the first step of this qualitative identification is based on an assessment of the macro prospect depending on a review of current industrial developments the utilising for: • value creation networks • business models • product life cycles Where the other stages considering the factor of value creation of industrial value creation. It also gave a review of sustainable value creation in I4.0 The paper concluded that there is a shortage of quantitative studies of the ecological and social effects of I4.0	Some negative effects might happen regarding with simplification of jobs via technological systems, substation or job transparency via ICT that may cause exploitation of employees

8 Conclusion

It is now clear that Industry 4.0 is a restructuring of society in philosophical terms. It is anticipated that this transition will lead as with all other technological revolutions to significant shifts in culture, education, economics, and commerce. This article offers some detail on this topic discussed in both science and industrial societies about Industry 4.0. There are two practical additions to the paper: Firstly, it aims to explain the basic principle for clinicians, providing the Industry 4.0 meaning. Second, eight architecture criteria of the Industry 4.0 scenario should be implemented in favour of implementation. It also defined in Industry 4.0 the possibilities of generating industrial value. It concluded that policies on investment in growth and science are also important to enhance the productivity of the workforce and to promote renewable energy use. In addition, comprehensive ecological and social impacts research analyses of industry 4.0 are sparse. There is a scarcity of evidence available.

Based on a content analysis study of the industry 4.0 management papers, we infer that most of these reports have restricted the significant role of technology in various fields in the fourth industrial revolution and in sustainability. Where the studies offered analysis, problems and ideas pertaining to the fourth industrial revolution. This research does not attempt to analyse problems and boundaries based on a small number of documents examined, the choice not to use statistical analysis and the likelihood that cross-disciplinary subjects be established from other fields of science. Future research should concentrate on realizing the multiple opportunities that enable companies to connect the two approaches to achieve complete Industry 4.0 objectives.

References

1. Alcácer, V., Cruz-Machado, V.: Scanning the industry 4.0: a literature review on technologies for manufacturing systems. Eng. Sci. Technol. Int. J. **22**(3), 899–919 (2019)

2. Andronie, M., Lăzăroiu, G., Iatagan, M., Hurloiu, I., Dijmărescu, I.: Sustainable cyber-physical production systems in big data-driven smart urban economy: a systematic literature review. Sustainability **13**(2), 751 (2021)
3. Ardito, L., Petruzzelli, A.M., Panniello, U., Garavelli, A.C.: Towards Industry 4.0: Mapping digital technologies for supply chain management-marketing integration. Bus. Process Manag. J. **25**(2), 323–346 (2019). https://doi.org/10.1108/BPMJ-04-2017-0088
4. Al-Shoqran, M., Shorman, S.: A review on smart universities and artificial intelligence. In: Hamdan, A., Hassanien, A.E., Razzaque, A., Alareeni, B. (eds.) The Fourth Industrial Revolution: Implementation of Artificial Intelligence for Growing Business Success. Studies in Computational Intelligence, vol. 935. Springer, Cham (2021). https://doi.org/10.1007/978-3-030-62796-6_16
5. Bai, C., Dallasega, P., Orzes, G., Sarkis, J.: Industry 4.0 technologies assessment: a sustainability perspective. Int. J. Prod. Econ. **229**, 107776 (2020)
6. Blunck, E., Werthmann, H.: Industry 4.0–an opportunity to realize sustainable manufacturing and its potential for a circular economy. In: DIEM: Dubrovnik International Economic Meeting, vol. 3, no. 1, pp. 644–666. Sveučilište u Dubrovniku (2017)
7. Dev, N.K., Shankar, R., Qaiser, F.H.: Industry 4.0 and circular economy: operational excellence for sustainable reverse supply chain performance. Resour. Conserv. Recycl. **153**, 104583 (2020)
8. Ejsmont, K., Gladysz, B., Kluczek, A.: Impact of industry 4.0 on sustainability—bibliometric literature review. Sustainability **12**(14), 5650 (2020)
9. Farsi, M.A., Zio, E.: Industry 4.0: some challenges and opportunities for reliability engineering. Int. J. Reliab. Risk Safety Theory Appl. **2**(1), 23–34 (2019)
10. Furstenau, L.B., Sott, M.K., Kipper, L.M., Machado, E.L., Lopez-Robles, J.R., Dohan, M.S., Cobo, M.J., Zahid, A., Abbasi, Q.H, Imran, M.A.: Link between sustainability and industry 4.0: trends, challenges and new perspectives. IEEE Access **8**, 140079–140096 (2020)
11. Haseeb, M., Hussain, H.I., Ślusarczyk, B., Jermsittiparsert, K.: Industry 4.0: a solution towards technology challenges of sustainable business performance. Soc. Sci. **8**(5), 154 (2019)
12. Hidayatno, A., Destyanto, A.R., Hulu, C.A.: Industry 4.0 technology implementation impact to industrial sustainable energy in Indonesia: a model conceptualization. Energy Procedia **156**, 227–233 (2019)
13. Ivanov, D., Dolgui, A., Sokolov, B.: The impact of digital technology and Industry 4.0 on the ripple effect and supply chain risk analytics. Int. J. Prod. Res. **57**(3), 829–846 (2019)
14. Machado, C.G., Winroth, M.P., Ribeiro da Silva, E.H.D.: Sustainable manufacturing in Industry 4.0: an emerging research agenda. Int. J. Prod. Res. **58**(5), 1462–1484 (2020)
15. Miranda, J., Ponce, P., Molina, A., Wright, P.: Sensing, smart and sustainable technologies for Agri-Food 4.0. Comput. Ind. **108**, 21–36 (2019)
16. Mustafa, M., Alzubi, S.: Factors affecting the success of internet of things for enhancing quality and efficiency implementation in hospitals sector in Jordan during the crises of Covid-19. In: Internet of Medical Things for Smart Healthcare, pp. 107–140. Springer, Singapore (2020)
17. Nara, E.O.B., da Costa, M.B., Baierle, I.C., Schaefer, J.L., Benitez, G.B., do Santos, L.M.A.L., Benitez, L.B.: Expected impact of industry 4.0 technologies on sustainable development: a study in the context of Brazil's plastic industry. Sustain. Prod. Consum. **25**, 102–122 (2021)
18. Phuyal, S., Bista, D., Bista, R.: Challenges, opportunities and future directions of smart manufacturing: a state of art review. Sustain. Futures **2**, 100023 (2020)
19. Oztemel, E., Gursev, S.: Literature review of Industry 4.0 and related technologies. J. Intell. Manuf. **31**(1), 127–182 (2020)
20. Reinhart, G. (ed.): Handbuch Industrie 4.0: Geschäftsmodelle, Prozesse, Technik. Carl Hanser Verlag GmbH Co KG (2017)

21. Sanders, A., Elangeswaran, C., Wulfsberg, J.P.: Industry 4.0 implies lean manufacturing: research activities in industry 4.0 function as enablers for lean manufacturing. J. Ind. Eng. Manage. (JIEM) **9**(3), 811–833 (2016)
22. Stock, T., Obenaus, M., Kunz, S., Kohl, H.: Industry 4.0 as enabler for a sustainable development: a qualitative assessment of its ecological and social potential. Process Safety Environ. Prot. **118**, 254–267 (2018)
23. Stock, T., Seliger, G.: Opportunities of sustainable manufacturing in industry 4.0. Procedia Cirp **40**, 536–541 (2016)

A Novel Hybrid Renewable Power Platform: A Sustainable Innovation

Amrita[1], Ramesh Chandra Panda[2](✉), and Lakshmi Dhandabani[3]

[1] Computer Science & Engineering, School of Engineering & Technology, Sharda University, Greater Noida, UP, India
amrita.prasad@sharda.ac.in

[2] Research and Development Cell, Mechanical Engineering, Synergy Institute of Engineering and Technology, Dhenkenal, Orissa, India

[3] School of Computer Science and Engineering, VIT Bhopal University, Bhopal, Madhya Pradesh, India
lakshmi.d@vitbhopal.ac.in

Abstract. Recent days, the Hybrid Renewable Power Platform (HRPP) is disseminated widely as stand-alone power systems. HRPP is an emergent power generation technology that comprises a combination of distinct renewable energy systems. It helps in the production of electricity in remote areas due to technological advancement in the field of renewable energy. Due to the digital revolution, the consumption of electricity has been increased and there is a steady increase in the prices of petroleum products. Therefore, it is necessary to discover the most suitable hybrid combinations to develop HRPP more efficiently and also environment friendly. This paper proposes a novel and stand- alone HRPP that combines wind turbines, solar panels and water current energy all three together. It supports hybrid offshore power generation platforms in order to produce off-grid electricity in remote areas. Each of these can generate up to 24 KW in ideal conditions and there's one at each corner of each 12 × 2-m floating unit. On top of that, 8 KW wind turbines can be placed at each junction point that covers the entire top surface with solar panels. The combined units could generate power up to 20 KW. The proposed HRPP unit can be deployed in order to produce off-grid electricity to remote areas unreachable to the regional or national electricity grid.

Keywords: Off-grid electricity · Hybrid system · Hybrid Renewable Power Platform · Wind turbines · Solar panels · Water current energy

1 Introduction

In recent decades, the demand for electrical energy has grown substantially due to increase in populations and rapid growth of technological development. Majority of electrical energy is produced from fossil fuels that are primary conventional energy sources [1]. It results in depletion of fossil fuel resources, and emissions of harmful gases caused by fossil fuel. There is a progressive increase in its price too. Across the globe many are living in remote areas without grid electricity even recent days [2]. These

A. M. A. Musleh Al-Sartawi et al. (Eds.): EAMMIS 2021, LNNS 239, pp. 218–225, 2021.
https://doi.org/10.1007/978-3-030-77246-8_22

situations are turning attention towards viable alternative solutions as renewable energy sources (RESs) like water, solar, and wind for universal and sustainable electrification to all especially to remote areas and enforce the environment preservation too [3, 4]. RESs replenish themselves naturally without being depleted in the earth. In spite of this, most of RESs are climate-dependent and suffer from discontinuity of generation due to variable climate conditions and seasonal variations. It brings unreliability and uncertainty issues when used individually to supply electricity demand. Its utilization requires complex design, planning and control methods. To overcome this problem, the hybrid renewable power platform (HRPP) or hybrid renewable energy systems (HRESs) has emerged as a promising technology for the generation of electricity for remote areas. It integrates more than one distinct RESs and energy storage systems (ESSs) that lead to improved system efficiency and economic power supply [5], either in grid connected or stand-alone architecture. The renewable energy share is up to 88.793 GW and Hydro energy is up to 45.70 GW out of total installed capacity 372.69 GW in India [6].

Solar power and wind power systems have strong intermittency, volatility and randomness. Large-scale grid connection of solar power and wind power has always posed challenges for the stability of the power system. As a large-scale, hydropower is a highly regulated power supply that can effectively regulate the instability of solar power and wind power. One of the solutions to this problem might be the exploitation of hybrid renewable energy technologies in off-grid areas or remote areas. However, there is a slow adoption of HRESs in many developing countries due to its high cost. Therefore, it is vital to discover the most suitable combinations of RESs that access electricity generation and reduce energy cost.

This paper proposes a novel stand-alone HRPP that combines wind turbines, solar panels and water current energy harvesters by employing hybrid offshore power generation platforms to produce off-grid electricity in remote areas. This paper is organized as follows. Section 2 presents the related work. The proposed HRPP and its technical specifications are detailed in Sect. 3, and finally, the conclusions are presented in Sect. 4.

2 Literature Review

Over the past decade, a huge research has been done in the area of HRPPs that integrate more than one RERs and ESSs in both grid connected and stand-alone mode [7–9].

A study of the approaches and strategies utilized to employ energy management both in grid-connected and in standalone HRES are presented in [10]. This review covers the research related to the various configurations of the HRES for electrification only.

An energy management strategy is proposed in [11] for a standalone HRES and storage systems as hydrogen and battery. An economic analysis is conducted for the decision of the selection of energy source or storage device. The parameters considered in the economic analysis are degradation of components, hours of operation, and lifetimes of the component. The study represents the reliable electricity support for the hybrid system.

Six different combinations are considered using wind turbines, photovoltaic (PV) system, fuel cell, electrolyzer, converter, batteries, and hydrogen tank for electricity generation in [12]. In this, the wind/PV/battery based system is found the most cost effective.

A Stand-alone HRES using multi-optimal combination is proposed in [13]. It encompasses wind, PV, battery, converter and diesel generator. The proposed HRES guarantees for the economical viability in terms of cost of energy and net present cost and substantial low GHG emissions and decrease in CO2.

The techno-economic analysis of HRES using off-grid is carried out under different climatic and geographic conditions [14]. It is observed that the cost of energy in grid is less than the off-grid electricity. In this, HRES combined wind energy conversion system (WECS) with fuel cell and battery as storage and PV system. The battery storage system is found to be more economical than a fuel cell.

A proper planning is required in the investment of RES-like wind-based systems and a model for WECS for expansion planning is proposed in [15].

A comprehensive feasibility analysis of a gird-off HRES for electricity generation is presented in [16]. A framework is proposed for appropriate planning and design of HRES, which employs techno-economic optimization analysis. Distinct hybridization combinations of a wind turbine, solar photovoltaic, battery storage, converter technologies, and diesel generator together with a diesel generator-based energy system as base case are modeled, evaluated and compared. The aim is to find the most viable and reliable solution with minimum system cost and realistic environmental impacts.

3 Research Methodology

Hydropower is the most verified and well-developed form of renewable electricity generation. Particularly low head hydropower plants are facing some challenges like reservoir restrictions, base load, water level etc. HRPP as depicted in Fig. 1 is a reliable hybrid solution for the hydropower industry, the design of HRPP is modelled by Solidworks Computer Aided Design (CAD) software.

It is a combination of a set of Turbine-Generator-Unit with a battery, solar cell and wind tunnel units. The control system and electrical power plant is integrated with batteries to optimize the revenue and extend the lifetime of the power plant. This unit can harvest green energy even with slow moving water.

According to a report, approximately 1.4 billion persons are living without electricity across the globe, the majority are living in distant and remote regions. They are unreachable to the regional or national electricity grid [17].

The conventional energy solution is accessible only through the utilization of gas generators. Gas generators produce air pollution and can be expensive for maintenance in the long term. The utilization of green technologies is swiftly growing nowadays as an alternative. RES solutions are becoming popular particularly for residential usage. The basic requirement of electric primary load is determined at 12.06 kW per day for single household appliances which includes the essential load demand such as lighting, and computers.

Every year, billions of dollars are invested on wind and solar systems. These renewable energy products do not provide optimal solutions for meeting the demand for electricity because intermittent production of RESs are conditioned by changes in weather. Consequently, these systems are utilized only upto 12–35% of their capacity implicating an over sizing of installations and storage systems to collect the produced energy. One

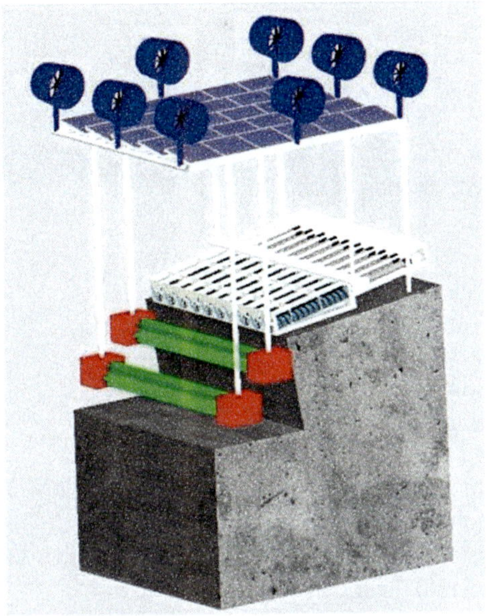

Fig. 1. Hybrid renewable power platform

predictable and regular RES unexploited is the river, which can impart a reliable and predictable energy source ceaseless among all RESs. Idénergie's river turbine (IRT) [18] is one of the solutions, which can produce continuous energy offered by the river. Among other benefits, it allows the use of less batteries than the standard norm, uninterrupted production of energy, and less dependence on changes in weather. IRT is the only water driven technology, which directly supplies electricity to remote areas. It is an outstanding emergency backup system because it provides endless energy supply and easy to install.

This turbine can easily be dismantled into pieces to make it easy for shipping to the remote areas and can easily be assembled on site and has low maintenance. It requires only 2 persons and less than a 24 h time for installation.

IRT has an embedded smart converter, which converts water current into electricity to recharge batteries and other power home appliances. In addition, it automatically regulates the rotational speed of the turbine and obtains maximum power from the current. This smart turbine can utilize the generator in motor mode to initiate enough movement to produce electricity in case water speed is very low.

IRT utilizes a novel kind of electric generator, which is very proficient at low water velocity, invented and manufactured by Idénergie. It produces energy between 2 and 6 kWh/day, which is equivalent to 4 to 10 solar panels.

The major components used are the eco-worthy 400 W Wind Turbine Generator and 100 W Monocrystalline Solar Panel for Off Grid 12 V Battery Charging is fitted along with sets of turbines and auto lifting linear actuation system for easy vertical locomotion.

The integral parts of Hybrid Renewable Power Platform are described in the following subsections.

3.1 Wind Turbine

The specification details of High Efficient Generator

- Start-up Wind Speed: 5.6 mph (2.5 m/s)
- Maximum wind speed: 78 mph (35 m/s)
- Rated Wind Speed: 23.5 mph (10.5 m/s)
- Rated Voltage: DC12-24 V
- Rated Rotate Speed: 800 r/min
- Fan in the weight: 17.6 lb (8 kg)
- Tower diameter: Suggest more than 3.1 in. (80 mm)
- Rotor diameter: 47.2 in. (1.2 M)
- Generator Material: carbon fiber composite
- Altitude of tower: 14.8–32.8 ft (4.5 M–10 M)
- Generator style: Permanent-Magento-Phase A.C
- Controller type: Wind & Solar Hybrid Controller
- High module conversion efficiency, ideal daily output: 2000 Wh per day (according to the availability of wind, the speed 35 m/s)
- Packing: 27.1 * 14.4 * 8.3 inches

3.2 100 W 12 V Mono Solar Panel

- Open-Circuit Voltage (Voc): 21.6 V
- Maximum Power at STC: 100 W
- Short circuit current (Isc): 6.13 A
- Working current (Iop): 5.78 A
- Temperature range: −40 °C to + 80 °C
- Size: 39.8 * 20.1 * 1.4 in.
- Weight: 18.96 lb
- Output Tolerance: ±3%

3.3 Wind and Solar Power Hybrid Controller

- Rated Solar Power: 500 W/1000 W
- Rated Wind Generator Capacity: 400 W/800 W
- Wind Generator Recover Voltage: 13.2 V/26.4 V
- Wind Generator Braking Voltage: 14.5 V/29 V
- Max Discharging Current: 20 A
- Rated Battery Voltage: DC 12 V/24 V

3.4 Solar Cable and Y Branch

- 1pc 16.4 ft black cable and 1pc 16.4 ft red cable with connector
- For extending the cables of panel, and connecting to controller

3.5 Linear Actuator: Pa-14 Specifications

The PA-14 Mini Linear Actuator as shown in Fig. 2 utilizes a fundamental linear motion design in a compact body. The low current, medium force actuator is ideal for home automation, robotics, agricultural industry, and cabinetry. It requires cords from the controller to the battery bank and a pole to install the wind turbine. It recommends using a gel battery or lead-acid, about 200–400 Ah. And the inverter depends on usage and battery setup 12 V/24 V, use a 1000 W/1500 W inverter for small appliances or 3500 W for a big appliance like small air-condition. First, connect the battery with the controller, then connect the Wind Turbine with the controller and finally, connect the solar panel with the controller.

3.6 Mounting Options

The BRK-14 bracket is designed for the Mini Linear Actuator PA-14. It allows for a simple installation process. The dimensional drawing for BRK-14 is shown in Fig. 2. It is a lightweight mounting bracket and allows the rotational direction up to 180°. It allows for a simple installation process.

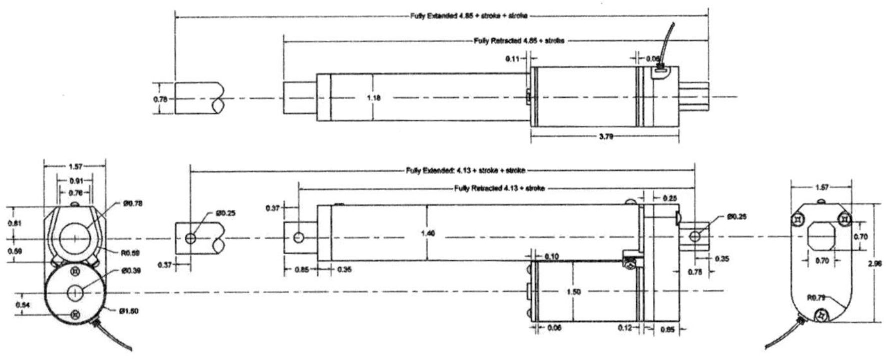

Fig. 2. Linear Actuator: PA-14

3.7 Control Systems

The control system supports plug and use methods for controlling linear actuators. The low current draw allows for connectivity with most of our control systems. It supports both of the power input options such as AC and DC. It supports generating DC output as well. Further it supports to attach micro controllers, relay and motor drivers.

4 Conclusion

The paucity of electricity access is declining the progression and obstructing the accomplishment of development objectives of any nation. As the country aims to accomplish trustworthy energy access for all, the selected hybrid renewable energy system might be the solution to electrification of remote areas, mitigate impacts of changes in weather, and substantial low GHG emissions and decrease in CO2. In this paper, a novel standalone Hybrid Renewable Power Platform (HRPP) has been proposed. It combines wind turbines, solar panels and water current energy harvesters by employing hybrid offshore power generation platforms to produce off-grid electricity. Each of these can generate up to 24 kW in ideal conditions and there's one at each corner of each 12 × 12-m floating unit. On top of that, 8 kWp wind turbines can be placed at each junction point, and cover the entire top surface with solar panels, which could contribute up to a total of 20 kW to the final output of the unit. It can stick units together to scale the whole thing up. It can produce continuous electricity supply to remote areas unreachable to the regional or national electricity grid. It can be reliable, convenient, cost effective, and robust power platform to mitigate existing power crises. It can also have a great impact on improving the socio-economic condition of remote areas people as well as will be a good sign of green energy technology.

References

1. IRENA: Global Energy, a Roadmap to 2050 by IRENA (2020). http://www.irena.org
2. Shouman, E.R.: International and national renewable energy for electricity with optimal cost effective for electricity in Egypt. Renew. Sustain. Energy Rev. **77**, 916–923 (2017). https://doi.org/10.1016/j.rser.2016.12.107
3. Senthil Kumar, J., Charles Raja, S., Srinivasan, D., Venkatesh, P.: Hybrid renewable energy-based distribution system for seasonal load variations. Int. J. Energy Res. **42,** 1066–1087 (2018). https://doi.org/10.1002/er.3902
4. Quek, T.Y.A., Alvin Ee, W.L., Chen, W., Ng, T.S.A.: Environmental impacts of transitioning to renewable electricity for Singapore and the surrounding region: a life cycle assessment. J. Clean. Prod. **214**, 1–11 (2019). https://doi.org/10.1016/J.JCLEPRO.2018.12.263
5. Akinyele, D.: Techno-economic design and performance analysis of nanogrid systems for households in energy-poor villages. Sustain. Cities Soc. **34**, 335–357 (2017). https://doi.org/10.1016/j.scs.2017.07.004
6. https://powermin.nic.in/en/content/power-sector-glance-all-india. Accessed 27 Sept 2020
7. Zia, M.F., Elbouchikhi, E., Benbouzid, M.: Microgrids energy management systems: a critical review on methods, solutions, and prospects. Appl. Energy **222**, 1033–1055 (2018). https://doi.org/10.1016/j.apenergy.2018.04.103
8. Twaha, S., Ramli, M.A.M.: A review of optimization approaches for hybrid distributed energy generation systems: off-grid and grid-connected systems. Sustain. Cities Soc. **41**, 320–31. https://doi.org/10.1016/j.scs.2018.05.027
9. Sawle, Y., Gupta, S.C., Bohre, A.K.: Review of hybrid renewable energy systems with comparative analysis of off-grid hybrid systems. Renew. Sustain. Energy Rev. **81**, 2217–35. https://doi.org/10.1016/j.rser.2017.06.033
10. Olatomiwa, L., Mekhilef, S., Ismail, M.S., Moghavvemi, M.: Energy management strategies in hybrid renewable energy systems: a review. Renew. Sustain. Energy Rev. **62**, 821–835 (2016). https://doi.org/10.1016/j.rser.2016.05.040

11. Torreglosa, J.P., García-Triviño, P., Fernández-Ramirez, L.M., Jurado, F.: Control based on techno-economic optimization of renewable hybrid energy systems for stand-alone applications. Expert Syst. Appl. **51**, 59–75 (2016)
12. Al-Sharafi, A., Sahin, A.Z., Ayar, T., Yilbas, B.S.: Techno-economic analysis and optimization of solar and wind energy systems for power generation and hydrogen production in Saudi Arabia. Renew. Sustain. Energy Rev. **69,** 33–49 (2017). https://doi.org/10.1016/j.rser.2016.11.157
13. Hossain, M., Mekhilef, S., Olatomiwa, L.: Performance evaluation of a stand-alone PV-wind-diesel-battery hybrid system feasible for a large resort center in South China Sea, Malaysia. Sustain. Cities Soc. **28**, 358–366 (2017). https://doi.org/10.1016/j.scs.2016.10.008
14. Duman, A.C., Güler, Ö.: Techno-economic analysis of off-grid PV/Wind/Fuel cell hybrid system combinations with a comparison of regularly and seasonally occupied households. Sustain. Cities Soc. **42**, 107–126 (2018). https://doi.org/10.1016/j.scs.2018.06.029
15. Valinejad, J., Marzband, M., Funsho Akorede, M., Elliott, I.D., Godina, R., Matias, J., Pouresmaeil, E.: Long-term decision on wind investment with considering different load ranges of power plant for sustainable electricity energy market. Sustain. **10**, 3811 (2018). https://doi.org/10.3390/su10103811
16. Elkadeem, M.R., Wang, S., Sharshir, S.W., Atia, E.G.: Feasibility analysis and techno-economic design of grid-isolated hybrid renewable energy system for electrification of agriculture and irrigation area: A case study in Dongola, Sudan. Energy. Convers. Manag. **196**, 1453–1478 (2019). https://doi.org/10.1016/j.enconman.2019.06.085
17. Azimoh, C.L., Klintenberg, P., Mbohwa, C., Wallin, F.: Replicability and scalability of mini-grid solution to rural electrification programs in sub-Saharan Africa. Renew. Energy. **106**, 222–231 (2017). https://doi.org/10.1016/j.renene.2017.01.017
18. Idenergie river turbine (2020). http://idenergie.ca/en/river-turbine/

Utilization of Blockchain, DLT, and Fintech for AI and IoT in the Digital Eras

The Usage of Artificial Intelligence in New Media

Abdulsadek Hassan(✉)

Ahlia University, Manama, Kingdom of Bahrain

Abstract. The study aims at identifying the usage of Artificial Intelligence in new media in the world. The results indicated that in Artificial intelligence, all the elements of the communication process play their role, whether in the presence of the human element in it, or in the absence of the human element and replaces it by one of the techniques of artificial intelligence and the fourth industrial revolution, the study also revealed that Artificial Intelligence will allow media professionals to analyze data, identify patterns and trends from multiple sources, see things that the naked eye cannot see.

Keywords: Artificial intelligence · New media · Media applications

1 Introduction

Traditional media faces a real challenge in the era of modern technologies and artificial intelligence, and the future is represented in its ability to keep pace with the new changes and focus on developing media according to two detailed words (change and honesty). We have to acknowledge that the power of the media outlet is determined today by the power of attracting by the media and advertising content it provides, and its speed in spreading and influencing [47, 50], which will continue in the future, but with the presence of sophisticated smart media outlets to spread the news and advertisements that take into account the desires and tendencies of the sender and the recipient at the same time, depending on the technique's Artificial intelligence [23]. It will have a clearer impact in the economic media sector than in the political, social and cultural media sector, due to the reliance of the first on the scientific analytical aspect, tables and digital market indicators convertible into more texts than the second, or what is known scientifically as text generation.

Artificial Intelligence applications invade our world, specifically the fields of social media. And the media organizations' use of so-called "chatbot" or automated spokes should be stopped, as Facebook Messenger contains 30,000 "bots" that respond to messages from members of groups or subscribers to pages [24, 62]. The same is the case on "Twitter". We all examine the results of It was carried out by Artificial Intelligence applications that reveal what is the trending news at the moment, the news or the most circulated name. Smart applications also recognize our faces through the images that we broadcast on "social media" and convert images into texts, texts into "infographics" or videos that support reports. The news that humans work on.

A. M. A. Musleh Al-Sartawi et al. (Eds.): EAMMIS 2021, LNNS 239, pp. 229–240, 2021.
https://doi.org/10.1007/978-3-030-77246-8_23

The announcement of Amazon, in December 2016, is the first store fully managed by artificial intelligence, one of the most prominent the daily applications of artificial intelligence [20], whatever the customer has to do it only when entering the store by logging in to his account on the website of Amazon, taking what they needed from the store and out quietly [24, 63], as many sensors switched on and cameras to recognize and identify the customer items purchased by the consumer, and at the end the bill of the account on the phone comes to him easily [2].

2 What Is Artificial Intelligence?

Artificial Intelligence (AI) is algorithms, programs, machines and systems explain intelligence [43] that Simulate the human Intelligence [18], and includes machines simulating human intelligence behavior [39].

Artificial Intelligence depends on many technologies such as: natural language processing, neutral networks, deep learning, physical robots and robotic process automation [12].

In the future, Artificial Intelligence is likely to change both media strategies and audience behaviors [11].

Artificial Intelligence performs many tasks without any human intervention as transferring data from Social Networks into recordkeeping system, replacing the traditional methods to new method such as reading documents to extract key provisions using natural language processing [24].

The use of Social Networks brought about a great change at all levels, as it affected all fields of life because of its important role in human communication, which took an approach that was continuously followed in all daily life practices, but this use differed based on individual and social differences [33]. These networks have played a role in opening new horizons in knowledge, the most important of which is Artificial Intelligence in their various life practices [45].

2.1 Stages of Development

The process of developing Artificial Intelligence went through five major stages that represent the life cycle of artificial intelligence, starting from the stage of understanding things, creating relationships between them [19], then realizing the entire surrounding environment, then full independence and decision-making alone [45], until we reach a fifth wave in which Artificial Intelligence excels at Human capabilities, and these stages are:

1- The First Stage - Artificial Intelligence Capable of Understanding: It is the simplest stage of Artificial Intelligence and is the ability to understand human needs through learning, where the algorithm monitors the human behavior of an individual through his usage to Social Media Applications & Internet sites and monitors the routine habits that he performs (such as: the news that he prefers, and the places he goes, and the books he likes to read [33, 61] This wave has greatly deepened in the lives of individuals, and has become a feature of human life, and it is an intuitive part of it, as the favorite news nominations that appear to users on various Internet sites, the ads that appear for products you need but you did not search for directly, and the recommendations of friends and pages on sites Social communication [48], suggested videos that appear on video sites, movies, books, and even your favorite meals that appear on the Internet; All of them are examples of the first wave of Artificial Intelligence internet [45], which became a major wave in human life nearly five years ago, that is, since 2012, and which was supported by the strong competition between social networking sites [42, 59].

2- The Second Stage - Artificial Intelligence Capable of Creating Relationships Between Variables: here, Artificial Intelligence has become able to collect a greater amount of information, arrange it, create relationships between them, and know the relationship between cause and effect. Many companies have started using smart algorithms capable of learning and creating relationships between direct and indirect variables that are often difficult to monitor [37]. This helped them to improve operating efficiency and accurately predict the future, which enabled them to better study market needs, and to predict their sales and earnings during the year [1, 58].

The companies were able to evaluate their products based on the opinions of customers on various websites and social media, and this wave became clear in 2013 with the emergence of many emerging companies that began providing Artificial Intelligence services to other companies such as Element (AI) in Canada and 4th Paradigm in China and others [6].

3- The Third Stage - Artificial Intelligence Characterized by Full Awareness: At this stage, Artificial Intelligence became able to fully awareness and perception of its surroundings and its environment, so it distinguished pictures, faces and sounds, and differentiated between different things [16], so smart phone applications appeared that solve the homework of students, by taking the student's picture for homework via the mobile application, which in turn analyzes the elements of this image and provides the required solution [46]. Artificial Intelligence may have multiple uses at this stage, so smart phones have used face and handprints as a safety element, and machines are able to distinguish voices [45], and even imitate and convert them into written text, and have also been distinguished by their ability to translate, so self-help programs that work via voice commands appeared, Such as: ALEXA for Amazon, Cortana for Microsoft, Google Assistant for Google, Jarvis for Facebook, and many others [22].

Augmented reality technologies have also appeared that can distinguish places [33], allowing individuals to take pictures of a tourist place or street and show the data on this site to the user in real time, and Artificial Intelligence is also able to filter the content, and find out if this content posted on the Internet and Social Media is prohibited or not, and then deal with it in an immediate way [25], and Amazon has used it in its (Amazon Go) store to distinguish customers and also distinguish what products they buy from their store, so an awareness of Artificial Intelligence that is able to distinguish individuals [13] things and sounds has begun to form, and the features of this wave have become clear since 2016 with the emergence of augmented reality applications and the spread of face and voice fingerprint phones [22].

4- Stage Four - Artificial Intelligence That is Independent of Itself and is Able to Make Its Own Decisions: In this stage, Artificial Intelligence is characterized by its ability to autonomy and the ability to make decisions independently, as all the previous stages merge, starting with understanding and then creating relationships between several variables that help it awareness and perception of the surrounding environment so that it becomes able to act and deal with it completely independently [22, 55], and this is the fourth wave, where Artificial Intelligence becomes independent in itself and is able to absorb all the matters around it, and take the appropriate decision [11]. So, the main wave of this stage may become with the year 2020 with the spread of Social Media we will find self-driving cars in the streets, drones and flying taxi projects in the sky [33], and the emergence of robots massively on the service institutions to provide their services to the public, and in hospitals to provide medical services and carry out operations. Surgical, and in military battles to fight instead of soldiers [22], where the machine turns into a semi-human capable of hearing, seeing, perceiving, distinguishing matters, analyzing them and making a decision about them [44, 52].

5- The Fifth Stage - Artificial Intelligence is Capable of Developing Itself: As a result of the unprecedented ability of Artificial Intelligence to collect information from all sources, whether from books, Social Media, websites or videos, analyze it and create relationships between them in record time, the Artificial Intelligence will become smarter, perhaps more [4, 62]. Intelligence from a human being made by himself as a result of his ability to learn that exceeds the ability of man, so that it becomes able to design another similar machine, or treat a technical defect that occurs with it, so it develops generations of himself [23], or finds solutions to problems that are difficult for a person to address, or realize that the sources energy is the secret of his existence, so he tends to control it [5], or he realizes that strength is the secret of survival, so he develops weapons that humans cannot confront, so that humans turn into followers of the machines that they invented, developed and taught [21, 57], so that humans find no solution except in entering into an inevitable war of fate against machines, and prophecies are fulfilled [27]. Science fiction movies and the prophecies of scientists who have acknowledged that Artificial Intelligence will be the cause of the death of humanity [13].

Therefore, it was necessary to prepare for a stage in which Artificial Intelligence would be controlling human life, and work must be done towards establishing a comprehensive legal, value [23] ethical and institutional system that preserves human rights as well as preserves the development of this industry that will change the form of human life [37]. but it must ensure that the machine does not control the human being. And that the human being remains the main control over it [10, 50].

From the previous view, Social Media jobs require some kind of innovation and creativity, and there are many roles that Media Professionals now play completely far from innovation, such as writing [23], reviewing texts, or even reading news bulletins. All are easy roles that the "robot" can perform without human intervention [35]. The world is now is heading to a new media, which is Artificial Intelligence media, a term that many have not mentioned, and coincides with the techniques of the Fourth Industrial Revolution [23], as the media seeks to employ the techniques of Artificial Intelligence and the fourth industrial revolution to the fullest [37]. The Artificial Intelligence media does not need all that cost, or camera crews, broadcasters, and informers, which will have a positive impact on the media industry and reduce its cost [2].

3 The Types of Artificial Intelligence

It has classified Artificial Intelligence technology in three categories: artificial narrow intelligence (ANI), general Artificial Intelligence (AGI), and artificial superintelligence (ASI).

3.1 Artificial Narrow Intelligence

It refers to Artificial Intelligence systems that can only perform specific tasks independently and with capabilities that mimic human capabilities. This means that these systems are limited in capabilities and their capabilities do not exceed the tasks for which they are designed [16, 56].

All contemporary applications of (AI), including the most complex and most capable applications, fall into this category. With reference to the first method of classification, it includes interactive machines with limited memory without exception for systems that use machine learning and deep learning [17, 59].

3.2 Artificial General Intelligence

It means the ability of Artificial Intelligence to learn, perceive, understand and act just like a human being by simulating human capabilities [47]. These systems will be able to build diverse capacities and reach connections and generalizations across several domains, which will greatly reduce the time required to train these systems [13].

3.3 Artificial Superintelligence

The development of super AI is the pinnacle of research in this area. It based on the fact that artificial general intelligence will be the most capable form of intelligence on Earth [23] when it is developed and advanced, the artificial superintelligence will be distinguished at all levels thanks to its superiority in terms of data processing and analysis speed, memory and decision-making capabilities [41].

3.4 Artificial Intelligence Tools

1- Reactive Machines
It is the oldest system of artificial intelligence, with a very limited ability, and it simulates the ability of the human mind to respond to different types of stimuli [49].

At the same time, interactive machines do not use memory, and therefore they cannot benefit from their previous experiences in guiding their subsequent steps, in other words, they lack the ability to learn [43].

This type of AI system also automatically responds to a limited set or combination of inputs only. An example of this type is the Deep Blue device, which was developed by the American company IBM in 1997 [34, 60].

2- Limited Memory Machines

In addition to having the capabilities of interactive machines, machines with limited memory can employ historical data in making decisions, and most of the currently known applications of Artificial Intelligence fall into this category [11] including deep learning systems that are trained with huge amounts of data that are stored in their memory as a reference for a solution to future problems [40].

For example, thousands of images and their description are being used to train (AI) based image recognition systems, and these systems learn to relate images to their names. Consequently [47], when other images are shown on these systems, they depend on the images used in the training to perceive the contents of the new images in many Social Media sites [39] The learning experience determines the accuracy of these systems in naming new images.

Examples of AI machines with limited memory range from chatbots in instant messaging and other apps, and virtual assistant apps in smartphones, all the way to self-driving cars [43, 52].

3- The Theory of Mind

In contrast to the previous two types, whose applications are widely spread, systems based on theory of mind are still in the development phase. It represents the next level of advancement of artificial intelligence [39]. By distinguishing their needs, emotions, beliefs, and thought processes, they will be able to better understand the entities with which they interact [22].

While artificial emotional intelligence is an emerging field that attracts the best researchers in the field of artificial intelligence, achieving the level of theory of mind necessarily requires an evolution in other branches of artificial intelligence [48], given that understanding human requirements needs to visualize human beings as individuals who need different factors in forming their minds, this method is used widely in many Social Media sites today [8].

4- Self-aware

This stage is considered the last stage in the development of Artificial Intelligence systems, as it has so far not exceeded the limits of assumptions, as it means self-conscious artificial intelligence [47], where this awareness reaches a point very close to the human mind to the point that allows it to realize its existence and develop self-awareness [29].

This type is the desired goal of all (AI) research, although it still takes decades and perhaps centuries before it is actually achieved. A self-aware will not only be able to understand and stimulate emotions in those with whom it interacts, but it will also have its own emotions, needs, beliefs, and possibly desires [21, 53].

Although the self-awareness of Artificial Intelligence and its development may support human civilization progress, it is likely to lead to a catastrophe that may threaten the existence of human race [32]. And through the awareness and perception of these machines themselves, they may reach the point of preserving their existence, which may mean a direct or indirect threat to human race. Consequently [31], it will be easy for them to compete with human minds, and they may reach to surpass them and to develop detailed plans that target humanity, this stage is the future stage of the use of new media which will revolutionize the entire communication process [15].

4 Media Applications of Artificial Intelligence

The technological development in the field of communication has achieved a tremendous revolution, the details of which have been reflected on the internet, Social Media, satellites, digital technology and digital economy [10, 15], and computer devices automation, and everything related to the outputs of artificial intelligence, and those effects led to forming a unique case, and crystallizing new media concepts and mechanisms that have been credited with the emergence of a modern contact map, which combines the visual, audio, and print, and can conveying content very quickly to groups and individuals, and a state of interaction also occurs [9].

In some cases, technology is assigned roles that are rather than humans. Artificial Intelligence techniques have provided an advanced concept in the field of media known as "Media Intelligence Artificial" or Media Robot [17], which is likely to lead to transformations media institutions. It is also expected to represent the Artificial Intelligence systems as a unique case in collecting News and writing them, apart from human effort, especially [36].

Artificial Intelligence simulates behavior Humanitarian, by understanding it and turning it into programs. A computer could make decisions and search for solutions to specific problems. The way of describing and inferring them, through the information fed into the computer [26].

There is no doubt that the techniques and applications of Artificial Intelligence will lead to a technical revolution in the ability of the media to influence and address the public and the formation of public opinion [43, 54] which is required of the various media, prepare early for this, which will increase the competition and the feverish race for leadership [7].

In the media, at the level of the Middle East and the world, Artificial Intelligence applications will provide Progress and speed in conveying news and interacting with it. The Social media have smarter tools It has become clear in recent years that automation has infiltrated news organizations [23].

International standards have succeeded in producing thousands of news stories without human intervention, starting with collecting information, classifying it, and then editing and publishing it [14], in addition to weather conditions, and it is expected to expand to include the areas of all news coverage [30]. It includes the uses of Artificial Intelligence in the media, according to global experiences [27].

The Tasks of Artificial Intelligence

- Automatically producing short news topics based on statistical data.
- Track breaking news and alert journalists of new information relevant to a topic.
- Conduct faster and more accurate searches, link information quickly and efficiently, and convert it into graphs [21].
- Automatic spelling, grammar and stylistic correction.
- Check facts quickly and reliably, and discover false news, such as the algorithms that Facebook began using it to get rid of fake news [20].

5 How Will Artificial Intelligence Affect the Social Media Sector?

Artificial Intelligence will have an enormous impact on the Social media sector, and it is already starting to do so. However, it is difficult to predict exactly how (AI) will affect this field because there are many possibilities that arise from this technology. Some of these opportunities for (AI) based media are exciting and innovative, while others may seem more alarming [3, 54].

(AI) media is a real thing and has already been used by many news organizations around the world. In 2017, journalists from the Associated Press used (AI) to write more than 2,400 stories about their quarterly corporate earnings reports [21]. The Associated Press's use of Artificial Intelligence represents an important step in the advancement of journalism as we know it [14].

Artificial Intelligence will help media professionals to perform their jobs better, that it will replace humans in media work in the coming years and Artificial Intelligence tools and techniques are already used in many media institutions in order to facilitate work, the media bias towards adopting technology is divided into two types [19], the first of which is unconscious bias, and the second is the adopted bias. On the editorial policies and political orientations of each media, as well as the inability of Artificial Intelligence to interact directly and directly with readers, not to mention collecting information from sources, which is a human effort par excellence [37].

In light of the progress witnessed by the automation sector and the entry of robots in the media field, those in charge of development processes should not neglect the ethical principles of the media as an important tool to educate and inform people about the issues around them [38] indicating the need to take into account the standards of Social Media and the validity of the published materials, and that the work of Artificial Intelligence applications is vulnerable. Errors are just like humans, and they are not the magic bullet that can do everything in the media [38].

Search results on famous search engines confirm the use of algorithms that guarantee objectivity are always subject to certain criteria, inputs and outputs, so Artificial Intelligence cannot provide us with the complete, absolute or objective truth [37].

6 Discussion

The study revealed that one of the most important uses of Artificial Intelligence technologies in the media industry; They are applications to limit the increasing spread of misleading information due to the capabilities it provides in smart handling of texts and language, as well as monitoring and verifying the authenticity of digital media.

The increasing use of Artificial Intelligence technologies in Social Networks to counter fake news; After these networks became one of the main sources for obtaining news, due to their ability to provide and update information quickly and in real time.

The global trend has increased towards the use of Artificial Intelligence algorithms in simultaneous translation, and the production of news materials using robots to transmit voice and images with high efficiency, explaining that Artificial Intelligence contributes to saving time and effort for media professionals in a manner that gives them the opportunity to be creative in writing news stories and articles and provide objective analysis of media content Submitted to the masses.

One of the most important digital technologies currently used in the media industry are virtual reality technologies, which depend on computer simulations of the real world in a way that enables online readers to experience a virtual experience that takes them from the real world to the world of the story in the virtual reality, which contributes to showing more interaction with the media message.

In the future, Artificial Intelligence will be able to do a lot more than direct news reporting. Artificial Intelligence will allow Media professionals to analyze data, identify patterns and trends from multiple sources, see things that the naked eye cannot see, in addition to converting data and spoken words into long narrative text, as well as analyzing scenes of objects, faces, text or colors. And so, we will enter the era of augmented Media.

However, this rapid expansion of Artificial Intelligence on a sensitive field such as Media, makes many fears the future of media, and Media ethical values that we may lose. Fortunately, it does not yet appear that automated software can play the role of media Professionals alone. It also does not appear, after the fake news scandal on Facebook that affected the 2016 US elections, that the press is an appropriate arena for playing artificial intelligence [51].

Recommendations

- Incorporating Artificial Intelligence technology in the media industry process, especially virtual reality technology, to save time and effort on media professionals through a format that allows them the opportunity for creativity, as well as exploiting the development in the emergence of video on demand and the spread of the fifth generation of communication networks.
- Creating partnerships to use Artificial Intelligence technology to develop the media industry, conducting comprehensive studies on the media needs to define an action plan for the use of digital alternatives, and updating the media policy in line with the technological development and the needs of the consumers in the media (based on public opinion studies), as well as updating Mechanisms and responsibilities of the media policy sender to serve the recipient.
- The necessity to follow the correct professional and ethical foundations in the media in general. And that by giving attention to training the various cadres working in the media field; Especially in times of crisis, such as the period of the Coronavirus, and journalists'demand for the importance of honesty and balance in providing information and removing the cloak of personal opinions, which affects gaining the public's confidence.
- The establishment of an international federation of media professionals working directly with Artificial Intelligence technologies.
- Finding the best ways to employ the technologies of the Fourth Industrial Revolution in the media.

References

1. Adami, C.: Artificial intelligence: robots with instincts. Nature **521**(7553), 426–427 (2015)

2. Akkılıç, S.: Media representation of AI: cases of US, UK, China and Turkey, Ph.D. thesis, Istanbul: Istanbul Bilgi University (2020)
3. Ali, W., Hassoun, M.: Artificial intelligence and automated journalism: contemporary challenges and new opportunities. Int. J. Media J. Mass Commun. **5**(1), 40–49 (2019)
4. Amazeen, M.A., Bucy, E.P.: Conferring resistance to digital disinformation: the inoculating influence of procedural news knowledge. J. Broadcast. Electron. Media **63**(3), 415–432 (2019)
5. Amazeen, M.A., Wojdynski, B.: Reducing native advertising deception: revisiting the antecedents and consequences of persuasion knowledge in digital news contexts. Mass Commun. Soci. **22**(2), 222–247 (2018)
6. Al-Sartawi, A.: Social media disclosure of intellectual capital and firm value. Int. J. Learn. Intellect. Cap. **17**(4), 312–323 (2020)
7. Al-Sartawi, A.: Does it pay to be socially responsible? Empirical evidence from the GCC countries. Int. J. Law Manage. **62**(5), 381–394 (2020)
8. Al-Sartawi, A.: Information technology governance and cybersecurity at the board level. Int. J. Crit. Infrastruct. **16**(2), 150–161 (2020)
9. Al-Sartawi, A.: Assessing the relationship between information transparency through social media disclosure and firm value. Manage. Account. Rev. **18**(2), 1–20 (2019)
10. Al-Sartawi, A., Sanad, Z.: Institutional ownership and corporate governance: evidence from Bahrain. Afro-Asian J. Finance Account. **9**(1), 101–115 (2019)
11. Al-Sartawi, A.: Ownership structure and intellectual capital: evidence from the GCC countries. Int. J. Learn. Intellect. Cap. **15**(3), 277–291 (2018)
12. Al-Sartawi, A.: Institutional ownership, social responsibility, corporate governance and online financial disclosure. Int. J. Crit. Account. **10**(3/4), 241–255 (2018)
13. Al-Sartawi, A.: The effect of corporate governance on the performance of the listed companies in the gulf cooperation council countries. Jordan J. Bus. Adm. **11**(3), 705–725 (2015)
14. Al-Sartawi, A.: Corporate governance and intellectual capital: evidence from gulf cooperation council countries. Acad. Acc. Financial Stud. J. **22**(1), 1–12 (2018)
15. Al-Sartawi, A. (ed.): The big data-driven digital economy: artificial and computational intelligence. In: Kacprzyk, J. (ed.) Studies in Computational Intelligence, vol. 974. Springer (2021)
16. Barro, S., Davenport, T.H.: People and machines: partners in innovation. MIT Sloan Manage. Rev. **60**(4), 22–28 (2019)
17. Kang, C.-M., Lin, S.-J., Lin, L.: The hybrid artificial intelligence model for analyzing the default risk of non-profit financial intermediation. J. Test. Eval. **44**(5), 2045–2058 (2018)
18. Ho, C.W.L., Soon, D., Caals, K., Kapur, J.: Governance of automated image analysis and artificial intelligence analytics in healthcare. Clin. Radiol. **74**(5), 329–337 (2019)
19. Choi, D., Lee, K.: An artificial intelligence approach to financial fraud detection under IoT environment: a survey and implementation. Secur. Commun. Netw. (2018)
20. Cukurova, M., Kent, C., Luckin, R.: Artificial intelligence and multimodal data in the service of human decision-making: a case study in debate tutoring. Br. J. Educ. Technol. **50**(6), 3032–3046 (2019)
21. Davenport, T., Guha, A., Grewa, D., Bressgot, T.: How artificial intelligence will change the future of marketing. J. Acad. Mark. Sci. **48**, 24–42 (2020)
22. Dedehayir, O., Steinert, M.: The hype cycle model: a review and future directions. Technol. Forecast. Soc. Change **108**, 28–41 (2016)
23. Edwards, C., Edwards, A., Spence, P.R., Westerman, D.: Initial interaction expectations with robots: testing the human-to-human interaction script. Commun. Studi. **67**, 227–238 (2016)
24. Gondwe, G.: News believability & trustworthiness on african online networks: an experimental design. Int. Commun. Res. J. **53**(2), 51–74 (2018)

25. Gong, B., Ordieres-Meré, J.: Prediction of daily maximum ozone threshold exceedances by preprocessing and ensemble artificial intelligence techniques: case study of Hong Kong. Environ. Model. Softw. **84**, 290–303 (2018)
26. Goodfellow, I., Bengio, Y., Courville, A.: Deep Learning. The MIT Press, Cambridge (2016)
27. Gupta, M., Sikarwar, T.S.: Modelling credit risk management and bank's profitability. Int. J. Electron. Bank. **2**(2), 170–183 (2020)
28. Gupta, N.: Influence of demographic variables on synchronisation between customer satisfaction and retail banking channels for customers of public sector banks of India. Int. J. Electron. Bank. **1**(3), 206–219 (2019)
29. Hicks, M.: Programmed Inequality: How Britain Discarded Women Technologists and Lost Its Edge in Computing. MIT Press, Cambridge (2017)
30. Huang, M.H., Rust, R.T.: Artificial intelligence in service. J. Serv. Res. **21**(2), 155–172 (2018)
31. Ivancsics, B., Hansen, M.: Actually, It's About Ethics, AI, and Journalism: Reporting on and with Computation and Data. Tow Center for Digital Journalism, New York, NY (2019)
32. Reis, J., Santo, P.E., Melão, N.: Artificial intelligence in government services: a systematic literature review. In: Proceedings of the World Conference on Information Systems and Technologies, pp. 241–252 (2019)
33. Corvalán, J.G.: Digital and intelligent public administration: transformations in the era of artificial intelligence. A&C Revista Direito Adm. Const. **18**(71), 55–87 (2018)
34. Jiang, Y., Jiang, Z., Hui, Y., Dong, H., Li, S.Ma., et al.: Artificial intelligence in healthcare: past, present and future. Stroke Vasc. Neurol. **2**(4), 230–243 (2017)
35. Gulson, K.N., Webb, P.T.: Mapping an emergent field of 'computational education policy': policy rationalities, prediction and data in the age of artificial intelligence. Res. Educ. **98**(1), 14–26 (2017)
36. Kaplan, A., Haenlein, M.: Siri, Siri, in my hand: who's the fairest in the land? On the interpretations, illustrations, and implications of artificial intelligence. Bus. Horiz. **62**(1), 15–25 (2019)
37. Kietzmann, J., Paschen, J., Treen, E.R.: Artificial intelligence in advertising: how marketers can leverage artificial intelligence along the consumer journey. J. Advert. Res. **58**(3), 263–267 (2018)
38. Wall, L.D.: Some financial regulatory implications of artificial intelligence. J. Econ. Bus. **100**, 55–63 (2018)
39. Halaweh, M.: Artificial intelligence government (Gov. 3.0): the UAE leading model. J. Artif. Intell. Res. **62**, 269–272 (2018)
40. Musleh Al-Sartawi, A.M.A.: E-learning improves accounting education: case of the higher education sector of Bahrain. In: Themistocleous, M., Papadaki, M., Kamal, M.M. (eds.) Information Systems. EMCIS 2020. Lecture Notes in Business Information Processing, vol. 402. Springer, Cham (2020)
41. McKelvey, F.: Internet Daemons: Digital Communications Possessed. University of Minnesota Press, Minneapolis (2018). p60
42. Noorbakhsh-Sabet, N., Zand, R., Zhang, Y., Abedi, V.: Artificial intelligence transforms the future of health care. Am. J. Med. **132**(7), 795–801 (2019)
43. Pedro, F., Subosa, M., Rivas, A., Valverde, P.: Artificial Intelligence in education: challenges and opportunities for sustainable development, UNESCO (2019)
44. Qin, X., Jiang, Z.: The impact of AI on the advertising process: the chinese experience. J. Advert. **48**(4), 338–346 (2019)
45. Rauch: AI for good — how artificial intelligence can help sustainable development (2018). https://medium.com/@C8215/ai-for-good-howartificial-intelligence-can-help-sustainable-development-58b47d1c289a
46. Russel, S.J., Norvig, P.: Artificial Intelligence. A Modern Approach, 3rd edn. Pearson Education Limited, Harlow (2016)

47. Mikhaylov, S.J., Esteve, M., Campion, A.: Artificial intelligence for the public sector: opportunities and challenges of cross-sector collaboration. Philos. Trans. R. Soc. Math. Phys. Eng. Sci. **376**, 1–21 (2018)
48. Scoular, C., Care, E., Hesse, F.W.: Designs for operationalizing collaborative problem solving for automated assessment. J. Educ. Meas. **54**(1), 12–35 (2017)
49. Sharafi, A., Iranmanesh, H., Amalnick, M.S., Abdollahzade, M.: Financial management of public private partnership projects using artificial intelligence and fuzzy model. Int. J. Energy Stat. **4**(2), 1650007 (2016)
50. Stray, J.: Making artificial intelligence work for investigative journalism. Digital Journal. **7**(8), 1076–1097 (2019)
51. Swartz, J., Wasko, J., Marvin, C., Logan, R.K., Coleman, B.: Philosophy of technology: who is in the saddle? Journal. Mass Commun. Q. **96**, 351–366 (2019)
52. Syam, N., Sharma, A.: Waiting for a sales renaissance in the fourth industrial revolution: machine learning and artificial intelligence in sales research and practice. Ind. Mark. Manage. **69**, 135–146 (2018)
53. Sanad, Z., Al-Sartawi, A.: Investigating the relationship between corporate governance and internet financial reporting (IFR): evidence from Bahrain Bourse. Jordan J. Bus. Adm. **12**(1), 239–269 (2016)
54. Sun, T.Q., Medaglia, R.: Mapping the challenges of artificial intelligence in the public sector: evidence from public healthcare. Gov. Inf. Q. **36**, 2368–383 (2019)
55. Tay, B.T., Low, S.C., Ko, K.H., Park, T.: Types of humor that robots can play. Comput. Hum. Behav. **60**, 19–28 (2016)
56. Tucker, C.: Privacy, algorithms, and artificial intelligence. In: The Economics of Artificial Intelligence: An Agenda. University of Chicago Press (2018)
57. Marda, V.: Artificial intelligence policy in India: a framework for engaging the limits of data-driven decision-making. Philos. Trans. R. Soc. Math. Phys. Eng. Sci. **376**, 1–19 (2018)
58. Van Dijck, J.: The Culture of Connectivity: A Critical History of Social Media. Oxford University Press, Oxford (2013)
59. Van Lente, H., Spitters, C., Peine, A.: Comparing technological hype cycles: towards a theory. Technol. Forecast. Soc. Change **80**(8), 1615–1628 (2013)
60. Wang, W., Deng, S., Tan, C.-W., Pan, Yu.: Smart generation system of personalized advertising copy and its application to advertising practice and research. J. Advert. **48**(4), 356–365 (2019)
61. Weigand, E.: Dialogue and artificial intelligence. Lang. Dialogue **9**(2), 294–315 (2019)
62. You, Q., Jin, H., Wang, Z., Fang, C., Luo, J.: Image captioning with semantic attention. In: Proceedings of the IEEE Conference on Computer Vision and Pattern Recognition, pp 4651–4659 (2016)
63. Zhavoronkov, A.: Artificial intelligence for drug discovery, biomarker development, and generation of novel chemistry. Mol. Pharm. **15**(10), 4311–4313 (2018)

Knowledge Management in Higher Education Institutions: Facts and Challenges

Rami Mohammad Abu Wadi[1](✉) and Lubna Sameer Khalf[2]

[1] Ahlia University, Manama, Bahrain
rwadi@ahlia.edu.bh
[2] Taibah University, Medina, Saudi Arabia
lkhalf@taibahu.edu.sa

Abstract. The Higher Education Institutions (HEIs) are the most appropriate institutions to adopt the knowledge management and its implementation because they are replete with massive strong knowledge infrastructure, while a knowledge society-imposed roles and functions, most notably the production and application of knowledge through investment in scientific research. Across the globe, the role of HEIs as knowledge providers has been scrutinized and challenged by the various stakeholders. Therefore, HEIs are putting significant efforts to manage their knowledge based strategic assets. This study aimed at exploring knowledge management at HEIs in reality and identifying the knowledge management processes of these institutions.

Keywords: Knowledge management · Knowledge sharing · Knowledge creation · Higher education institutions

1 Introduction

Knowledge have become a key driver for the economy and social progress and become an effective weapon for any institution if well managed, thus the importance of knowledge management has emerged, which is working to organize institution's experiences, skills, and knowledge to take advantage in achieving their goals, support in the decision-making process and achieve creativity and innovation, which results in achieving excellence and competitive advantage. The HEIs are the most appropriate institutions to adopt the knowledge management and its implementation because they are replete with massive strong knowledge infrastructure, while a knowledge society-imposed roles and functions, most notably the production and application of knowledge through investment in scientific research.

Although the importance of educational research as a fundamental pillar of human development in the community, and a necessity for the development of education, modernization and solving its problems, but it has seen a noticeable decline in recent times, which makes it most in need to apply knowledge management to development and improvement. Therefore, it emerged the HEI's need for knowledge management to be able to move from traditional management to more effectively management which

A. M. A. Musleh Al-Sartawi et al. (Eds.): EAMMIS 2021, LNNS 239, pp. 241–248, 2021.
https://doi.org/10.1007/978-3-030-77246-8_24

encourage the operations of creation and dissemination of knowledge and its application in order to improve the performance of educational research and productivity and access to a degree of excellence and innovation. This study therefore reviews the articles and research writing from peer reviewed Journals to understand the scholarly works on the topic of KM practices, enablers, and barriers in HEIs.

2 Knowledge Management

Knowledge is most valuable intangible assets for the society. Any society with strong base offers competitive advantage to the society. Whether it is society or organization, the decision making is based on the knowledge which must be sufficient, reliable, accessible, and independent. The knowledge-based decision-making offers lots of the advantages to the organizations and society in general.

Liebowitz (1999) defines knowledge in unique way by combining numbers with alphabets which compel the user to learn new things [1]. According to the definition, data forming from number and alphabet are used by different people, organizations, and society in different way in different context as per their requirements. The knowledge can be acquired by observations, interpretations, and actions.

Higher education inherently is used to assume as knowledge-intensive entity because these are always focused on knowledge creation, maintenance, and dissemination. Knowledge is always top agenda in HEIs [2, 3]. HEIs have traditionally used as an entity that will transfer the knowledge to the students that will enable students to contribute to society growth upon graduation [4].

KM in higher education institutions have always been involved in knowledge management. The three missions of HEIs, research, education, and service to society, are closely related with knowledge dissemination, knowledge transfer and knowledge creation, respectively [5, 6]. However, to remain competitive in the knowledge economy HEIs need to manage their knowledge processes within the context of a deliberate KM strategy.

2.1 Knowledge Management in HEIs

According to Lahijanian (2012) there is a lot of evidence that demonstrates the possibility of benefiting from theories and practices of knowledge management in higher education institutions, from looking at universities as representing the most important centers of knowledge production and dissemination [7]. Gill (2009) adds that the educational organization represents in its essence an integrated system for the production, dissemination, and distribution of knowledge, which is one of the basic processes in knowledge management [8].

Kidwell et al. (2000) argue that KM is vital to HEI, bringing crucial benefits to educational institution processes such as research, curriculum development, student and alumni services, administrative services, and strategic planning [9]. The challenge is convert the knowledge that currently resides in each individual/service and make it widely and easily available to any faculty member, staff person or other actors. Bhusry and Ranjan (2011) pointed that the main challenge in HEI is to create a knowledge environment, and the recognition of knowledge as intellectual capital [10]. Effective KM requires significant change in the culture and values, organizational structures, practices, and systems. In addition, several studies reveal the nascent nature of KM in HEI and the urgent need to adopt information technologies that address the needs of the initiatives and practices [11]. According to some authors [12, 13] HEI have many particularities and specificities, which should be taken into consideration, since they impact the knowledge management issues:

- Knowledge management is characterized by the nature of sustainability, which means that knowledge management is not just a transient, time-bound project, but rather it must be a philosophy adopted by higher education institutions on a permanent basis that entails an integrated and lasting knowledge management strategy.
- Knowledge management processes are expensive. Therefore, applying knowledge management in universities requires investing extensive resources, and for this reason knowledge administrators must pay attention to the return realized from knowledge management and be a justification for investing in the necessary financial and human resources.
- Knowledge management requires the availability of knowledge managers, so the application of knowledge management in the university requires the presence of a specialized, self-contained administration affiliated with the higher management of the university.

2.2 KM Practices

Knowledge management generally consists of a set of key practices that should be available and working together in an integrated manner in order to achieve the desired goals behind the application of knowledge management in different organizations. In HEIs, databases, directories, procedural handbooks, and e-mail messages are amongst the examples of knowledge codification practice. By reviewing different works in the field of KM, the study noticed many KM practices [14, 16] reported that there are six KM practices in HEIs as follow:

2.3 Knowledge Creation

This process is characterized by the creation of knowledge through the use of investigative tools, and the discovery that enables members of the university institution to access new innovative ways to perform the required tasks and processes or bring them in from outside from their external sources. Examples of knowledge creation processes: scientific research activities and development [15]. Despite the acknowledgement that universities should be regarded as knowledge- creating institutions on account of their heavy involvement in research, only a few studies have investigated the knowledge creation role of universities.

2.4 Knowledge Capture

Knowledge capture occurs when you gain valuable knowledge and are closely related to the current or future needs of the university. Usually, a group of methods are used to capture the knowledge obtained from the university's external sources, such as: making benchmarking comparisons of best practices used in other HEIs, attending conferences, seminars, and workshops, contracting with consultants, and monitoring economic, social and technical trends of society and the surrounding environment, collecting data from faculty, students and other competitors, exploiting available resources, contracting new faculty members, and building knowledge links with other higher education institutions.

2.5 Knowledge Organization

This process is characterized by arranging, revising, and then organizing the new knowledge that is created by performing filtering operations with the aim of identifying the positive dimensions of knowledge of the products and different services provided by HEIs. It also includes placing knowledge that is created in a context that allows for practical application, and constantly reviewing it to ensure its modernity and its relevance to reality.

2.6 Knowledge Storage

It is characterized by classification, storage, and retention of knowledge in appropriate formats and templates, enabling others in HEIs to access them quickly and easily. In this framework, it is possible to take advantage of database management systems, digital storage repository techniques for data and information, as well as the use of expert people, practical guidebooks, and web applications such as: e-mail messages and other applications.

2.7 Knowledge Dissemination

This process includes the distribution and dissemination of knowledge in procedurally useful formulas and templates that meet the special needs, specific to HEIs, knowledge is usually published and shared using a common language that uses standardized tools that can be understood by all users without exception. Examples of procedural mechanisms usually used by HEIs to transfer and exchange knowledge include books, published periodicals, presentations, Web sites available on the Internet, teaching and education activities, and official published policies and reports.

2.8 Knowledge Application

It is characterized by applying the acquired knowledge to new situations and contexts that enable HEIs to learn and generate new knowledge. During this learning process, a variety of processes of analysis and critical evaluation should be undertaken with the aim of generating new patterns of knowledge that can be benefited from their use and applied in the future. Examples of procedural tools used by universities and HEIs in this regard include: applying organizational decision-making systems at the organizational level, encouraging creativity and innovation, managing relationships that link the university with stakeholders, and the practical application of the acquired knowledge may result in generating other new knowledge, or updating and developing existing knowledge that will then need to be stored and retained In the databases of HEIs.

3 Knowledge Management Enablers and Barriers

Knowledge management enablers are factors or institutional mechanisms that stimulate knowledge creation, facilitate knowledge sharing [16], and promote knowledge transfer [17]. Barriers, on the other hand, are factors that have a negative effect on KM and the likelihood of its being beneficial. The literature on KM in higher education identifies a wide range of enablers and barriers, such as organisational culture, rewards and incentives, community partnership, leadership, technology, and organizational structure.

3.1 Organisational Culture

It refers to the prevailing organizational climate, or patterns and templates for participation in the circulation of knowledge related to behaviors, perceptions, degree of openness and incentives provided to members of the university community to encourage them to effectively apply knowledge management systems and initiatives. However, the role of culture is complex and contested. For example, researchers have suggested that the culture in universities is individualistic and can be self-serving and instrumental [5] and that academics selfgovern and tend to work independently [18]. Tsui et al. (2009) (a) argue that a knowledge-sharing culture needs to be built [19], but this may require significant change in the culture and values of HE.

3.2 Technology

It refers to the basic infrastructure available from devices, equipment and technical systems that enhance knowledge building and distribution efforts at the HEIs level. Further, there is a consensus that whilst the use of appropriate information communication technologies can help universities to move towards a knowledge-based learning organization, a 'good fit' between information technology, socio-organizational factors and a sustainable organizational culture is also required [20]. On the other hand, some studies disagree on the importance of IT in knowledge sharing. Fullwood et al. (2013) found that academics were neutral as regards the importance of technology, possibly due to their high level of autonomy and engagement in disciplinary communities [5].

3.3 Rewards and Incentives

It usually aims to involve all employees, and members of the university community at all levels in promoting cultural development efforts and initiatives, providing them with tangible material incentives, and enhancing their motivation for development, as well as building an academic environment capable of advancing knowledge management efforts, and therefore appreciation and rewards for the contributions of workers in the development of knowledge management processes based on accurate procedural indicators of their performance in knowledge management. This poses a challenge for universities who need to ensure that incentives recognize academics' contributions to any knowledge sharing system [21] and fulfill their expectations of positive outcomes of knowledge sharing, both in terms of extrinsic rewards and in terms of the development of relationships [5]. Academics expect their engagement in knowledge sharing to improve and extend their relationships with colleagues, and to offer opportunities for internal promotion and career development in other universities [5, 22].

3.4 Leadership

It includes the university's ability to link knowledge management behaviors and its organizational strategies, as well as identifying opportunities for growth, upgrading the value of knowledge management, publishing, trading and sharing the best strategies suitable for practical application, facilitating the task of organizational learning occurrence, and providing codified performance indicators that can be used to assess the impact of management Knowledge. Previous studies have concluded that leaders can play an important role in initiating KM [6, 8], since they control the knowledge processing environment [23]. However, Fullwood et. al. (2013) did not find leadership to be central to knowledge-sharing [5].

3.5 Organizational Structure

The organizational structure affects the mechanism and method of transferring, exchanging, storing, processing, retrieving and accumulating knowledge. Therefore, there must be a flexible organizational structure that helps to perform knowledge management operations efficiently and effectively, which is reflected in the performance of workers. Adhikari (2010) suggests that both formal and informal organisational structures can be important, with factors such the physical layout of offices facilitating social interaction and communities of practice [20].

3.6 Community Partnership

In essence, universities are organizational institutions that do not operate in a vacuum, as they are closely linked to strong and solid relations with their local communities by entering with them in strategic partnership relations that depend on the exchange of political, economic, social, and cultural interests. Hence, the importance of the local community participation in pushing the university's efforts towards applying effective knowledge management systems. Universities in developing countries generally face greater challenges in such alliances because they look to the Government to provide the overall framework for developing these linkages, which requires the formulation of policy directions and reward systems [24].

4 Conclusion

Higher education contributes to preparing human resources in society, which is a strategic investment for every country, which contributes to providing manpower and skilled hands that the labor market needs. The higher education system around the world has witnessed many of the changes required by these contemporary scientific, civilizational, and informational developments. The performance of higher education institutions depends on applying administrative ideas and theories in them, where knowledge management contributes to developing the performance of public and private institutions and businesses for the better.

HEIs have a very large knowledge compared to other business sector institutions, and knowledge is available in its human resources from its employees as faculty, employees and students, and knowledge is also available in its material assets from libraries, research centers and the information network, and this in turn needs an effective management that invests knowledge In solving the problems facing HEIs and the development of administrative processes in them, improving performance and developing their role in society.

The application of knowledge management requires the availability of a set of important elements such as that the organizational structure be more suitable for knowledge management so that it leads to more independence in decision-making, and helps to work with team spirit, and the organizational culture must expand to contain the many aspects of knowledge management, and besides that There must be leadership that encourages the adoption of knowledge management as it must have the ability to explain the vision to others, and set an example for them. In addition to the capabilities provided by modern information technology for knowledge management, such as: the information network, data stores, the internal network, and knowledge work systems, which facilitates knowledge management in organizations.

References

1. Liebowitz, J.: Building Organizational Intelligence: A Knowledge Management Primer. CRC Press (1999)
2. Cronin, B., Davenport, E.: Knowledge management in higher education. In: Bernbom, G., (ed.) Information Alchemy: The Art and Science of Knowledge Management, vol. 3, EDUCASE Leadership Strategies (2001)
3. Rowley, J.: Is higher education ready for knowledge management? Int. J. Educ. Manage. (2000)
4. Tippins, M.J.: Implementing knowledge management in academia: teaching the teachers. Int. J. Educ. Manage. (2003)
5. Fullwood, R., Rowley, J., Delbridge, R.: Knowledge sharing amongst academics in UK universities. J. Knowl. Manage. (2013)
6. Ramachandran, S.D., Chong, S.C., Wong, K.Y.: Knowledge management practices and enablers in public universities: a gap analysis. Campus-Wide Information Systems (2013)
7. Lahijanian, A.: Knowledge Analytical Approach on Promoting Environmental Management (2012)
8. Gill, A.: Knowledge management initiatives at a small university. Int. J. Educ. Manage. (2009)
9. Kidwell, J.J., Vander Linde, K., Johnson, S.L.: Applying corporate knowledge management practices in higher education. Educ. Q. **23**(4), 28–33 (2000)
10. Bhusry, M., Ranjan, J.: Knowledge collaboration in higher educational institutions in India: charting a knowledge management solution. Int. J. Comput. Sci. Issues (IJCSI) **8**(5), 332 (2011)
11. Ramakrishnan, K., Yasin, N.M.: Knowledge management system and higher education institutions. International Proceedings of Computer Science and Information Technology **37**(1), 67–71 (2012)

12. Kok, A.: Intellectual capital management as part of knowledge management initiatives at institutions of higher learning. Electron. J. Knowl. Manage. **5**(2), 181–192 (2007)
13. Van Deuren, R.: Capacity development in higher education institutions in developing countries. In: Massachusetts School of Management Working Papers, (2013/30) (2013)
14. Matin, E.K., Nakhchian, A., Kashani, B.H.: Effect of employees' entrepreneurial orientations on knowledge management in small and medium enterprises in Iran. J. Basic Appl. Sci. Res. **3**(3), 608–617 (2013)
15. Alexandropoulou, D.A., Angelis, V.A., Mavri, M.: Knowledge management and higher education: present state and future trends. Int. J. Knowl. Learn. **5**(1), 96–106 (2009)
16. Lee, H., Choi, B.: Knowledge management enablers, processes, and organizational performance: an integrative view and empirical examination. J. Manage. Inf. Syst. **20**(1), 179–228 (2003)
17. Ranjan, J., Gera, R.: Bridging the gap in knowledge transfer between academia and practitioners. Int. J. Educ. Manage. (2012)
18. Goh, S.K., Sandhu, M.S.: Knowledge sharing among malaysian academics: influence of affective commitment and trust. Electron. J. Knowl. Manage. **11**(1) (2013)
19. Tsui, E., Tian, J., Nakamori, Y., Wierzbicki, A.P.: (a). Knowledge management and knowledge creation in academia: a study based on surveys in a Japanese research university. J. Knowl. Manage. (2009)
20. Adhikari, D.R.: Knowledge management in academic institutions. Int. J. Educ. Manage. (2010)
21. Tsui, E., Arntzen, A.A.B., Worasinchai, L., Ribière, V.M.: (b). An insight into knowledge management practices at Bangkok University. J. Knowl. Manage. (2009)
22. Cheng, M.Y., Ho, J.S.Y., Lau, P.M.: Knowledge sharing in academic institutions: a study of Multimedia University Malaysia. Electron. J. Knowl. Manage. **7**(3) (2009)
23. Martin, J.S., Marion, R.: Higher education leadership roles in knowledge processing. Learn. Organ. Int. J. **12**(2), 140–151 (2005)
24. Bano, S., Taylor, J.: Universities and the knowledge-based economy: perceptions from a developing country. Higher Educ. Res. Develop. **34**(2), 242–255 (2015)

Artificial Intelligence and the Covid-19 Challenge

Ismail Noori Mseer(✉)

Ahlia University, Manama, Bahrain

Abstract. The exploration paper looks to recognize how uses of human-made reasoning were used during the flare-up of the Coronavirus. The realities demonstrated to the states of increasing speed and assistance to which these applications contributed significantly to information and data investigation and drew precise outcomes. A gigantic measure of data is straightforwardly identified with the Coronavirus, notwithstanding the colossal effort of data relating to sicknesses near it, as it gives a valuable perusing dependent on correlation and stock. It has been known as the utilizations of human-made brainpower and A.I. in lessening the component of time regardless of whether finding the hereditary strain of the infection or at the degree of antibody made. Simultaneously, these applications have demonstrated their most extreme advantage in observing the pestilence's development and restricting the speed of its spread.

Keywords: Covid19 · Pandemic · Artificial intelligence

1 Introduction

The entire world has focused on the Microsoft Bing application, where a guide [1] of the spread of the COVID-19 infection and the awful dangerous conditions influencing everybody have raised. The strikes proceed from the total and complete conclusion of the Chinese monster to the disease that no longer recognizes the first class and the general, pioneers, pioneers, or essential individuals. Everybody lives in the fixation on sterilizers and disinfectants, making a point to wear defensive covers and utilizing sterile palms. The motto that humankind settled upon today depends on "Be at home." Considering keeping up the most significant measure of segregation. As the roads of the cosmopolitan urban areas were purged of them and the minimal ones, they became apparition urban communities. Until every other person disregarded everybody, no handshakes, no arms, the suggested dividing was at least one meter. Everybody lives in a group of dread of contamination, the dad cautions of his child, and the family is quick to keep a protected distance, evading the evil of Coronavirus as all of the humankind got represented with their eyes and center towards web-based media, those that were not without energy on occasion and anticipation now and again. Where the cover between current realities and gossipy tidbits.

It was not without the development and the development of the paranoid fear about what is befalling the world. Particularly when increasingly more online media pioneers

A. M. A. Musleh Al-Sartawi et al. (Eds.): EAMMIS 2021, LNNS 239, pp. 249–262, 2021.
https://doi.org/10.1007/978-3-030-77246-8_25

are citing writings having a place with prominent scholars, and worse than the spread of the content that has a place with Martin Rees in his book "Our Last Hour" gave in 2003, in which he says in a real sense (I wager on 1,000 dollars That there will be a natural assault in 2020, wherein 1,000,000 individuals will be killed ?! [2] Where the detail that raises more doubt a lot about the exactness in deciding the year 2020 as though the writer composes content for an artistic film. Cynical dreams about the eventual fate of humanity, what is as yet given by more recognized scholastics having a place with numerous old colleges and various [3], where he is fixated on dread and tension about the quick advancement states of major logical tests, and what can result from calamitous mistakes, which can put the whole fate of the planet Earth in danger of end and demise. Inside the setting of cynicism, a CNN satellite station went toward reviewing the 2011 Contagion film [4] "Disease" (Hollywood), whose occasions are like what is strikingly occurring in the momentum world, mainly when the pandemic beginnings from China to spread the world later.

1.1 Significance

Human-made reasoning applications have demonstrated their significance even with the most serious pandemic confronting humanity in the fallout, after the spread of the Coronavirus, a scourge of all locales of the Earth. These applications were portrayed by precision, association, thorough correlations, and dependable outcomes. The data that arrives at a vast number of records were used in light of the grouping and arranging gave by the components of applying human-made consciousness. Something else, controlling or in any event, profiting by this pile of data was viewed as an unimaginable blow. If humankind has the advantage of This store of information, the administration stays restricted.

1.2 Tricky

The exploration endeavors to show a critical issue on which most human efforts are based, notwithstanding the danger that Coronavirus stances to human existence on the planet. Be that as it may, a portion of the foundations working in the field of computerized reasoning or medication producers are as yet contemplating how to keep their proprietary innovations. Subsequently, transparency, collaboration, trade of encounters, and serious cooperation in joining endeavors are considered the doorway to creating answers for humanity's emergency.

1.3 Targets

The investigation looks to uncover the dangers presented by the Coronavirus and how official and administrative specialists manage the infection toward the start of its development. The objective here depends on the significance of recognizing the game-plan of private and nonlegislative associations, as realities have demonstrated their energy and their work proceeds away from the traditional everyday practice and the conventional progressive system. While the facts saw during the times of the pandemic that cleared the human network because of the Coronavirus, the significance of computerized reasoning applications as a logical application equipped for improvement, association, and alteration and guarantees more certain outcomes in the field of encounters that humanity is experiencing in its vast future. Information and dream. Humankind has made significant steps in the field of logical exploration. Indeed, even the logical and epistemological advancements made by contemporary researchers, science groups, research organizations, and renowned colleges have made individuals live in states of consolation about the clinical and wellbeing fate of humanity. Particularly after the examination moved toward the admittance to unravel the charms and images of the human hereditary guide [5], since the year 2001, which is viewed as a stage that summed up the whole clinical history since the times of Hippocrates, promising endeavors that made everybody extraordinary and adherents live in states of lack of concern and enjoying a feeling of hopefulness about the imminent wellbeing hazards, while they took a gander at the past with distress over the scourges and pandemics that encompassed humanity and brought down a great many individuals until it was known as a pandemic The Plague (Black Death) [6]. Still, fate needs to make the offspring of the second decade of the twenty-first century (7) live in circumstances of cover among science and fantasy until they recover the number twenty condition, which goes with the start of every century spread out, killing 100,000 individuals. Furthermore, the 1820 cholera slaughtered many individuals—also, Spanish influenza in 1920, which guaranteed a great many lives. Simultaneously, the oddity arrives at its most extreme to be human despite the Coronavirus in 2020.

Vulnerability conditions are currently covering the world from east to West; Coronavirus will influence everybody, reliable and feeble, late and progressed, north and south. Indeed, even everybody has been calling out misleadingly, uses of human-made consciousness and the legitimacy of its investigations in the field of hereditary designing, trying to locate a solid perusing, add to opening the specialty towards considering this infection and destroying its substance [7]. This is by figuring the desire for the most compelling logical strategy in removing successful treatment equipped for confronting this infection [8]. This perplexed the psyches of trained professionals, having demonstrated its capacity to change until it got subject to hypotheses and assumptions. This is obvious through the positions taken by the nations of administration and logical greatness. After he emigrated from China, which remained by him through coordinated approaches, he started to extend in the terrains of the industrialist majority rule West with a terrifying soul. Indeed, even the crazy shout [9] from the heads of the created world. At the same time, trust stays on computerized reasoning and its promising applications in the field of shortening more stages towards finding the fitting antibody, and fierce and valuable commitment to drawing orderly plans fit for retaining how this infection works and spread, through the undertaking to contemplate the hereditary guide of the disease and work to build up a duplicate of it for study, assessment, and recognizability.

1.4 Past Encounters

It isn't the best proof of the means taken by the computerized reasoning research facilities in China, which were needed to remain on the hereditary chain of the infection during one month of the emergency disturbance in the year 2020 when the labs in 2003 during the SARS infection went up against more Months. The issue was not restricted to the

Chinese labs. Yet, the Australian commitment is apparent in the creation of a duplicate of the infection, which experts are hopeful about, as the guide that will be rehearsed by the laborers in the field of clinical preliminaries, where the assumption that depending on this duplicate will have a viable impact in decreasing the period time to acquire the fitting antibody in a period that doesn't surpass three months [10].

There is still trust in logical endeavors, as there is outright trust in the means that researchers had some expertise in their labs. As the severe collaboration by foundations and organizations working in human-made reasoning is featured, to ask all endeavors towards confronting this pandemic that compromises the destiny of humanity. This is apparent from the truth of the exertion made by the different organizations and organizations working in the field of human-made consciousness. The Longevity Vision Fund put forth no attempt to seek to activate capital and send solicitations to establishments operating in the field of biotechnology to discover a fix fit for halting the invasion of Virus Corona. One of its branches operating in the region of industry The medication is to quicken its exercises, so it has had the option to arrive at promising outcomes [11]. The issue didn't stop there; instead, the feeling of obligation caused this organization to give the good results that arrived close enough for everybody. Decisively, she distributed all the information and data on the organization's site, out of a forward-looking vision towards facing the pestilence that is clearing the world, and that confronting this savage plague assault, all endeavors require to be joined no matter what—taking into account that the truth requires a firm stand and an undertaking dependent on offering free assistance, away from rivalry or coordinating structures towards the issue of material benefit. The circumstance is more huge than material or serious increase, particularly regarding the eventual fate of the person on Earth [12]. The most pivotal issue is that mainstream researchers 3 have solid trust in the estimation of the field of human-made brainpower, which is a significant commitment to dealing with the Covid scourge emergency, particularly since past encounters have exhibited the capacity of this field to improve and create execution notwithstanding illnesses. It saves exertion, cash, and time in overseeing information and data and adds to characterizing subject matter experts' obligations and characterizing ways and utilization of medical services concerning the overall circumstance looked for by the wellbeing foundation despite the emergency [13]. The Republic of China has had productive utilization of uses of human-made consciousness during the new Coronavirus emergency and on different tracks. On the immediate clinical level, these applications were utilized to arrange crafted by clinical staff in clinics and to give a classified arrangement dependent on direct data, which contains the course of the connection between emergency clinic laborers (specialists, attendants, and administration laborers) until this association has created extreme outcomes Positive, it was to lessen the dangers to clinical staff presented as far as possible. With the assistance of the robot and ambulances controlled through the human-made reasoning framework, in offering direct clinical types of assistance to patients, if it was fair and the square of finding and potential dangers that could straightforwardly influence the clinical staff or work to quicken the consequences of the analysis, or even in the field Transfer of crisis and basic cases. As coordinated to utilize the robot's innovation, trying to speak with individuals and send alerts and give appropriate directions to anticipation or the best approach to manage crises [14]. Truly, direct experience demonstrates the achievement that has been accomplished by China in confining the scourge. However, it has spread quickly all through the Earth. The most specific issue here is identified with the strategy for the encounter. The technique for social detachment, and segregation dependent on the fourteen-day rule, might have been applied in China and a few nations of the world dependent on the direction strategy dependent on the state's centralization. As the numbers expanded in the United States [15], Considering the hesitance six conditions in settling the choices identified with the best approach to confront the emergency, generally, as the judgment stays identified with vital business interests, and this is apparent in President Trump's delay about segregating New York City for instance, and the monetary misfortunes that can result from this choice wide. In any case, the unequivocal showdown stays in the circle of human participation, as the desire gave by everybody towards defying this impending risk [16].

1.5 Cautioning and Alarm

A.I.'s. Immediate, major commitment arises around the COVID-19 flare-up, as Blue Dot has been the first to uncover the infection that compromises Wuhan's Chinese city. The admonition gave by this organization, which works in the field of human-made reasoning, amazingly went before the World Health Organization. The organization's work, which is situated in Toronto, Canada, depends on controlling irresistible infections by depending on information examination and A.I. [17], which is done through the affirmation of calculations that add to 8 extrapolations and derive the principles that manage the assortment work Phenomenal data. This organization had the option to arrive at critical outcomes, which was to screen the activities of an infection prompting intense pneumonia, the development of which is revolved around Wuhan in China [18]. The organization's revelation was not the first, as it had a few effective exercises nine during the 2003 SARS emergency and the years 2014 and 2016 [19]. Blue Dot's way of 0 thinking depends on the significance of attention to the quickening change covering the whole world's subtleties.

To the degree that innovation, examination, and experience develop, what undermines humanity should develop, and the extent of its danger will be more noteworthy and riskier [20]. The organization's work framework relies upon gathering more than 100,000 distributions for each day from different dialects, which number 65. Indeed, even the admonition was connected to what exactly is going on in the Chinese city of Wuhan, where the observation of the development of Coronavirus [21], and what happened already during 2003 during the emergency of (SARS) infection. Here it should be perceived that human-made reasoning has not arrived at the sought after reach, as it is as yet during the time spent experimentation, yet the outcomes identifying with it have not been appeared to demonstrate encouraging expectations. What identified with the Blue Dot organization depends on observing and determining the chance of the rise, spread, and spread of the infection on the planet? In this, the inquiry is about the adequacy of this work did by the organization. Information assortment and handling, and dependence on a variety of sources from; Social systems administration locales, Internet destinations, articles gave by legislative and nonadministrative associations and foundations, reports gave my research focuses, particular logical exploration, checking the advancement of carrier organizations, attempting to screen traveler thickness examples, and zeroing in on data gave by medical clinics, and Reports identified with epidemiological observing administrations around the globe, made this organization remain on a particular instrument, through which the strategy and example where the plague is spread across the globe [22].

1.6 Exploration in the Causes

From the truth of the focal point of the pandemic identified with respiratory sicknesses, COVID-19, where the world viewed the misfortune in China, how much theory about the birthplace and reasons for the scourge, and what amount slamming and accuse the overall population has been telling the Chinese public, about the absence of reservation in food propensities and the best approach to eat. Notably, a portion of the reports demonstrated that the wellspring of the scourge began from a bat, and another perusing related it to one of the kinds of snakes that the Chinese public used to eat. The significant thing in the entirety of this is that the scourge immediately spread and started to influence the whole planet Earth until it left no spot on the Earth aside from its heavy shadow. While the world keeps on experiencing hesitance to manage the new infection, regardless of its spread over 58 nations, the World Health Organization deferred its depiction of the scourge [23] (Pandemic). At the same time, the discussion stayed four while in transit to take a gander at the sickness dependent on assumptions and questions, however instead That the attention depended on the truth of correlation with past encounters, where the SARS infection, which killed 774 individuals and spread in 29 nations, gone on for a half year, and cost the world monetary misfortunes of forty million dollars. The administrators of Blue Dot are driven by the possibility of the significance of depending on human-made reasoning frameworks as the most encouraging methods notwithstanding irresistible sicknesses. In light of the actuation, everything being equal, and making them serve the down to earth target exercise, away from government standard and institutional conventions, this figures that the Internet has given the broadest scope of open [24] data. Simultaneously, the states of productive and valuable managing arose through opening up to numerous establishments that didn't stop for a second to give financing and liberal help by Horizons Ventures to this organization to finish its examination [25]. This depends on the variety of six and multidisciplinary fortes, as the organization depends on specialists and experts' endeavors in the field of the study of disease transmission, software engineers, and experts in the field of human-made consciousness and veterinary medication.

The errands of this organization don't stop at notice and notice of the rise or spread of a As the hover of interest, pandemic reaches out to incorporate more exercises, those identified with medical services, and general wellbeing. It is the pattern towards extending the scope of A.I. applications while giving a functioning and straightforward job for human insight. [26] The issue of man-made brainpower doesn't stop at the lines of early recognition of either illness. The correct activity depends on the significance of heading towards seeing how the illness is spreading. When Blue Dot analyzed Wuhan's Chinese city, it encouraged endeavors to connect the data gave via aircraft tickets and intercity travel objections to start the suspicions of injury and spread around the world [27]. Based on the calculation of man-made brainpower, in light of variety in observing creature, plant, and human infections [28], The desire 9 is for how the scourge will spread, its speed and sway, and what is the following objective for its pandemic. The entirety of these remaining parts is subject to the information and data; human-made consciousness is not an enchantment wand. To the extent that it speaks to exact, target treatment depends on the moral principles set in the framework [29]. Eventually, you manage unique and assorted societies. When China 0 moved in the direction of immovability in segregation measures and carefully towards shutting urban communities and sanctioning the development to a base, a few nations didn't spare a moment to coordinate the fault and fault for such activities, which were depicted as converging common freedoms. However, the conundrum arrived at its degree when it began. Those nations that have brought their hindrances up in reprimanding the Chinese experience surged towards its utilization and imitated it, having wound up at the core of an extreme pandemic, calling for an examination into all arrangements.

1.7 Potential Arrangements

Here, the central inquiry emerges about the potential arrangements that human-made reasoning can offer to confront this pandemic, which is spreading at an alarming pace. The appropriate response comes here from specialists who intervene in the combat zone in the Coruna infection as the sign that the capacities gave by the foundation of human-made reasoning have demonstrated power and proficiency regarding early location [30] of the scourge. The joined one endeavor of organizations working in this field to recognize the infection early, in light of information assortment and examination, advantage from witness records, and picture investigation, notwithstanding admonitions given by experts from the core of the occasion. Where the admonition sound was raised by a specialist working in Wuhan Chinese Hospital, that a diligent peril is the developing indications of infection [31] influencing the respiratory framework, be that two as it may, it remains a short time, particularly when managing another kind of infection. Yet, the ebb and flow advantage gave by human-made reasoning frameworks may center its job in the utilization of data, investigation, and extrapolation, and work on creating gadgets identified with observing and identification of wounds, and planning of infection, and the conceivable outcomes of [32] extension. Truly, human-made reasoning has a real and significant job in giving meaning. help, yet the choice remaining parts identified with the organization that animates and moves it [33]. Humans. For instance, China got an admonition from man-made consciousness stages working in observing the spread of many transmittable sicknesses and had the option to embrace a thorough crisis plan dependent on segregation and conclusion of schools and industrial facilities authorizing shopping exercises for people. As the contamination spread in New York City in a later period, regardless of the worldwide collaboration with the awfulness of the sickness and the risks it left under the sight and becoming aware of the entire world [34]. Artificial insight is as yet 5 in the improvement stage, and from this, it can't give the last answers for an emergency, for example, the coronary plague. At the same time, trust stays unblemished by how the investigation can be embraced for information and data identified with the general subtleties, exercises, and areas that recognize crafted by the current individual On planet earth.

This is the thing that can presently be utilized by human-made brainpower. It doesn't stop at a particular area, for example, sickness investigation, for instance, however, stretches out the course of work to incorporate more businesses and orders. Where wellbeing, the travel industry, avionics, shopping, food, poultry, veterinary, and fishing, the entirety of this are dissected by depending on exceptionally advanced and efficient calculations [35]. The showdown with these six infection remains a perplexing and burdensome cycle that calls for deliberate endeavors and the appropriation of choice dependent on the intensity and mental fortitude. Notwithstanding the pandemic's deteriorating states and the expanding number of casualties, it arrived at 1,000,000 contaminated with this infection. In any case, a few governments are as yet contrasting the monetary misfortunes that can result from the strategy of social segregation, halting work in production lines and business focuses, or receiving the group insusceptibility strategy [36]. Natural selection thinking about the pandemic focuses on the old who experience the ill effects of frail resistance and constant sicknesses. They are at first a monetary and administration trouble on the state.

1.8 Participation and Joint Effort

The significant organizations working in the field of human-made consciousness went towards initiating the strategy, embracing a technique for purposeful endeavors, and trading encounters and information, considering the epidemiological danger that influences humankind. Where Alibaba went toward building up the instrument of C.T. Picture Analytics Solutions, in China, specifically, it is a technique that helps examine C.T. examines, which work with profound learning calculations, have been tried in clinics all over China. Fit for foreseeing the chance of contamination with COVID-19 and basic pneumonia dependent on C.T. filters [37], and this strategy had a basic job in accelerating the contamination location measure multiple times over the past rates [38]. While the use of Alibaba cloud was coordinated towards considering the quality nine identified with Coronavirus within a couple of hours [39]. The Chinese organization Baidu didn't 0 stop for a second to find its partners working in the field of human-made reasoning, as it went towards building up its route framework working through satellite by zeroing in on pictures and sensors. Thermodynamics and work to screen the possible setback inside a brief timeframe and this innovation will significantly affect air terminal zones and public spots [40]. This is notwithstanding one the early location of the maltreatments that bandits can take, particularly criminal mafias attempting to exploit the emergency clearing the world to pass dubious arrangements [41]. The most basic test confronting organizations working in the field of human-made the reasoning is presently zeroing in on the conceivable commitment field in decreasing the period identified with creating a clinical immunization that adds to wiping out the infection, on the off chance that it is at the degree of tests or add to the instruments of creative techniques. Furthermore, because of the severe contamination on which this infection is based. The commitment of human-made brainpower was available at the degree of emotional investment in high-hazard confinement territories, where the robot was utilized in the fruition of the assignments too near patients conveying the infection. With striking proficiency and capacity, generally, as operations were of a deep level of exactness, speed, and association by tidying up and coordinating patient rooms and conveying drug portions, and chipping away at an immediate assessment of them and taking examples to catch up on debilitating conditions.

To follow the methodology of clean disengagement, automated airplane innovation was utilized to screen populated territories and to track the strategy for consenting to the guidelines gave by wellbeing specialists. What has had the most effect on limiting the infection and recognizing the system where it is spread? Notwithstanding utilizing four facial-acknowledgment innovations, the pattern is to screen people who are not dedicated to applying for clean allows, notwithstanding identifying fever that individuals might be presented to. The occasion to give the experience to catch up on potential instances of disease or help the infection using collaboration between the motorized stage and the cell phone organization. To diminish the spread of the scourge.

1.9 Perceptions and Estimates

The worry of experts today is to consider the conduct of the Coronavirus and the manner in which it spreads, to comprehend the instrument of development, and, as needs are, set up the suitable organization to restrict the spread. This methodology depends on capitalizing on the colossal measure of data and information identified with transportation and correspondence and the choices taken by neighborhood specialists around the globe. Also, don't dismiss every single imaginable detail, for example, interpersonal interaction destinations, and urge endeavors towards participation and correspondence with all foundation of human-made consciousness. The present reality has become a workshop whose essential mission is to develop the Coronavirus and endeavor to comprehend its development and spread and set potential plans to screen how it is applying [42] world. Accordingly, a large portion of the renowned eight examination organizations with long-standing experience has been making a decent attempt to utilize research systems. It implies that it is not founded on human-made brainpower. Yet, instead, the uniqueness of the immense region towards A.I. execution applications, given the problematic and distressing race wherein humankind got itself, concerning Coronavirus, which has been spreading at an alarming rate, is spreading everywhere in the world. Ohn S. Brownstein of Harvard University calls attention to the significance that data and information play in giving the capacity to foresee Covid conditions. The sign passed by it contains the essentialness of the story, whatever its source. In this showdown, there is not all that much, and the other is auxiliary; everything is fundamental as long as it is identified with Coronavirus; from here, the online journals, news, and hadiths gave by informal communication destinations all should be coordinated, arranged, and renamed. There is a promise for collaboration and participation between the foundation of human-made reasoning, which through this collaboration can decide the conceivable system wherein irresistible sicknesses are broke down, notwithstanding the help of governments [43]. This is with the significance of cautious thought of the best approach to organize and misuse the possibilities between AI ML and A.I., about the ideal method to accomplish explicit and exact bearings despite the changed Covid infection.

A.I. offers a wide chance in observing the manner in which this pandemic is spread, and computerized reasoning gives scientists occasions to put resources into accessible medications and the possibility of profiting by them even with this scourge. It is the data war and the manner in which this war is overseen in a way that gives the capacity to decide the model that can be utilized to restrict the spread of the pestilence. The central inquiry here depends on how the medications accessible and right now accessible can be picked and how they can be utilized in a way that doesn't meet the interests of the monster drug organizations, those that are between the choices of benefit and misfortune. At the same time, it would be useful for these organizations to reuse past restorative mixes, rather than Wasting time on tests needing in any event eighteen months, to arrive at the proper antibody, given that dependent on models that evade humanity, the results of results, and the risks that emerge from Errors in estimations because of speed. This is with the significance of mulling over that the medication advancement area in human-made brainpower is positioned better when analyzed than other clinical fields that involve the worries of human-made reasoning. As the interest in the drug, the region is multiple times that of the closest contending field [44]. Notwithstanding the far and wide goal of new companies that have gotten themselves and have entered a ton of rivalry and in the general test, until they have been endeavoring with every one of their endeavors and capacities to balance out their feet, beginning from the effort to put resources into the scourge emergency [45].

1.10 Discover an Arrangement

The idea of the arrangement shows up and spreads during the emergency that is clearing the world. Such a truism might be stunning?! Yet, this is the truth depends on the components of work and its guidelines that include numerous variables; speculation, business, regulatory, monetary, and aptitude. From this, the undertakings are observed depending on the organization's drug strength and their capacity to join in and contend. Little organizations working in the field of medication are looking for energetically towards introducing their advancement as an approach to draw in consideration of setting up and monster organizations as the more evolved organizations go towards invigorating the field of association and participation among them, to break the imposing business model of huge goliaths. Hence, the scourge of COVID-19 has become a field of limited testing and an open door for development and rise for drug organizations. The fundamental thought behind the proposed treatment depends on exploration in the decay of the Covid protein synthesis. Consequently, the Google Deep psyche automated reasoning motor has coordinated to consider six proteins identified with Coronavirus. To quicken admittance to an antibody that can be embraced as a pandemic treatment [46]. As the Hong Kong-based Insilico Medicine Company went to build up a plan that mimics Coronavirus to restrict its spread, expanding on the past experience the organization obtained during the spread of SARS in 2003 [47. In Germany, Innoplexus went to zero in on an investigation protein-related protein by depending on a medication motor plan that recognizes the medication [48]. These and different activities utilized by all the drug business, and imaginative and creative participation with human-made consciousness, need, best case scenario, an entire year, as per the assessment concurred by all specialists working in this specialization. From this, the pattern today depends on the significance of looking for accessible medications, chipping away at testing them, and the chance of utilizing them despite Coronavirus. From this reality, the Benevolent A.I. organization centered its endeavors on utilizing A.I. and computerized reasoning in the quest for accessible medications, which can decrease disease and spread [49].

2 Conclusion

The test confronting humanity depends on the inescapable of this infection, which has caused worldwide medication research centers to enter a hot competition to arrive at a flash of expectation through which humankind can recuperate. What's more, to the degree that experts in the assembling of writing and immunizations affirm that the revelation of another medication needs at any rate eighteen months, the pattern and concentrate today is coordinated towards utilizing what is accessible from past antibodies that have demonstrated to be compelling in the territory of pneumonia [50]. As the inquiry that stresses all of humanity spreads over the reasonable six times when the Coronavirus immunization is accessible and usable? Global news shows that the endeavors of more than 35 medication organizations and esteemed scholastic examination establishments have united in endeavoring industriously towards the sought after immunization industry. The indication of expectation here is hung on four investigations that have figured out how to test the antibody in creatures, and it is looking for, right away, to apply these examinations to people. This incredible shorthand is expected at that point, based on the novel Chinese collaboration that began in January 2020 to uncover the hereditary chain of the infection, given the advantage from past readings of the SARS infection. What's more, work to put the outcomes at removing all [51]. Interestingly, the infection is as yet slaughtering lives. The desire is right now centered around controlling the spread of contamination using past immunizations that have demonstrated viability in fighting numerous sicknesses [52]. To feature the devastating crossing point and the shortened rationale that describes the states of these organizations, which meets with the experience of logical participation between the worldwide medication organizations during the encounter with the SARS infection in 2003, where the SARS immunization was found in record time, humanity confronted more cataclysms and misfortunes if the reference is here to a portion of the drug organizations trying to utilize emergencies to accomplish interests. The coalition was set up in June 2019 between ten drug organizations to create anti-microbials [53] through dependence on the utilization of man-made consciousness. One of the products of this open collaboration was that the outcomes arrived at an anti-infection creation that battles 35 microbes, without bias to the organizations' proprietary advantages partaking in the test [54].

References

1. Wolfson, E.: Updated: March 27, 2020 2:53 Pm Edt | Originally Published: March 11, 2020 1:01 Pm Edt, Mapping the Spread of the Coronavirus Outbreak Around the U.S. and the World, Time Magazine
2. Rees, M.: Our Final Hour: A Scientist's Warning, p. 74. Basic Books, New York (2003)
3. Bostrom, N.: The Future of Humanity, New Waves in Philosophy of Technology Journal. In: Olsen, J.-K.B., Selinger, E., Riis, S. (eds.) p. 186 Palgrave McMillan, New York (2009)
4. Lowry, B.: The movies that predicted an outbreak like Coronavirus, CNN, 13 March 2020. https://edition.cnn.com/2020/01/30/entertainment/contagion-and-pandemics-in-movies/index.html
5. Frazer, K.A.: Decoding the human genome, Genome Research Journal, p. 1599. Cold Spring Harbor Laboratory Press (2012)

6. Benedictow, O.J.: The Black Death, 1346–1353: The Complete History, The Boydell Press, U.K., 7 American Universities Field Staff, 1974, Reports: West Europe series, vol. 9–12, p. 1 (2004)
7. Ali, M., Dapoigny, R. (eds.) Advances in Applied Artificial Intelligence. In: 19th International Conference on Industrial engineering and other Applied Intelligent System Annecy, France, June 2006, Proceeding, Springer, Switzerland, p. 972 (2006)
8. Al-Rodhan, N.R.F.: The Politics of Emerging Strategic Technologies, p. 81. Palgrave Macmillan, London (2011)
9. Amaro, S.: Italy asks for help from E.U. crisis fund as death toll spikes, 20 Mar 2020. https://www.cnbc.com/2020/03/20/italy-conte-calls-for-eu-crisis-fund-as-coronavirus-death-toll-rises.html
10. Peiris, M., Anderson, L.J., Osterhaus, A.D.M.E., Stohr, K., Yuen, K.-Y. (eds.): Severe Acute Respiratory Syndrome, p. 7. Blackwell Publishing, USA (2008)
11. Khatib, H.: Exclusive: Longevity Vision Fund invests into record-breakingA.I.DrugDiscoveryCompany, 2–13, September 2019. https://www.ameinfo.com/industry/healthcare/exclusive-longevityvision-fund-invests-into-insilico-medicine-a-record-breaking-ai-drug-discovery-company
12. Hayden Field, on March 26, 2020, inside a startup's pivot to have A.I. fight coronavirus, Insilico Medicine repurposed three of its machine learning systems to dream up treatments for Coronavirus. Its CEO says it's going well — but the process isn't without hiccups. https://www.protocol.com/ai-machine-learning-coronavirus-covid19
13. Acton, A.: Advances in Risk and Prevention Research and Treatment: 2012 Edition, p 804. Scholarly Editions, Atlanta, Georgia (2012)
14. Higgins-Dunn, N., Feuer, W.: MAR 17 2020, Coronavirus reaches all 50 states; stock futures fall, Pentagon sending respirators and ventilators, https://www.cnbc.com/2020/03/17/coronavirus-live-updates.html.
15. German Lopez March 25, 2020, The Trump admini6 station's botched coronavirus response explained, from insufficient testing to a lack of coordination, Trump's Covid-19 response has been a disaster year in the making. https://www.vox.com/policy-and-politics/2020/3/14/21177509/coronavirus-trump-covid-19-pandemic-response
16. Wu, J.: How Artificial Intelligence Can Help Fight Coronavirus, COGNITIVE WORLD, March 2020. https://www.forbes.com/sites/cognitiveworld/2020/03/19/how-artificial-intelligence-can-help-fightcoronavirus/#43da26a04d3a
17. Mena, J.: Data Mining Your Website, Digital Press, Boston (1999)
18. Simpson, M.: Canada commits $192 million to covid9 -19 sif stream; signs deal with bluedot to help track, make decisions on virus (2020). https://betakit.com/canada-commits192-million-to-covid-19-sif-stream-signs-deal-with-bluedot-to-help-track-make-decisions-on-virus
19. Mollman, S.: How artificial intelligence provided early warnings of COVID-19, 26 January 2020. https://qz.com/1791222/how-artificial-intelligence-provided-early-warning-of-wuhan-virus/
20. Covid-19: Artificial intelligence may be a pandemic lifesaver... one day, March 26, 2020, The Star. https://www.thestar.com.my/tech/tech-news/2020/03/26/covid-19-artificial-intelligence-may-be-pandemic-lifesaverone-day
21. Pressman, A.: How A.I. is aiding the coronavirus fight, magazine digital health, 16 Mar 2020. https://fortune.com/2020/03/16/ai-coronavirus-health-technology-pandemic-prediction/
22. Heaven, W.D.: A.I. could help with the n3 ext pandemic—but not with this one, 12 March 2020. https://www.technologyreview.com/s/615351/ai-could-help-with-the-next-pandemicbut-not-with-this-one
23. Ducharme, J.: World Health Organization Declares COVID4-19 a 'Pandemic'. Here's What That Means,11 March 2020. https://time.com/5791661/who-coronavirus-pandemic-declaration/

24. Abubakar, I., Stagg, H.R., Cohen, T., Rodrigues, L.C.: Infectious Disease 5 Epidemiology, p. 29. Oxford University Press (2016)
25. Bowles, J.: How Canadian A.I. startup Blue Dot spotted Coronavirus before anyone else had a six clue, 10 March 2020. https://diginomica.com/how-canadian-ai-start-bluedot-spotted-coronavirus-anyone-else-had-clue
26. Teo, D.: Two Cents: Why Artificial Intelligence can't replace us, 2 Jan 2020. https://hrmasia.com/twocents-why-artificial-intelligence-cant-replace-us/
27. Stieg, C.: How this Canadian startup spotted Coronavirus before everyone else knew about it, 3 March 2020. https://www.cnbc.com/2020/03/03/bluedot-used-artificial-intelligence-to-predict-coronavirus-spread.html
28. Chen, T.-M., Rui, J., Wang, Q.-P., Zhao, Z.-Y., Cui, J.-A., Yin, L.: A mathematical model for simulating the phase-based transmissibility of a novel coronavirus. Infect. Diseas. Poverty J. **9** (2020). Article number: 24
29. Davenport, T., Kalakota, R.: The potential for artificial intelligence in healthcare. Future Health care J. **6**(2), p. 95 (2019)
30. Ferran, M.: Should I Be Tested For Coronavirus? Here's What You Need to Know 1 About COVID-19 Tests, March 2020. https://www.sciencealert.com/how-does-the-coronavirus-test-work
31. Allen-Ebrahimian, B.: Timeline: The early days of China's coronavirus outbreak and cover2 up, 18 March 2020. https://www.axios.com/timeline-the-early-days-of-chinas-coronavirus-outbreak-and-cover-up-ee65211a-afb64641-97b8-353718a5faab.html
32. Kent, J.: Is Artificial Intelligence Mature Enough to Combat COVID -19? 18 March 2020. https://healthitanalytics.com/news/is-artificial-intelligence-mature-enough-to-combat-covid-19
33. Gouveia, S.S. (ed.): The Age of Artificial Intelligence: An Exploration, p. 260. Vernon Press, USA (2020)
34. U.S. Now Leads the World in Confirmed Cases, March 26, 2020, The New York Times, five https://www.nytimes.com/2020/03/26/world/coronavirus-news.html
35. Bhopalwala, B.: March 26, 2026, 0, COVID-19 & Machine Learning A potential Machine Learning approach that can help stop COVID-19. https://towardsdatascience.com/covid-19-machine-learning-4f064df53c43
36. Regalado, A.: What is herd immunity, and can it stop the Coronavirus? 17 March 2020. https://www.technologyreview.com/s/615375/what-is-herd-immunity-and-can-it-stop-the-coronavirus/bing.com/covid
37. C.T. Image Analytics for COVID-19. https://www.alibabacloud.com/solutions/ct-image-analytics
38. Heymann, D.L., Shindom, N.: COVID-19: what is next for public health? Lancet J. **395**(10224), 542 (2020)
39. Yu, E.: Alibaba Cloud offers an A.I. platform to support coronavirus medical efforts, 19 March 2020. https://www.zdnet.com/article/alibaba-cloud-offers-ai-platform-to-support-coronavirus-medical-research/
40. McFarland, A.: Baidu's A.I. Technology Being Used to Combat Coronavirus, one, 13 March 2020. https://www.unite.ai/baidus-ai-technology-being-used-to-combat-coronavirus/
41. Johnson, K.: How people are using A.I. to detect and fight the Coronavirus, 3 March 2020. https://venturebeat.com/2020/03/03/how-people-are-using-ai-to-detect-and-fight-the-coronavirus/
42. Godfried, I.: Machine Learning methods to aid in Coronavirus Response, 18 March 2020. https://towardsdatascience.com/machine-learning-methods-to-aid-in-coronavirus-response-70df8bfc7861

43. Ma, V.L.: Crimson Staff Writer, Harvard Medical School Professor's Artificial Intelligence System Maps Coronavirus Outbreak, 3 March 2020. https://www.thecrimson.com/article/2020/3/3/hms-coronavirustracking-tool/
44. Terry, M.: V.C. Funding for Artificial Intelligence in Drug Development Climbs to $5.2 Billion, 6 January 2020. https://www.biospace.com/article/vc-funding-for-artificial-intelligence-in-drug-development-climbs-to-5-2-billion/
45. Aspan, M.: This tech giant says A.I. has already helped it save $1 billion, 24 January 2020. https://fortune.com/2020/01/24/ai-ibm-human-resources/
46. Chowdhury, H.: DeepMind's A.I. could speed up research into a coronavirus vaccine, 6 March 2020. https://www.telegraph.co.uk/technology/2020/03/06/deepminds-ai-could-speed-research-coronavirus-vaccine, In 1720 plague, 1820 cholera outbreak, 1920 Spanish Flu, 2020 COVID-19–What is happening. https://www.thenews.com.pk/print/629878-in-1720-plague-1820-cholera-outbreak-1920-spanish-flu-2020-chinesecoronavirus-what-is-happenin
47. Ng, E.: Can A.I. speed up a cure for Coronavirus? This Hong Kong startup opens its resources to global drug firms for free, 17 February 2020. https://www.scmp.com/business/article/3050832/can-ai-speed-cure-coronavirus-hongkong-start-opens-its-resources-global
48. Innoplexus Opens Its Ontosight® AI Search Platform to COVID-19 Research, 15 March 2020. https://www.plantautomation-technology.com/viewprnewswire
49. Kristensen, U.: Analysis: 5 Applying A.I. in the Fight Against the Coronavirus, 01 March 2020. https://hitconsultant.net/2020/04/01/analysis-applying-ai-in-the-fight-against-the-coronavirus/#.XogueYgzb4Y
50. Arnold, C.: How soon will we have a coronavirus vaccine? The race against covid-19, 18 March 2020. https://www.newscientist.com/article/2237742-how-soon-will-we-have-a-coronavirus-vaccine-the-race-againstcovid-19/
51. Spinney, L.: Will a coronavirus vaccine be ready? 3–7 April 2020. https://www.theguardian.com/world/2020/apr/03/when-will-a-coronavirus-vaccine-be-ready
52. Wetsman, N.: Scientists are racing to find the best drugs to treat COVID-19, 23 March 2020. https://www.theverge.com/2020/3/23/21188167/coronavirus-treatment-clinical-trials-drugs-remdesivir-chloroquinecovid
53. Rees, V.: Ten big pharma companies collaborate on data sharing A.I. 7–9 March 2019. https://www.europeanpharmaceuticalreview.com/news/89540/ten-big-pharma-companies-collaborate-on-datasharing-ai/
54. Harris, E.: Scientists at MIT have Discovered a New Antibiotic Using Artificial Intelligence, 25 February 2020. https://towardsdatascience.com/scientists-at-mit-have-discovered-a-new-antibiotic-using-artificial-intelligencee3a3921420fb

A Research for Computerized Maintenance Management System and Predictive Maintenance Investments in Turkish Industry

Arafat Salih Aydıner(✉) and Burak Kubilay

İstanbul Medeniyet University, İstanbul, Turkey
arafat.aydiner@medeniyet.edu.tr

Abstract. The enterprises which are used in industries that capital-intensive investments take place such as aviation, maritime, manufacturing, and energy, become more complex and their total cost of ownership has risen gradually in parallel with the technological developments. To use these expensive technological instruments properly, organizations use systems that enable a much better maintenance efficiency like computerized maintenance management and predictive maintenance systems. Especially with Industry 4.0, predictive maintenance's ability for data analytics on certain equipment has become a very popular topic in literature. However, this popularity has some missing essential points which are directly connected with system and cost efficiency theories. It is assessed that making investments for these systems without a holistic approach could cause a potential risk and could lead organizations to catastrophic financial losses. Turkey as a developing country with its organizations in a turbulent financial environment should be very sensitive about properly designed and implemented maintenance information systems investment projects which are directly connected to their total efficiency. This research aims at providing a descriptive overview of the maintenance information technology investments in Turkish companies and creating a path for future research.

Keywords: Computerized maintenance management systems · Predictive maintenance · Organizational characteristics

1 Introduction

Increasing demand for quality, availability, and productivity has pushed institutions to have more complex and capital-intensive assets [1]. In parallel with this issue, maintenance, which had been seen as a "Necessary Evil" has begun to be considered as a strategic element that directly supports business objectives [2]. The increasing importance of maintenance and advances in information technologies have led to the emergence of systems that can increase the effectiveness of maintenance management such as Computerized Maintenance Management Systems (CMMS) and Predictive Maintenance (PdM). While computerized maintenance management systems accelerate the flow and control of information in the maintenance organization, modern PdM 4.0 [3]

A. M. A. Musleh Al-Sartawi et al. (Eds.): EAMMIS 2021, LNNS 239, pp. 263–277, 2021.
https://doi.org/10.1007/978-3-030-77246-8_26

systems analyze the data received from the equipment through sensors and notify the user of the possible future failure time [2, 4]. In the literature review, it has been observed that such systems are frequently mentioned individually [5–7]. However, there is a black box about the whole system (maintenance organization), which is composed of subsystems, behavior in terms of organizational performance and efficiency. Based on the researches conducted for the optimization of information technology investments, it is considered that the evaluation of this topic with system and cost-effective approaches would make a significant contribution to filling this gap. For this purpose, as a first step, it is aimed at this work to show the overall picture of CMMS and PdM investments, to see the system modules' usage efficiency, and to create a starting point that aims to improve maintenance information technology investment performance in Turkish companies. In this descriptive empirical study, scales derived from the literature were used to collect the data from large-scale organizations to understand the usage rate of CMMS and PdM. The remainder of the study is cultivated as follows. In the next part, information about CMMS, PdM, and organizational characteristics is given from the literature. In the third section, the methodology is explained, and the main questions of the research are presented. In the fourth part, the field research findings are given and finally, the findings and future research areas are evaluated.

2 Literature Review

2.1 Computerized Maintenance Management Systems (CMMS)

A CMMS is an asset that supports maintenance strategy based on an information system and a set of functions aimed to produce indicators by processing information to support maintenance activities [8]. In this context; CMMS enables maintenance management to control a wide variety of data on maintenance workforce, spare parts inventories, repair schedules, and equipment histories; and it can be used for planning & scheduling work orders, accelerating the sending of breakdown calls, and managing overall maintenance workload and providing refined data at the right time to facilitate decision-making for maintenance [9, 10]. Besides, CMMS helps maintenance departments to increase their general productivity by automating maintenance management [11] as well as to improve the communication and coordination skills of maintenance departments by strengthening vertical communication channels [12]. CMMS also supports a team approach in problem-solving between maintenance personnel/vendors/service providers which are separated by distances by providing effective communication between maintenance departments and external service providers with advanced information and communication technology [13]. In literature, common characteristics and set of functions which CMMS have can be found and it is not limited to the following:

- Asset Management: It contains recording all assets (or equipment) and keeping a historical record of maintenance & repair interventions and equipment parts list;
- Work Orders Management: This ability allows maintenance management to control the workforce by planning & scheduling and enables users to give orders to the maintenance technicians;

- Preventive Maintenance Management: That supports the planning & scheduling and control of activities;
- Inventory Control: This ability allows users to get information about spare parts availability;
- Report Management: CMMS processes, analyses, and controls data flow and helps maintenance management to produce maintenance performance indicators;
- Control of Costs and Budget: CMMS provides users to monitor and control maintenance budget and expenses [8–12, 14, 15].

In a complex organization like current production systems which is composed of many different machine parks with thousands of components and hundreds of maintenance workers, it is a really hard task to control the whole system manually and to get success in collecting the appropriate data to measure the efficiency and to get an oversight for maintenance performance indicators. A truly applied CMMS that is aligned with other systems/subsystems could allow maintenance management to do it so. In their work, Accorsi et al. [16] proposed a maintenance management system composed of several computerized tools which save more than 20% of the costs and time required for maintenance.

2.2 Predictive Maintenance (PdM)

In recent years, the predictive maintenance approach, which is the developed version of the condition-based maintenance strategy, has attracted great interest in theory and practice. PdM enables users to maximize the equipment's current usefulness as well as its residual usefulness by allowing service and repair interventions to be undertaken at exactly the right time when an item is about to fail [17]. A report published in 2017 by Pwc and Mainnovation companies stated that the application of big data analytics in maintenance represents the fourth level in the predictive maintenance growth model and this level is called "Predictive Maintenance 4.0" [3]. In the same report, predictive maintenance 4.0 is affiliated with predicting future failures in assets and ultimately determining the most effective predictive measure by applying advanced analytical techniques to technical condition, usage, environmental factors, maintenance history, data of similar equipment, and indeed all big data that may be related to the performance of an asset. The use of large volumes of data (Big Data) in Industry 4.0 and the increasing development of cognitive systems greatly strengthen the field of predictive maintenance in decision making and early fault detection based on real-time data [18]. According to Jimenez et al. [19], predictive maintenance has two main approaches: diagnostic methods that aim to determine the root causes and current health status of the system by detecting failures to prevent unexpected failures and prognostics methods used to predict the future state and remaining useful life of the system. From a different point of view, Mathews et al. [20] proposed a simulation-based prediction model of coal-fired power plant condenser maintenance that provides considerable savings. The organizations which use PdM systems and approaches have some characteristics. In their empirical studies, Bottani et al. [21:158] and Alsyouf [22:220] use the below variables to measure the predictive approach in Italian and Swedish industries:

- Investments are made for the improvement of skills and competencies of the maintenance employees,
- CMMS systems are commonly adopted
- Off-line tests are taken place to monitor possible criticalities of the asset,
- On-line tests are performed to monitor possible criticalities of the asset,
- Maintenance activities are planned based on some specific performance metrics of the asset,
- Maintenance activities are planned based on statistical failure data.

Organizations that are looking for a world-class maintenance level should benefit from information technology opportunities like PdM and CMMS. In addition, whole maintenance system optimization is a vital need for the sustainability of organizations' efficiency.

2.3 Organizational Characteristics

In general, the public sector is less willing to take risks and take a proactive approach than private sector companies [23]. Furthermore, some sectors are more active in the use of information technology applications [24]. Accordingly, examining the applications in different sectors will contribute to the analysis of industries' behavior based on IT investments in maintenance. In parallel with this purpose, since most of the maintenance managers, technical service managers, and mechanical engineers in organizations are authorized to make decisions on maintenance strategies and therefore have a more comprehensive and holistic overview about maintenance information technologies investments, they provide useful sources of industrial data for researches. The organizational scale could play an important role in maintenance IT investments: large companies generally use more financial resources and manpower to reach higher information capacity [25]. The scale of a company is connected with the number of its employees [24]. From a different point of view, older companies could be more advantageous than young companies due to their experience advantages by means of sustaining their development [26] but they may also avoid investing in new technologies by creating social inaction [24].

In Swanson's work, a useful descriptive framework for analyzing maintenance organizations is provided including characteristics of plants, plant maintenance departments, and CMMS [2]. In parallel with this study, the annual maintenance budget, the average market value, the average age of machines, the number of personnel working in maintenance department, and the predominant use of maintenance field personnel are considered to be useful in terms of comparing the organizations and used in this paper.

In line with the aforementioned literature with the combination of CMMS and PdM, this study proposes three research questions to analyze the characteristics of maintenance practices in an emergent country, Turkey. The set research questions are presented below: What are the usage level and characteristics of CMMS in subjects? What are the usage levels and characteristics of PdM system in subjects? And What are the organizational characteristics of the subjects?

3 Research Methodology

Since Turkey is in the category of developing countries and has a positive approach to information technology investments, it is considered to be an attractive area of research. The focus of this study is the organizations that operate maintenance activities and their maintenance personnel. On-line and mailing survey techniques were used as data collection tools (Because of the Covid-19 period, face-to-face surveys could not be conducted.). The Qualtrics online software was used in the creation of the questionnaire, social media, and e-mail channels were used for its distribution. There are five sections in the content of the survey questionnaire, which is derived from literature and shaped by the contributions of academicians and maintenance sector professionals. The first section consists of a clarification text prepared in accordance with the GDPR law-6698 and a preface explaining the research. In the second, third, and fourth sections, there are CMMS and PdM questions respectively. CMMS and PdM scales derived from the previous literature [8–12, 14, 15, 21, 22]. In the last section of the survey, scales of Swanson [2:12] and Aydıner & Tatoğlu [24:61]'s are used. During the data collection phase, questionnaires were sent to 62 institutions/organizations and 112 maintenance professionals. These organizations are private companies, state-owned enterprises operating in maritime, aviation, and production industries and engineering chambers (Mechanical, Electrical, Marine Engineering, and Petroleum). Besides, distributions were made especially for the senior maintenance managers in parallel with the study's approach. Survey responded by 127 participants. In the examination phase, it was understood that 72 observations were partly done while 55 of them were completely answered and suitable for analysis. It indicates a 43% survey completion rate. It is considered that the requested information in the survey (Especially annual budget, annual maintenance budget, number of employees, and number of maintenance employees) are acknowledged as a commercial/professional secret and therefore the response rate for the last part of the survey is low.

4 Findings

4.1 CMMS

The rate of the organizations which have a CMMS between 55 respondents is only 61%. The average of the organizations which implement CMMS for 7 years or more composes 33% of the respondents of this research. It implies that older organizations are not very aggressive about having maintenance information technology capabilities. This situation could be in accordance with social inaction theory. Even though the literature shows that CMMS has been used for decades, however, the analysis indicate that CMMS applications are relatively new among companies in Turkey which is based on the fact that 67% of the respondents' CMMS is younger than 7 years (Table 1). CMMS usage could be connected with the organizational scale and be seen as a useful way to deal with complexity [12]. Unlikely, although 60% of the respondent organizations have 100 or more employees and an annual budget which is more than 5 million Turkish Liras (TL), respondents have relatively low CMMS ability usage overall with 55% (Table 2).

Table 1. Characteristics of plant/firm CMMS

Item	Average %
Does your plant/firm have a CMMS?	
Yes	61
No	39
Average number of years CMMS has been in operation	
One year or less	11
1–3years	26
3–5 years	15
5–7 years	15
7 years or more	33

Furthermore, 64% of the respondent organizations‘ total equipment value is more than 1 million (TL) and 61% of these assets are older than 10 years (Table 9).

There are many products of CMMS in the market and every product has its capabilities. Table 3 gives an overview of the CMMS capabilities of the organizations and their usage rates. The ability of historical tracking is common and hasextended usage relative to other abilities (Work orders, maintenance activities, and equipment repair history). Additionally, technical documentation and procedure storing ability is another common characteristic of CMMS in Turkey. Noticeably, having material and spare parts purchasing rate is 45% and the lowest one between CMMS capabilities of the users. And the usage rate of this capability is relatively low. It is possible to mention that this module of CMMS is attracted to have limited attention by users, and they use different ways in purchasing processes to contact their vendors or service providers.

This is parallel with the outside of service call/dispatch capability which has the highest ability percentage (%63%), while it has the lowest usage rate (only two point one percent). The second-lowest ability rate belongs to maintenance budgeting and spare parts requirement planning (49%) which enables users to control expenses of the plant/firm. It should be noted that maintenance spare parts requirements planning and maintenance budgeting are among the main characteristics of CMMS in the literature. According to Table 3, the most common abilities of CMMS that the organizations have, and use are work order management and preventive maintenance management. This could be a sign that CMMS is seen as a maintenance personnel control tool, mainly. On the other hand, connected with the PdM capacity of organizations, having equipment failure diagnosis system ability is 51% which is the fourth-lowest ability between other CMMS characteristics and its extent rate is only 3,3 out of 5 which could be a sign that CMMS support to PdM is relatively low and this condition could affect the whole systems' efficiency in a negative way.

Table 2. Characteristics of respondent plants/firms

Item	Average
Sector engagement	
Public	10%
Private	87%
Unknown	3%
Annual budget (Turkish Lira)	
Less than 100k	0%
100k–500k	5%
500k–1 million	10%
1 million–5 million	18%
More than 5 million	60%
Unknown	7%
Age of plant/firm	
Less than 5 years	6%
5–10 years	16%
10–15 years	14%
15–25 years	22%
More than 25 years	40%
Unknown	2%
Number of employees	
Less than 50	27%
50–100	11%
100–500	25%
500–1000	15%
More than 1000	20%
Unknown	2%

The CMMS factor loadings can be seen in Table 4. It shows that all variables have sufficient loading scores. The lowest score belongs to "Outside service call/dispatch capabilities" variable while the highest score belongs to "Historical tracking of work orders" variable.

4.2 Predictive Maintenance

According to Table 5, having PdM applications/abilities is 89% higher than possession of CMMS. On the other hand, the extent of these applications/abilities is lower than the extent of CMMS applications in organizations. It should be taken into account that

Table 3. Use of CMMS Abilities

Item	System ability %	Extent of use*
Workforce planning and producing preventive work orders	56	4,2
Work order generation, prioritization and tracking by equipment/component	58	4,2
Historical tracking of work orders generated which become sortable by date, person responding etc.	61	4,5
Tracking of scheduled and unscheduled maintenance activities	61	4,2
Storing of maintenance procedures as well as all warranty information by component	58	4,1
Storing all technical documentation or procedures by component	60	4
Equipment repair history	60	4,4
Equipment parts list	54	3,7
Real-time reports of ongoing work activity	58	3,4
Mobile device interface to streamline input and work order generation	52	3,4
Inventory control	52	3,3
Material and spare parts purchasing	45	3,4
Spare parts requirements planning	49	3,6
Equipment failure diagnosis	51	3,4
Outside service call/dispatch capabilities	63	2,1
Maintenance budgeting	49	3,6
Integration with management information systems (inside/outside) like ERP, SAP, etc.	56	3,3

*1 = Rarely 5 = Common

commonly adapted CMMS is an important fact for PdM applications as mentioned before. In this context, it could be mentioned that the situation in Turkey is not parallel with the literature. This could be connected with IT maturity levels of these organizations and be a sign of ineffective maintenance IT investment policy without a holistic approach.

The most common and used PdM applications are investments of maintenance staff and maintenance planning on specific performance parameters which have a higher extent rate relative to other PdM applications. On-line and off-line test capabilities applications and their extent rates are relatively low. Since they are connected with identifying possible future criticalities, it could be mentioned that utilizing diagnostic capabilities of PdM is relatively low to other capabilities between organizations in Turkey. As emphasized by Pwc&Mainnovation [3] and Jimenez et al. [19], prognostic capabilities and data-driven models are at the center of focus in publications in recent years. Based on

Table 4. CMMS factor

Items	Loading
How long have you been using CMMS?	0,763
Workforce planning and producing preventive work orders	0,856
Work order generation, prioritization and tracking by equipment/component	0,898
Historical tracking of work orders generated which become sortable by date, person responding, etc.	0,954
Tracking of scheduled and unscheduled maintenance activities	0,941
Storing of maintenance procedures as well as all warranty information by component	0,914
Storing all technical documentation or procedures by component	0,922
Equipment repair history	0,933
Equipment parts list	0,837
Real-time reports of ongoing work activity	0,819
Mobile device interface to streamline input and work order generation	0,805
Inventory control	0,809
Material and spare parts purchasing	0,753
Spare parts requirements planning	0,837
Equipment failure diagnosis	0,829
Outside service call/dispatch capabilities	0,703
Maintenance budgeting	0,825
Integration with management information systems (inside/outside) like ERP, SAP, etc.	0,829

this, it could be stated that prognostic PdM capabilities should be more concentrated in Turkey. In Table 6, PdM factor variables can be seen and it is evident that all of them are at acceptable levels and useful for future research in Turkey.

4.3 Organizational Characteristics

Identifying the respondent organizational characteristics is an important fact for analyzing the behavior of Turkish industries about maintenance IT investments. According to Table 6, 87% of the respondents engaged in the private sector. 60% of the organizations' annual budget is more than 5 million Turkish Liras. Besides, the number of employees is above 100 in 60% of the organizations which means mostly, middle and large-scale institutions participated in our survey. Connected with this fact, 76% of the respondents' plants/firms are older than 10 years and it is possible to mention that most of the respondents are not young organizations.

In Tables 7 and 8 characteristics of respondent staff can be seen which indicates manufacturing industry dominance over other participants and 83% of the respondents have managerial roles.

Table 5. Use of PdM

Item	Plants/firms application/ability %	Application/ability*
Investments are made to improve skills and competences of the maintenance staff	96	3,6
Real-time data flow is made by using IoT(Internet of Things)	90	3,2
Off-line tests are performed to identify possible criticalities of the manufacturing plant	87	3,2
Online tests are performed to identify possible criticalities of the manufacturing plant	80	3,2
Maintenance interventions are planned on the basis of some specific performance parameters of the plant	92	3,8
Predictive maintenance interventions are planned by analyzing the data of statistical failures	89	3,5

* 1 = rare 5 = common

Table 6. PdM factor

Item	Loading
Investments are made to improve skills and competences of the maintenance staff	0,668
Real-time data flow is made by using IoT (Internet of Things)	0,723
Off-line tests are performed to identify possible criticalities of the manufacturing plant	0,898
Online tests are performed to identify possible criticalities of the manufacturing plant	0,954
Maintenance interventions are planned on the basis of some specific performance parameters of the plant	0,941
Maintenance interventions are planned on the basis of statistical failures data	0,914

Table 7. Characteristics of respondent staff

Item	Average
What field/sector do you work in?	
Marine industry	7%
Aviation industry	7%
Manufacturing industry	27%

(*continued*)

Table 7. (*continued*)

Item	Average
Information technology industry	6%
Petrochemical industry	7%
Medical industry	3%
Energy industry	3%
Telecommunications industry	2%
Construction industry	2%
Plastic injection molding industry	2%
Transportation industry	6%
Service industry	3%
Railroad industry	6%
Municipal services	2%
Other industries	10%
Unknown	7%

Table 8. Characteristics of respondent staff

Item	Average
What is your role in the organization?	
Technical service manager	36%
Maintenance manager	34%
Technician	2%
General manager	4%
Coordinator	4%
Engineer	5%
Other positions	11%
Unknown	4%

Table 9. Characteristics of plan/firm maintenance department

Item	Percent
Annual maintenance budget (Turkish Lira)	
Less than 50k	%7
50k–100k	%11

(*continued*)

Table 9. (*continued*)

Item	Percent
100k–500k	%18
500k–1 million	%21
More than 1 million	%36
Unknown	%7
Value of equipment (TL)	
Less than 100k	%7
100k–500k	%9
500k–1 million	%16
1 million–5 million	%9
More than 5 million	%55
Unknown	%4
Average equipment age	
Less than 5 years	%20
5–10 years	%27
10–15 years	%33
15–25 years	%14
More than 25 years	%4
Unknown	%2
Number of maintenance employees	
Less than 20	%62
20–50	%9
50–100	%18
100–500	%7
More than 500	%2
Unknown	%2
Predominant maintenance man-hour utilization	
Emergency work	%31
Planned work	%45
Standing work orders	%18
Others	%4
Unknown	%2

According to Table 9, 57% of the participant organizations has annual maintenance budgets above 500k (TL) and 64% of the participants' value of the equipment is above 1 million (TL). Average equipment age data signifies 47% of the respondents have equipment younger than 10 years. The number of maintenance employees is less than 20 in 62% of the respondents' organizations. Predominant maintenance man-hour utilization is emergency work in 31% of the organizations which is not in accordance with common PdM capability usage (89%) which enables users to have a more proactive and planned work schedule.

5 Discussion

With this descriptive study CMMS, PdM, and characteristics of maintenance organizations in Turkey were analyzed and, in conclusion, three points should be stated about maintenance investments. First, the difference between CMMS and PdM possession rates are unlike the expectations; although commonly adopted CMMS is an important need for PdM, as mentioned, there is a noticeable gap between CMMS and PdM application/ability overall rates with 55% and 89% (Tables 3 and 5). Second, organizations in Turkey do not seem to take advantage of prognostic capabilities of PdM. On-line and off-line tests for detecting possible criticalities of equipment are relatively low compared to other applications/abilities usage in Table 5. This fact is not parallel with recent developments in PdM theory and practice. Third, the level of emergency works in Turkish organizations is approximately one-third of the average maintenance workload which should be a deep concern for companies in the context of effectiveness and efficiency (Table 9). These three points should be focused to have a good understanding of maintenance information system structure effectiveness level in Turkey.

6 Conclusion

IT investments are vital and strategic elements for organizations and worth deep analyzing. CMMS and PdM ability gap of maintenance organizations in Turkey could be negatively connected with this issue and refer to maintenance IT investment decision analyzing problems. Besides, prognostic capabilities usage of PdM is an important fact in terms of having a proactive maintenance organization. It is noticeable in our data that although organizations have prognostic PdM abilities, they do not prefer using them. Accordingly, a lack of concentration for these capabilities could stem from the low IT maturity levels of organizations. From another point of view, having a one-third emergency weighted workload in the presence of PdM tools could probably be a sign of holistic approach lack in Turkey in terms of organizational maintenance efficiency and effectiveness. Since it is directly inconsistent with literature and practice, the reason for CMMS and PdM usage difference in Turkey is worth analyzing and should be further examined. Contextualizing this topic with maintenance performance indicators could be a good way in terms of searching for its potential effects on organizational efficiency and effectiveness. On the other hand and secondly, prognostic PdM capabilities usage extent is lower than other capabilities and, as mentioned, PdM prognostics has a high potential for improving the efficiency of PdM. Based on this fact, the lack of prognostics PdM capability usage extent in Turkey could be a beneficial research area in the context of IT maturity levels of maintenance organizations Third, the case that emergency works establish approximately one-third of the average maintenance workload and it could be a tackling position to reach a "World Class Maintenance " level for organizations in Turkey and it should be further analyzed for having a more efficient and proactive maintenance system. Efficient and effective maintenance information systems are the cornerstones of organizations which means that they are essential for industries and the whole economy of a country. It is evaluated that emphasized facts in this study would be an appropriate starting point for further research to analyze and improve maintenance systems in Turkey.

References

1. Labib, A.W.: World-class maintenance using a computerized maintenance management system. J. Qual. Main. Eng. **4**(1), 66–75 (1998)
2. Laura, S.: Computerized maintenance management systems: a study of system design and use. Prod. Inven. Manage. J. Second Q. **38**(2), 11–15 (1997)
3. Pwc & Mainnovation. Predictive maintenance 4.0: predict the unpredictable. Report, June 1–32 (2017)
4. Labib, A.W.: A decision analysis model for maintenance policy selection using a CMMS. J. Qual. Main. Eng. **10**(3), 191–202 (2004)
5. Dhall, R., Solanki, V.: An IoT based predictive connected car maintenance approach. Int. J. Interact. Multimed. Artif. Intell. **4**(3), 16–22 (2017)
6. Olalere, I.O., Dewa, M.: Early fault detection of elevators using remote condition monitoring through IoT technology. S. Afr. J. Indus. Eng. **29**(4), 17–32 (2018)
7. Kızrak, M.A., Bolat, B.: Predictive maintenance of aircraft motor health with long-short term memory method. J. Inf. Tech. **12**(2), 103–109 (2019)
8. Lopes, I., Senra, P., Vilarinho, S., Sá, V., Teixeira, C., Lopes, J., Alves, A., Oliveira, J.A., Figueiredo, M.: Requirements specification of a computerized maintenance management system – a case study. Procedia CIRP **52**, 268–273 (2016)
9. Ahuja, P.S.: Handbook of maintenance management and engineering. chapter 17, pp. 417–459 (2009)
10. Abbas, M., Mahmood, S.: An overview of maintenance management strategies for the corroded steel structures in extreme marine environments. Mar. Struct. **71**, 1–23 (2020)
11. Majeed Ali, A.A.: Application of computerized maintenance management system in industry. In: Proceedings of the 37th International Conference on Computers and Industrial Engineering, pp. 984–995 (2007)
12. Swanson, L.: An information-processing model of maintenance management. Int. J. Prod. Econ. **83**, 45–64 (2003)
13. Velmuguran, R.S., Dhingra, T.: Maintenance strategy selection and its impact in maintenance function. Int. J. Oper. Prod. Manage. **35**(12), 1622–1661 (2014)
14. Fumagalli, L., Macchi, M., Rapaccini, M.: Computerized maintenance management systems in SMEs: a survey in Italy and some remarks for the implementation condition based maintenance. In: 13th IFAC INCOM(INCOM 2009), Moscow, Russia, June 3–5, 1615–1619 (2009)
15. Azahar, N.F., Othuman Mydin, M.A.: Potential of computerized maintenance management system in facilities management. Anal. Univ. Eftimie J. **21**(1), 51–59 (2014)

16. Accorsi, R., Gallo, A., Tufano, A., Bortolini, M., Penazzi, S., Manzini, R.: A tailored maintenance management system to control spare parts life cycle. Proc. Manuf. **38**, 92–99 (2019)
17. Tran Anh, D., Dabrowski, K., Skrzypek, K.: The predictive maintenance concept in maintenance department of the industry 4.0 production enterprise. Found. Manage. **10**, 283–292 (2018)
18. Rousopoulou, V., Nizamis, A., Vafeiadis, T., Ioannidis, D., Tzovaras, D.: Predictive maintenance for injection molding machines enabled by cognitive analytics for industry 4.0. Front. Artif. Intell. **3**, 1–12 (2020)
19. Jimenez, J.J.M., Schwartz, S., Vingerhoeds, R., Grabot, B., Salaün, M.: Towards multimodal approaches to predictive maintenance: a systematic literature survey on diagnostics and prognostics. J. Manuf. Syst. **56**, 539–557 (2020)
20. Mathews, I., Mathews, E.H., Van Laar, J.H., Hamer, W., Kleingeld, M.: A simulation-based prediction model for coal-fired power plant condenser maintenance. Appl. Therm. Eng. **174**, 1–10 (2020)
21. Bottani, E., Ferretti, G., Montanari, R., Vignali, G.: An empirical study on the relationships between maintenance policies and approaches among Italian companies. J. Qual. Main. Eng. **20**(2), 135–162 (2014)
22. Alsyouf, I.: Maintenance practices in Swedish industries: survey results. Int. J. Prod. Econ. **121**, 212–223 (2009)
23. Zhou, K.Z., Li, J.J., Zhou, N., Su, C.: Market orientation, job satisfaction, product quality, and firm performance: evidence from China. Strat. Manage. J. **29**(9), 984–1000 (2008)
24. Aydıner, A.S., Tatoğlu, E.: A field study on information systems in business applications in Turkey. Information Technologies Journal **12**(1), 59–73 (2019)
25. Mani, D., Barua, A., Whinston, A.: An empirical analysis of the impact of information capabilities design on business process outsorcing performance. MIS Q. **34**, 39–62 (2010)
26. Chen, Y., Wang, Y., Nevo, S., Benitez-Amado, J., Kou, G.: IT capabilities and product innovation performance: the roles of corporate entrepreneurship and competitive intensity. Inf. Manage. **52**, 643–657 (2015)

Artificial Intelligence in Personnel Management: Opportunities and Challenges to the Higher Education Sector (HES)

Kirk Chang, Yousif Abdelbagi Abdalla(✉), and Alhashmi Aboubaker Lasyoud

College of Business Administration, University of Sharjah, Sharjah, United Arab Emirates
{kchang,yomer,alasyoud}@sharjah.ac.ae

Abstract. Drawing on multi-disciplinary literature, this conceptual article provides an integrative review of AI's applicability in personnel management. Prior studies indicate that AI has the ability to change the ownership and responsibility for managerial decision making, information accumulation and strategic management. Following this line of research, we review the literature of AI and discuss its application in personnel management, particularly from the context of Higher Education Sector (HES). We draw insights from three theoretical perspectives (job replacement, psychological contract and demands-resources), aiming to understand the influence of AI-driven management on the organizations, managers and employees. Specifically, two questions are proposed and critically discussed. First, what might be the opportunities and challenges of AI-driven management? Second, how AI-driven policies could be implemented to the personnel management? Based on the analysis of theoretical perspectives and empirical research findings, we have learnt that managers' attitude is crucial to the implementation of AI in personnel management, leading to different organizational outcomes. If not handled well, AI-driven management may cause job insecurity and dissatisfaction to the employees. Research findings has brought new insights into the AI-management literature and clarified AI's applicability in HES. Implications for future studies are discussed.

Keywords: AI · AI-driven · Artificial intelligence · Higher education · Personnel management

1 Introduction

Artificial intelligence (AI) is often regarded as innovative and smart technology, helping the organizations and employees to deliver quality service and performance (Smith and Anderson 2014). AI has been widely applied to marketing strategies, energy saving, climate forecast, risk prevention, customer supporting services, and problem solving in business operations (Brown et al. 2017). Recently, AI has started to exert its influence into the management field, such as performance-tracking software (Ernst and Young 2018), holistic process management (Chang 2019), development of automatable jobs (Lima et al. 2021), employee retention and attrition (Sivathanu and Pillai 2018), employee

A. M. A. Musleh Al-Sartawi et al. (Eds.): EAMMIS 2021, LNNS 239, pp. 278–289, 2021.
https://doi.org/10.1007/978-3-030-77246-8_27

rights and ethics (Jeske and Calvard 2020), employee emotion management (Henkel et al. 2020) and burnout prevention (Kong et al. 2021).

Following the development of AI-driven management and aforementioned studies, the current research is keen to understand how AI-driven management affects the organizations and their employees in the Higher Education Sector (HES), with the following rationale: i). AI has the ability to change the ownership and responsibility for decision making, information accumulation, strategic management, but its applicability on personnel management is still unknown (Grace et al. 2018; Malik et al. 2019); ii). Scholars have proposed different theories and viewpoints to predict the impact of AI-driven policies on organizational outcomes and employee behaviors (Chang 2020; Farndale et al. 2019). However, how these theoretical perspectives could be implemented into the actual managerial practices is still unclear; iii). AI applications are crucial to all fields of life, particularly for the sector of education. Education providers nowadays are no longer limited to education provision but also contribute to the development of communities and societies (Aldosari 2020); and, finally, iv). Higher Education Sector (HES) is an important driver to the growth performance, prosperity, and competitiveness both in national and global economies (Zhu et al. 2017). HES also equips students and apprentices (i.e., future employees) with the latest knowledge, such as digital technology, e-service and products (Lucena et al. 2019).

In the current research, we first draw insights from the literature and critically discuss the advantage and disadvantage of AI-driven policies in management. Three theoretical perspectives are consulted in evaluating the impact of AI-driven management on the organizations, managers and employees. We then analyze how AI-driven management could be implemented to the personnel management. In closing, we respond to the knowledge gap of AI's applicability in HES, clarifying its implementation opportunities and challenges. It is our hope that HES staff, management scholars and HR practitioners will benefit from our research.

2 Literature Review

Artificial intelligence was first proposed in 1960s to study the heuristic- and cognitive-related themes, including idea generation, problem solving and symbolic representation (Brunette et al. 2009; SAS 2018). Continuing this line of research, the US Department of Defense designed a series of computational modules to learn human mind and predict its possible behavioral outcomes (SAS 2018). These preliminary findings have built the foundation for the automatic reasoning and information accumulation that we see in electronic devices and products nowadays, such as decision support services and smart search systems that are designed to complement and augment human abilities (Russell and Norvig 2009). AI seems omnipotent and offers different benefits in modern life, shaping the future of virtually every industry and every human being. For instance, AI has increased the efficiency of marketing strategies, improved the efficacy of energy saving, enhanced the accuracy of weather forecast, and facilitated problem solving in business activities (Brown et al. 2017). AI also acts as intelligent and advanced technology, assisting business and organizations in delivering better-quality service and more performance (Smith and Anderson 2014).

In the field of management, more specifically, scholars have scrutinized AI's characters and praised its applicability. Grace et al. (2018) have indicated that AI is better than human intelligence in two characteristics. These are: *cognition* (e.g., translating jargons into plain languages and writing analytic reports) and *coordination* (e.g., selling products, customer-interaction skills and operating surgeries). These two characteristics are crucial to management, particularly when the organizations today tend to have complex structure, in which departments and teams share numerous tasks and distributed responsibilities (Mullins 2016). In view of what has preceded, it is therefore reasonable to state that AI has great potential in the personnel management; and, actually, recent studies have provided some credence to support the statement.

For instance, team leaders and managers have improved individual performance through the AI-driven techniques, such as KPI monitoring software and performance-tracking applications (Ernst and Young 2018). These AI-driven applications are informative to both managers and their subordinates, helping both parties "stay on the same page" during the annual appraisals. Different from the conventional management approach that focuses on the performance outcome, AI-driven management approach enables both managers (e.g., leaders, appraisers) and subordinates (e.g., followers, appraisees) to monitor the performance through a holistic perspective, from the initial stage of objective-setting, to the final stage of completion (Chang 2019). Managers may apply the concept of "Big-Data" into their managerial practices, with a view that AI-driven analyses offers better insights into how to execute and operate in staff succession planning and job analysis (Wang et al. 2017). Through AI-based software and hardware, organizational leaders, personnel managers and HR practitioners become more capable of understanding the employee retention and attrition (Sivathanu and Pillai 2018), employees' emotion management (Henkel et al. 2020) and their burnout prevention (Kong et al. 2021). Ernst and Young (2018) also indicate that "*Human resources executives have faith that merging AI into HR administration functions will benefit and improve the overall employee experience. This will provide more capacity, more time and budget, and more accurate information for decisive people management, p.3*".

Following the discussion of AI's applicability aforementioned, one may assume that AI is powerful and has salient potential in management. Yet, not all managers acclaim AI's applicability in the practice, and the association between managers and AI's applicability is a bit mysterious. A recent study has shown that managers may develop opposite attitudes of AI' applicability in management, subject to the interpretation of AI's role. Chang (2019) explains that some managers are willing to apply AI-driven software in their recruitment practices, and they benefit from AI's swift and accurate selection of appropriate applicants. These managers evaluate AI's applicability and recognize AI's merits from a more objective and positive manner. Interestingly, some managers are reluctant to accept AI, fearing that AI may compromise their job roles and dwindle their influence in the workplace. These managers tend to interpret AI as a career threat and evaluate AI's influence from a more subjective and negative manner. Following this logic, we can conclude that AI's applicability in management is not necessarily related to the function of AI, and managers' attitude speaks louder on this occasion (See Table 1 for details).

Table 1. AI's applicability in management: from the eyes of managers.

Attitude	Evaluation	Consequence
In the eyes of managers, AI acts as support	AI's function and merits are appreciated and recognized by the managers	Managers are more likely to adopt AI-driven devices/software in their managerial policies and practices. Managers receive the benefits of AI-driven software and devices
	The evaluation tends to be more objective and positive	Managers treat AI as support, assisting their decision-making and managerial practices
In the eyes of managers, AI acts as threat	AI's function and merits are either depreciated or ignored by the managers	Managers are less likely to adopt AI-driven devices/software in their managerial policies and practices
	The evaluation tends to be more subjective and negative	Managers worry about the possibility that AI may compromise their job roles and dwindles their influence in the workplace

To our knowledge, AI-driven management and its development in the real world is still at its infancy stage. Scholars from the fields of human resource management and organizational psychology jointly have called more studies to be carried out (e.g., Farndale et al. 2019 Malik et al. 2019), aiming to understand the influence of AI-driven management on both organizations and employees. In responding to this call, the current article is keen to advance the knowledge; specifically, we have reviewed relevant literature and three theoretical views are selected for discussion. These are: *job replacement*, *psychological contract* and *demands-resources* (See Table 2 for details). These theoretical perspectives are different in nature, explaining the influence of AI on management in three unique ways. Details are outlined below:

To begin with, Huang and Rust (2018) propose that AI may re-shape employees' job tasks and hence make them exposed to the redundancy risk, and that AI acts like as amalgamation of innovation and efficiency in implementing tasks. That is, the more tasks AI can take over, the more risk employees are exposed to. Huang and Rust have specified four intelligences required for service tasks (i.e. mechanical, analytical, intuitive & empathetic) and outlines the way managers should decide between labor and machines for accomplishing those tasks. According to Huang and Rust, AI and its influence evolves in a staged manner, with mechanical mostly preceding analytical, analytical mostly preceding intuitive, and intuitive mostly preceding empathetic intelligence. Huang and Rust (2018) argue that AI job replacement occurs fundamentally at the task level, rather than

Table 2. AI's applicability in management: from different theoretical views.

Views	Authors	Main argument	AI's applicability in management
Job replace approach	Huang and Rust (2018)	AI may re-shape employees' job tasks and hence make them exposed to the redundancy risk. The more tasks AI can take over, the more risk employees are exposed to	If unavoidable, the replacement may occur at the task level, rather than the job level, and for lower intelligence tasks first. Tasks with heavy-cognition-component are less likely to be replaced at present, although more advanced AI devices (hardware, software) may develop over time and gradually change the course of replacement
	Lima et al. (2021)	AI has great potential to streamline the production lines and minimize the labor needs. One demonstration of the AI's potential is the emerging phenomenon of automation in Brazil's industry, aiming to enhance the efficacy of existing production technology, such as better quality and more performance	If AI is to be applied in the next decade, 60% of the employment in Brazil will be highly impacted by the practice of automation, due to the possibility that eight out of the ten occupations with the biggest workforce are highly automatable; that is, AI replaces human-dependent tasks and jobs. Automation is favored by the investors and business owners, but the employee sides think differently
Psychological contract approach	Rousseau (1989)	Employees are keen to negotiate what they must do to satisfy their side of the bargain, and what they can expect in return. A health give-and-take relationship makes employees satisfied at job	A balanced give-and-take relationship is imperative to the employees. For the same reason, implementing AI requires great attention, and managers have the responsibility to learn AI, avoiding its side effects to the employees

(continued)

Table 2. (*continued*)

Views	Authors	Main argument	AI's applicability in management
	George (2010)	Psychological contract helps people to understand what the parties should, or should not do and how it should be done. When the parties' expectations match each other, objectives are likely to be accomplished and job satisfaction obtained	To ensure employees maintain healthy psychological contract, AI must be adopted in an appropriate manner, at least in a way that empowers employees, not threatens them. When threatened, employees may lose confidence and feel disengaged, leading to the negative outcomes
	Farndale et al. (2019)	HRM practices and ICT systems have crucial relationship with employee attitudes and behaviors, highlighting the importance of individual differences	AI-driven management should not be conducted without the consideration of individual characteristics and preferences. Employees may interpret AI's impact along with their individual values and therefore engage with their jobs differently. Different consequences may follow
Demands-resources approach	Bakker and Demerouti (2007)	Whereas every occupation may have its own specific working characters, these characters are generally divided in two categories (job demands vs. job resources). People feel stressed when the demand-resource relationship is not balanced, for instance, when job demands are high but job resources are low	AI affects managerial practices, particularly when AI has potential to become either job demand or job resource, and the absence of either potential may put employees and employers at a disadvantageous position. Managers have the responsibility to prevent the demand-deficiency and resource-deficiency caused by the AI-driven management. A precautious and preventive approach is encouraged

(*continued*)

Table 2. (*continued*)

Views	Authors	Main argument	AI's applicability in management
	Chang (2019)	AI's applicability in management is related to managers' interpretation of AI's role. AI-driven management may improve the efficiency of existing practice (e.g., more resourceful) or reframing the ownership of existing practice (e.g., manager's input is less demanding)	Manager with positive attitude of AI are keen to probe AI's function and receive its benefits, whereas managers with negative attitude of AI are reluctant to adopt AI and regard AI as career threat, such as compromising their job roles and dwindling their influence in the workplace. Whether AI-driven management is adopted may depends on managers' evaluation of AI's function and influence

the job level, and usually for lower intelligence tasks initially. Based on their argument, AI may first replace some of a service job's tasks, a transition stage seen as augmentation, and then progresses to replace human labor entirely when it has the ability to take over all of a job's tasks. The progression of AI task replacement from lower to higher intelligences results in predictable shifts over time in the relative importance of the intelligences for service employees. Following this line of logic, analytical skills will become trivial in future jobs, as AI may take over more analytical and complex tasks, giving the intuitive and empathetic skills more weight and credits in the job market. Eventually, AI will have the capability to perform not just mechanical and routine tasks of a job, but also in its entirety, even jobs that have high intuitive and empathetic content.

Next, human beings are a social species and rely on relationships in every aspect of their life. Take employment for instance, workplace relationship often emerges through the inter-personal relationships formed at work. Rousseau (1989) states that employees slowly negotiate what they must do to satisfy their side of the bargain, and what they can expect in return, and this give-take relationship has formed the sense of psychological contract. The aforementioned negotiation appears in different ways, such as conducting appraisal and performance review sessions, exchanging expectations between assessors and assesses, or acting/reacting to organizational policies. Scholars have also found that psychological contract helps people to understand what the parties should, or should not do and how it should be done. When the parties' expectations match each other, objectives are likely to be accomplished and satisfaction obtained (George 2010). Following the concept of psychological contract, managers and leaders need to be prudent in their introduction as well as implementation of AI strategies. Organizational leaders and senior executives also have the responsibility to learn AI and understand how to use it.

AI must be adopted in an appropriate manner, at least in a way that empowers employees, not threatens them.

The influence of AI-driven management may also be interpreted through the concept of demands and resources. Bakker and Demerouti (2007) indicate that whereas every occupation may have its own unique characters, these characters are generally divided in two categories (job demands vs. job resources), thus constituting an overarching model that may be applied to various occupational settings, irrespective of the particular demands and resources involved. Bakker and Demerouti continue that people may feel stressed when the demand-resource relationship is not balanced (e.g. when job demands are high and when job resources are limited). From the perspective of employees, when AI is regarded as job demands (AI often functions as an innovative and assistive tool; Wisskirchen et al. 2017), individuals without AI knowledge/skills are more likely to miss job opportunities, whereas those with AI skills may secure more opportunities instead. From the perspective of employers, when AI is regarded as job resource (as AI helps to efficiently automate many back-office functions for reliable transactions and service delivery; Wang et al. 2017), firms with AI resources are more likely to gain competitive advantage, compared to those without resources.

Overall the aforementioned three theoretical views (*job replacement*, *psychological contract* and *demands-resources*.) have attempted to explain the potential influence of AI-driven management on the organizations, managers and employees from their respective perspectives. Jointly, the three perspectives have offered a comprehensive analysis on how employers and employees interact with (or are affected by) AI-driven management. These perspectives also help clarify how AI-driven management may be interpreted by the managers and employees, leading to different outcomes.

3 AI and Higher Education Sector

The current research is intrigued to know how AI exercises its influence in the Higher Education Sector (HES) and how HES should respond to such influence. For the same reason, we have reviewed the latest empirical studies and their findings are outlined below.

By investigating the impact of AI in HES, Ma and Siau (2018) have discovered that mundane jobs, routine duties and structured tasks are likely to be replaced by AI-driven applications, indicating that managers shall carefully evaluate AI's implication and make remedying plans to alleviate its consequence, such as job relocation and employee redundancy. Similarly, Siau (2017) claims that with the advancement of AI, robotics and automation, higher education institutions need to be reshaped, as these institutions may no longer be viewed as a means to employment or career development. Higher education institutions need to equip students with the necessary skill sets to compete in the AI era (Ma and Siau 2018).

In terms of student enrollment, liberal arts and humanities majors are less impacted by "AI-invasion.", whereas areas such as accounting and financial analysis are more impacted (Ma and Siau 2018). AI-driven applications facilitate academic practices through the provision of virtual teaching assistant, intelligent tutoring systems, and smart education, and scholars have found positive feedback in AI's application, such as improving students' learning, promoting academic guidance, enhancing student assessment and scoring process, and improving programs and quality assurance (Aldosari 2020; Ma and Siau 2018). AI also shows great potential on curriculum development, as AI is renowned for its advantageous characters, such as speed, accuracy and consistency (Brown et al. 2017).

In terms of personnel management, more specifically, there are limited studies analyzing the application of AI-driven management in HES. Scholars claim that there is no theoretical basis within HRM literature for understanding how, and to what extent, will AI impact on employees and their management (Malik et al. 2019). Following this line of research, Chang (2020) proposes an APM model, evaluating the potential outcome of AI-driven management at the organizational-, managerial-, and individual job-level. The model is meaningful and informative, as it helps clarify the opportunity and constraint of AI-driven management. Yet, the applicability of the model cannot be analyzed and affirmed until an empirical research is conducted. In a similar vein, Czarnowski and Pszxzolkowski (2020) claim that many managers still treat AI as information collector, which does not help promote the application of AI nor clarifying its advanced role in management. Overall, although AI has demonstrated its excellence and exerted its influence ubiquitously (*as per* literature review), how AI affects HES employees is still unclear, and how AI-driven management could be implemented into the personnel management in HES also requires further research and discussion. Having said this, however, we believe that managers and leaders from the HES can still benefit from our analysis (see Tables 1 and 2 for the detailed comment and suggestions). Where appropriate, we also believe that managers and leaders from the HES can incorporate our remarks into their personnel policies and employee management strategies.

4 Discussion

Regardless of its side-effect and threats, AI has possessed numerous merits and outperformed human intelligence in different aspects of human life. AI-based applications have strong potential to raise employee productivity and help managers become knowledgeable agents that boost employee performance (Ernst and Young 2018). The success of any organization depends on how effectively it combines people, process and technology to deliver transformational value at optimized cost, and AI will help to efficiently automate many back-office functions for reliable HR transactions and service delivery (Wang et al. 2017). Our proposition to the HES leaders and managers is: AI is like a technology tsunami, and it might be too late to action if there is no action strategy prepared in advance. Rather than concerning the possibility that AI-driven management may replace jobs and harm the employees, HES leaders and managers should adopt a more proactive attitude to evaluate the opportunity and constraint in using AI in personnel management. Taking a precautious approach and engaging with AI's development shall be more realistic.

As aforementioned in the literature, the main reason for the popularity of AI in management is its great potential. With the AI's assistance, organizational leaders and managers can easily digest massive quantity of information, which helps refining their decision-making and devising practical and timing managerial strategies (Brunette et al.

2009; Chang 2020). The advancement in AI can also empower business owners and personnel managers in understanding and supporting their employees more easily, such as performance management (Ernst and Young 2018; Hughes et al. 2019), emotional regulation (Henkel et al. 2020), training and development programmes (Maity 2019) and intervention of job burnout (Kong et al. 2021).

It is evident that the Higher Education Sector (HES) is different from the conventional business-oriented sectors and does not endorse the culture of profit-generating and customer-service ethos (Aldosari 2020; Ma and Siau 2018). Having said this, however, we believe that an educational institution is still an organizational entity, in which people are recruited as employees and units are formed to accommodate with the organizational needs and objectives. For the same reason, we believe that the analysis and remarks from Tables 1 and 2 are transferable to the HES and its employee management. We also believe that three theoretical perspectives have offered HES managers an integrative review on how managers and employees could be affected by AI-driven management in the workplace.

5 Conclusion

Based on the multi-disciplinary literature, this conceptual article has provided an integrative review of AI's applicability in personnel management. We draw insights from three theoretical perspectives (job replacement, psychological contract and demands-resources), analyzing the role of AI-driven management and its influence on the organizations, managers and employees. After reviewing both theoretical perspectives and empirical studies, we have learnt that managers' attitude is closely linked to the implementation of AI in personnel management. AI's applicability in management is not necessarily related to the function of AI, as managers' attitude plays a more important role in judging whether the AI should be adopted in management. Moreover, if not handled well, AI-driven management may cause job insecurity and dissatisfaction to the employees. Our research findings have brought new insights into the AI-management literature, advanced the knowledge of AI's opportunity and constraint in the personnel management, and clarified AI's applicability in HES.

In terms of managerial implications, three theoretical perspectives have offered different but practical suggestions. The perspective of job replacement first suggests that managers and leaders need to plan themselves in training of AI management; in addition to traditional labor management, they also need to know how to manage AI or at least equip themselves with skills in managing AI software/hardware. Next, the perspective of psychological contract suggests that, even if AI challenges managerial paradigms, managers shall appreciate the fundamental and human questions, such as how AI may be deployed to maximize the optimum efficacy of managerial policy? And, probably more importantly, in order to ensure employees maintain healthy psychological contract, how do managers engage the employees (e.g. those who are affected by AI), so they do not feel being left behind? Finally, the perspective of resource-demands suggests that managers shall remain cautious of AI's influence in their managerial practices, as AI has potential to become job demand or job resource, and the absence of either potential may put the employees and/or employer at a disadvantageous position.

In terms of research limitation and suggestions for future studies, our research focuses on the employees and relevant managerial issues, rather than the students in HES. Future research may consider qualitative methods (e.g., semi-structured interviews with AI-policy makers) and quantitative methods (e.g., questionnaire surveys distributed to AI-users), aiming to understand the role of AI and its influences in the campus, such as students' recruitment, engagement and their academic performance. Next, the current research regard administrative- and academic-staff as the same category of employees. Yet, administrative- and academic-staff may have different job duties and professional characteristics, whether such differences affect the implementation of AI-driven management requires further examination. Both cross-sectional and longitudinal research design shall be considered, so the full picture of AI and its implication on different categories of employees can be further analyzed. Finally, AI-driven management is found to enhance the organizational productivity and employee performance, particularly in the business organizations (Ernst and Young 2018). Whether the same findings are transferrable to the HES and its employees requires further examination and discussion.

References

Aldosari, S.A.: The future of higher education in the light of artificial intelligence transformations. Int. J. High. Educ. **9**(3), 145–151 (2020)

Bakker, A.B., Demerouti, E.: The job demands-resources model: state of the art. J. Manag. Psychol. **22**(3), 309–328 (2007)

Brown, J., Ling, T., Gurdeniz, E.: Managing Next Generation of AI in Banking. Oliver Wyman - Marsh & McLenna, New York (2017)

Brunette, E.S., Flemmer, R., Flemmer, C.: A review of artificial intelligence. In: 4th International Conference on Autonomous Robots and Agents, ICARA 2009, Wellington, New Zealand, February 10–12 (2009). https://doi.org/10.1109/icara.2000.4804025

Chang, K.: Has the Era of Artificial Intelligence (AI)-Driven Management Arrived Yet: The Case of Fudan University Student Canteen (in Chinese). Fudan Development Institute (FDDI), Fudan University, Shanghai, China (2019)

Chang, K.: Artificial intelligence in personnel management: the development of APM model. Bottom Line **33**(4), 377–388 (2020)

Czarnowski, I., Pszxzolkowski, P.: A novel framework for decision support system in human resource management. Procedia Comput. Sci. **176**, 1548–1556 (2020)

Ernst and Young: The New Age: Artificial Intelligence for Human Resource Opportunities and Functions. Ernst & Young LLP, London (2018)

Farndale, E., McDonnell, A., Scholarios, D., Wilkinson, A.: A look to the past, present, and future of the journal and HRM scholarship. Hum. Resour. Manage. J. **30**(1), 1–12 (2019)

George, C.: Psychological Contract: Managing and Developing Professional Groups. Open University Press, Maidenhead, G.B. (2010)

Grace, K., Salvatier, J., Dafoe, A., Zhang, B., Evans, O.: When will AI Exceed Human Performance? Evidence from AI Experts. Future of Humanity Institute, Oxford University, Oxford (2018)

Henkel, A.P., Bromuri, S., Iren, D., Urovi, V.: Half human, half machine – augmenting service employees with AI for interpersonal emotion regulation. J. Serv. Manage. **31**(2), 247–265 (2020). https://doi.org/10.1108/JOSM-05-2019-0160

Huang, M.H., Rust, R.T.: AI in service. J. Serv. Res. **21**(2), 155–172 (2018)

Hughes, C., Robert, L., Frady, K., Arroyos, A.: Artificial intelligence, employee engagement, fairness, and job outcomes. Managing technology and middle- and low-skilled employees. In: The Changing Context of Managing People, pp. 61–68. Emerald Publishing Limited, London (2019). https://doi.org/10.1108/978-1-78973-077-720191005

Jeske, D., Calvard, T.: Big data: lessons for employers and employees. Empl. Relat. **42**(1), 248–261 (2020). https://doi.org/10.1108/ER-06-2018-0159

Kong, H., Yuan, Y., Baruch, Y., Bu, N., Jiang, X., Wang, K.: Influences of artificial intelligence (AI) awareness on career competency and job burnout. Int. J. Contemp. Hosp. Manage. (2021). https://doi.org/10.1108/IJCHM-07-2020-0789

Lima, Y., Strauch, J.C.M., Esteves, M.G.P., de Souza, J.M., Chaves, M.B., Gomes, D.T.: Exploring the future impact of automation in Brazil. Empl. Relat. (2021). https://doi.org/10.1108/ER-08-2020-0364

Lucena, F.J.H., Diaz, I.A., Reche, M.P.C., Rodriguez, J.M.R.: Artificial intelligence in higher education: a bibliometric study on its impact in the scientific literature. Educ. Sci. **9**(1), 51–70 (2019)

Ma, Y., Siau, K.L.: Artificial intelligence impacts on higher education. In: MWAIS 2018 Proceedings, vol. 42 (2018). http://aisel.aisnet.org/mwais2018/42

Maity, S.: Identifying opportunities for artificial intelligence in the evolution of training and development practices. J. Manage. Dev. **38**(8), 651–663 (2019). https://doi.org/10.1108/JMD-03-2019-0069

Malik, A., Budhwar, P., Srikanth, N.R., Varma, A.: May the bots be with you! Opportunities and challenges of artificial intelligence for rethinking human resource management practices. In: Paper Accepted for Presentation at the Annual Conference, British Academy of Management 2019 (2019)

Mullins, L.J.: Management and Organizational Behavior, 11th edn. Pearson, Essex, England (2016)

Rousseau, D.M.: Psychological and implied contracts in organizations. Empl. Responsib. Rights J. **2**(2), 121–139 (1989)

Russell, S., Norvig, P.: Artificial Intelligence: A Modern Approach, 3rd edn. Prentice Hall Press, Upper Saddle River (2009)

SAS: Artificial Intelligence: What it is and Why it Matters. SAS, North Carolina, US (2018). https://www.sas.com/en_us/insights/analytics/what-is-artificial-intelligence.html#history

Siau, K.: Impact of artificial intelligence, robotics, and automation on higher education. In: Americas Conference on Information Systems (AMCIS 2017), Boston, MA, August 10–12 (2017)

Sivathanu, B., Pillai, R.: Smart HR 4.0 – how industry 4.0 is disrupting HR. Hum. Resour. Manage. Int. Digest **26**(4), 7–11 (2018). https://doi.org/10.1108/HRMID-04-2018-0059

Smith, A., Anderson, J.: AI, robotics, and the future of jobs. Pew Research Center, Washington (2014). http://www.pewinternet.org/2014/08/06/future-of-jobs/

Wang, A.M., Wang, T.L., Huang, C.J.: Big Data on Human Resource Management: Model, Theory and Application. Tsinghua University Press, Beijing, China (2017)

Wisskirchen, G., Biacabe, B.T., Bormann, U., Muntz, A., et al.: Artificial Intelligence and Robotics and Their Impact on the Workplace. IBA Global Employment Institute, London (2017)

Zhu, T.T., Peng, H.R., Zhang, Y.J.: The influence of higher education development on economic growth: evidence from Central China. High. Educ. Policy **31**(1), 139–157 (2017)

Blockchain Technology for Reinforcing Sustainability Practices Across Complex Multi-tier Supply Networks

Mohammad Najjar(✉)

An-Najah National University, Nablus, Palestine
m.najjar@najah.edu

Abstract. Complex supply networks feature many inter-connected and geographically distributed partners. Many of these partners are beyond the direct realm of the company, making the management of sustainability a difficult task. Inevitably, this research aims to explore how blockchain technology can facilitate sustainability initiatives across multi-tier supply networks. The findings suggest that blockchain can enhance the visibility of suppliers' sustainability practices. Managers will have increased transparency and traceability of their multi-tier supply networks, which will eventually reduce the cost of monitoring a large number of geographically distributed suppliers. Furthermore, information can be shared with different stakeholders to increase credibility and legitimacy.

Keywords: Blockchain · Complex supply networks · Multi-tier supply networks · Sustainable supply chain management · Sustainability

1 Introduction

Recent technological advancements are provoking significant disruptions to existing supply chain management models [1]. Blockchain is one of these technologies that offer several principal benefits for supply chain management. The technology consists of decentralized ledgers and distributed processing, which increases the immutability of data, higher information transparency, and further traceability [2–6]. These benefits have stimulated many global companies, e.g. Unilever and WalMart, to integrate the blockchain technology into their supply networks in order to promote efficiency and performance [6]. Inevitably, the technology will impact on existing sustainable supply chain management frameworks, specifically frameworks that strive to embed sustainability practices across multi-tier, complex, and global supply networks. During the last three decades, both the social and environmental dimensions of sustainability have been gaining increasing attention among different stakeholder groups. These dimensions have become undoubtedly an important driver for customer loyalty [6]. Customers are requesting more access to information about products and suppliers' sustainability operations [7–10]. This has put an increasing pressure on focal companies to monitor the operations of their extended supply networks [11, 12].

A. M. A. Musleh Al-Sartawi et al. (Eds.): EAMMIS 2021, LNNS 239, pp. 290–299, 2021.
https://doi.org/10.1007/978-3-030-77246-8_28

Today's global supply networks are complex and feature many inter-connected [13] and geographically distributed partners that work across different industrial sectors and cultural norms [11, 14], with the aim to provide competitive and sustainable products to customers [12, 13]. Focal companies have adopted several management systems, e.g. certification systems and codes of conduct, to track and control the sustainability practices across downstream suppliers [7, 8, 15, 16]. However, Choi et al. [17, p. 359] questioned the effectiveness of these practices and highlighted that "what happens to tertiary-level suppliers is often unknown to the final assemblers, and this is outside of the realm of their control". Furthermore, it is difficult to recognize "who is responsible for what aspect" in lengthy and complex supply networks [18]. Indeed, managers usually struggle with the inevitable lack of prediction and control that exist across their supply networks [17], making the management of sustainability a difficult task. Inevitably, blockchain can be a promising technology in these situations; indeed, it still at its early stages of development [1]. It has the potential revolutionize supply network and sustainability models [3] by facilitating the transfer of sustainability information across the different tiers, as well as to different stakeholder groups. Therefore, rigorous exploration of blockchain is required in order to understand the extent to which it may help focal companies in creating value [3].

Following the suggestion of Cole et al. [19], this research strives to understand the adoption of blockchain technology across complex multi-tier supply networks. It attempts to explore the role of blockchain technology in supporting sustainability practices across global multi-tier supply networks [1]. In addition, the research seeks to investigate supply network structural barriers that may impede blockchain adoption, responding to increasing call from researchers to investigate the main challenges that are associated with blockchain technology adoption [1, 4–6, 20]. In particular, this research attempts to answer two questions:

1. How can blockchain enhance the visibility and tractability of suppliers' sustainability practices across complex multi-tier supply networks?
2. What are the main challenges that are associated with adoption of blockchain technology in complex and global supply networks?

The study contributes to the body of literature focusing on the utilization of blockchain to improve multi-tier supply networks' sustainability. It provides a framework to facilitate the adoption of blockchain technology across multi-tier supply networks and identify supply network structural complexities that might hinder effective adoption of this disruptive technology. The remainder of this research is organized as follow. The following section discusses the literature on blockchain technology and sustainable supply chain management. It gives attention to the complexity of global multi-tier supply networks in order to thoroughly investigate the adoption of blockchain technology. The third sections proposes a framework that describes the role of blockchain in facilitating the diffusion of sustainability practices across multi-tier supply networks and the different challenges that global companies may encounter during the adoption process. The final section is a conclusion and contribution of this research.

2 Literature Review

Blockchain has gained increasing attention during the last few years due to its potential to revolutionize supply chain management [5, 6]. The literature has posited several principal advantages such as improved visibility, traceability, transparency, efficiency and cost saving, as well as improved brand recognition [2, 5, 21]. In addition, several global companies (such as Unilever) have pursued to adopt this disruptive technology (e.g. Unilever) in order to enhance the sustainability performance of their supply networks. Blockchain is a digital decentralized ledgers that include transactions as blocks [6]. The blocks are connected to each other through a cryptographic hash [1]. The connectivity feature enhance data immutability because data transactions are replicated and shared among all member on the chain [22]; therefore data cannot be altered or deleted [3]. Since blockchain is decentralized, the computational power in the blockchain is distributed across the different partners. Two popular blockchain technologies exist, i.e. private and public. In private blockchain, only invited members are authenticated to participate in the chain and use the chain data. In contrast, in public blockchain, any member/entity can join the chain, access and use the data and ledgers. In the context of supply chain management, the connectivity feature and the ability to trace information and products in private blockchain have stimulated many companies to integrate the technology into their supply network models, with the aim to improve sustainability performance of downstream suppliers.

Seuring and Muller [23, p. 1700] defined sustainable supply chain management as "the management of material, information and capital flows as well as cooperation among companies along the supply chain while taking goals from all three dimensions of sustainable development, i.e., economic, environmental and social, into account which are derived from customer and stakeholder requirements". Different stakeholder groups have started to recognize the role played by global companies in the promotion of sustainability practices across their extended multi-tier supply networks. As a consequence, companies have adopted several management practices to ensure suppliers' acquiescence to sustainability standards, while gaining enough credibility and legitimacy among different stakeholders, e.g. through codes of conduct, certification systems as well as auditing practices [8, 13, 15, 16, 24]. However, the literature still questions the validity and credibility of these management systems and how much they can reach further down the supply network [8, 25]. Hannibal and Kauppi [8, p. 78] stated that "information asymmetries between supply network actors and stakeholders can result in uncertainty about how a good has been produced and traded, resulting in sustainability uncertainty". Furthermore, unethical practices usually originate from lower-tier suppliers that are beyond the preview of the focal company [15]. Global multi-tier supply networks are complex settings with many suppliers distributed across different geographical locations and operate under different cultural norms [7, 13–16]. Depending on the length of the supply network, it is sometimes difficult trace and monitor the social and environmental impact of products from inception to final use by end consumers [6, 8, 26] using the traditional and costly management systems. In addition, stakeholders are still requesting further transparency and accessibility to sustainability information across the entire supply network [6, 7, 16]. Inevitably, blockchain can play an important role in these situations. Azzi et al. [2] stated that "blockchain can be integrated into the supply chain architecture to create a reliable, transparent, authentic and secure system".

Despite the potential benefits of blockchain technology on improving the sustainability practices of suppliers across multi-tier supply networks, the number of companies applying the technology is still limited. This might be due to the barriers that exist across the multi-tier supply networks [4–6]. Kouhizadeh et al. [6] examined the different challenges that may play a role in the slow adoption of blockchain technology. In addition, Saberi et al. [5] have categorized barriers into three main groups; technological barriers are related to security, accessibility and immaturity of technology, organizational barriers include issues related to commitment, policies and culture, and environmental barriers related to information disclosure across supply network members, lack of collaboration, communication, and involvement of different stakeholder groups (for a more inclusive list of barriers, please refer to [5, 6]. Although several research paper have addressed the different barriers that may impede the adoption of blockchain technology [2, 4–6], this research gives attention on the complexities that reside within multi-tier supply networks and explores how blockchain can aid the integration of sustainability practices across complex supply networks. Supply networks as a complex system consists of many interconnected partners that exchange products and information [17, 27, 28]. In complex and global supply networks, suppliers are heterogeneous; meaning that they work in different industries, operate under different cultural norms, and located in different areas [14, 15, 19, 27]. Due to the complexity of global supply networks, Choi and Krause [29, p. 639] stated that "much of what occurs in supply networks is outside the managerial purview of the focal company". In addition, Choi et al. [17, p. 359] highlighted that "what happens to tertiary-level suppliers is often unknown to the final assemblers, and this is outside of the realm of their control". Furthermore, it is arduous to recognize "who is responsible for what aspect" in lengthy and complex supply networks [18]. Indeed, managers usually struggle with the inevitable lack of prediction and control that exist across their supply networks [17], making the management of sustainability a difficult task.

In summary, managers adopt different practices to manage the sustainability aspects of their downstream suppliers. However, their efforts may stagnate due to the complexity of their multi-tier supply networks where many suppliers are beyond their direct visibility domain. Therefore, this research strives to understand the adoption of blockchain technology across complex multi-tier supply networks. It attempts to introduce a framework that utilized blockchain as approach to facilitate the integration of sustainability across the different tiers of complex supply network.

3 Research Method

In this research, we sought to how blockchain technology can enhance complex and multi-tier supply networks sustainability. Therefore, three main areas have been investigated in the literature. Firstly, the literature related to blockchain technology has been explored to understand how this technology operates as well as the implementation issues and benefits that might be gained from the adoption of this technology. Secondly, the literature on complex supply networks has been investigated to explore the different barriers and complexities that may hinder the adoption of blockchain technology.

Finally, the literature on multi-tier sustainable supply networks and the different governance structures have been explored in order to understand how blockchain technology can promote sustainability across the different layers of the supply network.

In order to gain enough knowledge on these topics, relevant and well cited papers were searched across different journals (e.g. International journal of production economics, Supply chain management: An international journal, International journal of information management, International journal of production research) using the online databases such as EBSCO, Science direct, and Emerald Group. The following keywords were utilized to collect relevant research papers: "Blockchain technology", "Blockchain and industry 4.0", "Complex supply networks", "Complexity theory", "Sustainable supply chain management" and "Multi-tier sustainable supply networks". A list of highly cited and peer-reviewed papers has been categorized and explored in more details (initial abstract investigation), filtering the articles based on their relevance and added-value to the research topic in this study. Finally, the list of selected articles was then screened again to remove articles that provide duplicate results, and to link the different articles to the different research questions proposed in this study. These steps have resulted in 34 relevant research papers that have been selected for full review and analysis.

4 Discussion: The Integration of Blockchain in Complex Multi-tier Supply Networks to Reinforce Sustainability Practices

Blockchain technology and its implications on business operations (specifically sustainable supply chain management) is still under-researched area. Companies operate within a complex network of interdependent partners. Therefore, companies' ability to gain competitive position in the industry is linked to the operations and sustainability performance of other supply network members. Mena et al. [30] proposed three governance approaches to manage multi-tier supply networks, i.e. open, closed, and transitional. According to them, in the open mechanism, the focal company and the second-tier suppliers are connected through the first-tier supplier, and information and material flow among those partners in a linear fashion [15, 30]. In the closed approach, the focal company and the second-tier suppliers interact directly and exchange information, whereas the transitional approach is between the open and closed in which the focal company and the second-tier suppliers start to build direct relationships [30]. In addition, Tachizawa and Wong [10] identified four governance mechanisms, as shown in Fig. 1, to manage the sustainability practices of downstream suppliers, i.e. indirect, direct, work with third parties, and don't bother. According to authors, the direct approach fosters lineal interaction between the focal company and the second-tier suppliers (which is similar to the closed approach in Mena et al. [30]), whereas the indirect approach the focal company depends on the first-tier suppliers to monitor the operations of their sub-suppliers [10] (which is similar to the open approach in Mena et al. [30]). In the third party approach, the focal company authorizes Third Party Organization (e.g. a certification body) to monitor the operations of second-tier suppliers [10, 15]. When the focal company's power is low, information on lower-tier suppliers is limited, and when suppliers are located in distant countries, the company may utilize the don't bother approach [10].

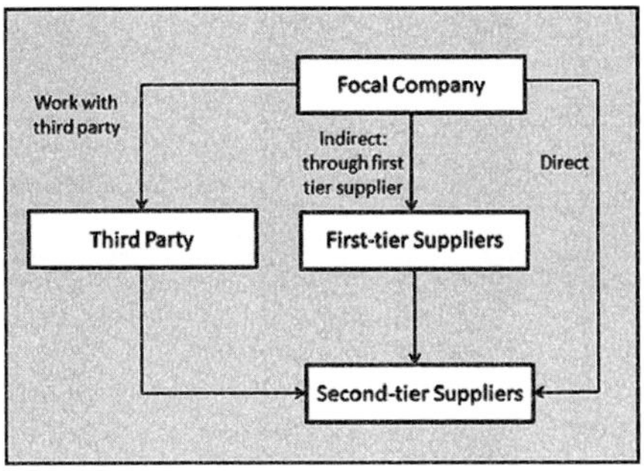

Fig. 1. The different governance mechanisms (adopted from Tachizawa and Wong [10])

It is sometimes costly to monitor the operations of lower-tier suppliers, in particular suppliers beyond the second-tier. In global, lengthy, and complex supply networks, many suppliers are outside the direct visibility of the focal company and located in different geographical areas. Therefore, company's efforts to control the operations of their suppliers across the different tiers may deteriorate. Blockchain technology can facilitate and support companies' sustainability initiatives through promoting traceability and transparency, and access to sensitive data across the entire supply network, even suppliers that are beyond the second tier.

The framework depicted in Fig. 2 describes how blockchain can facilitate and integrate sustainability practices across the multi-tier supply networks. The automated nature of blockchain enables trusted information flow and automatic updates of records between geographically dispersed, distant, and disconnected partners at different tiers. Blockchain can facilitate the transfer and sharing of credible, up-to-date, and correct sustainability information to network members, and when necessary can be shared with different stakeholder groups. The technology can stimulate confidence in the operations of the supply network partners by monitoring and following up the flow of products and materials across the supply network [6, 31].

Blockchain can track and increase the visibility of sustainability practices across the different tiers. It provides an approach for increased transparency and smooth information flow among supply network partners [1, 19]. Indeed, the utilization of blockchain technology may change the governance structures and focal companies may reduce the utilization of third party approaches since information is inevitably available to all supply network partners without the need for further intervention of external parties to monitor lower-tier suppliers (specifically suppliers at n-tier).

In our evaluation of the complex supply networks, several structural barriers have been identified. These obstacles should be addressed for effective adoption of blockchain in sustainable multi-tier supply networks. Firstly, in complex supply networks, suppliers are inter-connected [17, 32, 33]. This means that transactions made by any member will influence the blocks of other members. As a result, the blockchain may require additional computing power to duplicate and maintain records on all connected nodes. This is specifically crucial for global and multi-national corporations that have a large number of suppliers. Furthermore, companies operating in global supply networks usually work with family businesses and small-size suppliers that may lack enough resources

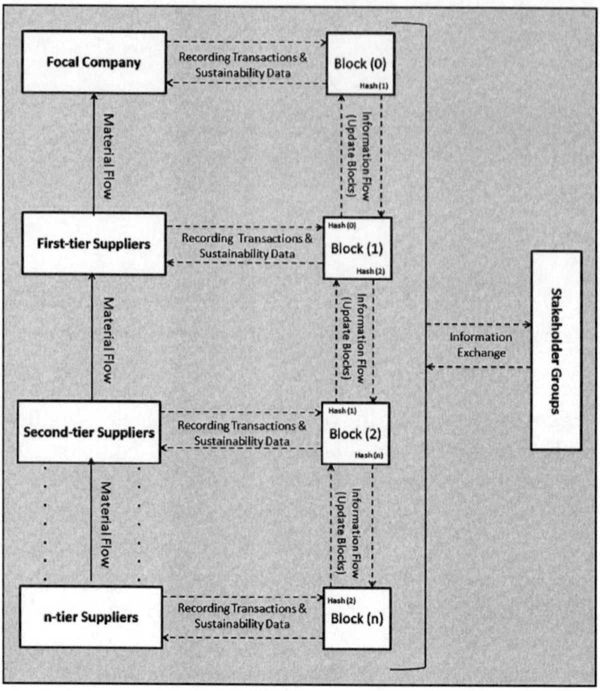

Fig. 2. The role of blockchain in facilitating information flow across multi-tier supply networks

and expertise, which may impede the adoption of block chain technology. Indeed, it is important to have both tangible and intangible resources for successful adoption of blockchain technology as well as the supporting policies to protect disclosure of sensitive information [6].

Environmental aspects of sustainability can be monitored and measured using the technological tools. However, many of the social aspects of sustainability are intangible (e.g. child and forced labor) and companies may find it difficult to trace the compliance of the distant suppliers. Park-Poaps and Rees [34] found that suppliers with limited resources may attempt to give incorrect sustainability data. Therefore, both academicians and practitioners should establish clear measures and systems to trance the intangible aspects of sustainability for effective utilization of blockchain technology.

5 Conclusion

Complex supply networks feature many inter-connected and geographically distributed partners that work across different industrial sectors. Focal companies have adopted several management systems to track and control the sustainability practices across downstream suppliers. However, Choi et al. [17, p. 359] questioned the effectiveness of these practices and highlighted that "what happens to tertiary-level suppliers is often unknown to the final assemblers, and this is outside of the realm of their control." Indeed, managers usually struggle with the lack of prediction and control that exist across their multi-tier supply networks [17], making the management of sustainability across the different tiers a difficult task. Inevitably, blockchain can be a promising technology in these situations; indeed, it still at its early stages of development [1]. The findings highlighted that blockchain technology can support the management of sustainability across complex multi-tier supply networks. This has important implication for manager since the blockchain technology can enhance the visibility and transparency of suppliers' sustainability practices across the different layers of the supply network. In addition, managers will have increased traceability of products to their origin, which will eventually reduce the cost of monitoring the large number of geographically distributed suppliers in global and complex supply networks. Blockchain enables efficient, transparent and secure information flow across the different tier. Information can be shared with different stakeholder groups to increase credibility and legitimacy, gaining better competitive position in the industry.

Despite the potential benefits enabled by the adoption of blockchain technology in sustainable multi-tier supply networks, there exist vital challenges that may hinder the adoption of the technology to facilities the management of sustainability. Firstly, according to the complexity theory, complex supply networks consist of a large number of inter-connected suppliers, which may require additional computing power to duplicate and maintain records on all connected nodes. In addition, global supply networks extend to developing countries, and many suppliers in these countries are small-size and family business. These suppliers lack enough resources and expertise to adopt the technology. Furthermore, in contrast to environmental issues, many of the social issues are intangible (e.g. child and forced labor), and academics and practitioners need to establish clear measurement systems to be able to trace the different dimensions of sustainability in an effective manner. The research has important implications for practitioners and researchers who are interested in a more thorough exploration of the impact of blockchain on supply chain management practices. The overall objective of this research is to understand blockchain technology and how it can facilitate sustainability practices across multi-tier supply networks. In this context, although blockchain is provoking disruptive changes in the industry, it is remodeling the sustainability monitoring practices across multi-tier supply networks through enhancing traceability, transparency, and visibility of the supply network members. The proposed framework can support managers and researchers in identifying major challenges in complex supply networks, while at the same time creates new research agenda to investigate how to overcome these complexities utilizing different theoretical lenses. Future research can investigate the proposed conceptual framework in different contexts. It can be applied to different supply network structures and industrial sectors with different features (e.g. labor vs. capital intensive industries) to understand the applicability of blockchain technology. Furthermore, the literature is lacking exploratory and survey based research that address the implications of blockchain technology on the sustainability practices of global and multi-tier supply networks. Future research can address this gap in order to have a more thorough and generalized exploration of blockchain technology across complex, global, and multi-tier supply networks.

References

1. Queiroz, M.M., Telles, R., Bonilla, S.H.: Blockchain and supply chain management integration: a systematic review of the literature. Supply Chain Manage. **25**, 241–254 (2019)
2. Azzi, R., Chamoun, R.K., Sokhn, M.: The power of a blockchain-based supply chain. Comput. Ind. Eng. **135**, 582–592 (2019)
3. Treiblmaier, H.: The impact of the blockchain on the supply chain: a theory-based research framework and a call for action. Supply Chain Manage. **23**, 545–559 (2018)
4. Queiroz, M.M., Wamba, S.F.: Blockchain adoption challenges in supply chain: an empirical investigation of the main drivers in India and the USA. Int. J. Inf. Manage. **46**, 70–82 (2019)
5. Saberi, S., Kouhizadeh, M., Sarkis, J., Shen, L.: Blockchain technology and its relationships to sustainable supply chain management. Int. J. Prod. Res. **57**, 2117–2135 (2019)
6. Kouhizadeh, M., Saberi, S., Sarkis, J.: Blockchain technology and the sustainable supply chain: theoretically exploring adoption barriers. Int. J. Prod. Econ. **231**, 107831 (2021)
7. Mejías, A.M., Bellas, R., Pardo, J.E., Paz, E.: Traceability management systems and capacity building as new approaches for improving sustainability in the fashion multi-tier supply chain. Int. J. Prod. Econ. **217**, 143–158 (2019)
8. Hannibal, C., Kauppi, K.: Third party social sustainability assessment: is it a multi-tier supply chain solution? Int. J. Prod. Econ. **217**, 78–87 (2019)
9. Kembro, J., Näslund, D., Olhager, J.: Information sharing across multiple supply chain tiers: a delphi study on antecedents. Int. J. Prod. Econ. **193**, 77–86 (2017)
10. Tachizawa, E.M., Wong, C.Y.: Towards a theory of multi-tier sustainable supply chains: a systematic literature review. Supply Chain Manage. **19**, 643–663 (2014)
11. Najjar, M., Small, M.H., Yasin, M.M.: Social sustainability strategy across the supply chain: a conceptual approach from the organisational perspective. Sustainability **12**, 10438 (2020)
12. Najjar, M., Small, M.H., Yasin, M.M.: Adopting and extending social sustainability practices across the supply chain. Compet. Forum **17**, 105–113 (2019)
13. Sauer, P.C., Seuring, S.: Extending the reach of multi-tier sustainable supply chain management–Insights from mineral supply chains. Int. J. Prod. Econ. **217**, 31–43 (2019)
14. Zorzini, M., Hendry, L.C., Huq, F.A., Stevenson, M.: Socially responsible sourcing: reviewing the literature and its use of theory. Int. J. Oper. Prod. Manage. **35**, 60–109 (2015). https://doi.org/10.1108/IJOPM-07-2013-0355
15. Govindan, K., Shaw, M., Majumdar, A.: Social sustainability tensions in multi-tier supply chain: a systematic literature review towards conceptual framework development. J. Clean. Prod. **279**, 123075 (2020)
16. Fraser, I.J., Müller, M., Schwarzkopf, J.: Transparency for multi-tier sustainable supply chain management: a case study of a multi-tier transparency approach for SSCM in the automotive industry. Sustainability **12**, 1814 (2020)
17. Choi, T.Y., Dooley, K.J., Rungtusanatham, M.: Supply networks and complex adaptive systems: control versus emergence. J. Oper. Manage. **19**, 351–366 (2001). https://doi.org/10.1016/S0272-6963(00)00068-1
18. van Bommel, H.W.M.: A conceptual framework for analyzing sustainability strategies in industrial supply networks from an innovation perspective. J. Clean. Prod. **19**, 895–904 (2011). https://doi.org/10.1016/j.jclepro.2010.12.015
19. Cole, R., Stevenson, M., Aitken, J.: Blockchain technology: implications for operations and supply chain management. Supply Chain Manage. **24**, 469–483 (2019)
20. Ray, P., Om Harsh, H., Daniel, A., Ray, A.: Incorporating block chain technology in food supply chain. Int. J. Manage. Stud. **6**, 5 (2019)
21. Angelis, J., da Silva, E.R.: Blockchain adoption: a value driver perspective. Bus. Horiz. **62**, 307–314 (2019)

22. Christidis, K., Devetsikiotis, M.: Blockchains and smart contracts for the internet of things. IEEE Access **4**, 2292–2303 (2016)
23. Seuring, S., Muller, M.: From a literature review to a conceptual framework for sustainable supply chain management. J. Clean. Prod. **16**, 1699–1710 (2008). https://doi.org/10.1016/j.jclepro.2008.04.020
24. Mena, C., Schoenherr, T.: The green contagion effect: an investigation into the propagation of environmental practices across multiple supply chains tiers. Int. J. Prod. Res. **58**, 1–18 (2020)
25. Jiang, B.: The effects of interorganizational governance on supplier's compliance with SCC: an empirical examination of compliant and non-compliant suppliers. J. Oper. Manage. **27**, 267–280 (2009). https://doi.org/10.1016/j.jom.2008.09.005
26. Ivanov, D., Dolgui, A., Sokolov, B.: The impact of digital technology and Industry 4.0 on the ripple effect and supply chain risk analytics. Int. J. Prod. Res. **57**, 829–846 (2019)
27. Frostenson, M., Prenkert, F.: Sustainable supply chain management when focal firms are complex: a network perspective. J. Clean. Prod. **107**, 85–94 (2015)
28. Li, G., Yang, H., Sun, L., Ji, P., Feng, L.: The evolutionary complexity of complex adaptive supply networks: a simulation and case study. Int. J. Prod. Econ. **124**, 310–330 (2010). https://doi.org/10.1016/j.ijpe.2009.11.027
29. Choi, T.Y., Krause, D.R.: The supply base and its complexity: implications for transaction costs, risks, responsiveness, and innovation. J. Oper. Manage. **24**, 637–652 (2006). https://doi.org/10.1016/j.jom.2005.07.002
30. Mena, C., Humphries, A., Choi, T.Y.: Toward a theory of multi-tier supply chain management. J. Supply Chain Manage. **49**, 58–77 (2013)
31. Dai, H.-N., Zheng, Z., Zhang, Y.: Blockchain for internet of things: a survey. IEEE Internet Things J. **6**, 8076–8094 (2019)
32. Pathak, S.D., Day, J.M., Nair, A., Sawaya, W.J., Kristal, M.M.: Complexity and adaptivity in supply networks: building supply network theory using a complex adaptive systems perspective*. Decis. Sci. **38**, 547–580 (2007). https://doi.org/10.1111/j.1540-5915.2007.00170.x
33. Serdarasan, S.: A review of supply chain complexity drivers. Comput. Ind. Eng. **66**, 533–540 (2013)
34. Park-Poaps, H., Rees, K.: Stakeholder forces of socially responsible supply chain management orientation. J. Bus. Ethics **92**, 305–322 (2010). https://doi.org/10.1007/s10551-009-0156-3

Segmenting Potential Customers with Kohonen Network: A Banking Sector Case Study

Tutku Tuncali Yaman(✉)

Beykent University, Istanbul, Turkey
tutkuyaman@beykent.edu.tr

Abstract. The main objective of this paper is to understand the characteristics of the new customers of a web-based lending bank via a segmentation methodology with Kohonen Network. The main problem involved here is the proposition of a data-driven segmentation approach in the prevention of the loss of remunerative customers' acquisition in the digital era's banking industry. The reason for choosing Kohonen Network is the ability to work on large volumes of data and the most appropriate determination of the cluster number, which is the most important decision for cluster analysis, by the technique itself. The well-known purpose of cluster analysis in business analytics is to provide the necessary foresight by revealing previously unknown critical customer characteristics and degrees of importance in order to assist the business in charge in strategic marketing decisions such as market segmentation and target market selection. As a result of the study, new marketing strategies regarding the loss of new customers' acquisition suggested through the findings of Kohonen clusters.

Keywords: Market segmentation · Kohonen network · Clustering · Lending bank · Data mining

1 Introduction

Market segmentation is also specified as a set of grouping consumers' identical or similar characteristics [1]. In the literature of business science, market segmentation has an essential role itself. In most cases, customization cannot be thought of without proper customer segmentation. Market segmentation allows a company to focus its limited resources on the most promising opportunities by segmenting customers into economically manageable and "priority" segments [2]. With its great advantage in taking different actions in marketing such as generating customized advertising, pricing, and promotions, several different analytical approaches have been generated and applied in the generation of the different customer segments in different sectors. In the first market segmentation criteria, a single demographic characteristic or the purchase and use of the product were prioritized. Over time, more features have been used together and post-hoc measurements have been developed. With these developing criteria, market segmentation has begun to be examined in different dimensions, general and specific to the domain [3]. One of the most basic approaches of market segmentation is determining the strategies

A. M. A. Musleh Al-Sartawi et al. (Eds.): EAMMIS 2021, LNNS 239, pp. 300–312, 2021.
https://doi.org/10.1007/978-3-030-77246-8_29

to be used by determining the main criteria and modeling them. Another approach is to make marketing strategies applicable and to ensure the transition from theory to practice [4]. The most important focus in segmentation is choosing the best analysis technique. In their well-known chapter, Wind and Bell offered six different types of analyzes used in market segmentation. These are classification methods (clustering), discriminant methods (discriminant and regression), concurrent evaluation (multidimensional scaling), simulation and optimization (conjoint), Bayesian methods (modeling), and customer interaction (data mining) [5].

The banking industry, with a huge amount of customer data in terms of demographical and transactional, is seen as suitable for exploratory data analysis. However, with the rise of the 'Big Data Era', as a customer-focused industry such as insurance and telecommunication, banks took one of the leading roles as an application area of big data analytics [6]. Without a doubt, another gain of the 'Big Data Era' could be seen that the contribution to the rise of artificial intelligence (AI) based recommender systems. Most of the up-to-date AI-based applications in the banking industry are offering intelligent systems in credit scoring in lending [7–9]. With the help of real-time recommender systems, the applicant's data are processed, and an objective decision upon the loan request. Except for credit scoring, in the banking sector, under-standing behavioral patterns of positively evaluated potential loan borrowers by an AI-based solution has seemed an open area for research. A dynamic segmentation approach by occupying AI that is supported by a set of relevant marketing activities would be advantageous in preventing the loss of new customers.

In this paper, a case study of an online-lending bank that is serving in Turkey is handled with the objective of understanding factors affecting the loss of acquisition of potential customers. By following this aim, a total of 5825 single loan applications of individual customers in 4 months (from May to September) of 2018 were analyzed. Although all of these applications were approved, only 1745 of them were completed. To create a meaningful segmentation for potential customers (loan applicants) and understand the characteristics of the 'quitters' who are the applicants that canceled their approved credit application, following a feature selection algorithm that is aimed to define important characteristics in customer behavior, a clustering analysis based on Kohonen Self Organizing Neural Network (SONN) was performed. With the help of the informative results of the analysis, a set of suggestions provided to the bank in terms of preventing quitting behavior of their approved loan applicants. The value of the research is being a pilot case study that its objective is associated with not only the understanding of the new customers but also the problem of new customer acquisition in the digital era's banking industry. Plus, the suggested analytical approach (Kohonen SONN) is open for AI-based real-time segmentation practice since being a member of the neural network's family.

The following sections of the paper are included a literature review, description of the examined bank case study, dataset definition, details of the followed analytical procedure in segmentation, explication of the findings, and discussion, respectively.

2 Literature Review

To the extent of a proper literature review, studies that both aimed to perform market segmentation and involved in the new customer acquisition problem in the banking industry, are reviewed. In one of the early applications of survey-based segmentation studies in the banking sector [10], authors conducted field research among 100 Swiss bank officials and performed a segmentation study according to production technology and normative coherence of attitudes and orientations of the bank's staff. According to the results, using separate production technology is a significant attribute for different market segments. An example study from Indonesia [11] in-tended to perform a segmentation study for the Indonesian banking industry by con-ducting a field study to understand expected customer benefits from a bank. A following quantitative study was initiated to identify relationships between desired consumer benefits and demographic characteristics such as age, gender, income, and marital status. According to their results that gathered via hierarchical and k-means cluster analysis, the use of promotional activities and benefit bundles was found beneficial to retain new customers. Surprisingly, they have found no significant association between desired benefits and the demographics of the surveyed customer group. Another paper [12] is performed an RFM, loan-to-value (LTV) ratio, and customer characteristics based segmentation methodology with the use of a combination of neural network analysis and k-means clustering algorithm. They used the Iranian bank customers' data in their practice. As a result, the existing customers were divided into nine segments according to their shared transactional behavior and characteristics. Profiles of customer groups were found convenient for marketing professionals to create both new acquisition and retention strategies. Another Iranian case study [13] realized a guild segmentation with the use of RFM clustering and customer lifetime value (CLV) analysis. Similar to previously discussed papers, this one has also used the K-means algorithm in clustering the guilds. In their paper, Khajvand and Tarokh [14] provided a market segmentation framework for estimating customer future value based on adapted weighted RFM analysis which is an ARIMA based CLV calculating model, for each segment of Iranian customer in retail banking scope. A hierarchical clustering-based segmentation study [15] proposed a methodology to identify similarities between the existing Taiwanese college library users. The library entry data was found similar to true transaction records in banks. Pursuing their aim, before clustering, the dissimilarity between transaction sequences was defined. According to the results, they obtained three clusters and they claimed that the proposed method better results than other traditional methods. There are similar studies that aimed to define groups of banks' customers by segmenting based on customers' financial transactions with the use of cluster analysis [16, 17]. Since Islamic banking has become a global phenomenon in both Islamic and western countries, a Nigerian example segmentation study benefits from secondary data sources exposed that the targeted Muslim segment is substantial [18]. Also, Islamic banking in Nigeria is worthwhile and has the potential to gain new customers. Another study that discussed an Islamic banking example from Bangladesh [19] aimed to examine the impact of demographic disparities on the bank selection criteria. The dataset was obtained from one of the Islamic banks of Bangladesh. According to the results of regression analysis, when demographic factors were controlled, along with having corporal efficiency, core-banking services, and confidence; adoption of e-banking solutions, offering advanced promotional activities, and having experienced staff were found important attributes by the majority of the respondents. In a Greek retail bank example [20], authors examined the importance of customer satisfaction that contains service quality and such other attributes as price, convenience, and innovation in customer segmentation. It is found that the facets of customer satisfaction as explanatory cues for the switching behavior of individual and business customers were tested successfully. In Loureiro and Miranda's research [21], a Finite Mixture Partial Least Squares-based market segmentation was proposed in the inter-net banking context. Their model was involved in customer trust, perceived risks and benefits of online banking, brand awareness/associations, perceived quality, and their impact on bank's brand equity and brand loyalty. According to the results that retrieved two customer segments from a Portuguese private bank's existing customers; the impact of online benefits on trust in the service provided is stronger for the first segment than for the second. In their study, Kahreh et al. [22] aimed to group an Iranian commercial bank's customers by a benefit segmentation based on customer lifetime value (CLV). According to CLV results all customers are divided into 6 segments with their characteristics. The findings implied that each segment requires different marketing plans and programs in terms of strategic marketing perspective. In a recent study, it was aimed to realize a benefit based segmentation study for the customers of Shahr bank of Iran. After conducting 165 interviews with the customers, survey data was analyzed and four customer segments of benefit-oriented, peace-oriented, interest-oriented, and moderate were identified [23]. In a classification-based predictive study [6], the authors aimed to estimate potential bank customers on the bank dataset obtained by telemarketing method in UCI Machine Learning Repository. In a conference paper [24], the authors investigated that whether there is a statistically significant connection between demand for consumer loans and the age of the customer with a logistic regression analysis.

As a member of the artificial neural network family, Kohonen SONN applications in the market segmentation are focused on clustering retail customers [25–39]. There are two papers were performed Kohonen SONN analysis in the detection of fraudulent activity in the finance sector [40, 41]. The use of Kohonen SONN in customer segmentation in the banking industry was realized in only two studies. The first one performed a hybrid methodology that combines Kohonen network and hierarchical clustering in segmentation study based on psychological characteristics of Korean internet banking customers [42]. The other was realized a survey among 850 Croatian firms and implemented a business customer segmentation via Kohonen network [43].

As a result of a comprehensive literature review, there is a lack of research in analyzing potential customer segments in the banking industry rather than segmenting the existing ones. Even though the Kohonen SONN has been performed in several segmentation studies and has been found as a robust technique, there is not any re-search that benefits from this technique and handling the segmentation of new individual customers in the banking industry. In the following section, the online lending bank case study will be discussed in detail.

3 Online Lending Bank Case Study

To gain a better understanding of the case study discussed in the paper, it is important to explain the online loan application process in detail. The case study will be explained under three headings: the loan lending process, marketing actions, and new customer acquisition problem.

The Loan Lending Process: The aforementioned bank provides consumer loans, which one only can apply online via the Internet. Any person can make the loan application by entering the requested information in the application form on the bank's website. This information covers citizenship number, demographics, requested loan amount (since the upper limit of the product offered by the bank to its customers is 30,000 TL, the system recommends this upper limit at most even if the customer's request an amount above this amount on the application page), and contact information such as address and phone number. In the evaluation phase after the application is made, the bank verifies the personal information entered via the citizenship number. Besides, the credit score of the person is questioned via the citizenship number. The bank first decides on the approval & rejection status of the loan application through the decision support algorithm it follows in the process. Then, with a credit risk score-based calculation, the amount of credit that can be allocated to the person is determined, taking into account the desired loan amount. For rejected applications, an information message is sent to the applicant. Information about the approved applications is forwarded to the call center. The call center staff calls the person whose application has been accepted and informs about the documents required for the acceptance of the application and the loan allocation. The person whose application is approved must prepare the necessary documents and confirm that he/she will send them to the bank via a courier on a day to be determined within 20 days from the date of application. On the confirmed date, the bank sends a courier with the contract to be signed at the customer's address. The customer transmits the necessary documents to the courier and signs the contract. After the documents brought by the courier are checked by the bank, if no deficiencies are detected, the loan amount allocated to the IBAN sent by the customer during the application process is deposited. In this process, if the customer does not fulfill his/her obligations within 20 days from the date of application, gives up the loan, or any deficiency is detected in the transmitted documents, the loan application will be canceled even though it was approved.

Marketing Actions: Internet ads are at the top of the marketing efforts for a bank that does not have any branches to gain new customers. Through these advertisements, the bank increases the awareness of the loan service it offers with advantageous interest rates and directs willing customers to the application page. Among the websites where it advertises, there are business pages of daily newspapers' websites, informative pages on credit and finance blogs, and e-commerce sites. Also, it uses Google Ads, publishes banners on social media sites, posts informative content on coupon pages, and makes promotion announcements via mailing. To monitor the effectiveness of the advertising activities related to new customer acquisition, the information on the type of advertisement that all customers who apply to the application page on the website are also kept.

New Customer Acquisition Problem: The performance criteria for new customer acquisition for their only consumer loan product offered by the bank are: the increase in the number of new loan applications and the completion of the loan allocation to customers who apply for and receive approval. It is known that increasing new loan applications is associated with offering advantageous interest rates and advertising expenditures. However, as explained in the above process, the reasons for taking a step back are unknown for customers who did not complete the loan application within 20 days even though the loan application was approved. This is the main problem addressed in this case study. A segmentation study will be conducted for all customers whose loan application has been approved to understand those who will be called 'Quitters' and to reduce their ratio among those whose loan applications are approved in the future. In this way, it will be possible to make suggestions to follow different marketing strategies by comparing them with 'Acquired' customers who have completed their approved application. In the following section, the information will be given about the methodology followed for market segmentation, which is presented as a solution to the problem explained in detail above.

4 Research Methodology

As explained in detail in the previous section, the data of 5825 approved applications belonging to the middle four months (from May to September) of 2018 were considered to understand the characteristics of the 'Quitter' customers of the bank in the case study. Out of these, only 1745 customers are 'Acquired', while the remaining 4080 are 'Quitter'. The data set includes the following information about customers: ID, application date, application expiry date, age, gender, marital status, education status, employment status, profession, province of residence, requested loan amount, amount of loan offered by the bank, over-limit request (whether the loan amount requested by the customer is over 30,000 TL), advertising channel directed to the loan application website (Google Ads, Banner, etc.), application status (Acquired/Quitter), allocation amount, loan allocation date for those whose application is completed, credit score, credit score group, interest rate. Since the distribution of Acquired/Quitter entries in the available data is not equal, Acquired customers are re-sampled and the ratio of the number of customers representing these two groups is balanced (46% vs. 54% respectively). Then, a likelihood ratio based feature selection algorithm has been applied to determine the important variables to be used in the segmentation study. Then, three available clustering algorithms (K-means, Two-step, and Kohonen) were run using variables that were found to be important, and the Kohonen SONN was chosen by considering the conceptual significance of the cluster structures tgether with the silhouette values. Before going through the evaluation of the analysis results, the fundamental information will be given about the methods used below.

4.1 Feature Selection

In business analytics studies that are involved in big data the high number of entries, as well as the high number of attributes is a factor affecting the significance of the analyses.

At this point, it is a preferred method to choose the important ones among the attributes to be used and to continue further analysis with these variables. Feature selection consists of three steps Scanning, Ranking and Selecting. For further details, please refer to [44].

4.2 Kohonen Self Organizing Network

The process of allocating data into different classes that have the most similarities in some characteristics is called clustering. Each cluster contains a dataset similar to another in its cluster and different from data in other ones [45]. Self-organizing network is suggested by Kohonen [46] is one of the data mining techniques that have a wide application area. Self-organizing maps are the most widely used technique after standard feed-forward networks, one of the artificial neural network techniques, and can be described as an uncontrolled two-layer artificial neural network used for clustering and size reduction [47]. As it is known, artificial neural network are parallel and distributed information processing structures, which are developed inspired by the human brain, are connected through weighted connections, and consist of processing elements each with its memory. Although all the other networks (namely Multilayer Perceptron and Radial Basis Function Networks) are designed for supervised learning tasks, Kohonen SONN is constructed essentially for unsupervised learning [27, 46]. The idea behind the SONN is to learn from the structure of the data and this feature makes Kohonen Network useful for explanatory data analysis. It is also mentioned as an advantage of Kohonen SONN is reducing the dimensionality of big data to improve its clustering analytic. Whereas it is not able to compare Kohonen SONN with other clustering methods, and it is difficult to measure the effectiveness of the SONN in clustering big data [48].

5 Results

Prior to the evaluation of the detailed analysis results, distribution of the balanced sample with 7571 observations are investigated. According to the descriptive statistics of the attributes covered in the dataset, most of the applicants are male (71.9%), married (63.2%) and high school or university educated. Employment ratio is the highest (88.6%) in terms of working status. The most preferred application channel is the webpage of the bank. The likelihood ratio-based feature selection algorithm was applied to all variables in the data set, as there were both continuous and categorical variables. In the scanning step, all fields are scanned to see if the maximum percentage of missing values is above 0.70, the maximum percentage of records in a single category is above 0.90, the maximum percentage of categories as a percentage of records is above 0.95, the minimum coefficient of variation is under 0.1, and the minimum standard deviation is under 0.00. According to the results, as the Acquired/Quitter variable is targeted, Interest Rate, Profession, Age, Education Status, Channel, Requested Amount, Marital Status, Employment Status, Offered Amount, Province, and Credit Score Category attributes are found as 'Important' with the 0.95 cut-off value and 'Marginal' with the 0.90 cut-off value. The Over Limit attribute found irrelevant in the scanning step due to its single category is too large and the Credit Score attribute is also removed because of having a coefficient of variation below the threshold. According to the cluster analysis results performed with variables determined as important, K-means and Two-Step algorithms both

Table 1. Cluster compositions

	Cluster 1	Cluster 2	Cluster 3	Cluster 4	Cluster 5	Cluster 6	Cluster 7	Cluster 8	Cluster 9	Cluster 10	Cluster 11	Cluster 12
Dimensions	X=3, Y=0	X=0, Y=0	X=3, Y=2	X=0, Y=2	X=2, Y=2	X=2, Y=0	X=1, Y=0	X=1, Y=2	X=0, Y=1	X=3, Y=1	X=1, Y=1	X=2, Y=1
Size	16%	14.40%	14%	12.70%	12.50%	8.30%	6.20%	5.50%	4.50%	2.60%	2.50%	0.80%
Acquired/ Quitter	Acquired: 52%	Quitter: 63.8%	Acquired: 94.2%	Acquired: 79.7%	Quitter: 98.3%	Quitter: 78.1%	Quitter: 69.2%	Quitter: 97.6%	Acquired: 57.2%	Acquired: 64%	Quitter: 66.3%	Quitter: 83.1%
Age	23.53	36.06	38.77	37.34	35.66	28.04	30.22	35.8	36.48	37.33	37.47	37.04
Channel	Web Page: 85.9%	Google Ads:99.5%	Web Page: 100%	Web Page:51.2%	Web Page:98.9%	Google Ads: 37.1%	Google Ads:97.9%	Web Page:89.4%	Google Ads: 92.3%	Web Page: 100%	Coupon: 56.3%	Web Page: 83.1%
Education Status	University: 68.8%	High School: 50.8%	High School: 60.5%	University: 98.8%	High School: 55%	High School: 50.2%	High School: 72.8%	University: 87.5%	University: 84.4%	High School: 45.2%	High School: 77.4%	High School: 67.8%
Marital Status	Single: 93.1%	Married: 96.1%	Married: 95.5%	Married: 92.7%	Married: 99.8%	Single: 98.7%	Single: 76.4%	Married: 97.4%	Married: 89.1%	Divorced: 75.6%	Married: 96.3%	Divorced: 91.5%
Interest Rate	1.18	1.23	1.35	1.13	1.18	1.11	1.23	1.15	1.27	1.29	1.02	1.22
Credit Score Category	1301-1350:23.5%	1351-1400:20.3%	1301-1350:16.8%	1451+: 30%	1251-1300: 20.8%	1251-1300:22.2%	1251-1300:20.8%	1451+:31%	1351-1400:20.4%	1251-1300:31.5%	1401-1450: 22.1%	1251-1300:28.8%
Offered Amount	8323.65	8855.54	9632.62	13771.79	8545.25	7287.33	6816.36	13090.11	10362.06	6415.34	10788.33	7078.43
Employment Status	Employed: 88.8%	Employed: 87.3%	Employed: 80.3%	Employed: 89.8%	Employed: 92.2%	Employed: 92.2%	Employed: 91.6%	Employed: 92.8%	Employed: 90%	Employed: 83.2%	Employed: 88.4%	Employed: 94.9%
Requested Amount	9136.95	11534.03	10420	11824.04	12065.16	9760.95	10017.24	15949.88	13592.33	8183.76	8814.74	16657.63

created 5 and Kohonen generated 12 clusters. Although the highest silhouette value is the k-means algorithm (0.151), it is important to propose actionable marketing strategies through the groups formed in the segmentation analysis, since the conceptual significance of the clusters is high, it seemed more appropriate to proceed with Kohonen results with a silhouette value of 0.135. The variables and their degrees of importance, which were found important as a result of the analysis and included in the network. According to the results, the most important attributes are channel, acquired/quitter, education status, marital status and age.

When the cluster compositions formed with Kohonen are examined (see Table 1), it is seen that approximately half of the volume of the clusters (50.2%) was represented by Quitters. In Clusters 1, 3, 4, 9, and 10 Acquired applicants are represented significantly higher than others. In almost all of the clusters where Acquired customers are represented (except Cluster 9), the bank's website was used as the application channel. Among the channels where Quitter segments except Cluster 5 are directed to application, some advertising channels and websites are frequently seen. It is observed that the applicants in the Quitter groups are at a lower education level compared to Acquired segments. Concurrently, the average interest rates and offered loan amounts of Quitter segments are lower than Acquired segments. However, customers in the Quitter segments demanded higher loan amounts compared to other groups. In four out of six Quitter segments, applicants' credit scores are belonged to the minimum category (1251–1300). In particular, there is a higher difference between the average loan amount requested by customers in these segments (5, 6, 7 and 12) and the average amount offered. In the light of these findings, suggestions regarding the marketing actions that can be taken for Quitter customers will be discussed in the following section.

6 Discussion

In parallel with developments in data storage and processing technologies, the volume of data collected in banking and finance has been increasing exponentially over the years. Many methods are evaluated in order to obtain meaningful and actionable results from the data collected in this age, which is called big data. In this paper, an example of a web-based lending bank was considered as a case study, and a segmentation analysis was carried out in the context of the characteristics of customers whose loan applications were approved. The main question here is to understand 'quitter' customers with an approved credit but not completed application by comparing the 'acquired' ones. In the methodology, firstly, a feature selection algorithm was followed to select the important and relevant variables among the available ones in the dataset, and then Kohonen SONN analysis, a heuristic method whose success was confirmed in large data sets, was performed. According to the results of the segmentation analysis, the most striking finding is that the application channels of quitter customer groups differ according to the acquired groups. Although acquired customers have made their applications through the bank's web page where they can search for detailed information about the application process lending conditions and required documents, most of the quitters are directed to the application page through the advertisement pages, so they may have applied without sufficient information about the conditions, but then found the conditions appropriate.

At this stage, it will be more beneficial to direct new customers who come through advertisements on other web pages to a front-page with detailed information text and/or video about the process before they are directed to the application page. Another finding was that quitter customers had lower education levels. In this context, it should be ensured that the informative contents provided from both the call center and the application page are more understandable and readable. Together with the interest rates, the loan amount offered is also lower in quitter customer groups. Although low loan rates are an attractive factor in making an application, the fact that the recommended loan amount after the evaluation is relatively lower than the amount requested at the time of application may be an important factor for these customers to take a step back. Again, since the average credit scores of the customers in this group are lower than the acquired customers, the amount of credit offered may have been suggested lower than desired. In this context, it may be recommended to inform customers about the evaluation process and to follow an open communication policy to avoid unexpected results after their loan applications. In addition to the recommendations listed above, different marketing campaigns and advertising content can be prepared by considering the characteristics of customer groups with different characteristics. Plus, differentiated communication methods can be developed based on predetermined customer segments in call center communication.

7 Conclusion

Prior to offering an analytical solution in customer acquisition for the aforementioned digital lending bank case study, a comprehensive literature review was realized by covering studies that both aimed to perform market segmentation and involved in the new customer acquisition problem in the banking industry. Even though there are plenty of studies that were tackled segmenting existing customers or were performed AI-based credit scoring modeling in the banking industry, it is found that there is a lack of clustering research involving behavioral patterns of potential customer segments. Being a pilot study that is investigating uncovered patterns in the behavior of new loan applicants, performing explanatory and neural networks-based Kohonen SONN technique, and attempting a contemporary issue in the digital era, the results of the study could be enlightening in future work. An obvious limitation of the study is being a case study that covers only a Turkish online lending bank's customers' data. As a suggestion for future studies of people working in the field, it can be given to evaluate the analyses to be made with equivalent methods and the data collected in a larger period for different case studies in different regions. Alternative case studies that are defined with different customer dynamics will also be vital in the development and notable evaluation of data mining methods.

References

1. Steenkamp, J.B.E., Ter Hofstede, F.: International market segmentation: issues and perspectives. Int. J. Res. Mark. **19**(3), 185–213 (2002). https://doi.org/10.1016/S0167-8116(02)00076-9

2. McDonald, M., Christopher, M., Bass, M.: Market segmentation. In: Marketing. Palgrave, London (2003). https://doi.org/10.1007/978-1-4039-3741-4_3
3. Wedel, M., Kamakura, W.A.: Market Segmentation: Conceptual and Methodological Foundations. Kluwer Academic Publishers, Dordrecht/Boston/London (2000)
4. Haghirian, P., Dickinger, A.: Identifying success factors of mobile marketing. Assoc. Consum. Res. **17**(1), 8–13 (2004)
5. Wind, J., Bell, D.: Market segmentation. In: The Marketing Book. Taylor & Francis, London (2007)
6. Başarslan, M.S., Argun, İ.D.: Prediction of potential bank customers: application on data mining. In: Hemanth, D., Kose, U. (eds.) Artificial Intelligence and Applied Mathematics in Engineering Problems. ICAIAME 2019. Lecture Notes on Data Engineering and Communications Technologies, vol. 43. Springer, Cham (2020). https://doi.org/10.1007/978-3-030-36178-5
7. Jakšič, M., Marinč, M.: Relationship banking and information technology: the role of artificial intelligence and FinTech. Risk Manage. **21**, 1–18 (2019). https://doi.org/10.1057/s41283-018-0039-y
8. Donepudi, P.K.: AI and machine learning in banking: a systematic literature review. Asian J. Appl. Sci. Eng. **6**(3), 157–162 (2017)
9. Duan, J.: Financial system modeling using deep neural networks (DNNs) for effective risk assessment and prediction. J. Franklin Inst. **356**(8), 4716–4731 (2019). https://doi.org/10.1016/j.jfranklin.2019.01.046
10. Eriksson, K., Mattsson, J.: Organising for market segmentation in banking: the impact from production technology and coherent bank norms. Serv. Ind. J. **16**(1), 35–45 (1996). https://doi.org/10.1080/02642069600000004
11. Alfansi, L., Sargeant, A.: Market segmentation in the Indonesian banking sector: the relationship between demographics and desired customer benefits. Int. J. Bank Mark. **18**(2), 64–74 (2000). https://doi.org/10.1108/02652320010322976
12. Namvar, M., Gholamian, M.R., KhakAbi, S.: A two phase clustering method for intelligent customer segmentation. In: 2010 International Conference on Intelligent Systems, Modelling and Simulation, Liverpool, pp. 215–219 (2010). https://doi.org/10.1109/isms.2010.48
13. Khobzi, H., Akhondzadeh-Noughabi, E., Minaei-Bidgoli, B.: A new application of RFM clustering for guild segmentation to mine the pattern of using banks' e-Payment services. J. Glob. Mark. **27**(3), 178–190 (2014). https://doi.org/10.1080/08911762.2013.878428
14. Khajvand, M., Tarokh, M.J.: Estimating customer future value of different customer segments based on adapted RFM model in retail banking context. Procedia Comput. Sci. **3**, 1327–1332 (2011). https://doi.org/10.1016/j.procs.2011.01.011
15. Hsu, F.-M., Lu, L.-P., Lin, C.-M.: Segmenting customers by transaction data with concept hierarchy. Expert Syst. Appl. **39**, 6221–6228 (2012). https://doi.org/10.1016/j.eswa.2011.12.005
16. Bijak, K., Thomas, L.C.: Does segmentation always improve model performance in credit scoring? Expert Syst. Appl. **39**, 2433–2442 (2012). https://doi.org/10.1016/j.eswa.2011.08.093
17. Mo, J., Kiang, M.Y., Zou, P., Li, Y.: A two-stage clustering approach for multi-region segmentation. Expert Syst. Appl. **37**, 7120–7131 (2010). https://doi.org/10.1016/j.eswa.2010.03.003
18. Mawoli, M.A., Abdulsalam, D.: Effective market segmentation and viability of Islamic banking in Nigeria. Aust. J. Bus. Manage. Res. **1**(10), 1–09 (2012)
19. Rashid, M., Kabir, H.M.: Customer demographics affecting bank selection criteria, preference, and market segmentation: study on domestic Islamic banks in Bangladesh (2009). Int. J. Bus. Manage. **4**(6) (2009). https://ssrn.com/abstract=3263078

20. Athanassopoulos, A.D.: Customer satisfaction cues to support market segmentation and explain switching behavior. J. Bus. Res. **47**(3), 191–207 (2000). https://doi.org/10.1016/S0148-2963(98)00060-5
21. Loureiro, S.M.C., Miranda, F.J.: Brand equity and brand loyalty in the internet banking context: FIMIX-PLS market segmentation. J. Serv. Sci. Manage. **4**(4), 476–485 (2011). https://doi.org/10.4236/jssm.2011.44054
22. Kahreh, M.S., Tive, M., Babania, A., Hesan, M.: Analyzing the applications of customer lifetime value (CLV) based on benefit segmentation for the banking sector. Procedia Soc. Behav. Sci. **109**(8), 590–594 (2014). https://doi.org/10.1016/j.sbspro.2013.12.511
23. Aghaei, M.: Market segmentation in the banking industry based on customers' expected benefits: a study of Shahr bank. Iran. J. Manage. Stud. (2021, in press). https://doi.org/10.22059/ijms.2021.305952.674132
24. Rasskazova, A., Potekhina, K.: Influence of age of the borrower on demand of a consumer loan. In: 2017 Tenth International Conference Management of Large-Scale System Development (MLSD), Moscow, pp. 1–5 (2017). https://doi.org/10.1109/mlsd.2017.8109673
25. Vellido, A., Lisboa, P.J.G., Meehan, K.: Segmentation of the on-line shopping market using neural networks. Expert Syst. Appl. **17**(4), 303–314 (1999). https://doi.org/10.1016/S0957-4174(99)00042-1
26. Taşkin, Ç., Emel, G.G.: Segmentation of online shoppers by means of an integrated data mining approach: a case study. J. Ind. Relat. Hum. Resour. **10**(3), 74–88 (2008)
27. Sharma, A., Chopra, A.: Artificial neural networks: applications in management. J. Bus. Manage. **12**(5), 32–40 (2013)
28. Hung, C., Tsai, C.F.: Market segmentation based on hierarchical self-organizing map for markets of multimedia on demand. Expert Syst. Appl. **34**(4), 780–787 (2008). https://doi.org/10.1016/j.eswa.2006.10.012
29. Vellido, A., Lisboa, P.J., Meehan, K.: Segmenting the e-commerce market using the generative topographic mapping. In: Mexican International Conference on Artificial Intelligence, pp. 470–481. Springer, Berlin, Heidelberg (2000)
30. Tiwari, R., Saxena, M.K., Mehendiratta, P., Vatsa, K., Srivastava, S., Gera, R.: Market segmentation using supervised and unsupervised learning techniques for E-commerce applications. J. Intell. Fuzzy Syst. **35**(5), 5353–5363 (2018). https://doi.org/10.3233/JIFS-169818
31. Clewley, N., Chen, S.Y., Liu, X.: Applications for data mining techniques in customer relationship management. In: Encyclopedia of Information Science and Technology, pp. 188–192, Second edn. IGI Global (2009)
32. Kim, K.J., Ahn, H.: Using a clustering genetic algorithm to support customer segmentation for personalized recommender systems. In: International Conference on AI, Simulation, and Planning in High Autonomy Systems, pp. 409–415. Springer, Berlin, Heidelberg (2004)
33. Decker, R.: A growing self-organizing neural network for lifestyle segmentation. J. Data Sci. **4**(2), 147–168 (2006)
34. Kim, K.J., Ahn, H.: A recommender system using GA K-means clustering in an online shopping market. Expert Syst. Appl. **34**(2), 1200–1209 (2008). https://doi.org/10.1016/j.eswa.2006.12.025
35. Vellido, A., Lisboa, P.J.G., Meehan, K.: A systematic quantitative methodology for characterizing the business-to-consumer e-commerce market. ACM SIGBIO Newsl. **20**(1), 24 (2000). https://doi.org/10.1145/346335.346352
36. Hiziroglu, A.: A neuro-fuzzy two-stage clustering approach to customer segmentation. J. Mark. Anal. **1**(4), 202–221 (2013). https://doi.org/10.1057/jma.2013.17
37. Liang, Y.H.: Customer relationship management and big data mining. In: Information Granularity, Big Data, and Computational Intelligence, pp. 349–360. Springer, Cham (2015)

38. Vohra, R., Pahareeya, J., Hussain, A., Ghali, F., Lui, A.: Using self organizing maps and K means clustering based on RFM model for customer segmentation in the online retail business. In: International Conference on Intelligent Computing, pp. 484–497. Springer, Cham (2020)
39. Gang, L., Lu, A.: Clustering analysis of e-commerce transactions with self-organizing map. Data Anal. Knowl. Disc. **24**(9), 70–77 (2008). https://doi.org/10.11925/infotech.1003-3513.2008.09.12
40. Bach, M.P., Vlahović, N., Pivar, J.: Self-organizing maps for fraud profiling in leasing. In: 2018 41st International Convention on Information and Communication Technology, Electronics and Microelectronics (MIPRO), pp. 1203–1208, IEEE (2018)
41. Bach, M.P., Vlahović, N., Pivar, J.: Fraud prevention in the leasing industry using the Kohonen self-organising maps. Organizacija **53**(2), 128–145 (2020). https://doi.org/10.2478/orga-2020-0009
42. Shin, T.S., Hong, T.H.: Customer segmentation model for internet banking using self-organizing neural networks and hierarchical clustering method. Asia Pac. J. Inf. Syst. **16**(3), 49–65 (2006)
43. Bach, M.P., Juković, S., Dumičić, K., Šarlija, N.: Business client segmentation in banking using self-organizing maps. South East Eur. J. Econ. Bus. **8**(2), 32–41 (2014). https://doi.org/10.2478/jeb-2013-0007
44. IBM SPSS Modeler 18.1.1 Modeling Nodes. ftp://public.dhe.ibm.com/software/analytics/spss/documentation/modeler/18.1.1/en/ModelerModelingNodes.pdf
45. Han, J., Kamber, M., Pei, J.: Data Mining: Concepts and Techniques (The Morgan Kaufmann series in data management systems). Morgan Kaufmann Publications, Burlington, MA (2011)
46. Kohonen, T.: Self-organized formation of topologically correct feature maps. Biol. Cybern. **43**, 59–69 (1982). https://doi.org/10.1007/BF00337288
47. Hammer, B., Micheli, A., Sperduti, A., Strickert, M.: Recursive self-organizing network models. Neural Netw. **17**(8–9), 1061–1085 (2004). https://doi.org/10.1016/j.neunet.2004.06.009
48. Chiroma, H., Abdullahi, U.A., Alarood, A.A., Gabralla, L.A., Rana, N., Shuib, L., Herawan, T.: Progress on artificial neural networks for big data analytics: a survey. IEEE Access **7**, 70535–70551 (2018). https://doi.org/10.1109/ACCESS.2018.2880694

Applications of Artificial Intelligence in COVID-19 Pandemic

Mohammad Al-Shoqran[1](✉) and Samer Shorman[2]

[1] Mathematical Sciences, Languages and General Studies Department, Ahlia University, Manama, Bahrain
malshoqran@ahlia.edu.bh

[2] Department of Computer Science, Applied Science University, Eker, Bahrain
samer.shorman@asu.edu.bh

Abstract. COVID-19 pandemic is the most notable world crisis in the modern era that originated from Wuhan, china in late 2019 has affected all the nations and triggered a global economic crisis. As well artificial Intelligence (AI) plays an important role in the estimate and expectation of the next phase and they are helpful for searches operation. Covide19 structure was studied in multiple research papers that studied the impact of artificial intelligence and how to apply its algorithms to predict Covid19 directions and cases. This study addresses significant issues on Covid19 which will study the impact of Total Cases and Serious-cases on Total Deaths using artificial intelligence language-python to implement linear regression on some countries data. Likewise used the content analysis to determine the strengths and advantages of using AI on COVID19. In addition to explaining the role of artificial intelligence on COVID19.

Keywords: Artificial intelligence (AI) · COVID-19 pandemic · Linear regression · Python

1 Introduction

Since the emergence of the COVID-19 pandemic at the end of 2019, 219 countries around the world affected 97,422,123 confirmed cases and 2,086,159 deaths, on the twenty-first of January 2021 (website, worldometers.info) see [16]. Most of them have taken a quick response and many varied Procedures to limit the spread of this disease and to try to find a treatment or a vaccine for it. And they used different methods in this regard. One of the most important things is the use of artificial intelligence against this pandemic because it is one of the most powerful methods for dealing with the "Covid-19" pandemic, whether to limit the spread of the pandemic or in diagnostic operations. also, when using artificial intelligence to produce a vaccine or in drug manufacturing methods.

In the literature, many researchers have been studied the role of artificial intelligence and its application in many reals life fields such as the role of artificial intelligence in smart universities and how it would be the education in the future based on AI applications [3]. The use of Artificial intelligence has already helped many countries to limit the

A. M. A. Musleh Al-Sartawi et al. (Eds.): EAMMIS 2021, LNNS 239, pp. 313–321, 2021.
https://doi.org/10.1007/978-3-030-77246-8_30

spread of the pandemic. In addition, Artificial intelligence has been used to provide a safer environment for doctors and nurses, and healthcare providers. Also, artificial intelligence has been used to detect the high temperature of people and to discover the gathering places in public places through special robots and so on [11]. Countries of the world have taken many procedures to limit the spread of the pandemic. Among these procedures for example social distancing, not leaving the house except for necessity, wearing masks, using sterilizers, keeping away from gatherings, and many other procedures.

Artificial intelligence played an important role in teaching and learning by providing an appropriate virtual environment in distance education, which helped a lot in the continuation of education and evaluation processes for students in various educational stages. Artificial intelligence applications around the world have also contributed to the development of medicine and health care systems by improving diagnostic processes and contributing to disease prevention. This great achievement would have not been without global and human efforts that had been combined in finding a solution to this pandemic.

Artificial intelligence is significant in efforts made in solving various problems facing the world, and it is expected that artificial intelligence will change the shape of the world in the future and will contribute to providing new job opportunities that were not previously known. Up to this moment, human efforts to confront this pandemic have not stopped, some vaccines have finally been distributed and many others are being developed, as the battle against COVID-19 is not over yet, so that artificial intelligence must be used more and more widely in this regard especially. This study presents a review for different articles related to the employment of AI to deal with COVID-19 such as (see Fig. 1) and to highlight the main method and techniques that are used and implemented, with advantages and disadvantages of those articles.

All the above confirms the importance of using artificial intelligence in solving various problems facing the world, and it is expected that artificial intelligence will change the shape of the world in the future and will contribute to providing new job opportunities that were not previously known. Up to this moment, human efforts to confront this pandemic have not stopped, some vaccines have finally been distributed and many others are being developed, as the battle against COVID-19 is not over yet, so that artificial intelligence must be used more and more widely in this regard especially. This investigation presents an analytical review for some papers related to applications of artificial intelligence in the covid-19 pandemic to highlight the main techniques that are implemented in this regard combined with the Strengths and Weaknesses of these papers.

This study arranged as follows: The first section includes a general introduction of the study. The second section contains the research methodology. The third section deals with a review of some scientific articles related to the topic of this study, as well the fourth section includes an analysis and interpretation of data related to artificial intelligence applications and the COVID-19 pandemic, the last section contains the most important results and recommendations.

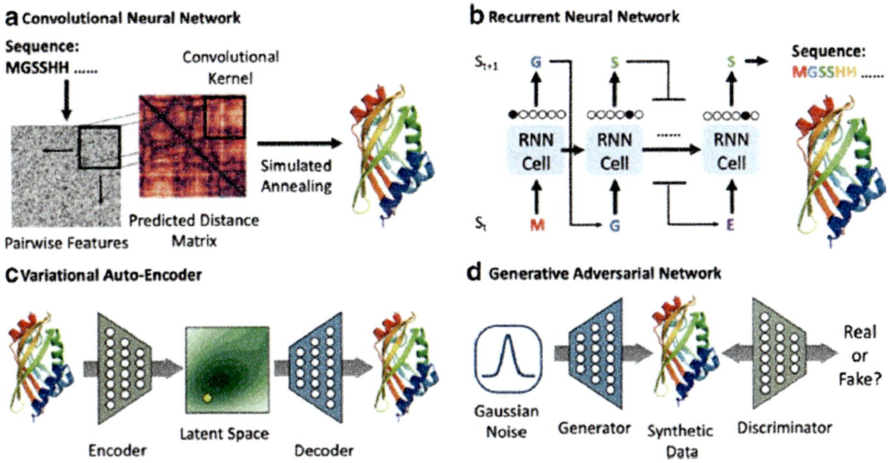

Fig. 1. Artificial intelligence algorithm neural network into Portion modeling [12].

2 Research Methodology

Research Methodology shows and explains the method of obtaining the scientific results in clear and regular steps by clarifying the research conducted method to achieve the objectives and to explain the methods of obtaining the results. This study is based on an analytical descriptive study by analysing the content of some previous studies related to the applications of artificial intelligence to find a solution to COVID-19 pandemic from aside, on the other hand, to present and analyse some related data. In addition to some graphical presentation with their explanations. Also, to present and analyse some related data using linear regression model [6, 11].

3 Articles Review

Artificial intelligence with Covid19 has been studied intensely by many researchers during 2020, this section contains the analysis for some articles related to the applications of artificial intelligence and its algorithms on the COVID-19 pandemic (see Table 1).

Based on the content analysis for some selected articles related to applications of artificial intelligence in a covid-19 pandemic, we noted that most of these articles have been focused on trying to how to minimize the effects of the COVID-19 pandemic, whether in the medical aspects related to doctors, nurses, and medical staff on the one hand, and on the other hand to mitigate the economic effects that affected on many people and countries around the world. where this is done by finding and using some artificial intelligence tools and applications to help in the following aspects especially, cases diagnosis, cases treatment, and contact tracking of the discovered cases to contribute to controlling this pandemic. In conclusion, through the results and numbers have been obtained in these studies, we can say that artificial intelligence applications played a good role in fighting against this pandemic.

Table 1. Literature review of applications of artificial intelligence in covid-19.

Articles	Strengths and advantages	Weaknesses and disadvantages
[1]	The paper discussed a computational method to discover new candidate drugs or vaccines. the authors focused on COVID-19 drug and vaccine development using AI and the potential of intelligent training for discovering therapeutics of it. Also, they presented (Corona DB-AI), which is a dataset of compounds, peptides, and epitopes discovered from silico or vitro that can be used in the training models regarding finding a treatment of COVID-19.in conclusion, the authors believe that computational models are capable to obtain a reliable therapy. also, they noticed that the main protease and the viral spike protein have been the most popular options of drug discovery and vaccine development. and they collected a dataset titled Corona DB-AI	There is no sufficient information or examples on how to extract mathematical models, in terms of finding a solution based on the data-driven
[2]	The study polled the opinions of experts to show existence gaps, useful technologies in terms of emergency management. Results indicated that various IT-systems like artificial intelligence, online service, computational methods, and IoT could have different applications to manage an outbreak. They assure on the role of IT in reducing the effect of crises effects. So that the authors classified these applications as follows: • diagnosis and prediction • treatment • protection • management goals In conclusion, the authors assure on the role of IT in minimization of the crisis harms and risk management. And the role of strategic planning in improving the public health responses to outbreaks, economic damages	The authors have some results without clear convenient evidence, so we suggest using more scientific research tools to reinforce their profs or results
[4]	The review discussed how AI and Big Data applications help in finding a solution to the pandemic such as: • A Quick Pandemic Alert • Diagnosis and Tracking of COVID-19 cases • Identification of a potential therapeutic treatment • Facilitating the Implementation of Public Health Interventions • Enhancing cities and favoring the development of healthy, smart, resilient cites the review classified the Possible applications of AI and Big Data in terms of COVID-19 outbreak management as sort, mid, and long terms time. in addition to examples on each like: • AI can aperient epidemiological data collection, decision-making, and risk assessment • Accomplishment of public health interventions • Identification of specific diagnostic features • Identification of the existing drugs The review concluded that AI and Big Data can be applied to follow the prevalence of coronavirus in real-time	It would be better if more examples of the AI and Big data applications against COVID-19, in addition, to explain the working mechanism

(continued)

Table 1. *(continued)*

Articles	Strengths and advantages	Weaknesses and disadvantages
[6]	The survey summarized the applications of AI in clinical management with fighting against COVID-19 and Big Data applications in terms of identification coronavirus the authors classified AI techniques in clinical data analysis, such as:neural systems, SVM, etc. Also, the paper discussed the role of using AI-oriented cloud computing in fighting against various like COVID-19. in addition to a discussion about how AI in fighting against similar viruses Artificial intelligence is a powerful tool in differentiating between Early diseases due to COVID-19 and Computer-based intelligence Computer-based intelligence is useful for suitable medical checkups and it Can follow COVID-19 emergencies by different scales	However, the survey is very strong and powerful. But it would be great if the authors give more explanation on how to overcome the difficulties in deep learning
[7]	This paper showed how to combat COVID-19 using Artificial Intelligence, by illustrating Deep Learning techniques, Generative Adversarial Networks, Long and Short-Term Memory, and Extreme Learning Machine And it showed that one of the most important advantages of AI- platforms are to speed up the process of treatment and diagnosis of the disease of Covid-19. these platforms help AI experts to analyze big datasets and they assist physicians in terms of Improving the data analyzed to deal with the virus speedily and accurately Also, the paper discussed a mechanism of choosing the suitable models of prediction and estimation for desired parameters using clinical and non-clinical datasets. The paper is desirable because of its potential in creating a workspace during physicians and AI experts could work together	The paper did not suggest mechanisms or ways to encourage the wider use of these methods, and how to increase the confidence in reliance on it and to increase its accuracy
[8]	This paper aimed to widely review the role of Machine Learning and AI regarding screening, forecasting, and drug development COVID-19. Using an information selective assessment was implemented on a database related to the applications of AI and ML on Covid-19. the paper discussed a few challenges while using AL and ML algorithms in real-life problems. The paper concluded that the development in ML and AI has improved screening, forecasting, and vaccine/drug development, etc. process against this pandemic	In data analysis(results) There is a need to show how the accuracy and the error have been obtained
[10]	This paper covered the role of Smart technology with the spread of COVID-19 such as robotics, and sensor technology Also, the paper discussed the drone technology-driven and robotics methods that have been used to assisting health systems, and disinfection procedures, etc. mask and sensor technologies are demonstrated, including a manufacturing method such as optical sensing and 3D printing The results illustrated the use of smartphone applications Connecting large masses can be configured to handling emergency situations. In addition, how these applications help to understand the impact of technology on society and how it can be tuned to respond in emergency cases. The paper gave a deep investigation of the new innovations in sensor technology, and mask during the Pandemic	It would be great if the authors suggest a solution to the expected threats like the impact of the pandemic on the economy And the technology impact on the daily real life, work, and audience psychology

(continued)

Table 1. (*continued*)

Articles	Strengths and advantages	Weaknesses and disadvantages
[14]	This investigation discussed the role of the Internet of Things (IoT) in increasing patient satisfaction and reducing the readmission rate in hospitals by employing interconnected networks The discussion of the results illustrates the effects of implementing IoT on reducing healthcare costs and improving treatment outcomes for the affected patient. This is done by clarifying the role of twelve IoT applications such as: • Rapid COVID-19 screening • Internet-connected hospital • Accurate forecasting of the virus The paper concluded that IoT is beneficial for a patient with COVID-19 in terms of symptoms identification and provides better treatment and IoT is helpful to predict the future situation of this illness	It would be great if the study suggested an explanation of the economic cost of these applications by measuring their economic impact, benefit, and ability to confront the pandemic

4 The Experiment

In this experiment, well apply Linear A multiple linear regression analysis was carried out to examine the relationship of the response variable (confirmed cases and serious-cases) with the regression (total deaths cases). The experiment variables distributed into X as two input variables (or independent variables, which represents in this experiment by total cases and serious cases). While the Y is the output/dependent variable (which represents in this experiment by total deaths). There are two main ways to perform linear regression in python with statsmodels and scikit-learn libraries, in this experiment will use both to validate the results. In this experiment, well adopt the Ordinary Least Squares (OLS) as a model and the Least Squares as a method, which means fit a regression line that would minimize the square of the distance from the regression line.

The assumptions include the following: if the total cases and serious-cases are increased then the dependent variable, which is total deaths, will be increased with excluding the other variables. In this study, data were extracted from the world health organization (WHO) (website, covid19.who.int.) see [15] and worldometers.info/coronavirus websites on the seventeenth of January - 2021. THE dataset was collected in an excel file (see Table 2) and analysed with the Spyder (python 3.7). The variables of interest include confirmed total cases and serious-cases and total deaths variable. As well consider the general linear regression model: The following OLS Regression Results (see Table 3).

Table 2. Dataset for countries and Total Cases Serious-cases Total Deaths

Country	Total cases	Serious cases	Total deaths
Iraq	603,739	128	12,906
Pakistan	506,701	2,340	10,717

(*continued*)

Table 2. (*continued*)

Country	Total cases	Serious cases	Total deaths
Saudi Arabia	363,949	312	6,295
Jordan	308,670	187	4,043
japan	286,752	864	4,044
UAE	232,982	191	711
Lebanon	222,391	585	1,629
Kuwait	154,841	50	945
Palestine	148,968	95	1,630
Qatar	146,068	26	246
Malaysia	141,533	190	559
Oman	130,944	23	1,508
Bahrain	95,879	15	356
china	87,591	18	4,634
Yemen	2,105	23	612

Table 3. OLS Regression Results based on statsmodels and scikit-learn libraries.

```
[*statsmodels]==============================================================
Dep. Variable:        Total_deaths              R-squared:             0.804
Model:                OLS                       Adj. R-squared:        0.771
Method:               Least Squares             F-statistic:           24.61
Date:                 Thu, 21 Jan 2021          Prob (F-statistic):    5.67e-05
Time:                 10:32:20                  Log-Likelihood:        -132.57
No. Observations:           15                  AIC:                   271.1
Df Residuals:               12                  BIC:                   273.3
Df Model:                    2
COVARIANCE TYPE:
============================================================================
                 coef        std err       t        P>|t|      [0.025    0.975]
----------------------------------------------------------------------------
constant              -1483.0481    860.275     -1.724    0.110         -3357.426
391.329
Total cases     0.0208      0.004         5.673     0.000        0.013     0.029
Serious-cases   0.3012      0.990         0.304     0.766       -1.857     2.459
============================================================================
Omnibus:               5.022        Durbin-Watson:        0.897
Prob (Omnibus):        0.081        Jarque-Bera (JB):     2.432
Skew:                  0.914        Prob (JB):            0.296
Kurtosis:              3.742        Cond. No.             4.96e+05
============================================================================
[*scikit-learn]
=================================================================
Intercept:  -1483.0481251317397
Coefficients: [0.0208442     0.3012144]
              nonrobust
```

5 Interpreting the Regression Results

Well highlighted several important components within the results from Table 3: standard errors assume that the covariance matrix of the errors is correctly specified. r-squared We can see here that this model has a much higher R-squared value (0.804) which meaning that this model explains 80.4% of the variance in our dependent variable. adjusted. As well, r-squared reflects the fit of the model. r-squared values (0.771) range from 0 to 1, where a higher value generally indicates a better fit, assuming certain conditions are met. constant coefficient is your y-intercept (Intercept: −1483.0481251317397). it means that if both the Total cases and Serious-cases coefficients are zero, then the expected output (i.e., the y Total deaths) would be equal to the constant coefficient. Total cases coefficient (0.0208442) represents the change in the output Total deaths (y) due to a change of one unit in the Total cases It means that as the Total cases variable increases by 1, the predicted value of Total deaths increase by 0.0208442. Serious-cases coefficient (0.3012144) represents the change in the output Total deaths (y) due to a change of one unit in the Serious-cases.

In addition, from Table 3, the standard deviation - std err reflects the level of accuracy of the coefficients. The lower it is, the higher is the level of accuracy. $p > |t|$ is your p-value. A p-value of less than 0.05 is statistically significant. Confidence interval represents the range in which our coefficients are likely to fall (with a likelihood of 95%). Moreover, the condition number is large, 4.96e+05. This might indicate that there are strong multicollinearity or other numerical problems. t and P-value there is 95% confidence intervals for the total cases (meaning we predict at a 95% confidence that the value of total cases is between 0.013 to 0.029). As well there is 95% confidence intervals for the Serious-cases (meaning we predict at a 95% percent confidence that the value of Serious-cases is between −1.857 to 2.459). Likewise, the skew value is (0.914) which considers if the value between −1 and – 0.5 or between 0.5 and 1, the data are moderately skewed. kurtosis value is (3.742) with >3 is called leptokurtic. Compared to a normal distribution, its tails are longer, and the peak is higher.

6 Conclusion

COVID-19 pandemic still outbreaks around the world with new strains of Coronavirus, furthermore, the positive news rise with vaccines is increased and some of the people vaccinated. This research has concluded some findings, based on the content analysis of some articles that are related to Artificial Intelligence techniques such as python. This work shows that Total Cases Serious-cases affect the Total Deaths of covid-19 and should follow the health sector instructions and personal and institutional responses to the pandemic until a vaccine is widely available. Finally, based on these results, we expect that there will be a significant change in the Total Cases Serious-cases scale, if the majority of people adherence to precautionary measures to prevent the spread of the disease, especially for people who have chronic diseases or respiratory diseases. The future work will focus on the other factors that may influence Total Deaths and how to avoid them.

References

1. Arshadi, A.K., Webb, J., Salem, M., Cruz, E., Calad-Thomson, S., Ghadirian, N., Collins, J., Diez-Cecilia, E., Kelly, B., Goodarzi, H., Yuan, J.S.: Artificial intelligence for COVID-19 drug discovery and vaccine development. Front. Artif. Intell. **3**, 65 (2020)
2. Asadzadeh, A., Pakkhoo, S., Saeidabad, M.M., Khezri, H., Ferdousi, R.: Information technology in emergency management of COVID-19 outbreak. Inform. Med. Unlocked 100475 (2020)
3. Al-Shoqran, M., Shorman, S.: A review on smart universities and artificial intelligence. In: The Fourth Industrial Revolution: Implementation of Artificial Intelligence for Growing Business Success, chap. 16, 1st edn. XII. Part of book series: Studies in Computational Intelligence, vol. 935 (2021). ISBN 978-3-030-62795-9. Series ISSN 1860-949X
4. Bragazzi, N.L., Dai, H., Damiani, G., Behzadifar, M., Martini, M., Wu, J.: How big data and artificial intelligence can help better manage the COVID-19 pandemic. Int. J. Environ. Res. Public Health **17**(9), 3176 (2020)
5. Hussain, A.A., Bouachir, O., Al-Turjman, F., Aloqaily, M.: AI techniques for COVID-19. IEEE Access **8**, 128776–128795 (2020)
6. Jamshidi, M., Lalbakhsh, A., Talla, J., Peroutka, Z., Hadjilooei, F., Lalbakhsh, P., Jamshidi, M., La Spada, L., Mirmozafari, M., Dehghani, M., Sabet, A.: Artificial intelligence and COVID-19: deep learning approaches for diagnosis and treatment. IEEE Access **8**, 109581–109595 (2020)
7. Lalmuanawma, S., Hussain, J., Chhakchhuak, L.: Applications of machine learning and artificial intelligence for Covid-19 (SARS-CoV-2) pandemic: a review. Chaos Solitons Fractals 110059 (2020)
8. Maurya, M.R., Sadasivuni, K.K., Al-Maadeed, S.A.S.: Smart Technologies Driven Approaches to Tackle COVID-19 Pandemic (2020)
9. Shino, E., Binder, M.: Defying the Rally During COVID-19 pandemic: a regression discontinuity approach. Soc. Sci. Q. **101**(5), 1979–1994 (2020)
10. Shorten, C., Khoshgoftaar, T.M., Furht, B.: Deep learning applications for COVID-19. J. Big Data **8**(1), 1–54 (2021). https://doi.org/10.1186/s40537-020-00392-9
11. Ting, D.S.W., Carin, L., Dzau, V., Wong, T.Y.: Digital technology and COVID-19. Nat. Med. **26**(4), 459–461 (2020)
12. Website of world health organization (WHO) Coronavirus Disease (COVID-19) Dashboard. https://covid19.who.int/. Accessed 17 Jan 2021
13. Website: https://www.worldometers.info/coronavirus/countries-where-coronavirus-has-spread/. Accessed 21 Jan 2021

An Approach for Document Clustering Using Semantic Similarity and Whale Optimization

Shivam Sawarn[1(✉)] and Gerard Deepak[2]

[1] School of Computing, DIT University, Dehradun, India
[2] Department of Computer Science and Engineering, National Institute of Technology, Tiruchirappalli, Tiruchirappalli, India

Abstract. Document ontology matching is one of the fastest-growing fields in the world of Semantic Web Technologies. Clustering documents based on their contents is a difficult task due to the under-optimized algorithms. In the proposed method, TF-IDF is used to select the features in order to analyse the documents and the word occurrences in the document. Multiple variations of TF-IDF are used as feature selection methods to analyse the best method clubbed with other measuring techniques and optimization algorithms to devise a high precision and accurate algorithm for Discovering Patterns in the documents. The Dataset used in the proposed method is CHIC heritage dataset in the English language which contains over 1000 documents. WebPMI and KL divergence is used as a measuring technique to measure semantic similarity between two queries. At last Whale Optimization Algorithm (WOA) is used to optimize the whole framework and produce high accuracy results.

Keywords: Clustering · KL divergence · Mini batch K-means · TF-IDF · Web PMI

1 Introduction

Documents play a very important role in any industry. In the early days, documents were the only storage medium on which people used to store their information. Big industries, corporations, government, everyone used to store information on papers. Nowadays, information can be stored digitally and even the documents are also typed in the computer and saved digitally.

But since we are in the age of "Data", the information that we are dealing with is huge. The number of documents are innumerable and are in need to get sorted and clustered in a specific manner. Since everything is getting automated, even clustering of documents can be done using Natural Language Processing (NLP) algorithms. The biggest obstacle in clustering digital documents is to achieve high precision with big datasets.

Bafna, P. et al. Document clustering: TF-IDF approach [1] which was focused on the large size of the dataset and presented an approach that helped tackle this problem.

A. M. A. Musleh Al-Sartawi et al. (Eds.): EAMMIS 2021, LNNS 239, pp. 322–333, 2021.
https://doi.org/10.1007/978-3-030-77246-8_31

The large size of the documents makes it very difficult to achieve high accuracy as the algorithms have to take the scaling of documents into account.

The proposed method uses various techniques to tackle all the issues that are generally seen in the clustering of documents. The framework uses TF-IDF scores to pre-process the documents and then calculates the WebPMI score and KL divergence are calculated between the documents and then the documents are clustered based on these scores by taking mean standard deviation as the threshold. To obtain clusters of the documents Mini K- means clustering algorithm is used which divides the dataset into mini-batches before the training which helps in reducing the statistical noise. The whole algorithm is optimized using whale optimization algorithm which makes the results of the proposed algorithm more accurate.

Motivation: Clustering documents is a hectic and tiresome job. There have been a lot of attempts in the past to automate the process and get the job done as easily and quickly as possible. Many methods have been proposed in the past for the same but Document clustering and Text Clustering on big datasets is simple as it seems, especially in the ancient documents with missing contents. Neto, J. L. et al. Document clustering and text summarization [3] presented an approach that aimed at clustering similar types of documents.

The proposed method aims to simplify this job and optimize it to a high accuracy so that the method can work on a large dataset also.

Contribution: The contribution of the proposed work is based on the following hypothesis:

1. The proposed method uses TF-IDF and different feature selection methods to extract the labels from the documents by statistically measuring their importance in the document and using them as a criterion for Clustering the documents.
   Zhang, W.et al. A comparative study of TF* IDF, LSI and multi-words for text clustering [2] used a similar approach in the TF IDF clustering methodology.
2. Using WebPMI, the similarity index of semantics between the two words is measured and then it is used in the increasing order to arrange the documents.
3. To calculate the association between probability distributions KL-Divergence measure is also used for the documents which are to be clustered and probability distributions in each category.
4. Whale Optimization Algorithm is used to cluster the documents based on the WebPMI Measure and KLD measure using Mean Standard deviation as the threshold.
5. Different feature selection methods along with TF-IDF is used and correlation is also measured to obtain the best feature selection method to be used in the pre-processing of the data.

Organization: The organisation of this paper is in the following format - Sect. 2 provides a comprehensive description of relevant works done in this field. Section 3 consists of the complete architecture of the model as well as the approach. Section 4 consists of the Implementation and the algorithm of the framework. In Sect. 5 performance of the model is discussed. In Sect. 6 the paper has been concluded.

2 Related Literature

Document clustering is a difficult job to do in the field of text categorization and document clustering. Many scholars have tried to propose several algorithms to cluster documents of different types. Many algorithms emphasize hybrid algorithms and propose an efficient solution. Bafna P et al. [1] presented an algorithm that uses TF-IDF and fuzzy k-means algorithm with the hierarchal algorithm to cluster documents based on content. Zhang W et al. [2] presented a paper that highlighted the analysis based on a comparison between LSI, multi-word and TF-IDF methods for text mining. Neto J et al. [3] presented an algorithm that aims at document clustering and Text categorization using the TF-IDF method which is an adaptation from TF-IDF and is providing satisfactory results. Popescu A et al. [4] presented a method for opinion mining using WebPMI measure and used semantic similarity measure for document clustering. Forman G et al. [5] used multiple feature selection and machine learning for text clustering and tested it on more than 200 problems. The tests yielded that the method Bi-Normal outperforms all the feature selection methods and optimizes the whole algorithm.

KL-Divergence is used as a statistical method used to measure the numerical gap between probability distributions of the document to be clustered and the document of each category. Goldberger J et al. [6] proposed a method based on approximating KL divergence associated with two Gaussian mixtures. The algorithm was tested on an image dataset and yielded better results than the previously mentioned methods. Yu D et al. [7] presented a method that shows the encompassment of Deep Learning methodology and use of KL- Divergence regularization method to realize the constraint to the adaptation criterion. Schneider KM et al. [8] presented a framework that used the KLD method as a feature selection method only and is used along with the TF-IDF weighting method along with another clustering method. KLD is also used as a clustering method. The corpus used in the testing of the framework is a real-world dataset and yields satisfactory results. Aljarah I et al. [9] presented a framework based on the WOA to train a deep learning neural network. The stochastic nature of the WOA helps to optimize the algorithm and makes it outperform all the current existing algorithms for the same problem.

3 Methodology

3.1 TF-IDF

TF-IDF is the acronym Term Frequency – Inverse Document Frequency. It is used to calculate the frequency of a specific word in a specific document but lower in the whole document dataset. At first, it calculates the term frequency of the word and then multiplies it with the IDF (Inverse Document Frequency) giving the whole TF-IDF of the dataset. To extract the labels TF-IDF measure is used based on the context. In this algorithm, TF-IDF is used as a text clustering algorithm and multiple feature selection methods which are the adaptation from TF-IDF are used for text clustering and the method with the best methodology is selected for data pre-processing and text clustering.

The feature selection methods used in this framework are TF-IDF and Logarithmic TF-IDF and for text clustering Augmented models are used. The equations for the used models are given in Eqs. (1), (2), and (3).

$$IDF(d,t) = log\left[\frac{(1+n)}{(1+df(d,t))}\right] + 1 \tag{1}$$

$$Tf(t,d) = log(1+ft,d) \tag{2}$$

$$f(x) = 0.5 + f\frac{(t,d)}{\max}f\left(t', d : t' \in d\right) \tag{3}$$

TF-IDF augmented models are used simultaneously and the result in each case is recorded. The best performing model based on the results is used for text categorization and document clustering.

A term document matrix is created and then these four TF-IDF models are applied to the dataset. These models are used to optimize the algorithm and find the best possible output to further cluster the documents based on their WebPMI score and KLD method taking mean standard deviation as the threshold value.

3.2 WebPMI – Semantic Similarity Clustering

WebPMI is a statistical measure to calculate the semantic gap between two queries. In this algorithm, the WebPMI measure is used to match the labels extracted by the TF-IDF method and use it to cluster the documents based on the content and context of the documents. For each document, a TF-IDF measure is calculated and the mean standard deviation of the whole dataset is calculated.

$$\mathrm{WebPMI}(a_1, a_2) = \left\{\begin{array}{ll} \text{if log match}(a_1 \text{ and } a_2) \leq x \leq 0 & \text{Otherwise} \\ \log_2[(\text{match}(a_1 \text{ and } a_2)/N)/(\text{match}(a_1)/N) * (\text{match}(a_2/N)] \end{array}\right\} \tag{4}$$

where a_1 and a_2 are the two queries or words or documents between which the similarity is to be measured. N stands for the total no. of queries or words or documents in the data and x is the threshold value which in this case is the mean standard deviation of the dataset.

Then using the WebPMI measure the semantic similarity of the documents are calculated. This process is implemented in two stages:

I. In the first phase, the WebPMI measure is calculated for each document and the semantic similarity is matched between the words in that document.
II. In the second phase the WebPMI measure of each document is calculated against the other documents of the dataset and the documents are clustered based on the similar content of the documents.

The advantage of using the WebPMI measure in two stages is that we ought to get more accurate results and better precision as the documents are clustered in the first phase based on the contents in them and then on the grounds of similar content that the other document in the same dataset has, the documents are clustered in the second phase. This is done after the TF-IDF calculation of the document dataset and which in turn gives more accurate results.

3.3 Kullback – Leibler Divergence Method – Probability Distributions Clustering

It is a statistical method utilized for the calculation of the numerical gap between the probability distributions on the same implantation. It is a well knows method in the field of Information Theory and is known as the relative entropy method. It tells us how two probability distributions are different over the same corpus.

It is defined as given in Eq. 5:

$$(A \vee B) = \sum\nolimits_{x \in X} A(x) log\left(\frac{A(x)}{B(x)}\right) \tag{5}$$

In the proposed methodology the KL divergence method calculates the distance between the probability distribution of similar words occurring in the same document. The implementation of this method is also done in two phases. The given phases are:

I. In the first phase, the difference in the probability distributions of different reoccurring words is calculated from the document matrix and is stored.
II. In the second phase, the difference is then calculated for each document against the whole dataset based on the document content and is stored.

In the Kullback Leibler Method the distance calculated is done using the symmetric method and is based on the following formula given by Eq. 6:

$$p(a \vee b) = \sum\nolimits_{x \in X} a(x) log\left(\frac{a(x)}{b(x)}\right) \tag{6}$$

The KLD method works on the assumption that the document is characterized by a vector of weights. Each document is represented by d and the category in which it is to be clustered is termed as c. The criterion used for the clustering in the KLD method is that the documents will be clustered based on decreasing order of the KLD score and increasing order of the WebPMI measure with mean standard deviation being the threshold value.

It should be noted that the Kullback Leibler Method is not a real or true metric. It fails to follow the Triangle Inequality and only some of its properties are satisfied when using it as a measuring unit.

3.4 Whale Optimization Algorithm

(WOA) Whale Optimization Algorithm is a Swarm bottomed metaheuristic technique that mimics the biological nature of the humpback whales and solves the problem in two phases i.e., at first chasing the prey on the grounds of the best possible search agent or any available random agent and then simulating the bubble net hunting approach.

To put in technical terms, in Whale Optimization Algorithm starts by assigning randomized values to each document and assumes the best optimal value for the objective function and the best value for each document is calculated concerning objective function then updates its location by the best possible value at each iteration or assigns a random value to it. The process of Randomization is a very important step in the process and it a very complex process and requires a lot of time. To avoid excess time consumption the values assigned to the documents will be based on the previous outputs received by the TF-IDF method, WebPMI, and KLD methods respectively. The assigned values will then be updated at each iteration optimizing the algorithm and will take lesser and lesser time to get implemented on the whole dataset.

The behavior of the Whale Optimization Algorithm can be represented in multiple equations:

Encircling the Prey: This step means finding the location of each target document or word.

$$\vec{X}(a+1) = \overrightarrow{X'}(t) - \vec{P}.\vec{Q}, \tag{7}$$

$$\vec{Q} = \left|\vec{C}.\vec{X}'(t) - \vec{X}(t)\right|, \tag{8}$$

$$\vec{P} = 2.\vec{a}.\vec{r} + \vec{a}, \tag{9}$$

$$\vec{C} = 2.\vec{r} \tag{10}$$

Spiral Updating Position: Here the distance is calculated between the prey and the whale. In other words, the distance between the word and the mechanism is calculated which is done using multiple functions based on the following formula:

$$\vec{X}(t+1) = \vec{D}.e^{bl}.cos(2\pi l) + \vec{X}(t+1) \tag{11}$$

The WOA is used as an optimizing technique in the presented method and is helping the clustering algorithm to achieve better accuracy in terms of results.

3.5 Mini Batch K-Means

Mini Batch K-means is an amplified model of K-means in which cluster centroids are updated in small batches rather than in large sample size or even the whole dataset due to which the whole process becomes faster and, in some cases, it becomes more robust to statistical noises. In this method clustering of the whole document, the dataset is done using the same algorithm. Parameters like TF-IDF scores, WebPMI score and KLD scores are calculated and using those scores the documents are clustered using Mini-Batch K-Means.

The number of batches created is random and the given algorithm combined with WOA performs very well on the mentioned dataset and gives a very accurate result.

3.6 Proposed Architecture

The model starts with implementing statistical analysis on the CHIC heritage corpus and calculating mean, median, mode, etc. parameters. After that TF-IDF score is calculated which gives inferences about the importance of words in the given corpus. The documents are clustered based on two functions namely WebPMI score and KL divergence.

In WebPMI the function calculates the semantic similarity of the documents and in KL divergence the probability distribution of the documents is calculated. Both the factors are taken into account while clustering the documents by taking mean standard deviation as the threshold. After the calculations, Mini K-batch means is used to cluster the documents based on the calculations. To optimize the whole model, the Whale Optimization algorithm is used which makes the model more sound to statistical noise and makes it more robust. The complete architecture has been shown in Fig. 1.

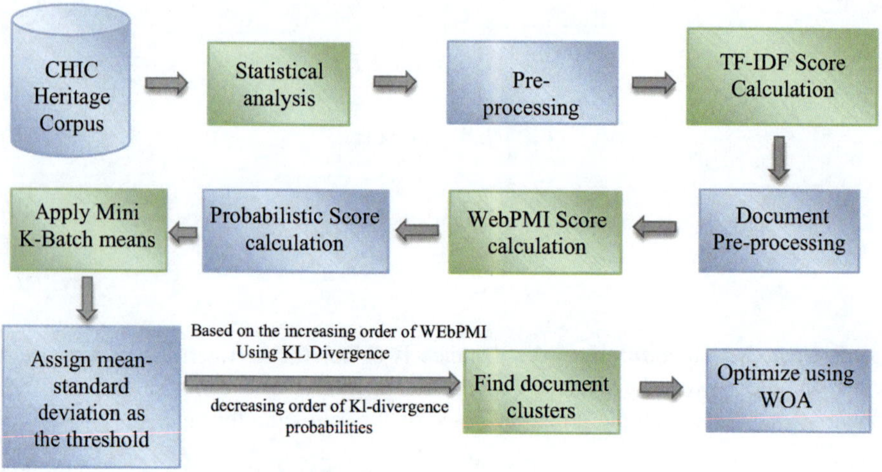

Fig. 1. The architecture for clustering of documents using feature selection and WOA

4 Implementation

The given algorithm starts with the statistical analysis of the dataset i.e., calculating the mean, median, mode, etc. parameters. After the statistical analysis, the TF-IDF score is calculated ad documents are pre-processed. Later, the WebPMI score and the KL Divergence of the documents are calculated and based on those scores the documents are clustered using the Mini batch K-means algorithm taking mean standard deviation as the threshold. To optimize the presented algorithm a metaheuristic algorithm is used, i.e., WOA, which refines the overall accuracy of the presented algorithm.

The framework has been tested on Linux Mint operating system having 16 GB RAM and an i7 processor. The IDE used for the development of the algorithm was Spyder under Anaconda Environment. The task was to cluster the CHIC heritage corpus. The pre-processing was done by removing stop words, stemming, lemmatization and tokenizing the documents. After that TF-IDF score calculation and statistical analysis helps in knowing the data better. After that WebPMI and KL divergence calculation based on which the documents are clustered and use of Mini batch- K means to find document clusters while using WOA to improve the performance and optimizing the algorithm as depicted in Table 1.

Table 1. Algorithm for clustering of documents using feature selection and WOA

Input- CHIC Heritage corpus

Output- Clustered documents based on semantic similarity and distance between probability distributions.

Begin

Step 1- Pre-processing of the CHIC heritage corpus is done. The techniques used stop words removal, lemmatization and stemming, tokenizing the documents separately.

Step 2- Statistical analysis and TF-IDF score calculation denoted by TF (t, d) and IDF (d, t).

Step 3- Computing WebPMI score and KL- Divergence.

Step 4- Clustering the corpus documents using Mini batch K-means algorithm using WebPMI and KL-Divergence scores as the factors.

Step 5- Optimizing the algorithm using WOA following the given approach:

Traversing and evaluating the semantic similarity between the documents and find the best match (BM) at each iteration.

```
while t<t_max
    Calculate the value using Equation (11)
    for each search query
        if h<0.5 then
            if A|<1 then X(t+1) =X*(t)- A.D
            else if |A|≥1 then X(t+1) = X_rand (t)- A.D*

            if end
        else if h≥0.5 then,

        X (t+1) = D.e^bl. cos(2πl) + X*(t)

        if end
        for end

        Evaluate the similarity of the document X(t+1) and update X*
```

End

5 Results

The algorithm put forth was tested on CHIC Heritage documents. To test the performance of the algorithm three metrics have been used namely GridSearchCV, Support Vector Regression Score and V-Measure. GridSearchCV is a multi-metric evaluation parameter that maps a dictionary with the name of the scorer, the scorer function, or the predefined name(s). The formula used to calculate the V-measure score is shown in Eq. 12.

$$V_\beta = \frac{(1+\beta)ab}{\beta a + b} \tag{12}$$

Table 2 defines the terms used in the V-measure Score calculation, where a is the class of the data point and b depicts the cluster. Calculation of V-measure gives the harmonic mean for completeness and homogeneity which helps to draw better inferences about the clustering performance of the model.

Table 2. Definition of the terms used in evaluation matrices

Vβ	"v_β" is the V-measure which measures the harmonic mean for completeness and homogeneity
a	"a" gives the belonging class of the data points
b	"b" is the belonging cluster of the data points

Further Support Vector regression has been used which is a statistical method that measures the linear relationship between two different variables. V-measure score; a very popular metric to measure the clustering performance has been used alongside SVR and GS.

To be used in machine learning models, sklearn module provides functions to calculate the SVR and GSCV in the python language. The use of WOA (Whale Optimization Algorithm) significantly optimizes the whole algorithm, the result of which can be seen by the better results obtained from the model. Table 2 gives a comparative analysis of the algorithm put forth with other baseline models in consideration.

Table 3. Performance measures of the presented approach in comparison with baseline models

Algorithm	GS	SVR	V-measure
Mausumi et al.	0.281	0.0081	0.3840
ENNPMI	0.321	0.0048	0.5627
Proposed	0.543	0	0.7166

The method put forth gives a better result for each evaluation metric, as can be seen from Table 3. The use of WebPMI and KL Divergence in association with WOA results in the improvement of the clustering process, which is then reflected in evaluation scores.

The Higher V-measure score achieved by the model is subjected to the procedure of clustering in phases using sophisticated feature selection algorithms. The absence of utilization of these sophisticated algorithms in the baseline models is reflected in the lower V-Measure score achieved.

The proposed methodology gives a V-Measure score of 0.7166 which demarcates it as a better performing model than the ones taken into consideration for comparison.

SVR score obtained by the proposed framework is 0 which further adds to the delineation of it giving better results than other models. The lower the SVR score, the higher would be the accuracy achieved by the model. The systematic use of two different clustering techniques shows significant improvement in the model performance.

It can be significantly marked out that this approach secures the highest GS score of 0.543 and hence, can be marked better performing than the other two.

Table 4. Comparison of the proposed approach with baseline models based on V-measure Score

No. of clusters (X-axis)	V-measure-proposed (Y-axis)	V-measure-Mausumi et al. (Y-axis)	V-measure-ENNPMI [20] (Y-axis)
20	0.4212	0.1271	0.3011
40	0.6585	0.2848	0.5131
60	0.7845	0.3211	0.5982
80	0.8463	0.5162	0.6894
100	0.8725	0.6712	0.7118

Table 3 gives the comparison report of the proposed framework with two similar modes where one model was presented by Mausumi et al. [10] and the other was CN, P et al. model [11]. Both the models focused on similar problems and based on the results shown in Table 4 it can be seen that this framework outperforms the performance of all the baseline models.

Similarly, based on the performance of the different number of clusters a similar comparison has been done by taking the V-measure score into account. This comparison has been depicted via the graph shown in Fig. 2.

On observing Fig. 2, it can be significantly observed that the model put forth outperforms the previous algorithms. The reason behind the high accuracy of the presented model is the use of two different functions to give the factors on which the corpus is clustered i.e., WebPMI score and KL divergence. The documents are clustered based on increasing order of WebPMI score and decreasing order of KL divergence which gives better accuracy. The use of Mini K-batch means makes the model more efficient during training as the corpus documents are given to the model in batches which improves the performance and reduces statistical noise. Finally, the use of the Whale optimisation algorithm provides a better result as by using this the model is independent of initial cluster central initialization which produces better clusters.

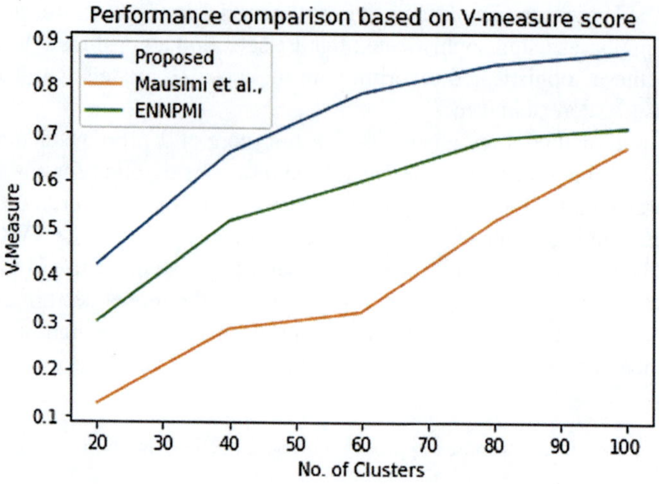

Fig. 2. Performance comparison based on V-measure

6 Conclusion

The presented model works on clustering heritage documents using machine learning methods. The model uses TF-IDF, WebPMI, KL-Divergence method to calculate probabilities on which they are to be clustered and then using Mini-Batch K-means are clustered while using the WOA optimizer algorithm. The presented framework yields highly accurate results and works very well on the mentioned dataset. It achieves a V-measure score of 0.7166, GS of 0.543 and SVR score of 0. Based on the results, it can be observed that the algorithm put forth performs better than the previous approaches and outperforms them all.

References

1. Bafna, P., Pramod, D., Vaidya, A.: Document clustering: TF-IDF approach. In: 2016 International Conference on Electrical, Electronics, and Optimization Techniques (ICEEOT), pp. 61–66. IEEE, March 2016
2. Zhang, W., Yoshida, T., Tang, X.: A comparative study of TF* IDF, LSI and multi-words for text clustering. Expert Syst. Appl. **38**(3), 2758–2765 (2011)
3. Neto, J.L., Santos, A.D., Kaestner, C.A., Alexandre, N., Santos, D.: Document clustering and text summarization (2000)
4. Popescu, A.M., Etzioni, O.: Extracting product features and opinions from reviews. In: Natural Language Processing and Text Mining, pp. 9–28. Springer, London (2007)
5. Forman, G.: An extensive empirical study of feature selection metrics for text clustering. J. Mach. Learn. Res. **3**(Mar), 1289–1305 (2003)
6. Goldberger, J., Gordon, S., Greenspan, H.: An efficient image similarity measure based on approximations of KL-divergence between two Gaussian mixtures, p. 487. IEEE, October 2003
7. Yu, D., Yao, K., Su, H., Li, G., Seide, F.: KL-divergence regularized deep neural network adaptation for improved large vocabulary speech recognition. In: 2013 IEEE International Conference on Acoustics, Speech and Signal Processing, pp. 7893–7897. IEEE, May 2013

8. Schneider, K.M.: A new feature selection score for multinomial naive Bayes text clustering based on KL-divergence. In: Proceedings of the ACL Interactive Poster and Demonstration Sessions, pp. 186–189, July 2004
9. Aljarah, I., Faris, H., Mirjalili, S.: Optimizing connection weights in neural networks using the whale optimization algorithm. Soft. Comput. **22**(1), 1–15 (2016). https://doi.org/10.1007/s00500-016-2442-1
10. Goswami, M., Purkayastha, B.S.: Discovering patterns using feature selection techniques and correlation. In: International Conference on Innovative Data Communication Technologies and Application, pp. 824–831. Springer, Cham, October 2019
11. Cn, P., Deepak, G., Zakir, M., Kr, V.: Enhanced neighborhood normalized pointwise mutual information algorithm for constraint aware data clustering. ICTACT (2016)

Check for updates

OWLW: Ontology Focused User Centric Architecture for Web Service Recommendation Based on LSTM and Whale Optimization

M. Arulmozhivarman[1] and Gerard Deepak[2](✉)

[1] Department of Electronics and Electrical Engineering, SASTRA Deemed University, Thanjavur, Tamil Nadu, India

[2] Department of Computer Science and Engineering, National Institute of Technology, Tiruchirappalli, Tiruchirappalli, India

Abstract. Web services are software packages proposed to facilitate machine-to-machine connectivity on an interoperable network. They are the web application elements, and it is possible to post, search, and use them on the Web. The use of the web service has grown and diversified in recent years as the Internet expands. This paper offers an innovative approach for recommending web services. The intention of the user is captured by the queries input by the user searches. The search queries are collected as input and preprocessed. Query topics are generated by formalizing the input queries. The formalized queries, along with the domain knowledge, are classified using LSTM. Upon classification, the top 10% of the search results' semantic similarity is computed using Jaccard and Lin similarity. Web service repositories such as UDDI and WSDL have been incorporated with similar terms in the dataset to recommend web services.

Keywords: LSTM · Ontological framework · Web service recommendation · Whale optimization

1 Introduction

As the 21st century becomes a digital book, the quantity of data accessible on the Internet is fascinating; filling in the information, doing tricky business achieve quality functions, and providing quality information to end-users by selecting the most appropriate web services. A web referral system related to service system operation services is designed to facilitate selecting a web service [1]. By using PCC, a user-based algorithm is employed for personalized web service recommendations [2]. Using ontology-based semantic analysis, the Jaccard and cosine similarity enables to identify with a very similar function related to web service [3, 4]. Any form of software that makes it accessible on the Internet and utilizes an organized XML informing framework is a web service. Web services allow various applications to connect and exchange content and services. Web Services uses a structured industry-standard communication protocol that offers the organization many advantages, such as a wide variety of options, cost savings due to competition, and quality enhancement.

A. M. A. Musleh Al-Sartawi et al. (Eds.): EAMMIS 2021, LNNS 239, pp. 334–344, 2021.
https://doi.org/10.1007/978-3-030-77246-8_32

Motivation: After receiving an input query from the user, the current user can click on server formulized user query topics. The formulated query topics are enhanced, and further domain knowledge is added by using static ontology. All these data are classified using LTSM. Furthermore, the data is stored in the Web service Dataset. Later, the top 10% items in each class are selected to be looked up and matched using Jaccard Similarity and cosine similarity with whale optimization [5]. This information is sent to a Real-World Web Service Repository like WSDL and UDDI, creating a list of recommendations and sent to the user.

Contribution: A novel framework to recommend web services are proposed in this work. The proposed approach's efficacy and accuracy are better than the baseline models as web service repositories such as UDDI and WSDL are encompassed to enhance the recommendation. Domain knowledge is incorporated with the query topics, and LSTM is employed to classify the dataset. The results obtained from the previous step are optimized using the Jaccard Similarity Optimization technique to recommend the most relevant web services for the user query.

Organization: The paper put out is structured in the following way. Related Works are in Sect. 2, and Sect. 3 encircles the proposed framework. Section 4 contains implementation and performance evaluation, while Sect. 5 carries the results. Conclusively, the paper is terminated in Sect. 6.

2 Related Work

Kang et al. [1] have combined functional with non-functional similarity measurements on comprehensive Qos of web service. Active wed service recommendation is useful because it recommends based on user' history. AWSR provides the consumer with a top-rated web service recommendation list. Jiang et al. [2] have improved the recommendation of web service on combing collaborative filtering with Pearson correlation coefficient similarity measurement, to sum up, personalized hybrid collaborative filtering (PHCF) which is useful by blending personalized user-based plus item-based algorithm. Bag et al. [3] have proposed a related Jaccard similarity that performs more accurately than another similar similarity model. By effective, two new models have been developed that consider all the user's evaluation vectors to arrange the suitable environments and create recommendations in the shortest computational time. Rana et al. [4] have combined collaborating filtering and Jaccard similarity to get more precise user recommendation. Mirjalili et al. [5] The bubble-net hunting strategy of whales stimulates the optimization algorithm that mimics the social behavior of humpback whales. The algorithms tested with 29 mathematical optimization problem and several problems in structural design. Huo et al. [6] By conducting widetrials on two real-world databases, the outcomeexhibit a definite ameliorationin the predictive accuracy of the knowledge tracing task compared to alive methods and performance in enhancing accuracy of the recommendation. Wang et al. [7] by capturing g the user data by LSTM and simultaneously encapsulating the movie title's feature with CNN, the model is further processed and improves the recommendations. Zhou et al. [8], by applying LSTM and serial filling at an equal ratio, the model can learn the user interest and changing trend in the steam dataset. Using a dimensionality reduction technique to improve the efficiency of the system is based on matrix factorization. Yu et al. [9] In the cell structure, by adding GATE functions, long short-term memory (LSTM) can take care of the problems of long-term dependencies. LSTM dominated networks, and LSTM integrated networks are two broad groups. Anithadevi et al. [10] Web service filtering is an indispensable method for suggesting QoS-based web services. Collaborative Filtering (CF) foretells the needing QoS values of web services. Neuro-Fuzzy Logic (NFL) improves classification to identify untrustworthy Web services. Wu et al. [11] the user finds it difficult to find the needed web service. Traditional approaches also tend to pay more attention to precision, but ignore diversity. The DivMTID approach increases performance by providing reliable and diversified suggestions. Using the TF-IDF algorithm to analyze the web service's weight vector, a form diversification is used to produce a top- recommendation. By these, we show DivMTID is a better method. Ayub et al. [12] Collaborative filtering (CF) is the typical method used in the recommender system. Similar approaches are used in CF to find user-to-user similarity. In this paper, they proposed an improved method, and it is compared to other similar methods in which the proposed method has shown improvement in predicting quality and accuracy. Dewi et al. [13] Global warming is a rising problem. Carbon pollution in the transport industry is the leading cause of this. The proposed algorithm is based on the Whale Optimization Algorithm (WOA) paired with the Tabu search algorithm. The result shows that HWOA is more efficient than other algorithms in minimizing the GVRP effect. [14] The approach relies on the development of a web archive of datasets from few known registry pages. Word-level semantics is used for clustering based on the NPMI value an ontology for the critical classes of web services is modeled to allow clustering based on the prior gathering of data and concepts. In [14–28] several Ontology based approaches in support of the proposed approach has been discussed.

Table 1. Analysis of frameworks that are intended to serve a similar purpose as that of the proposed model.

Authors	Techniques/Model incorporated	Disadvantages
Jiang et al.	Personalized Hybrid Collaborative Filtering	Data scarcity and ratings are required and cold start problem
Wang et al.	Combined recommendation model of LSTM and CNN	CNN model requires more high construction cost
Zhou et al.	Recurrent neural network based on LSTM	Computationally expensive and low relevance rate
Yu et al.	LSTM-dominated networks and integrated LSTM networks	By using LSTM the expense of the model is high and it has low relevance rate

(*continued*)

Table 1. (*continued*)

Authors	Techniques/Model incorporated	Disadvantages
Ayub et al.	The ratio between absolute rating values and number of commonly rated items by collaborative filtering and Jaccard similarity	There is a cold start problem in collaborative filtering and ratings are required

The Table 1 mentions a comparison of 5 models from the literature. The previously developed models are either expensive in building the model or have a low relevance rate because of using RNN and LSTM in their proposed model. By using collaborative filtering, there will be data scarcity, and ratings are required. In collaborative filtering, there is also a cold start problem for the above proposed models. These problems are rectified and overcame, and an alternative model OWLW is discussed in this paper.

3 Proposed Work

Queries are collected from the user, and the collected queries are preprocessed to generate query words. The query words obtained undergoes tokenization and lemmatization process. Tokenization is a series of actions of turning a meaningful part of data into a random string of characters called tokens. These tokens are not related to the original string, but they are used as references. Lemmatization is the algorithmic process of finding out the lemma based on the intentional meaning of the word. With the help of NER, queries are linked with the database. NER stands for Named Entity Recognition, is a process of information extraction from the input of the user queries into preprocessed categories or tags. NER digs out the predefined tags or keywords using natural language processing in any article or file. Semantic similarity is a measure identified by a collection of documents or words. The definition of distance between objects is based on the similarity of their importance or semantic content. These are mathematical functions used to approximate the semantic relationship's strength between units of language, concepts, and instances.

The user queries obtained in the previous process are formalized to generate query topics. Domain knowledge is merged with the query topics, and the LSTM technique is employed to classify the dataset. In this sense, the LSTM algorithm has been used to record information over arbitrary time intervals, and the gates control the flow of information in and out of the unit. Compared to the typical recurrent unit cell, LSTM cells have a better memory unit. LSTM has been exploited over RNN classifying the web service dataset owing to the advantages mentioned above. We can use as many LSTM

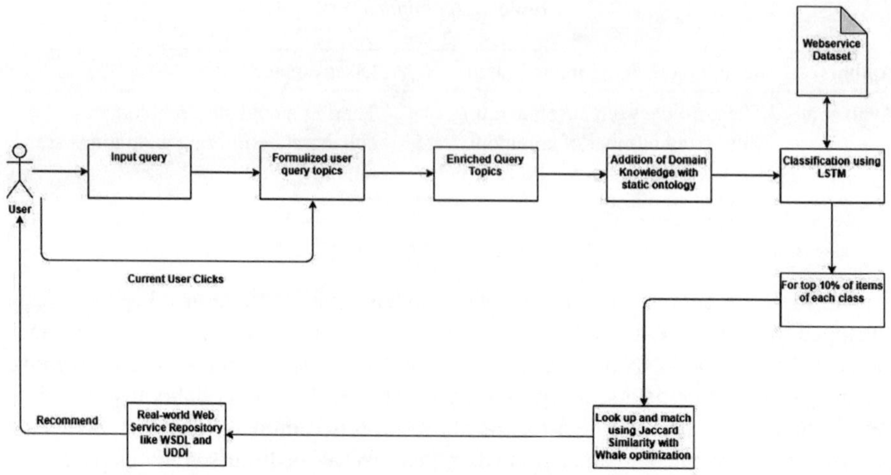

Fig. 1. Proposed OWLW architecture

cells for this process but using a group of LSTM cells, which reduces computational power. The output of the last LSTM cell is taken to the output layer. The top 10% of the items on each cluster are matched using Jaccard similarity with whale optimization. Web services repository like WSDL and UDDI are utilized to recommend web services for the user. The proposed architecture is illustrated in Fig. 1.

4 Implementation and Performance Evaluation

4.1 Dataset Preparation

The QWS Dataset admits over a list of 2,507 web services and computations of their Quality of Web Service (QWS) performed using our Web Service Broker (WSB) system during 2008. Web service and its associated nine QWS each row represents computation in this dataset. QWS metrics calculated using various Web service benchmark methods over six days are the initial nine components. Averages of the measurements gathered during that time are represented in the QWS values. A service name and a reference to a WSDL document are defined in the final two parameters.

Algorithm 1:The proposed Algorithm of the OWLW Framework

Input: User queries, Webservice Dataset
Output: Web Service that are mostappropriate to the need of the user.

Start
Step 1: The input queries are formalized to obtain enriched query terms.
Step 2: Domain knowledge with static ontology is incorporated with the enriched query terms.
Step 3: Classification using LSTM on the webservice dataset and enriched queries.
Initialize the algorithm for classification
RNN is operated for updating weights
Error percentage is calculated
Causing some disturbance to the weights
RNN is used again to update weights
Error percentage is calculated after the disturbance caused
Final weights is updated
Step 4: Calculation of semantic similarity using Jaccard measure
Step 5: Employ Whale Optimization on the classified data set.
Randomly launch the population of the whales
Consider the whale fitness principles to identify the right search agent.
if w<0.5
if |w| < 1 then $F(g+1) = F^*(g) - A.D$
else if |w|≥1 then$F(g+1) = F^*_{rand}(g) - A.D''$
end if
else if w≥0.5 then:
$F(g+1) = D'.e^{bl}.\cos(2\pi l) + F^*(g)$
end if
end for
Calculate the fitness of F(t+1) and renew F*
end while
Calculate the value of "w"
For each search agent
Calculate the fitness of F(t+1) and renewF*
Step 7: Real world information is obtained from web repositories such as UPPI and WSDL along with the results obtained in the previous step to facilitate web service recommendation
End

The optimization algorithm for whales begins with a bunch of random alternatives. Search engines change their positions on a randomly assigned search agent or the best-obtained alternative for each design change. To provide discovery and exploitation, the "w" parameter is reduced from 2 to 0, respectively. The Whale Optimization Algorithm is concluded once the termination criterion is met. There are three phases in total in whale optimization. The three phases are prey encircling, bubble-net attacking technique, and prey search. In the following Equations, the encircling of prey activity is represented.

$$\vec{D} = |\vec{C}.\overrightarrow{F_P} - \vec{F}(g) \tag{1}$$

$$\vec{F}(g+1) = \overrightarrow{F_P} - \vec{A}.\vec{D} \tag{2}$$

$$\vec{A} = 2\vec{w}.\overrightarrow{r1} - \vec{w} \tag{3}$$

$$\vec{C} = 2.\overrightarrow{r2} \tag{4}$$

$$F(g+1) = \vec{D}e^{bt}\cos(2\pi t) + \overrightarrow{F^*}(g) \tag{5}$$

$$\vec{D} = \left|\overrightarrow{F^*}(g) - \vec{F}(g)\right| \tag{6}$$

$$\vec{F}(g+1) = \begin{cases} \overrightarrow{F^*}(g) - \vec{A}\vec{D}p < 0.5 \\ \vec{D}e^{bg}\cos(2\pi g) + \overrightarrow{F^*}p \geq 0.5 \end{cases} \tag{7}$$

$$\vec{D} = \left|\vec{C}\overrightarrow{F_{rand}} - \vec{F}\right| \tag{8}$$

$$\vec{F}(g+1) = \overrightarrow{F_{rand}} - \vec{A}\vec{D} \tag{9}$$

Here, t, specify the current iteration, while $\vec{A}$ and $\vec{C}$ represents the coefficient vectors. The whale's position is denoted by $\overrightarrow{F_p}$, $\vec{F}$.In Eq. (3) and (4), r1 and r2 are vectors of random value from [0, 1]. The bubble-net attacking method is represented in the following Equations. Inaddition to bubble-net attacking, humpback whale randomly searches the prey and is coined as the exploration phase. The mathematical Equations are represented below. In Eq. (9), $\overrightarrow{F_{rand}}$ is the random position vector.

The semantic similarity between two documents or dataset is checked frequently, and the Jaccard similarity method is employed on the dataset to measure the semantic similarity. In mathematical terms, Jaccard similarity means the standard terms' ratio in two data sets concerning the total number of unique terms present in both the data set. The mathematical Equation for Jaccard similarity is given in Eq. (10).

$$J(X,Y) = \frac{|\mathrm{X} \cap \mathrm{Y}|}{|\mathrm{X} \cup \mathrm{Y}|} = \frac{|\mathrm{X} \cap \mathrm{Y}|}{|\mathrm{X}| + |\mathrm{Y}| - |\mathrm{X} \cap \mathrm{Y}|} \tag{10}$$

$$Cos(X,Y) = \frac{\mathrm{XY}}{|\mathrm{X}| * |\mathrm{Y}|} \tag{11}$$

In an inner product space, the measurement of similarity between two non-zero vectors is called cosine similarity. This method has been used in this approach to determine the similarity of the data irrespective of the data size. In cosine similarity, data objects are used in the dataset as vectors. The mathematical Equation for cosine similarity is illustrated in Eq. (11). In Eq. (11), X and Y are vectors found in the dataset. The top 10% values from both are combined and taken into consideration.

5 Results and Performance Evaluation

The queries which capture the requirement of the user are collected as input and preprocessed. The related terms among the preprocessed queries, static Ontologies are aggregated. Concept similarity among the classes and terms are measured by employing Jaccard similarity. Then LSTM is used to arrange the terms. The dataset is interpreted, and the Whale Optimization algorithm is operated on the annotated dataset to facilitate web service recommendation. Web service repositories like WSDL and UDDI, accompanying the annotated dataset, are employed to recommend the most suitable web services.

Table 2. Observation of the proposed approach with the baseline model and other alternatives

Metrics	OntoDisco	INFCF (Anita Devi et al.)	Proposed approach
Precision %	91.78	92.43	95.74
Recall %	89.45	90.12	93.78
Accuracy %	90.72	91.37	94.79
F-Measure %	90.6	91.26	94.74
FDR	0.11	0.10	0.07
nDGC	0.90	0.81	0.93

$$F - measure = \frac{2 * Precision * Recall}{Precision + Recall} \tag{12}$$

$$FDR = 1 - precision \tag{13}$$

$$Precision\% = \frac{True\ Number\ of\ positive}{True\ Number\ of\ positives + False\ Number\ Of\ Positives} \tag{14}$$

$$Recall\% = \frac{True\ Number\ of\ positive}{True\ Number\ of\ positives + False\ Number\ Of\ Negatives} \tag{15}$$

$$Accuracy\% = \frac{(Precision + Recall)}{2} \tag{16}$$

From the Table 2, OWLW's proposed architecture performs better than OntoDisco [13] and INFCF [9]. The OWLW has a better score for precision, F-measure, FDR, and accuracy. The F-measure value of OWLW for web service recommendation is better than OntoDisco and INFCF by 4.14% and 3.48%, respectively. The accuracy percentage of OWLW for web service recommendation is more remarkable than OntoDisco by 4.07% and INFCF by 3.42%. The OWLW's precision to recommend web service is 95.74, while that of OntoDisco and INFCF are 91.78 and 92.43, respectively. The recall of the proposed approach for web service recommendation is better than OntoDisco and

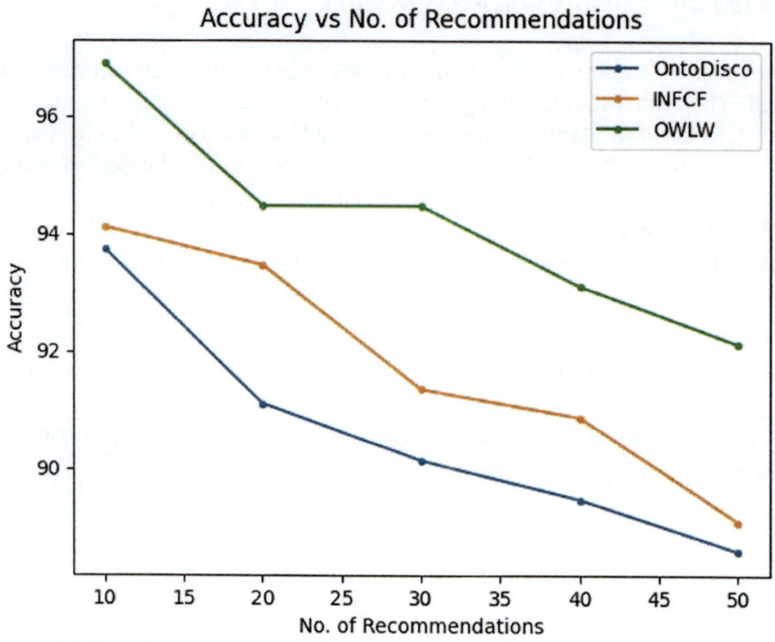

Fig. 2. Graphical representation of the proposed work and other baseline models

INFCF in percentage by 4.33 and 3.66. The FDR values of OntoDisco and INFCF are larger than OWLW by 0.04 and 0.03 for the web service recommendation.

The graphical representation of the base paper values and other alternatives are depicted in Figure 2. The F-measure and FDR are enumerated as portrayed in Eqs. (12) and (13). The Precision and Recall values are calculated using the Eqs. (14) and (15). The Accuracy percentage is calculated through the Eq. (16).

6 Conclusions

The results obtained employing the proposed OWLW framework proves its potential to be utilized for web service recommendations. The queries collected are pre-processed to receive the query terms. Domain Knowledge with static Ontology and Web service repositories like UDDI and WSDL are consolidated into the dataset to maximize classification accuracy. Upon classification, the Whale Optimization algorithm is applied to the sorted results for web services recommendations. The proposed OWLW framework generates an average accuracy of 94.79% with a significantly lower FDR of 0.07, which ensures that OWLW is the advanced approach for a recommendation of web services. The high accuracy obtained owes to the incorporation of domain ontologies and optimizing the classified data.As part of future work, further hybridization of meta-heuristic algorithms will be made to improve the model's efficacy.

References

1. Kang, G., Liu, J., Tang, M., Liu, X., Cao, B., Xu, Y.: AWSR: active web service recommendation based on usage history. In: 2012 IEEE 19th International Conference on Web Services, pp. 186–193. IEEE, June 2012
2. Jiang, Y., Liu, J., Tang, M., Liu, X.: An effective web service recommendation method based on personalized collaborative filtering. In: 2011 IEEE International Conference on Web Services, pp. 211–218. IEEE, July 2011
3. Bag, S., Kumar, S.K., Tiwari, M.K.: An efficient recommendation generation using relevant Jaccard similarity. Inf. Sci. **483**, 53–64 (2019)
4. Rana, A., Deeba, K.: Online book recommendation system using collaborative filtering (with Jaccard similarity). In: Journal of Physics: Conference Series, vol. 1362, no. 1, p. 012130. IOP Publishing, November 2019
5. Mirjalili, S., Lewis, A.: The whale optimization algorithm. Adv. Eng. Softw. **95**, 51–67 (2016)
6. Huo, Y., Wong, D.F., Ni, L.M., Chao, L.S., Zhang, J.: Knowledge modeling via contextualized representations for LSTM-based personalized exercise recommendation. Inf. Sci. **523**, 266–278 (2020)
7. Wang, W., Ye, C., Yang, P., Miao, Z.: Research on movie recommendation model based on LSTM and CNN. In: 2020 5th International Conference on Computational Intelligence and Applications (ICCIA), pp. 28–32. IEEE, June 2020
8. Zhou, Y., Ushiama, T.: LSTM-based recommendation approach for interaction records. In: International Conference on Ubiquitous Information Management and Communication, pp. 950–962. Springer, Cham, January 2019
9. Yu, Y., Si, X., Hu, C., Zhang, J.: A review of recurrent neural networks: LSTM cells and network architectures. Neural Comput. **31**(7), 1235–1270 (2019)
10. Anithadevi, N., Sundarambal, M.: A design of intelligent QoS aware web service recommendation system. Clust. Comput. **22**(6), 14231–14240 (2018)
11. Wu, S., Kou, H., Lv, C., Huang, W., Qi, L., Wang, H.: Service recommendation with high accuracy and diversity. Wirel. Commun. Mob. Comput. (2020)
12. Ayub, M., Ghazanfar, M.A., Khan, T., Saleem, A.: An effective model for Jaccard coefficient to increase the performance of collaborative filtering. Arab. J. Sci. Eng. (2020)
13. Dewi, S.K., Utama, D.M.: A new hybrid whale optimization algorithm for green vehicle routing problem. Syst. Sci. Control Eng. **9**(1), 61–72 (2021)
14. Pushpa, C.N., Deepak, G., Kumar, A., Thriveni, J., Venugopal, K.R.: OntoDisco: improving web service discovery by hybridization of ontology focused concept clustering and interface semantics. In: 2020 IEEE International Conference on Electronics, Computing and Communication Technologies (CONECCT), pp. 1–5. IEEE, July 2020
15. Deepak, G., Santhanavijayan, A.: OntoBestFit: a best-fit occurrence estimation strategy for RDF driven faceted semantic search. Comput. Commun. **160**, 284–298 (2020)
16. Kumar, A., Deepak, G., Santhanavijayan, A.: HeTOnto: a novel ap-proach for conceptualization, modeling, visualization, and formalization of domain centric ontologies for heat transfer. In: 2020 IEEE International Conference on Electronics, Computing and Communication Technologies (CONECCT), pp. 1–6. IEEE, July 2020
17. Deepak, G., Kasaraneni, D.: OntoCommerce: an ontology focused se-mantic framework for personalised product recommendation for user targeted e-commerce. Int. J. Comput. Aided Eng. Technol. **11**(4–5), 449–466 (2019)
18. Gulzar, Z., Leema, A.A., Deepak, G.: Pcrs: personalized course rec-ommender system based on hybrid approach. Procedia Comput. Sci. **125**, 518–524 (2018)
19. Deepak, G., Teja, V., Santhanavijayan, A.: A novel firefly driven scheme for resume parsing and matching based on entity linking paradigm. J. Discrete Math. Sci. Crypt. **23**(1), 157–165 (2020)

20. Haribabu, S., Kumar, P.S.S., Padhy, S., Deepak, G., Santhanavijayan, A., Kumar, N.: A novel approach for ontology focused inter-domain personalized search based on semantic set expansion. In: 2019 Fifteenth International Conference on Information Processing (ICINPRO), Bengaluru, India, pp. 1–5 (2019). https://doi.org/10.1109/ICInPro47689.2019.9092155
21. Deepak, G., Kumar, N., Bharadwaj, G.V.S.Y., Santhanavijayan, A.: OntoQuest: an ontological strategy for automatic question generation for e-assessment using static and dynamic knowledge. In: 2019 Fifteenth International Conference on Information Processing (ICINPRO), pp. 1–6. IEEE (2019)
22. Santhanavijayan, A., Kumar, D.N., Deepak, G.: A semantic-aware strategy for automatic speech recognition incorporating deep learning models. In: Intelligent System Design, pp. 247–254. Springer, Singapore (2021)
23. Deepak, G., et al.: Design and evaluation of conceptual ontologies for electro-chemistry as a domain. In: 2019 IEEE International WIE Conference on Electrical and Computer Engineering (WIECON-ECE). IEEE (2019)
24. Deepak, G., Priyadarshini, J.S.: Personalized and enhanced hybridized semantic algorithm for web image retrieval incorporating ontology classification, strategic query expansion, and content-based analysis. Comput. Electr. Eng. **72**, 14–25 (2018)
25. Kaushik, I.S., Deepak, G., Santhanavijayan, A.: QuantQueryEXP: a novel strategic approach for query expansion based on quantum computing principles. J. Discrete Math. Sci. Crypt. **23**(2), 573–584 (2020)
26. Varghese, L., Deepak, G., Santhanavijayan, A.: An IoT analytics approach for weather forecasting using raspberry Pi 3 model B+. In: 2019 Fifteenth International Conference on Information Processing (ICINPRO), pp. 1–5. IEEE, December 2019
27. Shreyas, K., Deepak, G., Santhanavijayan, A.: GenMOnto: a strategic do-main ontology modelling approach for conceptualisation and evaluation of collective knowledge for mapping genomes. J. Stat. Manag. Syst. **23**(2), 445–452 (2020)
28. Cn, P., Deepak, G., Zakir, M., Kr, V.: Enhanced neighborhood normalized pointwise mutual information algorithm for constraint aware data clustering. ICTACT J. Soft Comput. **6**(4) (2016)

KnowGen: A Knowledge Generation Approach for Tag Recommendation Using Ontology and Honey Bee Algorithm

N. Roopak[1] and Gerard Deepak[2(✉)]

[1] Department of Computer Science Engineering, SRM Institute of Science and Technology, Chennai, India

[2] Department of Computer Science Engineering, National Institute of Technology, Tiruchirappalli, India

Abstract. Diversity has recently been recognized as one of the main contributors in enhancing the efficiency of the recommendation system regarding user satisfaction. In Social Tagging-based systems, tag suggestion is used to suggest tags to an user for a resource. So far, diversity in tag recommendations has been disregarded in conventional tag recommendation methodologies. The proposed KnowGen model uses Synonymization with Word Embeddings and Ontology with data driven Knowledge Exploration and semantically aware by Honey Bee Algorithm. Hybridizing all these helps in yielding better accuracy. To assimilate users preferences for the recommendations, the MIcrosoft News Dataset (MIND) dataset is used. Efficiency of the model is estimated using Normalized Discounted Cumulative Gain (NDCG). The proposed model has achieved an Accuracy of 94.17% and tends to perform better than the other baseline models by attenuating the weakness of the common tag recommender systems.

Keywords: Knowledge exploration · Ontology · Synonymization · Tag recommendation · Word Embeddings

1 Introduction

People ought to be carefully led so that the products can be found according to their interests. The Recommendation System (RS) helps an user to effortlessly find their favorite products in a vast system where a huge volume of data is accessible on the web, among other items. Traditionally, recommendations study has centered on only to increase the accuracy of the Recommender Systems. However, it has been noticed that the accuracy and precision of the predictions alone could not always meet user needs. Beyond accuracy, there are metrics such as diversity, novelty, etc. [1] that are considered to be important dimensions for assessing the efficiency of a recommendation system to genuinely satisfy an user. Tagging helps the users to identify other users of a given group with common interests. Users with similar interests can put homogeneous tags and corresponding resources may be posted to them with similar tags. The recommendation of tags to end-users is an elemental feature of systems focused on Social Tagging. Users

A. M. A. Musleh Al-Sartawi et al. (Eds.): EAMMIS 2021, LNNS 239, pp. 345–357, 2021.
https://doi.org/10.1007/978-3-030-77246-8_33

may contribute to Social Tagging Based Structures (STS), also known as the Participative Network. Interest resources are annotated and categorized with freely selected keywords called tagging. It is possible to treat these tags as meta-data for the resources. Therefore, proper tag recommendation is important because it can play potential roles in the fields of Information Retrieval and Data Mining. It introduces the challenge of information overload as users are able to associate any tags to a resource. In order to overcome this situation, it has become essential to improvise Tag recommendations. Unlike resource recommendations or user recommendations, the small number of tag recommendations is a significant consideration. It is crucial that the collection includes as many different types as possible, because consumers are not interested in selecting from a long list of tags, which covers different facets without any repetitive suggestions. The proposed methodology is based on a binomial paradigm that uses a Synonymization with Word Embeddings and Ontology with data driven Knowledge Exploration and semantically aware by Honey Bee Algorithm to optimize an impartial function that considers both coverage and redundancy in order to diversify the recommendations provided by the method of tag recommendation [26].

Motivation: The current generation of people are very keen on gaining knowledge through the internet since everything is available on their hands. People always have a habit of reading through the news every day. Since there is abundant news available on the internet with different areas of topics, the people find it difficult to choose their area of interest in news and read through it. Here plays a major role by the recommender systems, where it helps people in recommending the news through the tags which are tagged with the news column that the people have already read through. The proposed hybrid Recommender System model provides the user with the news available in the internet's database and recommends it to the user according to the user's area of interest. This paper presents a new approach for recommending news based on the tags to the user who wants to explore the news topic by using tags he/she searches for the news topic by the keywords the user used.

Contribution: The proposed model is structured with various concepts like Synonymization with Word Embeddings, Ontology with data driven Knowledge Exploration, and Honey Bee Algorithm which helped in producing better accuracy than the baseline models. Data driven Knowledge generation helps in getting more accurate recommendations since Synonymization, Word Embeddings and the linking the terms with the static ontology paves a way for better recommendations. Experiments were conducted on the data crawled from MIcrosoft News Dataset (MIND) and hybridization of various algorithms and concepts into one helped in achieving a higher precision, recall, accuracy, F-Measure percentage, and very low FDR.

Organization: The remaining sections of the paper are structured as follows. Section 2 provides a transient overview of Related Work. Section 3 describes the Proposed System Architecture. Section 4 discusses the Implementation of the KnowGen model. Section 5 depicts the Results and Performance Evaluation. Finally, the paper is concluded in Sect. 6.

2 Related Work

Jayeeta Chakraborty et al. [2] have proposed a novel tag recommendation method that suggests a diverse but appropriate range of tags to users using Greedy-based algorithms. Babak Maleki Shoja et al. [3] have proposed a study that analyzed tag-aware recommendation systems that was systematically retrieved and reviewed between 2011 and 2018. This paper concluded that tag-aware recommendation systems are being studied and there are challenges and limitations to previous studies, which was seen in the review paper, and expected this pattern to be relatively steady or probably increasing over the next few years. Punit Gupta et al. [4] have proposed an emerging recommendation-based tagging method for the maintenance of digital libraries for users The approach of that paper would offer an easy way to buy a selective and usable book rather than an unused book. This will provide a guide to the book that is academically required by books referred by teachers in various fields or groups of subjects. Fatemeh Ghiyafeh Davoodi et al. [5] have suggested Open Directory Project (ODP) data as external awareness of web sites, In addition to tagging user behaviors on a social bookmarking platform. Their paper developed a recommendation framework focused on content that would recommend the most appropriate websites. Christian Wartena et al. [6] introduced topic aware recommendation algorithms, that detects multiple preferences of the user profile and then produces suggestions for each of these interests by first distinguishing different topics and then generating a list of similar items. Yang Song et al. [7] addresses the issue of tag recommendation from a machine learning perspective of view. Their paper suggests two novel, document-based methods that are capable of delivering accurate and efficient tag suggestions. Pijitra Jomsri et al. [8] exploits the use of sets of tags for recommending research papers to each user. It aims to propose a new paradigm using social tagging for a community-based research paper recommendation method. Yuan Zhang et al. [9] have proposed an algorithm called FolkRank, which adapts the famous Page Rank and evaluates and compares those approaches and shows that hybridization of those two methods will yield better results. In [10–24] several ontological models in support of the existing literature are discussed.

3 Proposed System Architecture

The proposed system architecture consists of knowledge generation and semantic awareness is computed by integrating semantic similarity models with honey bee algorithm for optimisation. Figure 1 depicts the proposed system architecture and the dataset used is MIcrosoft News Dataset (MIND) [25]. The dataset is then pre-processed with various techniques as per the requirements for this proposed model. Data cleansing is done so that only the data which is required for the model is taken. Data Transformation is done where removal of capital words from the available dataset, punctuation removal, etc. which helps in deforming the structure of the sentence. Word stemming is also done which helps in reducing the number of words that have the same stem words in it. By word stemming the number of words is limited. After the pre-processing of data, the available terms are loaded in the model for yielding the output.

After the loading of initial terms, Data driven Knowledge Generation happens. The first step of Knowledge generation is Synonymization of the loaded terms. Synonymization is achieved using WordNet 3.0. WordNet [1] is a lexical dictionary of words' semantic connections. WordNet connects terms such as synonyms, hyponyms, and meronyms into semantic relationships. The synonyms are organized into polysemes with brief meanings and descriptions of use [26]. While it is accessible through a web browser to human users, its primary use is in automatic text processing and applications for artificial intelligence. In order to understand the context and the meaning of the terms, There is a need for enrichment of the initial term set. Further the Word Embeddings are linked [26]. Word Embeddings are knowledge terms with respect to the dataset and are often related to the synonymized initial term set from the dataset. Word embedding is a type of word representation that allows for a common representation of words with a similar meaning. For each term, the key to the method is the idea of using a densely distributed representation. Each word, often in tens or hundreds of dimensions, is represented by a real-value vector. This is contrasted with the thousands or millions of dimensions, such as a one-hot encoding, needed for sparse word representations. Further the ontological terms are to be linked with the initial term set [27].

An ontology is a formal description of knowledge as a set of concepts within a domain and the relationships that hold between them. Here static ontologies are used for this model. Static ontologies are domain ontologies with respect to the initial term-set and are crawled to generate dynamic ontologies. These ontological terms are linked with the Word Embeddings in order to increase the information density as well as the semantic incorporation of words.

The next step is Knowledge exploration through SPARQL. Knowledge exploration is linking existing local knowledge with the global real-world knowledge. Global real-world knowledge is the knowledge from Linked Open Data Cloud (LODC), SPARQL, WikiData Endpoints, DBpedia and Google Knowledge base API. In this model SPARQL is used for knowledge exploration. SPARQL is an RDF query language, i.e., a database semantic query language capable of retrieving and manipulating RDF format data. SPARQL allows users to write queries against key-value data.

The main step in the model is semantic similarity computation using Honey Bee Algorithm. For the similarity calculations, Normalized compression distance (NCD) [28] is used. The formula for NCD is given in Eq. (8) [29]. The semantic knowledge is created from the initial terms in the dataset and the related ontological terms with the aid of Honey Bee calculations. Honey Bee algorithms that utilize genetic and behavioral mechanisms underlying the bee's mating behaviour, and algorithms that take their inspiration from the bee's foraging behaviour [30]. A population-based search algorithm that mimics the food foraging behavior of colonies of honey bees is the Honey Bees algorithm. The algorithm conducts a sort of neighborhood search in its simple version, combined with global search, which can be used both for combinatorial optimization and continuous optimization. For the implementation of the bee algorithm, the only condition is that some measure of distance is specified between the solutions. The bee algorithm consists of an initialization process and a key search cycle that is iterated several times for a given number T, or until an effective fitness solution is found. Each search cycle consists of five procedures namely recruiting, local search, diminishing neighborhood, abandonment of

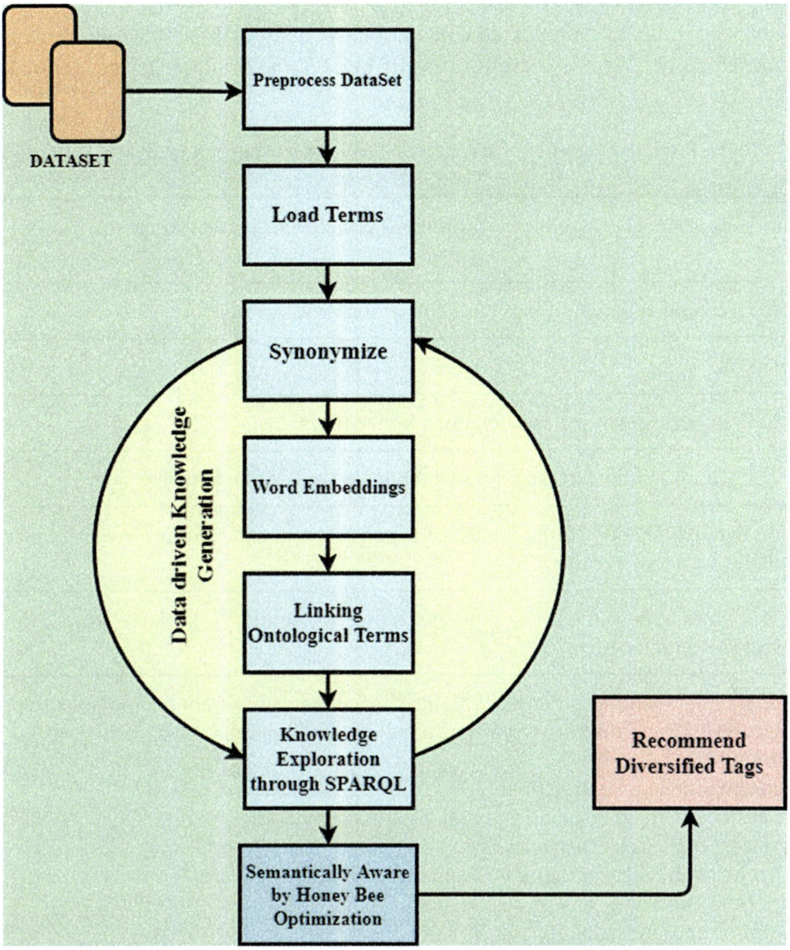

Fig. 1. Proposed system architecture of KnowGen.

the web, and global search. Then the tags are recommended to the user according to the users interests and needs. The recommended tags are classified into various classes which is also based on the users priority of interests [1].

4 Implementation

The proposed model is implemented using Python 3.9.0 in Windows 10 Operating system. The implementation process is realized in an i5 Intel Core processor with 32GB of RAM and 8GB of Nvidia graphics card. Google-Collaboratory is preferred for the implementation of this proposed model. The data set used in this model is taken from a large-scale MIcrosoft News Dataset (MIND) for news recommendation research, which is collected from the user behavior logs of Microsoft News3. OntoColab was used for data driven Knowledge Generation. The proposed model algorithm is depicted in Table 1.

The dataset is then pre-processed with various techniques as per the requirements for this proposed model. After the pre-processing of data, the available terms are loaded in

Table 1. Proposed algorithm for tag recommendation using Honey Bee Algorithm

Input: 75% of the Datas collected from the user behavior logs of Microsoft News 3 in MIND.
Output: Recommendation of tags based on user needs.
Begin
Step 1: Preprocessing of Dataset scrapped from MIND.
Step 2: Loading Terms into the proposed model in the form of tokens.
Step 3: Synonymization of the loaded terms as token is done. From here Data driven Knowledge Generation starts.
Step 4: Word Embeddings is done to represent similar tokenized words to be represented in similar ways.

Step 5: Linking Ontological terms with the Embedded words from the previous step, to increase the information density for the knowledge exploration.
Algorithm [31]:
1. Algorithm Standardization (name)
2. For each non-empty concepts of ontology o1
    Result = creates an empty set
    Tokenization tokens = (name) n = number of tokens in tokens
3. For i from 1 to n
    Term= Minusculisation (term)
    setbacks = add (term , result)
4. To end

Step 6: Knowledge Exploration is done with the help of SPARQL for linking the existing local knowledge with the global real-world knowledge. Data driven Knowledge Generation is created.

Step 7: Semantically aware by Honey Bee Algorithm is calculated between the initial term dataset and the linked ontological terms.
Algorithm [32]:
1. for i=1,…,ns
    i scout[i]=Initialise_scout()
    ii flower_patch[i]=Initialise_flower_patch(scout[i])
2. do until stopping_condition=TRUE
    i Recruitment()
    ii for i =1,...,nb

(*continued*)

Table 1. (*continued*)

1 flower_patch[i]=Local_search(flower_patch[i]) 2 flower_patch[i]=Site_abandonment(flower_patch[i] 3 flower_patch[i]=Neighbourhood_shrinking(flower_patch[i]) iii for i = nb,...,ns 1 flower_patch[i]=Global_search(flower_patch[i])}
Step 8: Recommending diversified tags to the user based on users interests. Diversification is done by classifying the results into various classes based on the user's needs.
End

the model for yielding the output. After the loading of initial terms, Data driven Knowledge Generation happens. The first step of Knowledge generation is Synonymization of the loaded terms. Synonymization is achieved by using WordNet 3.0. The synonyms are organized into polysemes with brief meanings and descriptions of use. While it is accessible through a web browser to human users, its primary use is in automatic text processing and applications for artificial intelligence. In Word Embeddings, the key to the method is the idea of using a densely distributed representation. The ontological terms are to be linked with the initial term set in order to increase the information density as well as the semantic incorporation of the words. Knowledge exploration is linking existing local knowledge with the global real-world knowledge. Finally, semantic similarity computation using Honey Bee Algorithm is done and then the tags are recommended to the users.

5 Results and Performance Evaluation

Any Recommender System aims to provide users with the best practicable and adequate recommendations. To accomplish this objective, the efficiency of this tag recommendation system is estimated. The metrics used for performance evaluation are Precision, Recall, Accuracy, F-Measure and False Discovery Rate (FDR). Equation (1) is used to calculate the precision percentage of the model. Equation (2) calculates the recall percentage of the KnowGen model. Equation (3) computes the percentage of accuracy. Equation (4) and Eq. (5) estimates the F-Measure and False Discovery Rate of the proposed approach. Accuracy is used to find the accuracy of the model predicted. F-Measure is used to evaluate a test's accuracy and is defined as the weighted harmonic mean of the precision and recall of the test [3]. False Discovery Rate (FDR) is the proportion of hypotheses that is true. Lesser the FDR, better the model. The proposed approach has been baselined with the base paper DTSRB which uses a Greedy based algorithm [1] for diversification of the tag recommendation. Since the proposed model is an incremental model which uses Synonymization with Word Embeddings and Ontology with

Knowledge Generation with honey bee optimization.

$$Precision\% = \frac{True\ Number\ of\ Positive}{True\ Number\ of\ Positives + False\ Number\ of\ Positives} \tag{1}$$

$$Recall\% = \frac{True\ Number\ of\ Positives}{True\ Number\ of\ Positives + False\ Number\ of\ Negatives} \tag{2}$$

$$Accuracy\% = \frac{(Precision + Recall)}{2} \tag{3}$$

$$F - Measure\% = \frac{2(Precision * Recall)}{(Precision + Recall)} \tag{4}$$

$$False\ Discovery\ Rate = 1 - Precision \tag{5}$$

$$nDCG = \frac{DCG\alpha}{IDCG\alpha} \tag{6}$$

$$DCG = \sum\nolimits_{i=1}^{\alpha} \frac{Rel_i}{log(i+1)} \tag{7}$$

$$NCD_z(x, y) = \frac{Z(xy) - min\{Z(x), Z(y)\}}{max\{K(x), K(y)\}} \tag{8}$$

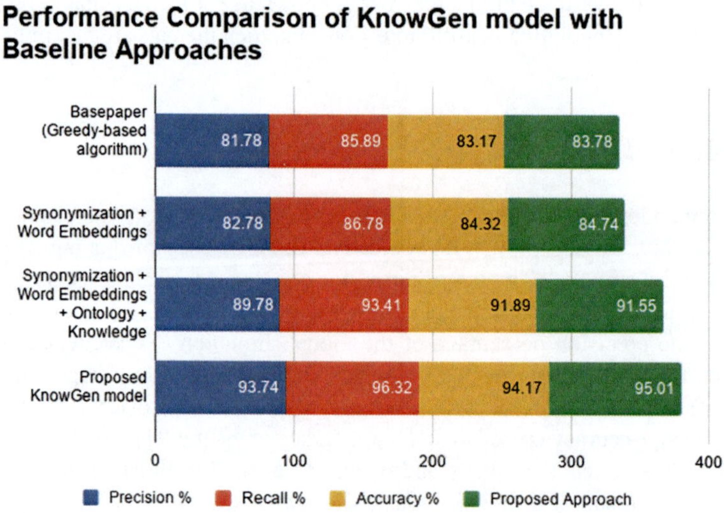

Fig. 2. Performance comparison of KnowGen model with baseline approaches

The comparison of the performance metrics in terms of Precision, Recall, Accuracy, F-Measure, and FDR of the proposed KnowGen model with the baseline models is depicted in Fig. 2. From Fig. 2, it is indicative that the proposed KnowGen model is more efficient than the baseline models. The proposed KnowGen model is composed of Synonymization, Word Embeddings, Ontology, Knowledge Exploration, and Semantic Aware Honey Bee Optimization. The baseline algorithms for diversification of tag recommendations are Greedy-based algorithm, Synonymization with Word Embeddings, and Synonymization with Word Embeddings and Ontology with data driven Knowledge Exploration. The tag recommendation using Greedy-based algorithm yields a precision of 81.78% and a recall of 85.89% and an accuracy percentage of 83.17 with an F-Measure of 83.78% and FDR of 0.13. The tag diversification using Synonymization with Word Embeddings furnishes a precision of 82.78%, a recall of 86.78%, an accuracy percentage of 84.32, and an F-Measure of 84.74% with an FDR of 0.18. The tag recommendation using Synonymization with Word Embeddings and Ontology with data driven Knowledge Exploration yields a precision of 89.78%, a recall of 93.41%, an accuracy of 91.89%, and F-Measure of 91.55% with an FDR of 0.11. The proposed KnowGen model yields a precision of 93.74%, a recall of 96.32%, with an accuracy of 94.17%, and an F-Measure of 95.01% with an FDR of 0.07. The goal of the research done in this paper is to assess the efficacy of tag recommendations using this hybrid methodology.

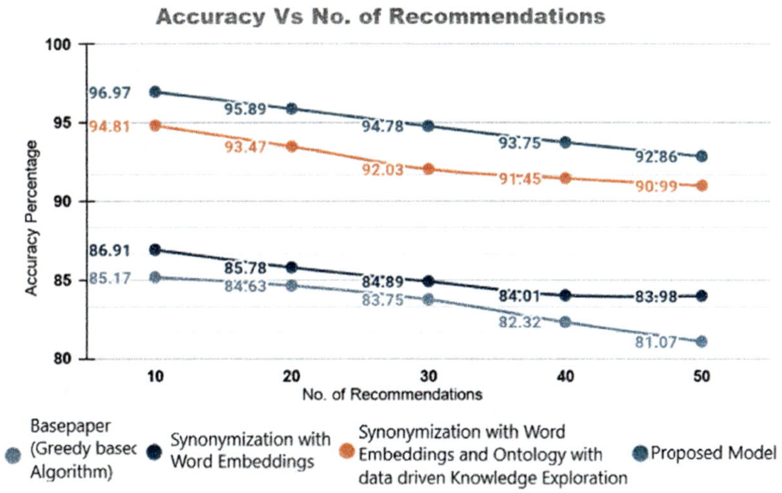

Fig. 3. Accuracy vs no. of recommendations

Figure 3 depicts the Accuracy Vs Number of Recommendations in steps of 10 up to 50 recommendations. From Fig. 3, it is evident that KnowGen has a higher Accuracy distribution when compared to the other baseline systems. The proposed KnowGen model has an accuracy of 96.97% for 10 numbers of recommendations but the other baseline models yield less accuracy for 10 numbers of recommendations. From Fig. 2, the FDR is the least for the proposed KnowGen model which shows that the other baseline models available are not efficient when compared to this fusion model. From Fig. 3, it is inferred that the proposed model has proved to be efficient in terms of Accuracy for an individual number of recommendations in numbers from 10 to 50. The KnowGen is efficient in comparison with the baseline models for diversification of tag recommendations for the reason that it uses Synonymization with Word Embeddings and Ontology with data driven Knowledge Exploration and semantically aware by Honey Bee Algorithm.

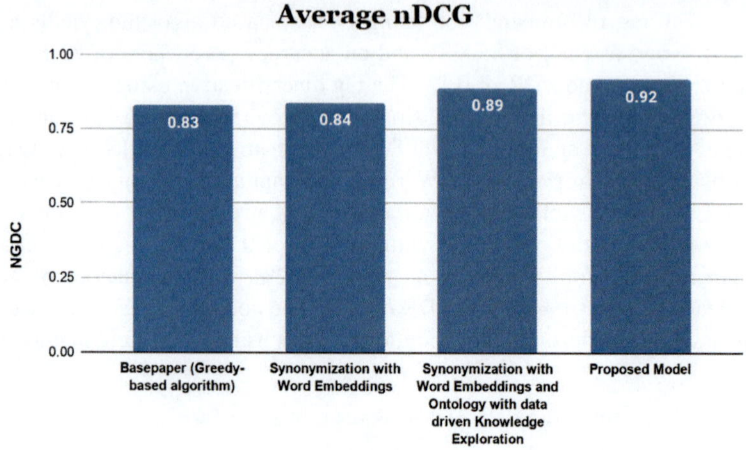

Fig. 4. Average nDCG for the baseline models and the proposed model.

From Fig. 4, it is clearly seen that the Normalized Discounted Cumulative Gain (nDCG) is higher for the proposed model then the other baseline models, which uses Synonymization with Word Embeddings and Ontology with data driven Knowledge Exploration and semantically aware by Honey Bee Algorithm. The proposed model has higher nDCG since we are hybridizing many concepts for tag recommendations. Since amalgamating various concepts into this paper the nDCG in Eq. (7) yields higher than other baseline models. Concepts like Honey Bee algorithm which is used for semantic awareness, Ontology and Synonymization with Word Embeddings helps in getting higher accuracy and nDCG. nDCG is used to evaluate the recommender systems. DCG depicted in Eq. (6) is calculated where k in the formula is answer options with more than 0 ground-truth relevance. With DCG the nDCG measures the ranking quality of the recommender systems by indicating the diversity of results.

The nDCG values ensure that there is a high diversity. So nDCG is directly proportional to the diversity of the recommendation. The diversity of tags is because of the usage of different concepts like Synonymization, Word Embeddings, Ontology with Knowledge Exploration that leads to dynamic Knowledge Generation and semantically aware by Honey Bee Algorithm. The main importance of this hybridization is yielding better accuracy and precision for the recommendation of tags. Usage of Honey Bee optimization helps in producing better recommendation of tags with Normalized compression distance which helps with the semantic similarity of the initial dataset terms and the linked ontological terms.

6 Conclusions

The proposed model employs the Deep Learning strategy for tag recommendation. This proposed paper firstly discusses with the literature survey of the other baseline papers that used different methodologies for tag recommendation. The recommendations are achieved based on the user's input queries and interests which are met perfectly to provide higher accuracy and F-Measure in comparison to other baseline models. The proposed model uses a hybrid model that consists of Synonymization with Word Embeddings and Ontology with data driven Knowledge Exploration and semantically aware by Honey Bee Algorithm when ensembled yields better precision, recall, accuracy, and F-Measure percentages with a low FDR. Experimentations have demonstrated that the usage of the Honey Bee Algorithm with Ontology and Synonymizations with Word Embeddings helps in bringing better recommendations. The proposed model has achieved an accuracy of 94.17% with a low FDR 0.07. Future Research on this topic can be done by experimenting this problem statement with different and new algorithms by hybridizing various concepts and yielding better accuracy.

References

1. Chakraborty, J., Verma, V.: Diversification in tag recommendation system using binomial framework. In: Mishra, D., Nayak, M., Joshi, A. (eds.) Information and Communication Technology for Sustainable Development. Lecture Notes in Networks and Systems, vol. 9. Springer, Singapore (2018). https://doi.org/10.1007/978-981-10-3932-4_44
2. Chakraborty, J., Verma, V.: Diversification in tag recommendation system using binomial framework. In: Information and Communication Technology for Sustainable Development, pp. 423–430, January 2018
3. Shoja, B.M., Tabrizi, N.: Tags-aware recommender systems: a systematic review. IEEE (2019)
4. Gupta, P., Jha, R.S.: Tagging based evolving recommendation system for digital library system. In: 2015 4th International Symposium on Emerging Trends and Technologies in Libraries and Information Services (2015)
5. Ghiyafeh Davoodi, F., Fatemi, O.: Tag based recommender system for social bookmarking sites. In: 2012 IEEE/ACM International Conference on Advances in Social Networks Analysis and Mining (2012)
6. Wartena, C., Wibbels, M.: Improving tag-based recommendation by topic diversification. In: Clough, P., et al. (eds.) ECIR 2011. LNCS, vol. 6611, pp. 43–54. Springer, Heidelberg (2011). https://doi.org/10.1007/978-3-642-20161-5_7
7. Song, Y., Zhang, L., Giles, C.L.: Automatic tag recommendation algorithm for social recommender systems. ACM Trans. Web Feb (2011)
8. Jomsri, P., Sanguansintukul, S., Choochaiwattana, W.: A framework for tag-based research paper recommender system: an IR approach. In: 2010 IEEE 24th International Conference on Advanced Information Networking and Applications Workshops (2010)
9. Zhang, Y., Zhang, N., Tang, J.: A Collaborative Filtering Tag Recommendation System based on Graph. IEEE (2010)
10. Kumar, A., Deepak, G., Santhanavijayan, A.: HeTOnto: a novel approach for conceptualization, modeling, visualization, and formalization of domain centric ontologies for heat transfer. In: 2020 IEEE International Conference on Electronics, Computing and Communication Technologies (CONECCT), pp. 1–6. IEEE, July 2020

11. Deepak, G., Kasaraneni, D.: OntoCommerce: an ontology focused semantic framework for personalised product recommendation for user targeted e-commerce. Int. J. Comput. Aided Eng. Technol. **11**(4–5), 449–466 (2019)
12. Gulzar, Z., Anny Leema, A., Deepak, G.: PCRS: personalized course recommender system based on hybrid approach. Procedia Comput. Sci. **125**, 518–524 (2018)
13. Deepak, G., Teja, V., Santhanavijayan, A.: A novel firefly driven scheme for resume parsing and matching based on entity linking paradigm. J. Discr. Math. Sci. Cryptogr. **23**(1), 157–165 (2020)
14. Haribabu, S., Sai Kumar, P.S., Padhy, S., Deepak, G., Santhanavijayan, A., Kumar D.N.: A novel approach for ontology focused inter- domain personalized search based on semantic set expansion. In: 2019 Fifteenth International Conference on Information Processing (ICINPRO), Bengaluru, India, 2019, pp. 1–5 (2019). https://doi.org/10.1109/ICInPro47689.2019.9092155
15. Deepak, G., Naresh Kumar, G., Sai Yashaswea Bharadwaj, V.S.N., Santhanavijayan, A.: OntoQuest: an ontological strategy for automatic question generation for e-assessment using static and dynamic knowledge. In: 2019 Fifteenth International Conference on Information Processing (ICINPRO), pp. 1–6. IEEE (2019)
16. Santhanavijayan, A., Naresh Kumar, D., Deepak, G.: A semantic-aware strategy for automatic speech recognition incorporating deep learning models. In: Satapathy, S., Bhateja, V., Janakiramaiah, B., Chen, Y.W. (eds.) Intelligent System Design. Advances in Intelligent Systems and Computing, vol. 1171. Springer, Singapore (2021). https://doi.org/10.1007/978-981-15-5400-1_25
17. Deepak, G., et al.: Design and evaluation of conceptual ontologies for electrochemistry as a domain. In: 2019 IEEE International WIE Conference on Electrical and Computer Engineering (WIECON-ECE). IEEE (2019)
18. Deepak, G., Priyadarshini, J.S.: Personalized and Enhanced Hybridized Semantic Algorithm for web image retrieval incorporating ontology classification, strategic query expansion, and content-based analysis. Comput. Electr. Eng. **72**, 14–25 (2018)
19. Deepak, G., Priyadarshini, J.S., Babu, M.H.: A differential semantic algorithm for query relevant web page recommendation. In: 2016 IEEE International Conference on Advances in Computer Applications (ICACA), pp. 44–49. IEEE, October 2016
20. Kaushik, I.S., Deepak, G., Santhanavijayan, A.: QuantQueryEXP: A novel strategic approach for query expansion based on quantum computing principles. J. Discr. Math. Sci. Cryptogr. **23**(2), 573–584 (2020)
21. Santhanavijayan, A., Kumar, D.N., Deepak, G.: A Novel hybridized strategy for machine translation of Indian languages. Soft Comput. Signal Process. 363
22. Deepak, G., Santhanavijayan, A.: OntoBestFit: a Best-Fit Occurrence Estimation strategy for RDF driven faceted semantic search. Comput. Commun. **160**, 284–298 (2020)
23. Pushpa, C.N., Deepak, G., Thriveni, J., Venugopal, K.R.: Onto Collab: Strategic review oriented collaborative knowledge modeling using ontologies (2015)
24. Leena Giri, G., Deepak, G., Manjula, S.H., Venugopal, K.R.: OntoYield: a semantic approach for context-based ontology recommendation based on structure preservation. In: Chaki, N., Cortesi, A., Devarakonda, N. (eds.) Proceedings of International Conference on Computational Intelligence and Data Engineering. LNDECT, vol. 9, pp. 265–275. Springer, Singapore (2018). https://doi.org/10.1007/978-981-10-6319-0_22
25. Wu, F., Qiao, Y., Chen, J.-H., Wu, C., Qi, T., Lian, J., Liu, D., Xie, X., Gao, J., Wu, W., Zhou, M.: MIND: A Large-scale Dataset for News Recommendation. Association for Computational Linguistics (2020)
26. Davoodi, F.G., Fatemi, O.: Tag based recommender system for social bookmarking sites. In: 2012 IEEE/ACM International Conference on Advances in Social Networks Analysis and Mining (2012)

27. Tuba, M., Akashe, S., Joshi, A. (eds.): Information and Communication Technology for Sustainable Development. AISC, vol. 933. Springer, Singapore (2020). https://doi.org/10.1007/978-981-13-7166-0
28. Ziv, J., Merhav, N.: A measure of relative entropy between individual sequences with application to universal classification. IEEE Trans. Inf. Theory **39**(4), 1270–1279 (1993). https://doi.org/10.1109/18.243444
29. Cilibrasi, R., Vitanyi, P.M.B.: R. Cilibrasi, P.M.B. Vitanyi, Clustering by compression. IEEE Trans. Inf. Theory, 27 October 2011
30. Diwold, K., Beekman, M., Middendorf, M.: Honeybee optimisation – an overview and a new bee inspired optimisation scheme. In: Panigrahi, B.K., Shi, Y., Lim, M.H. (eds.) Handbook of Swarm Intelligence. Adaptation, Learning, and Optimization, vol. 8. Springer, Heidelberg (2011). https://doi.org/10.1007/978-3-642-17390-5_13
31. Ontology Matching Algorithms: Bochra Brahma and Allaoua Refoufi. IPAC 2015, November 23–25, 2015, Batna, Algeria © 2015 ACM (2015)
32. Pham, D.T., Castellani, M.: The bees algorithm – modelling foraging behaviour to solve continuous optimisation problems. Proc. ImechE, Part C **223**(12), 2919–2938 (2009)

OntoReq: An Ontology Focused Collective Knowledge Approach for Requirement Traceability Modelling

V. Adithya[1,3] and Gerard Deepak[1,2](✉)

[1] Department of Computer Science and Engineering, Chennai, India
[2] National Institute of Technology, Tiruchirappalli, India
[3] SRM Institute of Science and Technology, Ramapuram, Chennai, India

Abstract. There is need for software-based applications in every domain and software developers offer high customizability to their client; it is very much important to deliver a defect free product that matches all the specification of the client and must pass all the testcases. So, to do this systematically, a requirement traceability matrix is used. Requirement traceability matrix is set of documents that consist of all the requirements and the test cases and the track of relation between the customer's requirements for the system. Manually creating a Requirement traceability matrix is a very time-consuming job and are always prone to errors. There is a lot of methods that are proposed in order to create a efficient traceability matrix using Language processing and machine learning techniques but these approaches require lot of data and does not take care of the real world knowledge and lead to errors. This paper proposes a knowledge engineering-based architecture to this problem with the use of ontology, machine learning and optimization algorithm which produces a dependability and steadiness of 97% and 95% respectively and the performance of the proposed model is compared with baseline approaches.

Keywords: Ontology · Requirement traceability matrix · Support vector classifier · Sunflower optimization · SRS (Software Requirement Specification) document

1 Introduction

In the recent times the world has moved so much towards software technology and web based or app based online platforms that most of the tasks that are done in a native way are now converted to e-platforms. Nowadays even the industries whose domain are in no way related to technology requires an application or website. Software Development companies nowadays needs to provide lots of customization for their software according to the customers' needs. In-order to have a defect free product, testers and software engineers must design and test products such that it matches all the specifications required by the client. In order to accomplish this objective of creating a defect free and error free uncompromising product, all the requirements and its true and false test cases must be created and designed and tested according. To do this in a more systematic way a

A. M. A. Musleh Al-Sartawi et al. (Eds.): EAMMIS 2021, LNNS 239, pp. 358–370, 2021.
https://doi.org/10.1007/978-3-030-77246-8_34

worksheet called as Requirement Traceability Matrix must be created. The reliability and efficiency of traceabilities in interdisciplinary industrial specifications management are becoming increasingly important [3].

Requirements traceability matrix is a document that contains track of the relation between the customer requirements for the system that one is developing and the work items that have been created to enforce and validate those requirements. These work items contain technical requirements, design specs, software code, test plans and other elements of the system development process. Specifications mapping allows the project team to consider which aspects of the design and technology meet the user's requirements and the checks are required to ensure that the user's requirements have been properly incorporated.

Motivation: Requirement traceability is the first and foremost important step that must be taken in-order to create a successful product. Requirements are subject to uncertainty, incompleteness and confusion, creating those manually takes lot of time and are prone to errors [6]. Traceability links generally get missed during the starting phase of the development because of the development pressure [1]. Requirement traceability matrix documents nowadays are mostly written and documented by people or they use software that uses simple query based answering system. Lot of research has been done on this and techniques have been proposed in order to create a highly efficient Requirement matrix. Since, the Requirement Traceability Matrix is a knowledge-based problem, this paper proposes a strategic method to create a requirement traceability matric using a knowledge engineering and machine learning based approach.

Contribution: This paper proposes a knowledge centric approach to create a Requirement Traceability Matrix. First the SRS document is pre-processed and the dynamically created requirements based on client meetings and secondary additional requirements are merged with it and real-world domain knowledge is added to it and it is classified as functional or non-functional requirements and from the domain ontology, entity graph is created. With this entity graph, the results yielded from classification is lined with Real World Knowledge using Wikidata and entity fact verification and aggregation and expansion of entity graph is done and entropy computation between the node of the entity and UML diagrams is done and semantic similarity is computed and optimized using Sunflower optimization and Requirement traceability matrix is created.

Organization: The latter part of the paper is organized as follows: Sect. 2 related research works done in this topic Sect. 3 explains the proposed system architecture. Section 4 explains the research methodology. The system's implementation is discussed and its performance is evaluated in Sect. 5 and Sect. 6 consist of the conclusion drawn from this experiment.

2 Literature Review

Creating an error free requirement traceability matrix is almost always difficult and people still rely on manual methods as still an efficient methodology hasn't into effect to provide a solution to this problem. Lot of research has been done related to this topic. Wang et al. [1] have Proposed a hybrid approach to the recover specifications They have used machine learning and logical thinking and have investigated the working of the use cases and code. they have also carried out a lot of studies to compare their proposed methodology to state-of-the-art approaches, and have said that their approach outperforms others.

Lucia et al. [2] have said that the existing system lacks automated or semi-automatic traceability connection creation and maintenance and have improved the artifact management system with a tool that uses latent similarity indexing which is an information retrieval tool. Their findings have also inferred that instruments like these can help to distinguish quality concerns in the textual description of the traced objects. Mahmood et al. [3] have presented a semantic based traceability link recovery (STLR) architecture which uses DBpedia knowledge base and Bablenet2.5 multilingual dictionary and semantic network to find similarities between requirement automation of retractability connection.

Zhang et al. [4] They have re-established traceability ties between current source code and program maintenance support documents. They have proposed an approach that establishes structured ontological representations for both documentary and source code objects. Guo et al. [5] have presented deep learning based solution that integrates the specification of artifact's semantics' and field awareness into the traceable solution. They have proposed a traceable networking architecture which utilizes Word Embeddings with RNN models to produce traceable connections. They have trained the BI-GRU with 360 different configurations of the tracking network using existing track ties in the Positive Train Control domain and also say that it is the best model.

Sundaram et al. [6] have examined the issues that have not been studied relating to this topic like the value of the language base used for tracing and testing and evaluating traceability linking and approaches using secondary steps and have performed empirical studies to understand the importance of this. Sultanov et al. [7] have used swarm technique, by using ant colony optimization for requirement traceability problem and have evaluated the approach with two real-world datasets. Klock et al. [8] Have proposed a Trace-eclipse, an Eclipse plug-in which is used incorporate may resonating attributes of traceability relation techniques into an easily usable suite. Zou et al. [9] have explored and found contrasts between three term-based optimization approaches intended to increase the efficiency of the probabilistic automatic monitoring tool. Ontological strategies are discussed in [10–23] in support of the literature of the proposed system by analyzing the existing literature it is very clear that the following are the gaps that are clearly evident:

- No hybridization approaches were used.
- There is no external knowledge incorporated.
- Many old traditional methodologies were used like latent similarity index, machine learning, only similarity index etc.

Whereas this problem requires hybridization, addition of external knowledge and also machine learning, so this proposed architecture amalgamates several approaches like knowledge engineering, machine learning and metaheuristics.

3 Proposed Architecture

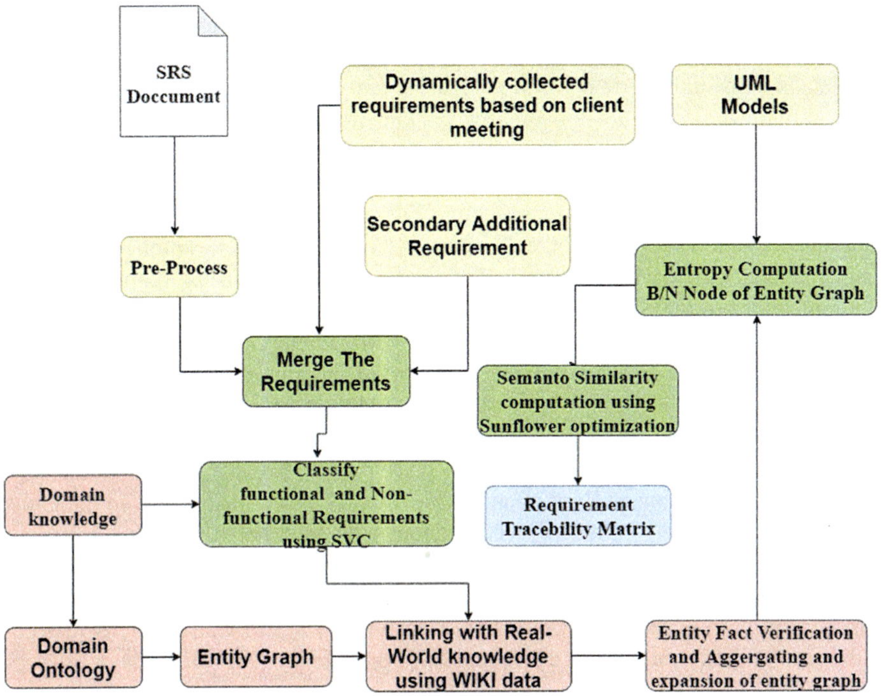

Fig. 1. Proposed OntoReq system architecture

The architecture of the proposed OntoReq is depicted in Fig. 1, which takes input as the Software Requirement Specification (SRS) document and Other Requirements Collected from the Clients apart from the Unified Modelling Language (UML) Repository from the Domain Relevant UML Repositories. The Software Requirement Specification (SRS) A software system description to be developed. It sets functional and non-functional requirements and is able to include a catalog of use cases the programs may have. The definition of software specifications sets down the framework for an arrangement between consumers and vendors or manufacturers on what the software system is to do and what it is not supposed to do. The specification of program specifications allows a rigorous evaluation of the requirements before the implementation can begin and avoids subsequent overhaul.

The requirements from the SRS document are subject to pre-processing, which is tokenized, lemmatized, and is further subject to Stop Word Removal. Further, the extracted requirements from the SRS document are subject to Named Entity Recognition (NER). The Dynamically Collected Requirements from the Client Meetings and the Secondary Additional Requirements are further pre-processed and merged with the Requirements from the document tree and is further classified into functional and non-functional requirements based on the Domain Knowledge Thesaurus which is annotated as functional and non-functional. The Domain Knowledge is further modelled into a larger Domain Ontology using OntoCollab which dynamically generates ontological concepts, axiomatizes and formulates into a taxonomy which is further transformed into an entity graph. The initially generated Entity Graph is linked with Real-World Entities using Wikidata as the Knowledge Store. The reason for using a Knowledge Store like Wikidata is to incorporate real-world knowledge and reduce the cognitive gap between the local framework and the global real-world knowledge.

The UML Repository comprises of Domain Relevant UML Models starting from Use-Case Models, Class Models, Sequence Models, Timing Diagram, Class Diagram, Component Diagram, Component Structure Diagram, Deployment Diagram, Object Diagram, Package Diagram, Profile Diagram, Activity Diagram, Communication Diagram, Interaction Overview Diagram for several possibilities of the domain relevant requirements from the SRS. However, the experimentations are stuck to only 10 specific inter-domains and thereby only 51 possible UML Models comprising mainly of the Use-Case, Sequence Diagrams, and the Interaction Diagrams are used in the approach. The Wikidata is queried by means of SPARQL Endpoints to populate the entity graph from the real-world entities and populate the entity graph into a fact verified enriched entity graph.

The approach computes Entropy at the first step between the Entities in the entity graph and the text labels from the UML Model by a threshold of 50% and the relevance between entities and the UML labels is computed. SemantoSim [24] is a metric that is an improvement of Pointwise Mutual Information (PMI) with a semantic property applied to it. SemantoSim is computed as shown in the Eq. (1) where x and y are two words whose similarity is computed and P () stands for probability function.

$$SemantoSim\,(x,\ y) = \frac{\text{pmi(x, y)} + p(x,y)\log[p(x,y)]}{[\text{p(x).p(y)}] + \log[\text{p(y,x)}]} \tag{1}$$

The entities that are relevant are retained in the first step, and further the Semantic Similarity is computed using the SemantoSim Measure under the purview of the Sunflower Optimization [25]. The reason for moderation of the SemantoSim with the Sunflower Optimization is mainly since the labelling in the UML Model is quite discrete and indirect. The Sunflower optimization is a modern meta-heuristic algorithm that is triggered by pushing sunflowers into sunlight, taking into account of pollination between neighboring sunflowers. Also, the Sunflower optimization is induced by inverse square law radiation and is a populace iterative recursive global algorithm inspired for multidisciplinary problems. Comparison with conventional algorithms, Sunflower optimization uses terminology such as root velocity and robust pollination.

SemantoSim alone will be termed insufficient and as a result, the Sunflower Optimization is integrated to govern the quality of the relevance of requirements that would be further integrated into the Requirement Traceability Matrix. The Requirements that have been infused into the structure of the Requirement Traceability Matrix are sequenced on the basis of their SemantoSim measure. If the value of the SemantoSim Measure is above 0.75 to 1, then it is prioritized as High Priority Requirements. However, if the value of the SemantoSim measure lies between 0.5 to 0.75, then the requirements are marked as Medium Priority Requirements. However, if the SemantoSim measure is between 0.35

and 0.5, then the Requirements are marked as low priority. If the SemantoSim measure is below 0.35. then the requirements are rejected and are not included in the Requirement Traceability Matrix. Figure 2 shows a small visual example of how requirement traceability matrix will look like.

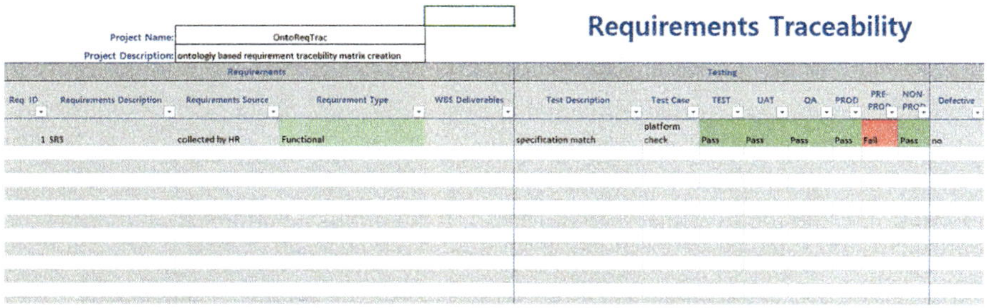

Fig. 2. Example visualization of a Requirement traceability Matrix

4 Research Methodology

It is almost always hard to construct a glitch free traceability matrix and using traditional machine learning and NLP approaches will not be much effective because of large requirement of datasets and all of them are learning based approaches. This problem can be solved using a semantically driven knowledge engineering paradigm. For this, a standard dataset from a standard repository has been used, A dataset called as public requirements dataset of 79 freely accessible natural language specifications records obtained from a number of sources. The dataset contains 34,268+ words has been utilized for this experimentation purpose 79 trials were run. Some of the data from the dataset were also split up and were used as input for the dynamically collected requirements.

The proposed architecture has been designed and implemented successfully and the implementation is explained in Sect. 5. The performance for 10 testcases where recorded and tabulated. Then the baseline approaches that resonate with the proposed technique are chosen and their performance are evaluated and bar graphs are plotted for the same and analyzed, and after analyzing, the paper is concluded with the reason why this novel approach is highly effective when compared to the baseline approaches.

5 Implementation and Performance Evaluation

The proposed system architecture was successfully implemented as shown in Table 1 and its performance was evaluated, using the Windows 10 operating system equipped with the Intel Core I5 8th generation processor with an 8 GB RAM. Python used by various NLP libraries to pre-process input data, namely NLTK, Re, Num2words and Sklearn. The Pellet Reasoner. Web Protégé a cloud-based ontology creator has been

Table 1. Algorithm of the proposed system architecture

Input: Heterogeneous collection of SRS document, dynamically collected requirements based on client meeting and secondary requirements and UML Diagrams. **Output:** Requirement Traceability matrix

Begin
Step 1: collect SRS documents
Step 2: if (SRS document! =NULL) {
Step 3: pre-process SRS document}
Step 4: else {
Step 5: goto step1}
Step 6: dynamically collect requirement from client meetings.
Step 7: if (dynamically collect requirement! =NULL) {
Step 8: go to step 11}
Step 9: else {
Step 10: goto step 6}
Step 11: secondary additional requirements to be collected.
Step 12: if (secondary additional requirements !=NULL) {
Step 13: goto step 15}
Step 14: else {
Step 15: goto step 16}
Step 16: merge the requirements {
Requirements=requirements(step1+step6+step11)
Step 17: Requirements=domain knowledge+ Requirements
Step 18: classification result = SVclassify(Requirements) //classified as functional and non functional
Step 19: entity graph=extract_entity_graph(domain ontology)
Step 20: linked =link(entitygraph,classification result,WIKI Data)
Step 21: verified= verify(linked) //entity graph verification and agreation is done to
Step 22: UML diagrams is collected
Step 23: if (UML !=NULL) {
Step 24: go to step 27}
Step 25: else{
Step 26: goto step 22}
Step 27: entropy= compute_entropy(UML,Verified)
Step 28:while(i!=79)
Step 29: { semantosim(entropy)
Step 30: i++}
Step 31: output Requirement Traceability Matrix
End

used to create and visualize ontologies. The algorithm of the proposed OntoReq system architecture is as shown in Table 1.

The performance metrics were drawn and compared with baseline approaches and they are tabulated as shown below in the Table 2.

Table 2. Performance comparison for 10 different test cases with baseline approaches

Test case	Keyword based traceability				Ontology based traceability creation				Proposed OntoReqTrac traceability recovery			
	P	R	F1	FNR	P	R	F1	FNR	P	R	F1	FNR
T1	0.32	0.4	0.36	0.6	0.54	0.74	0.62	0.26	0.87	0.9	0.90	0.06
T2	0.18	0.2	0.20	0.8	0.22	0.48	0.30	0.52	0.54	0.9	0.67	0.13
T3	0.36	0.4	0.39	0.6	0.56	0.68	0.61	0.32	0.87	0.9	0.89	0.08
T4	0.28	0.4	0.33	0.6	0.76	0.74	0.75	0.26	0.79	0.9	0.84	0.11
T5	0.52	0.5	0.52	0.5	0.44	0.54	0.48	0.46	0.85	0.7	0.79	0.26
T6	0.28	0.4	0.33	0.6	0.32	0.42	0.36	0.58	1	0.8	0.88	0.22
T7	0.32	0.5	0.38	0.5	0.34	0.54	0.42	0.46	0.87	1	0.93	0
T8	0.44	0.5	0.48	0.5	0.74	0.62	0.67	0.38	0.84	0.9	0.86	0.12
T9	0.48	0.6	0.53	0.4	0.52	0.64	0.57	0.36	0.94	1	0.96	0.02
T10	0.48	0.6	0.53	0.4	0.74	0.64	0.69	0.36	0.84	0.9	0.84	0.15

$$\text{Precision(P)} = \frac{\text{True number of Positives}}{\text{True number of Positives} + \text{False number of Positives}} \tag{2}$$

$$\text{Recall(R)} = \frac{\text{True number of Positives}}{\text{True number of Positives} + \text{False number of Negatives}} \tag{3}$$

$$\text{F} - 1\ \text{score(F1)} = \frac{2(\text{Precision} \times \text{Recall})}{(\text{Precision} + \text{Recall})} \tag{4}$$

$$\text{False Negative Rate (FNR)} = 1 - \text{Recall} \tag{5}$$

The performance metrices were drawn and compared with baseline approaches and they are tabulated as shown below in the Table 1 using the formulas as shown in the Eqs. (2), (3), (4), (5). The last 10 iterations metrics are recorded in the table. Three baseline approaches are tabulated accordingly. The first approach uses Keyword Traceability were the model uses only keywords-based linking, the second approach uses only ontology based linking and the proposed approach which uses ontology with machine learning as well as sunflower optimization algorithm. It can be observed that the proposed OntoReq is far superior in terms of performance from the start to the end of the test cases. The final test cases T10 was where the model testing was finished. T10 the final test case records a precision, Recall, F1 score, False Negative Rate of 0.48, 0.58, 0.53, 0.42 respectively for the keyword-based traceability. 0.74, 0.64, 0.69, 0.36 respectively for Ontology based traceability and 0.84, 0.85, 0.84, 0.15 for proposed OntoReq Traceability.

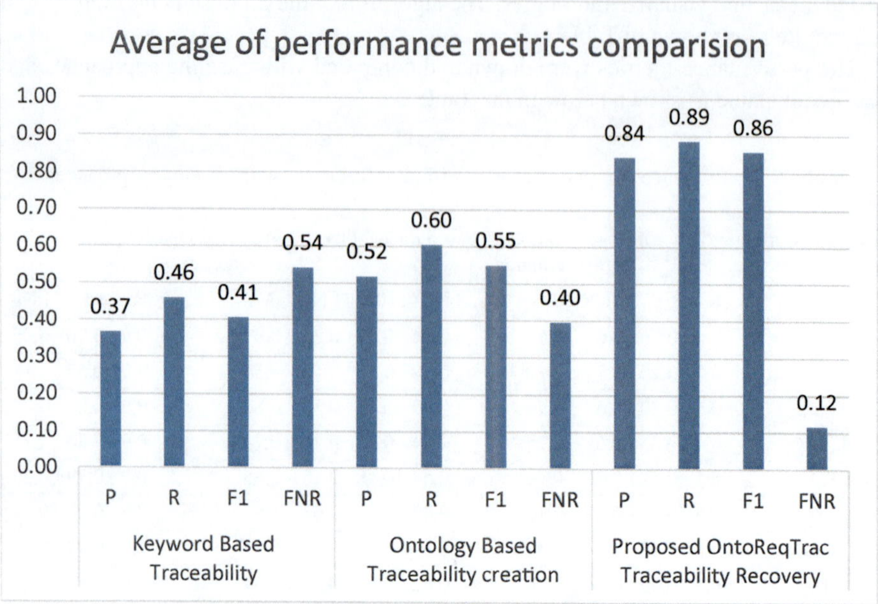

Fig. 3. Average performance metrics comparison

Figure 3 shows the average performance evaluation of baseline models comparison. The average precision, Recall, F1 score, False Negative Rate of the proposed model are 0.84, 0.89, 0.86, 0.12 respectively. The superiority of the proposed model in terms of performance is mainly because of the sunflower optimization that is being done that reduces the error so much.

The Fig. 4 shows the Dependability comparison of the baseline approaches to the proposed approach. It can be observed that the Keyword Based traceability observes a dependability of 77% and the Ontology based traceability approach records a dependability of 87% and the OntoReq observes a dependability of 97%. The dependability of the proposed OntoReq approach is high when compared to the base line approach because of the knowledge addition that is being done where real-world knowledge is used so the result becomes more practical and logical.

Figure 5 shows the steadiness comparison of the proposed OntoReq model with the baseline approaches. The proposed model achieves a steadiness of 95% and the ontology-based traceability model achieves a steadiness of 83% and the key word-based traceability approach achieves a steadiness of 74%. The OntoReq model's high steadiness is only because of both the use of ontology as well as optimization algorithm which increases the performance as well as takes care of real-world knowledge.

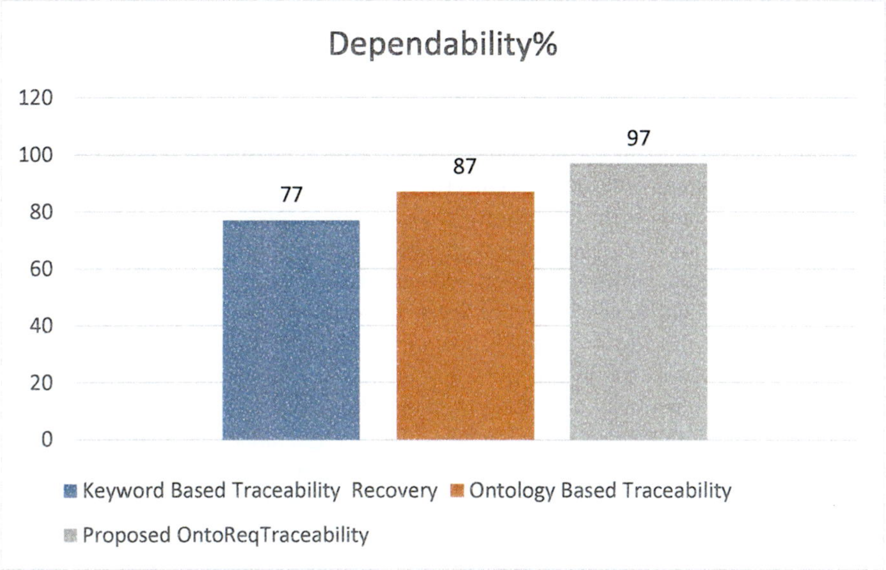

Fig. 4. Comparison of dependability

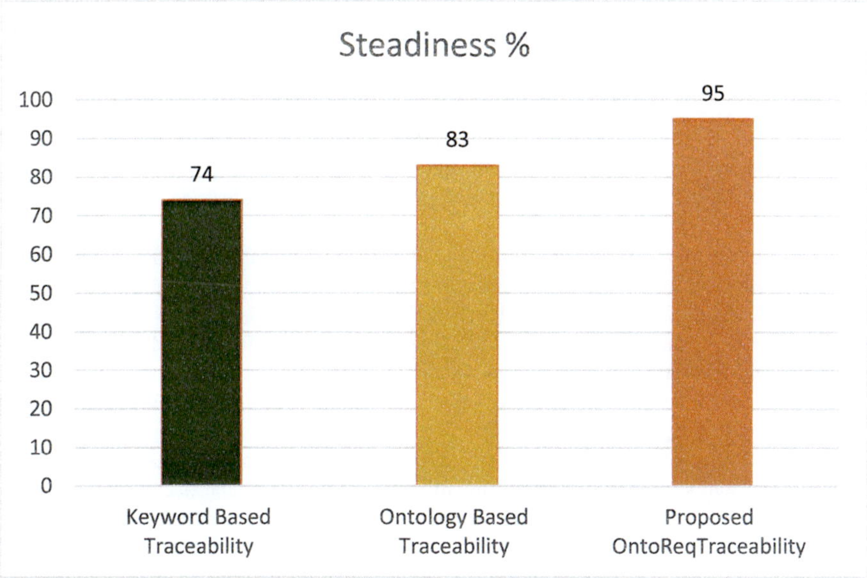

Fig. 5. Comparison of steadiness

6 Conclusion

Creation of requirement traceability matrix is the best way to systematically design a defect free and high-quality software product without confusion that matches the client's specifications. This paper proposes an architecture called as OntoReq which uses machine learning, Knowledge engineering, and, sunflower optimization algorithm that can create a requirement traceability matrix of high quality with a steadiness of 95% and Dependability of 97% tested with various test cases. This implies that using an ontology-based Knowledge approach adds auxiliary knowledge and gives solution with high stability and dependency. However, since this type of semantically driven approaches are still under research and is not much used at the same time API's and tools with better user interface aren't available so it is time consuming and takes much time to implement. The superiority of the proposed OntoReq model in terms of performance is because of the sunflower optimization algorithm and the model's high dependability is because of the ontology that is being used which leads to static knowledge addition. From this it can be concluded that use of hybrid knowledge engineering-based solution combined with a optimization algorithm can solve a knowledge-based problem with high efficacy and can provide practical and logical solution when compared to the traditional approaches. As a future work, this algorithm can be made into an application with better user interface where people can drop the required documents and a traceability matrix will be given as an output.

References

1. Wang, S., Li, T., Yang, Z.: Exploring semantics of software artifacts to improve requirements traceability recovery: a hybrid approach. In: 2019 26th Asia-Pacific Software Engineering Conference (APSEC), Putrajaya, Malaysia, pp. 39–46 (2019). https://doi.org/10.1109/APSEC48747.2019.00015
2. Lucia, A., Fasano, F., Oliveto, R., Tortora, G.: Recovering traceability links in software artifact management systems using information retrieval methods. ACM Trans. Softw. Eng. Methodol. 16 (2007). https://doi.org/10.1145/1276933.1276934
3. Mahmood, K., Takahashi, H., Alobaidi, M.: A semantic approach for traceability link recovery in aerospace requirements management system. In: 2015 IEEE Twelfth International Symposium on Autonomous Decentralized Systems, Taichung, 2015, pp. 217–222 (2015). https://doi.org/10.1109/ISADS.2015.33
4. Zhang, Y., Witte, R., Rilling, J., Haarslev, V.: An ontological approach for the semantic recovery of traceability links between software artifacts an ontological approach for the semantic recovery of traceability links between software artifacts. Software, IET. **2**, 185–203 (2008). https://doi.org/10.1049/iet-sen:20070062
5. Guo, J.L.C., Cheng, J., Cleland-Huang, J.: Semantically Enhanced Software Traceability Using Deep Learning Techniques (2018). https://doi.org/10.1109/ICSE.2017.9
6. Sundaram, S., Hayes, J., Dekhtyar, A., Holbrook, E.: Assessing traceability of software engineering artifacts. Requir. Eng. **15**, 313–335 (2010). https://doi.org/10.1007/s00766-009-0096-6
7. Sultanov, H., Hayes, J.H.: Application of swarm techniques to requirements engineering: requirements tracing. In: 2010 18th IEEE International Requirements Engineering Conference, Sydney, NSW, 2010, pp. 211–220 (2010). https://doi.org/10.1109/RE.2010.33

8. Klock, S., Gethers, M., Dit, B., Poshyvanyk, D.: Traceclipse: An eclipse plug-in for traceability link recovery and management. In: Proceedings - International Conference on Software Engineering (2011). https://doi.org/10.1145/1987856.1987862
9. Zou, X., Settimi, R., Cleland-Huang, J.: Improving automated requirements trace retrieval: a study of term-based enhancement methods. Empir Software Eng **15**, 119–146 (2010). https://doi.org/10.1007/s10664-009-9114-z
10. Kumar, A., Deepak, G., Santhanavijayan, A.:. HeTOnto: a novel approach for conceptualization, modeling, visualization, and formalization of domain centric ontologies for heat transfer. In 2020 IEEE International Conference on Electronics, Computing and Communication Technologies (CONECCT), pp. 1–6. IEEE, July 2020
11. Deepak, G., Kasaraneni, D.: OntoCommerce: an ontology focused semantic framework for personalised product recommendation for user targeted e-commerce. Int. J. Comput. Aided Eng. Technol. **11**(4–5), 449–466 (2019)
12. Gulzar, Z., Anny Leema, A., Deepak, G.: PCRS: personalized course recommender system based on hybrid approach. Procedia Comput. Sci. **125**, 518–524(2018)
13. Deepak, G., Teja, V., Santhanavijayan, A.: A novel firefly driven scheme for resume parsing and matching based on entity linking paradigm. J. Discr. Math. Sci. Cryptography **23**(1), 157–165 (2020)
14. Haribabu, S., Sai Kumar, P.S., Padhy, S., Deepak, G., Santhanavijayan, A., N. Kumar D.: A novel approach for ontology focused inter- domain personalized search based on semantic set expansion. In: 2019 Fifteenth International Conference on Information Processing (ICINPRO), Bengaluru, India, 2019, pp. 1–5 (2019). https://doi.org/10.1109/ICInPro47689.2019.9092155
15. Deepak, G., Naresh Kumar, G., Sai Yashaswea Bharadwaj, V.S.N., Santhanavijayan, A.: OntoQuest: an ontological strategy for automatic question generation for e-assessment using static and dynamic knowledge. In: 2019 Fifteenth International Conference on Information Processing (ICINPRO), pp. 1–6. IEEE (2019)
16. Santhanavijayan, A., Naresh Kumar, D., Deepak, G.: A semantic-aware strategy for automatic speech recognition incorporating deep learning models. In: Satapathy, S., Bhateja, V., Janakiramaiah, B., Chen, Y.W. (eds.) Intelligent System Design. Advances in Intelligent Systems and Computing, vol. 1171. Springer, Singapore (2021). https://doi.org/10.1007/978-981-15-5400-1_25
17. Deepak, G., et al.: Design and evaluation of conceptual ontologies for electrochemistry as a domain. In: 2019 IEEE International WIE Conference on Electrical and Computer Engineering (WIECON-ECE). IEEE (2019)
18. Deepak, G., Priyadarshini, J.S.: Personalized and Enhanced Hybridized Semantic Algorithm for web image retrieval incorporating ontology classification, strategic query expansion, and content-based analysis. Comput. Electr. Eng. **72**, 14–25 (2018)
19. Deepak, G., Priyadarshini, J.S., Babu, M.H.: A differential semantic algorithm for query relevant web page recommendation. In: 2016 IEEE International Conference on Advances in Computer Applications (ICACA) (pp. 44–49). IEEE, October 2016
20. Kaushik, I.S., Deepak, G., Santhanavijayan, A.: QuantQueryEXP: a novel strategic approach for query expansion based on quantum computing principles. J. Discr. Math. Sci. Cryptogr. **23**(2), 573–584 (2020)
21. Santhanavijayan, A., Kumar, D.N., Deepak, G.: A novel hybridized strategy for machine translation of indian languages. Soft Comput. Signal Processing **363**
22. Deepak, G., Santhanavijayan, A.: OntoBestFit: a Best-Fit Occurrence Estimation strategy for RDF driven faceted semantic search. Comput. Commun. **160**, 284–298 (2020)
23. Pushpa, C. N., Deepak, G., Thriveni, J., Venugopal, K.R.: Onto Collab: Strategic review oriented collaborative knowledge modeling using ontologies (2015)

24. Leena Giri, G., Deepak, G., Manjula, S.H., Venugopal, K.R.: OntoYield: a semantic approach for context-based ontology recommendation based on structure preservation. In: Chaki, N., Cortesi, A., Devarakonda, N. (eds.) Proceedings of International Conference on Computational Intelligence and Data Engineering. LNDECT, vol. 9, pp. 265–275. Springer, Singapore (2018). https://doi.org/10.1007/978-981-10-6319-0_22
25. Duong, T., Nguyen, T.: Application of sunflower optimization algorithm for solving the security constrained optimal power flow problem. Eng. Technol. Appl. Sci. Res. **10**, 5700–5705 (2020). https://doi.org/10.48084/etasr.3511.

KnowCrawler: AI Classification Cloud-Driven Framework for Web Crawling Using Collective Knowledge

N. Krishnan[1] and Gerard Deepak[2](✉)

[1] Department of Computer Science and Engineering, SRM Institute of Science and Technology, Ramapuram, Chennai, India

[2] Department of Computer Science and Engineering, National Institute of Technology, Tiruchirappalli, Chennai, India

Abstract. The need for Web Crawler in search engine optimization is very important, As the World Wide Web is huge and contains large amount of URL's it is important to rank these URL's and display more relevant web pages when a topic is searched in the search engine. AI cloud is a classification system which has a scheduler, dispatcher and a classification unit which uses bagging classifier and does parallel processing of data. The implication of the proposed approach is to build a Knowledge driven paradigm for crawling of web pages from the Web. Also, the approach focuses on integrating AI Classification Cloud and reduce the processing time for classification. The existing systems is not fully Semantic and is not compatible with standards of the Semantic Web. This paper solves the problem of lack of diversity and clod start problem in web crawling. This paper proposes an AI classification cloud driven framework to this problem with a use of Collective knowledge and ontology, and Cuckoo search. The architectures performance is evaluated and compared with the baseline approaches and it is seen that the proposed solution is higher in terms of performance and recorded a Harvest Rate of 94.41.

Keywords: AI cloud · Bagging classifier · Cuckoo optimization · Ontology and web crawling

1 Introduction

Web crawling is a technique which uses a web crawler to scrape a large amount of data from world wide web pages and process the unstructured data, perform analytics and make it into structured data. web crawler is an automated bot which helps browse the world wide web for indexing websites. Every crawler has a specific set of resources [1] assigned to its bandwidth. It is possible to efficiently use these tools to crawl the maximum number of web pages. When you go to a search engine and search for a term all the relevant URL's are displayed. The Web Crawler gives particular attention to new pages, modifications sites and dead links. Web crawler begins with the seed URL's to visit. Once the crawler browses the URL's it discovers all the hyperlinks present [5] in

A. M. A. Musleh Al-Sartawi et al. (Eds.): EAMMIS 2021, LNNS 239, pp. 371–382, 2021.
https://doi.org/10.1007/978-3-030-77246-8_35

the website and adds them to the visit next queue. While deciding the relevancy of the website meta tags [2] can also be taken into consideration. URL's are browsed recursively with respect to condition. The data crawled is copied and saved offline.

AI cloud is a data classifier model which increases the throughput by parallelly processing the data. As time taken decreases the throughput increases. Before a crawler is started the operations of the crawlers need to be initialized. In this way, order of execution is specified in advance and we can automate the process by using the same algorithm recursively [6] with the specified set of instructions. This decreases the time of execution. Then these initialized operations are automatically done by the crawler. From the data of the crawler the index is generated. This index will help search engines by giving most relevant URL's from the overall World Wide Web. We are adding Auxiliary Knowledge bases such as wiki data is added. Several community-based knowledges are also added and these are clustered as a collective ontology set.

Motivation: As we have enormous amounts of web pages in the world wide web, we need proper indexing to support the search engine to find the relevant URL's. Crawlers have the ability to read many types of data such as sitemaps, hyperlinks, contents in the page, HTML code of the site. Web crawlers help when you add a new relevant content to your website by changing the ranking of your site in the search engine. Here we need cloud-based architecture for classifying the data because we have a huge amount of data to be classified so by using AI based cloud, we can do parallel processing and classify the data chunks simultaneously by assigning n number of schedulers and dispatchers so that the throughput is more. In era of semantic web however there are so many web crawlers which exist there is a requirement for a highly focused semantically driven web crawler there is a need for hybridization.

Contribution: Here we create an initial set of seed URL data base and send it to the world wide web. Seed topics has been extracted the extracted seed topics is then synonymized as they will also contain relevant content. Along with auxiliary knowledge collective crowd source ontology is used to increase the quality of the relevant content from the processed seed topics. The above processed data is then sent to AI cloud which has 3 components a scheduler, a dispatcher, and a classification unit. The Ai cloud splits the entire data into data modules. Scheduler schedules the data modules for dispatching and the data will be classified parallelly. For classification bagging classifier is used which is a combination of decision tree along with random forest. Top 50% of the classified data Is then taken from the dataset after classification. Semantic similarity is calculated and with cuckoo search the more optimized solutions of URL's are sent to next step. The related URL's are then stored in the priority queue.

Organization: The structure of the publication is as follows: Section 2 consists of relevant research which has been previously done in the topics. Section 3 explains the proposed architecture. Section 4 gives insight on implementation. Session 5 is composed of performance evaluation and observed results. Section 6 is based on the conclusion.

2 Related Works

Manish Kumar et al. [1] Applied the idea of targeted crawling for information retrieval of Indian-based academics working abroad, which aims to build a database of such academics working in, and interacting with, universities abroad by developing a targeted crawler that can combine all such information. Manish Kumar et al. [2] Proposed a query-based oriented crawler utilizing searchable webpage interface as it does not entail following a route to enter web pages of relevance where crawler scans queries on collection from seed web pages using keyword list by maintaining and optimizing relevant keywords using learning method and updating the list dynamically. Varun Teja et al. [3] A new Natural Language Processing methodology has been suggested to illustrate the potential of data-driven HR to greatly increase the efficiency and pace of the overall recruitment process. Mareli et al. [4] evaluated various types of cuckoo search and on how it impacts when the values are switched dynamically. John et al. [5] Proposed a link structure by combining the relation between the hub pages set and relevant adorative pages. Suel et al. [6] proposed an architecture for high performance distributed web crawler by implementing the distributed crawler on the network of the workstation. This increases the rates of scaling and adoptability to various applications. Junghoo et al. [7] devised a method of execution of crawlers even when the entire web page is not accessible by using proper ordering scheme. Jalilian et al. [8] proposed a new method to weigh the related topics and added it as a key component to the schematic crawler and then relevance is checked. Here fuzzy interface is used to calculate the weights. Mahajan et al. [9] used ontology concepts and relevancy decision mechanism along with domain engineering to increase the crawler's performance. Fister et al. [10] summarized various research on cuckoo search how scheduling and optimization is done in various applications of this algorithm. In [11–25] several ontological models in support of the existing literature are discussed.

By analyzing the literature discussed in the section the following are the gaps identified Static Architecture, Ontology or Knowledge Driven Approach is not used, no specified strategy is used for data classification. The proposed system should contain knowledge-based architecture and should have a proper AI based Cloud architecture for classifying data quickly and decreasing the time complexity thus increasing the throughput of the system.

3 Proposed Methodology

3.1 Research Methodology

The information and URLs in the world wide web is massive. We need a proper web crawler that can support the search engine by filtering the relevant web pages and increasing the diversity of the data. So, this problem can be solved using Ai classification cloud driven framework for web crawling. Instead of collecting data individually existing standardized URL categorization Dataset have been used. Data preprocessing has been done. The proposed architecture has been designed and implemented successfully.

3.2 Proposed Architecture

Architecture Diagram consists of 3 phases. Phase 1 involves collection of seed URLs and creating a seed URL database. Seed URLs are then fetched to the World Wide Web through any search engines. The seed URLs are indexed by the search engine and gives the seed topics as output. Once the topics are prepared it is processed using WordNet 3.0 to synonymize. For the synonymized data auxiliary knowledge is added from Wikidata. The collective crowd source ontology is used for ontology driven enrichment. As ontology model is quite efficient in modeling the crowdsourced knowledge, it provides a scope for improving the Web Crawler.

Figure 1 illustrates the AI Cloud Architecture. In phase 2 Artificial intelligence-based classification cloud is configured according to the model. The cloud-based model functions by scheduling the classification and parallelizing the classification inturn reducing the response time taken for the classification process. Here the scheduler, dispatcher and then classification unit is used in the architecture. The classification unit follows the principle of bagging which uses decision tree along with random forest.

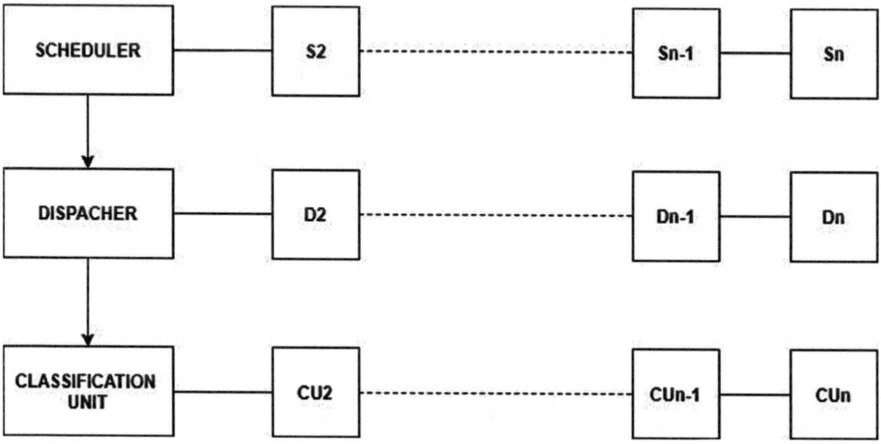

Fig. 1. Illustration of the AI Cloud Architecture

Instead of classifying the entire data, the data is sampled into smaller data modules so that we can increase the accuracy and throughput. Scheduler assigns all the incoming data modules, the data include topics, their synonyms, auxiliary knowledge and the data from collective crowdsourced ontologies along with the collected web pages dataset. The dispatcher sends the scheduled data from the scheduler to the assigned classification unit. A Bagging Classifier is just a meta-estimator set what maps the base classifiers in the random initialization sub-groups of the data and then collects their individual predictions to make a final prediction. Typically, by incorporating randomization into the construction process and then constructing an ensemble out of it, such a meta-estimator can be used as a way to reduce the variance of a black-box estimator like a decision tree. This classification unit uses bagging which uses decision tree which use tree-like branches to derive optimal solutions by reducing the cost. We use random

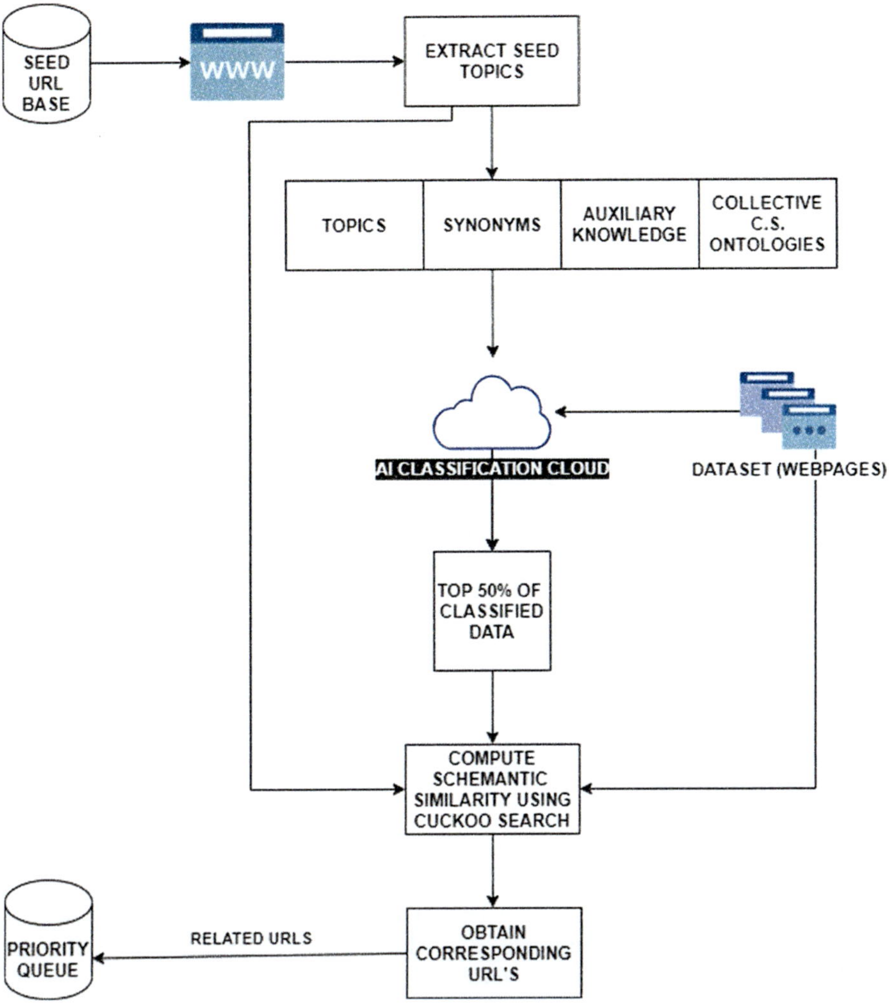

Fig. 2. Architecture Diagram of the proposed system

forest to reduce the variance of it. Top 50% of the classified data is then sent to compute semantic similarity.

In phase 3 the top 50% of the classified data from the AI cloud is passed to compute semantic similarity because only the most relevant data is required from the classification along with the reference of extracted seed topics and websites. Here we have Jaccard Similarity to find the similarity index between the functions. Jaccard Similarity between two functions u and v can be calculated using Eq. (1).

$$J(u, v) = \frac{|u \cap v|}{|u \cup v|} = \frac{|u \cap v|}{|u| + |v| - |u \cap v|} \tag{1}$$

Cuckoo search is used for selecting the optimized solution from the data after getting the values of the schematic similarity. Algorithm is inspired from the nature every nest containing an egg represents a solution. In our solution each data can either be stored in corresponding URL or not be considered for final evaluation. Our objective is optimizing the function so that most similar URL are chosen. The Eqs. (2) (3) represent the local and global random walks respectively.

$$Y_i^{c+1} = Y_i^c + \beta e \otimes h(S_a - \phi) \otimes (Y_p^c - Y_q^c) \tag{2}$$

$$Y_i^{c+1} = Y_i^c + \beta l(e, \lambda) \tag{3}$$

In the Eq. (2) Y_i^c is the position generated by random permutation, β is the positive scaling factor, step size is e, h is the high side function, S_a here is used to swift between local and global walk, ϕ is the random number chosen from a uniform distribution, $\otimes$ is the step value-based product of two vectors, $l(e, \lambda)$ is the Levy distribution function. Like a cuckoo bird lays maximum number of high-quality eggs in the most similar host so that the survival rate is increased for egg to become a cuckoo. Here in web crawling objective function is designed to select more similar URL's where each data modules quality is determined using the value of schematic similarity. More the similarity more is the chance of getting in the host's nest. The corresponding URL's are obtained. The related URL's are then sent to the priority queue. Figure 2 illustrates Architecture Diagram of the proposed system.

4 Implementation

The KnowCrawler system architecture was designed and implemented on Windows 10 Operating System. The implementation has been performed using Intel Core i7 9th Gen processor and 8 GB RAM. NLTK, Re and sklearn are the Python NLP tool libraries that was used for pre-processing data. The AI Cloud is Implemented using Google Cloud platform. In cloud storage cloud bucket is created. Pub/Sub topic is created. Cloud functions are created and invoked when an object is uploaded in the cloud storage bucket or when a message is received in Pub/Sub queue. Data loss prevention API is used. Cloud DLP inspects and classifies the data and moves the data to the appropriate bucket. URL categorization Dataset (https://data.world/crowdflower/url-categorization) is used. Under each categories of data 30% of the data has been taken by random sampling. For seed URL 5% data has been taken from the similar categories from the dataset. The proposed KnowCrawler algorithm is depicted as Algorithm 1.

Algorithm 1: Proposed KnowCrawler Algorithm for web crawling using the collective knowledge.

```
Input: Seed URL
Output: Recommended URL's
Begin
  Input SEED URL Base
  Generate population of Relevant set with z URL's.
  While (z < URL's generated) {
      Extract SEED topics
      From SEED topics generate Synonyms.
      initialize Dataarray
      Add synonyms to Dataarray
      Add Auxiliary knowledge to Dataarray
      Add Collective crowd Source ontologies
      AI classification Cloud (Dataarray) {
        schedule (Dataarray)
        Dispatch (Dataarray[i])
        Create Bucket Storage
        Generate Cloud Function (F1, F2)
        Call F1 when New Data is added in cloud
        Perform Classification using Bagging Classification
        Call F2 When Pub/sub message is received}
      Start cuckoo search
      pop top classified data y(j) Y = (Y_1, Y_2 ……..Y_m)
      initial population host nests xp (p = 1, 2...m)
     While (Y < h) or (end query) {
         generate p randomly for Cuckoo using Levy distribution
         Fitness Function F_p is Evaluated
         generate Q nest chosen among m random nests
         Evaluate its quality/fitness F_q,
         If (F_p > F_q) (
               Swap q by New Solution
            query end)
         check S_a if fraction abandon;
         Build next nest via Levy flights;
        Store Best Solutions
        Rank and select best solutions}}
  Obtain corresponding URL's
  Priority Queue with relevant URL's
End
```

5 Performance Evaluation and Result

The Precision, Recall, Accuracy, F-Measure and Harvest rate are calculated, and used as performance evaluation metrics. The Accuracy, F-Measure, and Harvest Rate are calculated and Plotted as graph 1 and graph 2. using the Eqs. (4), (5), (6), (7) respectively.

$$\text{Precision \%} = \frac{\text{True number of Positives}}{\text{True number of Positives} + \text{False number of Positives}} \quad (4)$$

$$\text{Recall \%} = \frac{\text{True number of Positives}}{\text{True number of Positives} + \text{False number of Negatives}} \quad (5)$$

$$\text{Accuracy \%} = \frac{\text{Precision} + \text{Recall}}{2} \quad (6)$$

$$\text{F-Measure \%} = \frac{2(\text{Precision} \times \text{Recall})}{(\text{Precision} + \text{Recall})} \quad (7)$$

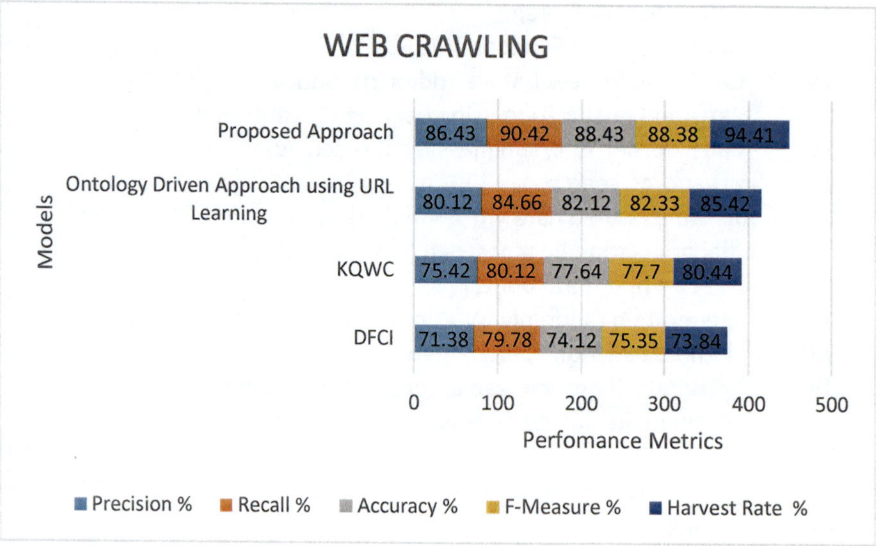

Fig. 3. The Performance metrics vs Percentage of performance measures of the KnowCrawler Architecture

In the Fig. 3 the various performance metrics marked on x axis are Precision, Recall, Accuracy, F-Measure and Harvest Rate and the models in the Y axis are KnowCrawler, Ontology Driven Approach using URL Learning, KQWC [2], DFCI [1] respectively. The values of the Precision, Recall, Accuracy, F-Measure and Harvest Rate of the proposed system is 86.43%, 90.42%, 88.43%,88.38% and 94.41%. For DFCI [1] value of Precision, Recall, Accuracy, F-Measure and Harvest Rate 71.38%, 79.78%,74.12%, 75.35% and 73.84%. For KQWC [2] 75.42%, 80.12%, 77.64%, 77.70% and 80.44%.

Ontology Driven Approach using URL Learning 80.12%, 84.66%, 82.12%, 82.33%, and 85.42% respectively. It is very evident that the proposed KnowCrawler approach has better performance metrics when compared to other models. This is because in the proposed system we are using the Ontology Driven Approach along with AI cloud and only top 50% data after classification is taken into consideration for final step. The other approaches lack in forming a highbred architecture between the classification model and knowledge driven approach.

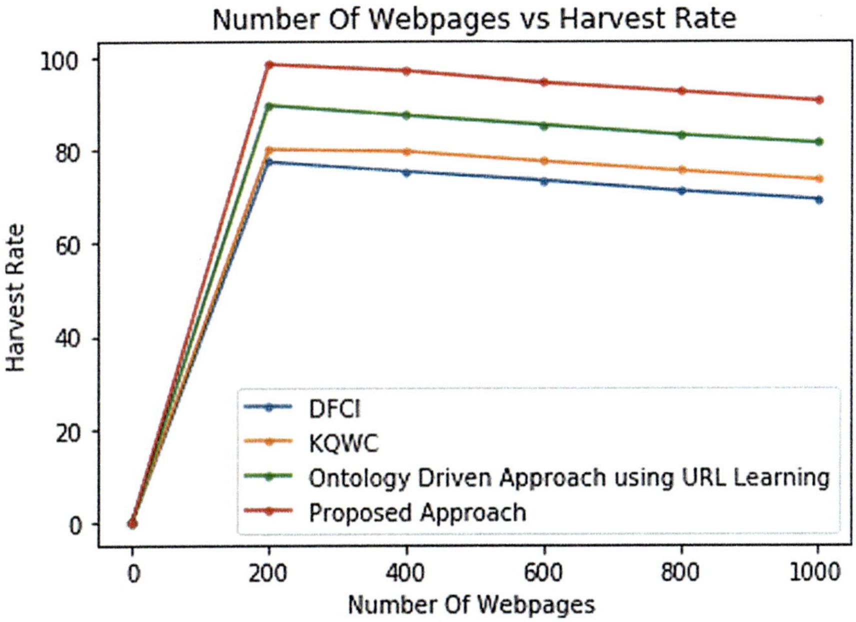

Fig. 4. Number of web pages vs Harvest Rate of various models.

The Fig. 4 depicts the Number of web pages vs Harvest Rate graph. Here KnowCrawler is baselined with DFCI [1], KQWC [2], Ontology Driven Approach using URL Learning. Analyzing the trend of the graph as Number of web pages increases the Harvest rate decreases. It is very clear that the Harvest rate of the proposed Know-Commerce system is higher consistently compared to other models. When the Number of web pages is 200, we obtain the peek Harvest rate which is 98.45. When number of web pages increased to 1000 the KnowCrawler system has a harvest rate of 90.45. which is comparatively higher when compared to other models DFCI [1], KQWC [2], Ontology Driven Approach using URL Learning whose harvest rate is 69.32%, 73.52%, 81.42% respectively. In the proposed approach we use cuckoo search for optimizing so the quality of the relevant content is increased so is the harvest rate.

In Fig. 5 the bar graph is plotted for each model vs Throughput Time. The Throughput of DFCI [1], KQWC [2], Ontology Driven Approach using URL Learning, Eliminating the AI Cloud in the Proposed Approach, KnowCrawler are 14.65, 16.18, 18.18, 15.48, 4.34 ms respectively. The difference between the throughput of KnowCrawler when

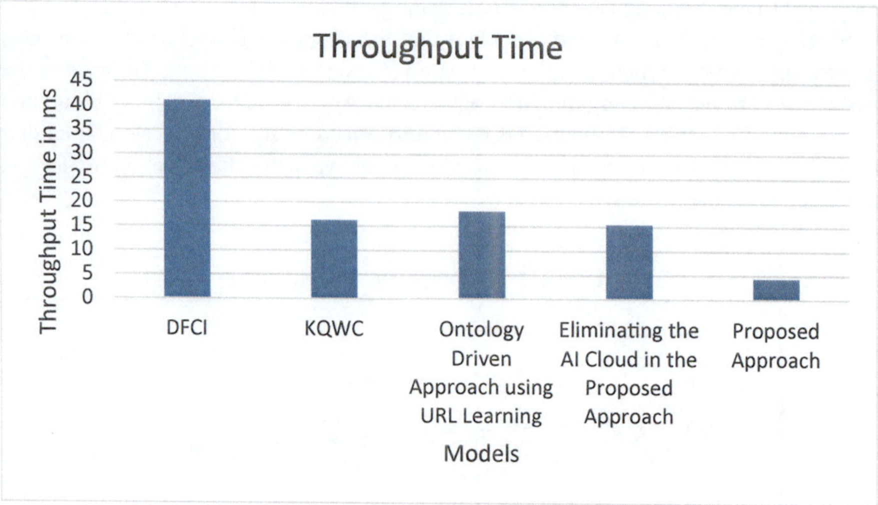

Fig. 5. Model vs Throughput Time

compared to Throughput of DFCI [1], KQWC [2], Ontology Driven Approach using URL Learning, Eliminating the AI Cloud in the Proposed Approach, KnowCrawler is 36.66,11.84,13.84 and 11.14 ms respectively. Throughput Time is very low for proposed KnowCrawler system because the time taken to execute the architecture is very less as the Ai cloud is using parallel processing to classify the data the entire data is split into modules and each module is processed simultaneously. When compared to other papers which use classical models for processing the data.

The DFCI [1] uses static architecture where only the depth of the webpage from seed URL varies. Recursive scraping with more data will result in more execution time and the accuracy will also be less. KQWC [2] architecture doesn't use any ontology or knowledge-based approach it passes keyword as a query in search boxes to search for relevant data. In Ontology Driven Approach using URL Learning even though it uses the ontology driven approach during URL learning huge amount of data is processed systematically which increases the time complexity. Even in the proposed system if the AI cloud is eliminated the time complexity increases and the throughput is increased. The proposed system uses crowd source ontology along with AI cloud for parallel processing so that the large datasets can be processed quickly. For classification bagging is used as it a combination of binary tree along with random forest the accuracy is more.

6 Conclusion

Web Crawling is one of the important parts of web indexing which is responsible for proper ranking and displaying relevant web URL's with respect to the seed topics generated from the seed URL's. This paper proposes an AI Classification cloud-driven framework for web crawling using the collective knowledge. The AI Cloud based model proved to be more effective than other models with a harvest rate of 94.41% and a throughput of 4.34 ms. The effectiveness of the proposed model is higher when compared to baseline approaches because in the proposed system a parallel processing AI classification cloud is used, as it leads to better computation lowering the time complexity. Cuckoo search is used to optimize the classified data by choosing the fittest data. From this it can be observed that in the future the knowledge engineering-based approach can produce better efficiency and can give practical and logical solution. Even though current implementation is optimized there is always a scope for improvement.

References

1. Kumar, M., Bhatia, R., Ohri, A., Kohli, A.: Design of focused crawler for information retrieval of Indian origin Academicians. IEEE (2016)
2. Kumar, M., Bindal, A., Gautam, R., Bhatia, R.: Keyword query based focused Web crawler. Procedia Comput. Sci. **125**, 584–590 (2018)
3. Deepak, G., Teja, V., Santhanavijayan, A.: A novel firefly driven scheme for resume parsing and matching based on entity linking paradigm. J. Discrete Math. Sci. Cryptogr. **23**, 157–165 (2020)
4. Mareli, M., Twala, B.: An adaptive Cuckoo search algorithm for optimization. Appl. Comput. Inform. **14**(2), 107–115 (2018). ISSN 2210-8327
5. Kleinberg, J.M.: Authoritative sources in a hyperlinked environment. J. ACM **46**(5), 604–632 (1999)
6. Shkapenyuk, V., Suel, T.: Design and implementation of a high-performance distributed Web crawler. In: Proceedings 18th International Conference on Data Engineering, San Jose, CA, USA, pp. 357–368 (2002). https://doi.org/10.1109/ICDE.2002.994750
7. Cho, J., Garcia-Molina, H., Page, L.: Efficient crawling through URL ordering. Comput. Netw. ISDN Syst. **30**(1–7), 161–172 (1988). ISSN 0169-7552
8. Jalilian, O., Khotanlou, H.: A new fuzzy-based method to weigh the related concepts in semantic focused web crawlers. In: 2011 3rd International Conference on Computer Research and Development, Shanghai, pp. 23–27 (2011). https://doi.org/10.1109/ICCRD.2011.5764237
9. Agre, G.H., Mahajan, N.V.: Keyword focused web crawler. In: 2015 2nd International Conference on Electronics and Communication Systems (ICECS), Coimbatore, pp. 1089–1092 (2015). https://doi.org/10.1109/ECS.2015.7124749
10. Fister, I., Yang, X.-S., Fister, D., Fister, I.: Cuckoo search: a brief literature review. In: Yang, X.-S. (ed.) Cuckoo Search and Firefly Algorithm. SCI, vol. 516, pp. 49–62. Springer, Cham (2014). https://doi.org/10.1007/978-3-319-02141-6_3
11. Kumar, A., Deepak, G., Santhanavijayan, A.: HeTOnto: a novel approach for conceptualization, modeling, visualization, and formalization of domain centric ontologies for heat transfer. In: 2020 IEEE International Conference on Electronics, Computing and Communication Technologies (CONECCT), pp. 1–6. IEEE (2020)
12. Deepak, G., Kasaraneni, D.: OntoCommerce: an ontology focused semantic framework for personalised product recommendation for user targeted e-commerce. Int. J. Comput. Aided Eng. Technol. **11**(4–5), 449–466 (2019)
13. Gulzar, Z., Anny Leema, A., Deepak, G.: Pcrs: Personalized course recommender system based on hybrid approach. Procedia Comput. Sci. **125**, 518–524 (2018)
14. Deepak, G., Teja, V., Santhanavijayan, A.: A novel firefly driven scheme for resume parsing and matching based on entity linking paradigm. J. Discrete Math. Sci. Cryptogr. **23**(1), 157–165 (2020)

15. Haribabu, S., Sai Kumar, P.S., Padhy, S., Deepak, G., Santhanavijayan, A., Kumar, N.D.: A novel approach for ontology focused inter-domain personalized search based on semantic set expansion. In: 2019 Fifteenth International Conference on Information Processing (ICINPRO), Bengaluru, India, pp. 1–5 (2019). https://doi.org/10.1109/ICInPro47689.2019.9092155
16. Deepak, G., Naresh Kumar, G., Sai Yashaswea Bharadwaj, V.S.N., Santhanavijayan, A.: OntoQuest: an ontological strategy for automatic question generation for e-assessment using static and dynamic knowledge. In: 2019 Fifteenth International Conference on Information Processing (ICINPRO), pp. 1–6. IEEE (2019)
17. Santhanavijayan, A., Naresh Kumar, D., Deepak, G.: A semantic-aware strategy for automatic speech recognition incorporating deep learning models. In: Satapathy, S.C., Bhateja, V., Janakiramaiah, B., Chen, Y.-W. (eds.) Intelligent System Design. AISC, vol. 1171, pp. 247–254. Springer, Singapore (2021). https://doi.org/10.1007/978-981-15-5400-1_25
18. Deepak, G., et al.: Design and evaluation of conceptual ontologies for electrochemistry as a domain. In: 2019 IEEE International WIE Conference on Electrical and Computer Engineering (WIECON-ECE). IEEE (2019)
19. Deepak, G., Priyadarshini, J.S.: Personalized and enhanced hybridized semantic algorithm for web image retrieval incorporating ontology classification, strategic query expansion, and content-based analysis. Comput. Electr. Eng. **72**, 14–25 (2018)
20. Deepak, G., Priyadarshini, J.S., Babu, M.H.: A differential semantic algorithm for query relevant web page recommendation. In: 2016 IEEE International Conference on Advances in Computer Applications (ICACA), pp. 44–49. IEEE (2016)
21. Kaushik, I.S., Deepak, G., Santhanavijayan, A.: QuantQueryEXP: a novel strategic approach for query expansion based on quantum computing principles. J. Discrete Math. Sci. Cryptogr. **23**(2), 573–584 (2020)
22. Santhanavijayan, A., Kumar, D.N., Deepak, G.: A novel hybridized strategy for machine translation of Indian languages. Soft Computing and Signal Processing, vol. 363
23. Deepak, G., Santhanavijayan, A.: OntoBestFit: a best-fit occurrence estimation strategy for rdf driven faceted semantic search. Comput. Commun. **160**, 284–298 (2020)
24. Pushpa, C.N., Deepak, G., Thriveni, J., Venugopal, K.R.: Onto Collab: Strategic review oriented collaborative knowledge modeling using ontologies (2015)
25. Leena-Giri , G., Deepak, G., Manjula, S.H., Venugopal, K.R.: OntoYield: a semantic approach for context-based ontology recommendation based on structure preservation. In: Chaki, N., Cortesi, A., Devarakonda, N. (eds.) Proceedings of International Conference on Computational Intelligence and Data Engineering. LNDECT, vol. 9, pp. 265–275. Springer, Singapore (2018). https://doi.org/10.1007/978-981-10-6319-0_22

Google's Use of Artificial Intelligence in a Two-Sided Advertising Market Causing Trust Loss, and Data Monopolization

Ben T. Yu[1(✉)] and Ekin Ulukok[2]

[1] International Regional Development, Bellingham, USA
Yudelin@irdfish.com
[2] New York Institute of Technology, New York, USA
Eulukok@nyit.edu

Abstract. Google, an American multinational technology company, offers internet products, e.g., digital advertising, search engine, cloud computing, software, and hardware. This paper describes Google search engine and Google Ads, two of its flagship products in relationship to the role of AI employed that may produce a network effect in a two-sided market. This is discussed through the structure of the search engine market and Google's position in this market. The digital use of AI robots is ubiquitous. This study brings about a new thinking, beyond the analogy between platform operations and city operations, notably along the dimension of private vs. public spaces since people are familiar with thinking of issues in that paradigm as the pre-internet days.

Keywords: Artificial intelligence · Google · Search engine · Google Ads

1 Introduction

We detect a loss of trust on the part of advertisers due to recent changes in the Google Partners program, as Google ads need to gain the trust of internet users. The paper comments on two recent antitrust lawsuits on Google Search and Advertising practices, which utilize indexes for approaching publicly accessible web pages content through the keywords and operators. The well-acknowledged (now famous) methodology matches the text on the web pages to the internet users' search of keywords. Google Ads is an online advertising platform that connects advertisers and internet users via publishing online text, image, video advertising on Google Search pages and display networks of web pages, mobile applications, videos [47, 48]. Google has powerful sources of artificial intelligence to develop its search engine and advertising platform. The usage of Google's artificial intelligence (AI) is for the purpose to strengthen the advertising's performance. Google advertising platform connects internet users and advertisers in a two-sided market environment. Data on both sides of this environment can be farmed by AI. The data so accumulated gives the company tremendous power to increase its success. This is usually done through both direct and indirect network effects, increasing

A. M. A. Musleh Al-Sartawi et al. (Eds.): EAMMIS 2021, LNNS 239, pp. 383–397, 2021.
https://doi.org/10.1007/978-3-030-77246-8_36

its primary income, the advertising revenue, in time to possibly create a data monopoly position. Google generates ad revenue directly with advertisers or through third parties that manage advertisers' ads. It offers third parties a partnership program with Google Partners. The company tries to improve its partners' advertising competencies with support such as training, activity, and certification. Google updates the terms of the partnership program, which raises queries about the overall benefits the program can do for its partners. Separately, Google faces antitrust lawsuits alleging monopolistic behavior in several areas where the company is leading.

2 Company Descriptions

Google is one of the five largest technology companies in the United States of America that can help us to understand the nature of a social contract between AI and the public for the 21st century. Often, these companies are being analyzed from a competitive theory perspective, emphasizing market concentration [15]. There can a more focused case-oriented approach to the inquiry. We choose this route for analyzing Google. The company was founded by Larry Page and Sergey Brin in 1998 in California, U.S. [44]. Google's initial public offering (IPO) took place in 2004, at a price of $85 per share, with a bid of 19,605,052 shares [17]. Through that process, the company generated $1.67 billion in sales, which increased Google's market value to over $23 billion [55]. Google's brand value has been in an upward trend since 2012. As of 2020, the brand value is 323.6 billion USD [27]. Google ranks second on Forbes' list of the world's most valuable brands in 2020 [53]. Its mission is organizing the world's information and making it universally accessible and useful [21]. The company believes, search makes it easy to discover a broad range of data from a wide variety of sources. Because Google does not only provide services in the search field and has many products and services, it competes with various companies in different sectors. Google acquired more than 100 companies, including several large organizations such as YouTube and Motorola Mobility, to expand its products and services [15]. Their list of products seems ever expanding Google Search, Google Ads, Google Analytics, Gmail, Google Maps, Google Drive, Google Docs Editors, Google Translate, Google Assistant, Google Shopping, YouTube, Google Pay, etc. Google is one of the desirable companies that people want to work with most. The number of full-time employees of Alphabet, Google's parent company, is 118,899 in 2019 [12, 13]. Alphabet's number of full-time employees is in a growth trend from 2013 to 2019 and has increased approximately five times since the beginning of the decade [18]. Google's 2017 diversity report states that its ethnicity is predominantly white at 56% and Asian at 35% [42, 43]. Thirty-one percent of the company's employees are women, 69% are men, and 20% of technology roles are women. Also, women make up 25% of leadership roles [42]. Google's parent company is Alphabet Inc. The annual report, which includes Google's shares, GOOGL and GOOG, on Nasdaq, is published by Alphabet Inc. In the year ended December 31, 2019, Alphabet Inc.'s revenue is $161,857 million; net income is $34,343 million [47]. Alphabet Inc.'s leading source of revenue is Google Ads, with a contribution of $134.8 billion in 2019 [10].

3 A Graphical Exposition of Two-Sided Markets with AI

Traditional market competitive analysis is neither here nor there for Google, as competition in the 21st century is marked by ecosystems going beyond a single market analysis. Two-sided markets as a matter of a business managing vendors and customers have been around since the early days of commerce, but two-sided markets with network effects are now a more recent phenomenon in the digital economy with AI now being augmented to the same [6]. [49] coined the term "two-sided markets" which the second author of the article was later rewarded for the 2014 Nobel Prize in Economics. The historical development of the concept of two-sided market is itself an interesting topic, it may be distracting for the purpose of this article to elaborate, suffice to say that it has its origin in what is called a Natural Monopoly problem, described in [23, 56]. Bandwagons, network externalities of various kind can be integrated in addition. More modern versions of this can be found in [14, 16]. Strip down to basics, the driving force behind the new form of competition is scale economies in the technology of production, typically characterized by a declining average total cost of production, as opposed to a U-shape average cost typical in perfect competition [39]. Thus, it is potential competition that is of relevance, which is basically Schumpeterian competition. We can use this as a starting benchmark.

Benchmark A: Natural Monopoly driven by Scale economies in Production (Fig. 1 – left chart). An industry which is characterised by a standard scale economy driven by technologies that have high fixed costs and low declining marginal costs. The situation is called Natural Monopoly as shown in Fig. 1 – left chart. D is the marginal valuation of a product that has scale economy technology. This marginal valuation curve is also known as the demand curve. Marginal valuation of a product can be cumulatively added up to a total that if divided by quantity is the average valuation, which is labelled as AD in the figure. A declining ATC curve denotes the scale economy property on the technology which together with average valuation, AD, gives a shaded yellow area that represent the total economic welfare in a situation satisfying what is called a Ramsey condition. The dotted yellow area represents the total fee net of costs awarded to the natural monopoly under fixed fee regulation, or the fees resulting from what is called the second-best Ramsey–Boiteux pricing [20].

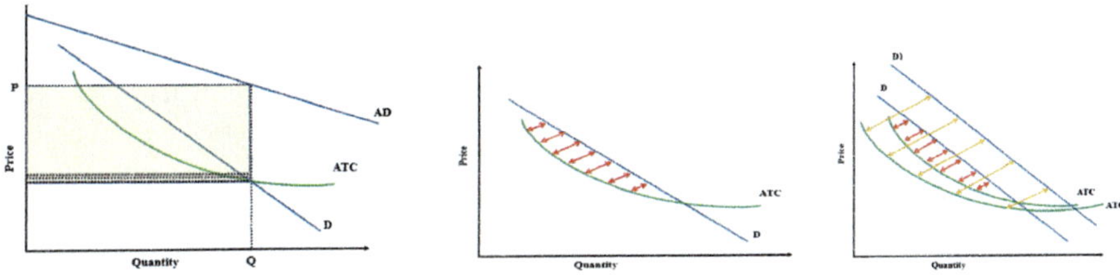

Fig. 1. Benchmark A (left chart), benchmark B (center chart), and benchmark C (right chart)

Benchmark B: A Two-sided market with network externalities (Fig. 1 – center chart). The model in Benchmark A is what is called an Amazon's linear-platform model, usually refers to as a traditional brick-and-mortar operation. (Johnson, n.d.). We now represent the Two-sided market model exhibiting network externalities. Some simplifying steps are used to modify Fig. 1 – left chart for a focus on the matching ability of the platform between the buyers and sellers. While the curve AD, shaded yellow box of total welfare and dotted yellow box of total fee net of costs hold true for Fig. 1 – center chart, they have been omitted for clarity purpose. The red arrows signify the interactive process of a Two-sided market which is the characteristic of many digital platforms. There can be direct effect of the two-sided market in that reviews of users of products posted by users generate a social media discussion on the buyer side of the market, increasing or decreasing demand in a way that vendors of the product can also adjust their product characteristics and methods of delivery as a matter of striving for survival. Also, there can be indirect effects that some type of a bandwagon effects can be generated by these social media discussions on either side of the market. Direct and indirect network effects generate value in a way which makes Amazon's role in this type of a market critical and is primarily the reason why many other players have not been able to disintermediate Amazon [2–6]. The point to note in Benchmark B is that the gap between demand and average cost in terms of the area denoted by the red arrow will be larger and larger. This questions whether the Ramsey-Boiteux pricing will continue to hold. Without sufficient and effective competition, the platform operator has no reasons not to extract the increase in this gap by engaging in monopoly pricing. Total economic welfare in that case will not conform to the Ramsey solution.

It might be interesting to comment on the early development of Microsoft in the context of Fig. 1 – center chart as well. Though not a platform company, Microsoft's products exhibit the property of scale economies in its production in that it involves high fixed costs of development and close to zero marginal cost. This is true both in terms of the different versions of its word processors as well as the development of internet browser, which they offered free to cross subsidize word processors development. Undoubtedly, the scope of development of Microsoft was much broader than the two products, which are just examples of the whole array of things they have developed. At the gist of all their development, it entails improving how individuals can better organize their private information (specifically in their PCs), and how they can add information to what they have for the purpose of improving the productivity or utility out of the information they have. The red arrow interactions in the figure captures this interaction. They are similar conceptually in that ATC of organizing information were being pushed down continuously over time by different versions of Word, while the users of Microsoft's product gain momentum as well as loyalty in the use of the products when users get additional information from the browser. Whether Ramsey-Boiteux price was charged or not has never been explored. In the context of a growing demand D, and a continuous lowering of ATC, that a high profit resulting from Ramsey-Boiteux pricing (rather than a pure monopoly price) is not implausible. The was the position articulated in [56].

Benchmark C: A Two-sided market with network effects implemented by AI (Fig. 1 – right chart). The yellow arrows are added in this diagram to denote the additional network effects. By employing AI robots, Google can increase its demand from D to D1 while simultaneously lowering its average total costs curve from ATC to ATC1. Total welfare in this situation will still be the maximum if the fee is subject to competition. This figure basically gives us a picture of how an efficient social contract can be formulated by a digital platform using AI in the absence of regulations. The formulation of AI effects in Fig. 1 – right chart goes beyond the original concept of the two-sided market suggested in [49]. It is notably different from that argued in [37] A field experiment of such nature has been attempted about the power of advertising, [54, 55].

4 Two-Sided Market in Advertising

Google's most significant source of income is online advertising products. The name of the self-service advertising platform is Google Ads. Google built this platform in 2000 as Google AdWords and changed its name to Google Ads in 2017. The platform, which serves with Pay Per Click (PPC) pricing model, displays advertisements on search engines such as Google Search or the web pages, mobile applications and videos of non-search engine members of the Google publishing network. Advertisers can publish advertisements as text on Google Search or as text, image, video on Google Display Network (GDN) publishers. Google's ads are divided into two types, performance ads and brand ads. Google Ads generate revenue because of the successful positioning of Google Search, which is Google's main product [50]. Google Search holds a large percentage of the search engine market. Google Search competes with general search engines, vertical search engines, social networks, other online and offline advertising forms, and digital video services in the advertising field. Baidu, Bing, Amazon, eBay, LinkedIn, Facebook, Snapchat, AppNexus, Criteo, Apple, AT&T companies can be given as an example to their competitors in the advertising field (Alphabet A, 2020).

Google Ads, a connector platform, helps advertisers prepare their ads most effectively because effective advertisements will be more successful in reaching the advertisers' goals. The more the advertisers achieve their goals, the more they prefer Google Ads to serve these ads. Therefore, the more successful Google Ads runs, the more revenue it generates. This is a characteristic of a two-sided market at the gist [49]. There is no price barrier for advertisers to use Google Ads. Advertisers can use Google Ads for low fees. To use the Google Ads platform, the advertiser must first determine their goals in advertising. For example, these goals could be to get more members to a friendship website or sell more products through an e-commerce platform. Advertisers need to determine where to show the ads. The location where the advertisement will be displayed will be suitable for the purpose of the ad. For example, a local market prefers to display advertisements in a local area, while a national brand can show its advertisements on a national scale. Advertisers should also choose the ad tour according to the target of the ad. For example, a beauty center operator may use photo ads, while a blog site may prefer video ads. Google Ads divides advertising placements into five groups: search, shopping, display, video, app (Google, 2020). Google Ads provides interaction between Google Search and Google Display Network (GDN) users and advertisers. Therefore, Google Ads is an intermediate platform that creates value for these two parties. In the Google Ads online advertising platform, internet users and advertisers are two different groups that interact with each other. The first group, Internet users, prefer to see ads that have been clicked too many times because their quality scores are high and because Internet users choose to click on ads that interest them. Ads with a high-quality score are ads that receive high clicks because they are more relevant to the users' interests.

The other group, advertisers, aim to have many users click on their ads. Advertisements are valuable if they reach the right target audience. Most certainly, there can be many research agenda in AI and advertising [34], but we shall concentrate on the network effect created by Google's data accumulation.

5 Artificial Intelligence for Network Effect

Google uses artificial intelligence solutions for the Google Ads platform. Google targets high performance in ads. Thus, it provides benefits for both internet users and advertisers in the double-sided market. Internet users want to see and click advertisements tailored to their needs. Search results can thus be distinguished by sponsored and organic [10]. The advertiser wants to reach internet users interested in the ads it publishes and click on those advertisements. Here, there is a win-win situation for both the internet user, the advertiser, and the Google Ads that brings the two parties together. Google Ads uses AI automation features that enable advertisers to target and deliver advertisements faster and easier. A Netflix documentary, Social Dilemma, captures the algorithm of AI used by Google succinctly. In the words of Tristan Harris, the Head of the Center for Humane Technology who was a Google design ethicist, the algorithm was designed to serve three main goals: "*There's the engagement goal: to drive up your usage, to keep you scrolling. There's the growth goal: to keep you coming back and inviting as many friends as possible and getting them to invite more friends. And then there's the advertising goal: to make sure that, as all that's happening, we're making as much money as possible from advertising.*"

These algorithms require as much data as possible to make the interactions of the three goals work together well. Through automation enabled by the AI, advertisers spend less effort to analyze data on the advertising platform provided to them. Advertisers use the information to create advertisements to develop more accurate business strategies. Google Ads AI helps advertisers with analysis, bidding, testing, reporting, following rules, and determining keywords. The automated analysis helps advertisers identify keywords with high conversion rates. Automatic bidding allows the advertiser to try more campaigns to better score while keeping current data and settings. The automated ad test makes it easy for large businesses that publish many advertisements to test their ads. Automatic reporting allows advertisers to monitor performance criteria that they can follow effortlessly. Advertisers can customize reports and view them according to their needs: this way, they can make decisions more easily and quickly. Advanced rules give advertisers the option to automate account settings and preset according to specific rules. This way, advertisers don't have to check metrics for each setting. The negative keyword selection feature enables the advertiser to automatically determine the negative keywords in their ad campaigns [57]. Consider the strategy of "growth hacking" among the three AI initiatives mentioned in previous paragraph. Google can collect data to reach users beyond what an advertiser original intention to serve. The more Google collects, and the more their AI can learn from the data it collects, and the more valuable it will become for its users. The more worthy the platform becomes for internet users; the more users prefer that platform. As the count of users increases, the size of data will increase, and as the number of data increases, the platform's learning ability will get stronger. This snowballing effect is more than the original network effect anticipated in the two-sided market [11, 12, 24].

The snowballing effect can grow so much that internet users rely more on Google's organic results driven by AI than on advertising results. Indeed, we observe cases where users trust in Google ads have been reduced because they encounter advertisements that they do not need to see. Yet, by using artificial intelligence, it can be ensured that the ads will fulfill success in terms of good performance in matrix if the client is willing to continue experiment with that methodology Thus, through organic growth, Google ads can hope to regain lost user trust [7]. Google Search Ads are text ads that are displayed as results on the search results page of Google search engine, depending on the keywords searched by internet users. Search ads are in various places on the Google search results page, Google maps, and Google shopping. Google prices its search ads based on the cost-per-click (CPC), related to the keywords, like an auction method. With the power of artificial intelligence, Google can prevent the same ad from being displayed again if you turn off the ad once. Thus, the insensitivity of internet users to Google ads will be reduced. In time, users will realize that when Google shows them an advertisement, it is an advertisement covering their needs. In this way, users will be more interested in Google ads than they currently are. It will increase the rate of ad clicking. The macro trend of how people obtain information is clear: more people are moving away from television to online video channels and streaming. Internet users use online videos for information, entertainment, education, etc. YouTube already has a big audience as a media channel, and YouTube ads are a part of Google ads revenue model [33]. Google video ads vary in length, and some of the ads take even longer than a minute. It means a reasonably long ad time for an internet user. Google sometimes requires that all of these long advertisements be watched until the end of it to show the page content to the internet user. In other words, these ads that last longer than one minute are compulsory to watch. Some Internet users prefer to leave the page instead of watching these advertisements. Google can detect those who leave the page with artificial intelligence. If a user has left the page after seeing an ad for more than a minute, Google should not show the same user the same length as a second time. It will increase users' interest in Google video ads and therefore increase the rate of ad clicks.

Irrespective of various positive intermittent reinforcement that Google adopts, the accumulation of data is an unquestionable advantage that if AI becomes the future of advertising, the business can be monopolized by monopolizing data. Indeed, the collection of data can be indiscriminating in that all data are relevant. Google assistant, by arrangement with cell phone manufactures, can be pre-installed on devices and record activities of the user without the users being aware of it. Youtube, owned by Google, bombards viewers with video ads mentioned in previous paragraph. Both Google assistant and Youtube ads can rob and interrupt users of various devices with or without viewers utilizing their free platforms. The experience of how viewing a free Youtube being interrupted by ads is common for many people. The distraction is a nuisance, obviously, but most people will accept that as the "price to pay" for using something "free". Yet, the AI of Google is monitoring every reaction to an ad, and in addition, possibly registering uninvited probing of private data by Google Assistant. A scene from Social Dilemma describes the interactive process by a "Yellow AI" in a vivid way: [yellow AI] narrates in the film: He's primed for an ad. Auction time. Sold! To Deep Fade hair wax. We had 468 interested bidders. We sold Ben at 3.262 cents for an impression.

6 Google Partners Program of Google Ads

Google allows Google Ads to be managed by third-party companies for brands, and for this, there is a marketing program called Google Partner. Third-party companies enrolled in this program can manage their Google Ads accounts of the brands. Google helps Google Partner improve itself in the field of ad management by organizing special training and events for companies that are Google Partners. It also offers Google Ads promotional offers to partners. Google has divided the partnership program into three groups: members, Google partners, and premier Google partners. Being a Google partner supposedly has many advantages. For example. Google offers product training, certifications, sales training, community membership, product updates, phone, chat, email support, promotional offers to all three groups. For partners and premier partners, Google provides badge assets, client event support, agency competitions. For premier partners, Google offers dedicated account management, new business strategy consulting, and annual partners summit participation. The advertising giant has plans for building this coalition, including the requirements for partner badge status and the way certificates are included in the badge status in the future. However, its plan might not satisfy many concerns from individual cases of this coalition. An example of this concern is that Google released an update to the requirements of the Google Partners program in early 2020. With this update, Google increased the minimum amount of advertising that the partner must make to be qualified, specifically, in the last 90 days of their business, from 10 thousand U.S. dollars to 20 thousand U.S. dollars [16]. Academy for ads certifications lost their validity, so everyone must take the skill shop exam. These will be more comprehensive tests that include Google ads search, Google ads display, Google ads video and shopping ads, which take more time to prepare than the previous exams. Another update is in the field of performance, and with this update, partner performance is to be gathered around with recommendations made by the Google advertising platform [30]. These requirements can always be stated in the name of enhancing quality. However, unlike other professions such as medical doctors or lawyers, it is not entirely clear that more stringent requirements are good for advertisers generally.

Some Google partners outright believe that Google's performance-enhancing recommendations are not effective. The recommendations do not make sense to Google partners, nor do they correlate with the goals of the advertisers they serve. For example, Google may suggest that advertisers in regulated industries use keywords that are not directly related to the business, and that will cause legal problems if applied [19]. These recommendations aim to increase ad spend and the adoption of new features. The imposition of these new updates and suggestions is, therefore, criticized by advertisers. Google partners are responsible for keeping the ad performance high on behalf of the customers of whose ad accounts they manage [25, 26, 58]. They try Google's suggestions to maintain ads' performance, but they also try many ways. With all these efforts, they find how they get the best efficiency from advertisements and apply them. However, Google uses OptiScore to measure ad performance. One of the issues that

Google partners are discussing is the accuracy of the scoring methodology. They allege that even though OptiScore may show a low score, there are ads that have performed very well. Some partners have also accused Google of misleading the ad performance of partners, therefore, misleading advertisers too, for the purpose to increase its own advertising revenue only.

Waves of defiance are now popping up everywhere. For example, Store Growers, an advertising agency and a Google partner for a long time, proclaimed openly on its web page why they are "no longer Google partner" and why other partners should leave the coalition also [40]. In an All Tech Considered radio show of NPR in 2019, host Audie Cornish interviewed Shannon Bond, an NPR Tech correspondent, mentioning software company Basecamp and Edible Arrangements expressing complaints about how their businesses have been handicapped, [11]. Even Nexstar Media Group Inc., the largest local news company in the U.S., has issues with Google's advertising business model [28, 29]. Perhaps we are only seeing the tip of the iceberg in a potential massive class action. It is not clear how Google will manage such unrests, although it did announce in response that it will use a new weight of 70% to evaluate the performance of their ad publications. [52], an intrinsic problem regarding how the coalition can be (or should be) effectively organized remains to be a big unknown. Our opinion is that unless an acknowledgement of how total welfare can be improved via the two-sided market described earlier in this paper, particularly in regard to how specifically Google AI aims to pull buyers and sellers together in a focused manner, rather than gathering data for the sake of gathering them, for the sole purpose to increase power for its AI robots, we do not believe minor adjustments in requirements will solve the intrinsic problem of satisfying the social contract that Google can unilaterally provide for the society as a whole.

7 Antitrust Lawsuits

Two links enlist the filing of complaints in legal format: (1) https://www.courtlistener.com/recap/gov.uscourts.txed.202878/gov.uscourts.txed.202878.1.0.pdf and (2) https://www.courtlistener.com/recap/gov.uscourts.txed.202878/gov.uscourts.txed.202878.1.0.pdf. Google operates in a highly competitive search engine market. Approximately 90% of internet searches are made through Google. Microsoft's Bing's market share, Google's closest competitor, is not even 3%. The leading position of Google is followed by Yahoo!, Baidu and Yandex. The shares of other search engines in the market, excluding these companies, are tiny [8]. Google Search claims to offer this service free of charge. One of the controversial issues regarding competition in the search engine market is that Google's privacy policy allows the company to cross-use the mass of big data it obtains from websites, applications, operating systems across different markets. Thus, Google is aggressively transforming data from one market into a cluster of other markets. This potentially can handicap competitors. This also means that the company poses obstacles to market entry, and these obstacles create a monopoly position that will be increasingly difficult for competitors to catch up [51]. The issue is related to Google's attempt to hinder its competitors' data acquisition because of its agreements with major companies. Google is provided as the default browser for internet users in Safari, Apple's web browser. There is an agreement between Google and Apple, and

Google makes payments to Apple every year as per the contract. Besides, Google is offered to users of the devices with the Android operating system preinstalled. Google also encourages companies like Verizon, AT&T, LG, to use the Google search engine through its agreements [9, 10]. There are allegations that Google uses anticompetitive practices to maintain its dominant position in the internet search and related advertising market. It is by monopolizing data that Google can protects it monopoly in the search engine and the advertising space.

The government has announced that it has stepped in to protect access to a free market for customers and Google's competitors [38]. A Justice Department spokesman explained that the government is trying to enforce Sherman Antitrust Laws to ensure competition, to make Google to keep their original promise in their mission of "doing the right thing". U.S. Department of Justice and 11 states initiated an investigation that continued for 14 months, and eventually in October 2020, filing a lawsuit against Google in a federal court [22]. It is anticipated that federal and state officials investigating how Google uses its power in the internet economy are preparing to open series of lawsuits. In the first case filed in Oct. 2020, the company was accused of making payments to maintain its leadership in the search engine area in cooperation with other companies, using anticompetitive tactics to protect and expand its monopoly, using the income it earned from its position as a monopolist in the search engine for these payments. Google's practices are anticompetitive under deep-rooted antitrust law [35, 36]. The alleged practices are hurting American consumers, advertisers, rival technology companies. The source of monopoly is not only based on pricing based on inelastic demand—economist's textbook version of monopoly. Cooperation with other companies will effectuate and magnify the search engine's efficiency because the efficiency of search results is affected by the monopolization of data. This efficiency prepares the ground for advertisers to be forced to choose Google [45].

At the time of writing of this article, another antitrust case has been filed by 10 state attorneys against Google specifically aiming at its ad technology by operating the largest electronically advertising exchange platform, ADX. The lawsuit alleges Google uses its market power to "extract a very high tax of the ad dollars otherwise flowing to the countless online publishers and content producers like online newspapers, cooking websites, and blogs who survive by selling advertisements on their websites and apps." (Dec. 16, 2020). Considering also the alleged advertising exchange practices described in [29], we cannot comfortably assure that Google has utilized its AI technology in a fixed rate market competition in the two-sided market bringing buyers and sellers together. The Google Antitrust cases are a continuation of political tensions between the U.S. government and major tech companies in USA. In 1998, a massive antitrust lawsuit was opened against Microsoft's tech company for making the Windows operating system a dominant and almost impossible default to be deleted. The Google lawsuit appears to be the second-largest antitrust lawsuit ever opened. The two cases filed so far focus on Google's monopoly in search engine domains and its advertising business model. Because Google uses its dominance to dominate areas such as advertising, smartphone operating system, web browsing. BASE. Google's search engine accounted for more than half of its total revenue, with about $21 billion per quarter. These series of cases against Google could pose an existential threat to the company (Razzaque & Hamdan, 2020).

In developed countries and the United States, almost all industries are not perfectly competitive or monopolies. However, most of the industries are positioned in between, either monopolistic competition or oligopoly [8]. Google holds over 90% of the United States search engine market and over 95% globally [41]. Only in China, South Korea and the Czech Republic are lower [20]. We do not believe addressing to concentration ratio is the significance of the Antitrust cases however, even though an above 90% indeed by conventional concept of monopoly will be considered to be too high; thus, the industry is highly concentrated [31, 32]. However, even if the relevant market for search engines is the world, market shares like concentrate ratios can be a misleading indicator of monopoly power [9]. Certainly, we expect country-specific, local, and regional methods of advertising such that worldwide market shares of advertising would be even less informative. The approach we suggest in this paper is not to ask the question of "market concentration", which could be a performance indicator of success, popularity, and economic efficiency in the form of scale economies, but to inquire into the monopolization of data, the method of getting them via AI, and to query whether "excessive" data collected were necessary for enabling the efficiency of a two-sided market to gain advantages in search engine, to gain advantages in advertising exchanges. That, in our opinion, is the crux of the question that Antitrust inquiries should focus in addressing, and not be confused with the result of the structural concept of market, which is a legitimate winning goal of dynamic competition. The variant of potential competition indeed will make ecosystems as a more relevant focus of analysis. The real question is whether Google has handicapped others in competition for the evolvement of ecosystems.

8 Conclusion

Google is one of the five largest technology companies in the U.S. In addition to Google's search engine, it has products and services in many different areas such as advertising, online streaming, artificial intelligence, and assistant services. There is controversy in the business world that the company has overgrown due to the vastness of Google's product range. Google Search is the most powerful search engine globally, and with a market share of over 90%, it is the market leader without exception. Although the search engine market structure is potential competition, and not a structural approach ingrained in the Industrial Organization tradition in economics, there are features of data monopolization in the case of Google that makes it different from earlier case of potential competition characterized by Microsoft in the 80s and 90s [57]. The demand structure of the technology ecosystem is evolving and increasingly becoming lock-in with some type of rigidity. Google has made its search engine power the company's primary source of revenue through advertising publishing. Google Search has a large percentage of the search. It uses artificial intelligence resources to expand on the advertising platform, bringing together the two-sided market of internet users and advertisers. Therefore, it is the AI strategy on big data collection by Google that can be questioned, and rightly, should be subjected to Antitrust scrutiny. Regarding the issue of monopolizing advertising, the reliability of any advertising program/platform for internet users/advertisers starts with a subjective guess, plans that someone thinks may work on others, for psychological or conspicuous consumption reasons. It is much weaker than the universality of a search engine, because users' interest level of passive-feed-form could be low with the advertisements competing for attentions, whereas a active-self-driven search is more focused and thus result more effective. Also, personalization by AI is limited. Advertising is a skill not comparable to professions such as legal practice and medical physicians. An over-requirement for qualification can be a danger. Google needs to harness the power of artificial intelligence and thus regain the lost trust of internet users. Despite updates Google Partners made in 2020 on the terms of the program and the scoring system of advertising efficiency, they are open to criticism. Google has also been accused of misleading business partners and advertisers about ad performance to increase ad revenue. Google will be facing many lawsuits in the years to come. The outcome of these lawsuits can directly affect both the search engine and the online advertising market, and looking at the bottom line, the company's well-being. This paper suggests a framework to evaluate how Google may want to have a more focused AI strategy for delivering a social contract that will be total welfare maximizing net of costs.

Acknowledgement. We thank Robert Donovan of DOM360 for motiving this study.

References

1. Al-Hashimi, M., Hamdan, A., Razzaque, A., Al-Sartawi, A., Reyad, S.: Skill Gaps in Management Information Systems Alumni. In: Hassanien, A.-E., Azar, A.T., Gaber, T., Oliva, D., Tolba, F.M. (eds.) AICV 2020. AISC, vol. 1153, pp. 773–782. Springer, Cham (2020). https://doi.org/10.1007/978-3-030-44289-7_72
2. Al-Sartawi, A.: Social media disclosure of intellectual capital and firm value. Int. J. Learn. Intell. Capital **17**(4), 312–323 (2020)
3. Al-Sartawi, A.: Does it pay to be socially responsible? empirical evidence from the GCC countries. Int. J. Law Manage. **62**(5), 381–394 (2020)
4. Al-Sartawi, A.: Information technology governance and cybersecurity at the board level. Int. J. Crit. Infrastruct. **16**(2), 150–161 (2020)
5. Al-Sartawi, A.: Assessing the relationship between information transparency through social media disclosure and firm value. Manage. Account. Rev. **18**(2), 1–20 (2019)
6. Al-Sartawi, A. (ed).:The big data-driven digital economy: artificial and computational intelligence In: Kacprzyk, J., (ed) Studies in Computational Intelligence, vol. 974. Springer International Publishing (2021)
7. Bartz, D., Paresh, D.: U.S. States Prepare Second Antitrust Lawsuit Against Google for December (2020). https://www.reuters.com/article/us-tech-antitrust-google-idUSKBN2832VR
8. Baum, D.: Google Ads Review 2020: A Brutally Honest Pros and Cons List. Impact Plus (2020). https://www.impactplus.com/blog/google-ads-review
9. Baye, M.R., Prince, J.T.: Managerial Economics and Business Strategy. McGraw-Hill/Irwin (2014)
10. Berr, J.: Google, Facebook, Reportedly Facing Multiple Antitrust Lawsuits (2020). https://www.forbes.com/sites/jonathanberr/2020/11/30/google-facebook-reportedly-facing-multiple-antitrust-lawsuits

11. Berry, S.: 2020 Search Market Share: 5 Hard Truths About Today's Market (2020). https://www.webfx.com/blog/seo/2019-search-market-share
12. Bond, S.: Google's Response to Antitrust Accusations (2019). https://www.npr.org/2019/10/14/770133480/googles-response-to-antitrust-accusations
13. Burguet, R., Caminal, R., Ellen, M.: In google we trust? Int. J. Ind. Organ. **39**, 44–55 (2015)
14. Clement, J.: Number of Full-time Alphabet Employees From 2007 to 2019 (2020). https://www.statista.com/statistics/273744/number-of-full-time-google-employees
15. Colander, D.: Microeconomics. McGraw-Hill Higher Education (2020)
16. Colt, J.: Google's Competitive Advantage and Strategy (2020). https://toughnickel.com/industries/Googles-Competitive-Advantage-Strategy
17. DeFazio, A.: Google Partner Badge: The New Timeline and What Advertisers Need to Know (2020). https://www.coywolf.news/paid/google-ads-partner-badge-update
18. Elgin, B.: Google: Whiz Kids or Naughty Boys? (2004). https://www.webcitation.org/5hT7Hw63l?url=http://www.businessweek.com/technology/content/aug2004/tc20040819_6843_tc120.htm
19. Elias, J.: Google Employees are Complaining the Company has Changed — This Chart Shows One Reason Why (2020). https://www.cnbc.com/2020/01/02/google-employee-growth-2001-through-2019.html
20. Finn, G.: What Google Partners Changes Mean for Agencies & Clients (2020). https://www.searchenginejournal.com/what-google-partners-changes-mean-for-agencies-clients-podcast/350014/#close
21. Fishkin, R.: As the Antitrust Case Against Google Kicks Off, Here's Where the DOJ Should Start (2019). https://sparktoro.com/blog/as-the-antitrust-case-against-google-kicks-off-heres-where-the-doj-should-start
22. Fry, M.: Google working to reach minority-owned businesses. NJBIZ, 28(31), 8 (2015)
23. Ghaffary, S., Molla R.: Why the US Government is Suing Google (2020). https://www.vox.com/recode/21524710/google-antitrust-lawsuit-doj-search-trump-bill-barr
24. Gilbert, R.: The role of potential competition in industrial organization. J. Econ. Perspect. **3**(3), 107–127 (1989). http://www.jstor.org/stable/1942764
25. Gregory, R.W., Henfridsson, O., Kaganer, E., Kyriakou, H.: The role of artificial intelligence and data network effects for creating user value. Acad. Manage. Rev. (2020). https://doi.org/10.5465/amr.2019.0178
26. Gupta, M., Sikarwar, T.S.: Modelling credit risk management and bank's profitability. Int. J. Electron. Bank. **2**(2), 170–183 (2020)
27. Gupta, N.: Influence of demographic variables on synchronisation between customer satisfaction and retail banking channels for customers' of public sector banks of India. Int. J. Electron. Bank. **1**(3), 206–219 (2019)
28. Guttmann, A.: Google's Global Brand Value From 2006 to 2020 (2020). https://www.statista.com/statistics/326046/google-brand-value
29. Haucap, J., Heimeshoff, U.: Google, facebook, amazon, ebay: is the internet driving competition or market monopolization? Int. Econ. Econ. Policy **11**(1), 49–61 (2014)
30. Hagey, K., Ngo, V.: How google edged out rivals and built the world's dominant Ad machine: a visual guide. Wall Street J. 7 (2019)
31. Henshaw, J.: Google Ads is Forcing Agencies to Use Its Recommendations or Risk Losing Their Google Partners Badge (2020). https://www.coywolf.news/paid/google-ads-partner-badge-controversy
32. Höffler, F.: Monopoly prices versus Ramsey-Boiteux prices: are they similar, and: does it matter?, Preprints of the Max Planck Institute for Research on Collective Goods, No. 2005,7, Max Planck Institute for Research on Collective Goods, Bonn (2004)
33. Jan, O.: Concentration Ratio (2019). https://xplaind.com/341367/concentration-ratio

34. Kuligowski, K.: YouTube for Business: Everything You Need to Know (2020). https://www.businessnewsdaily.com/9854-youtube-for-business.html
35. Li, H.: Special section introduction: artificial intelligence and advertising. J. Advert. **48**(4), 333–337 (2019)
36. Lorne, F.: Data in Cities and Digital Platforms: Some Implications for Digital Social Contracts with AI Robots. April 15, 2019 Mimeo (2019)
37. Lohr, S.: What Is Happening with the Antitrust Suit Against Google? (2020). https://www.nytimes.com/2020/10/20/technology/antitrust-google.html
38. Luchetta, G.: Is the google platform a two-sided market? J. Competition Law Econ. **10**(1), 185–207 (2013)
39. McKinnon, J.D.: Facebook, Google to Face New Antitrust Suits in U.S (2020). https://www.wsj.com/articles/facebook-google-to-face-new-antitrust-suits-in-u-s-11606742163
40. Memdani, L.: Demonetisation: a move towards cashless economy in India. Int. J. Electron. Bank. **2**(3), 205–211 (2020)
41. Moons, D.: Why I'm No Longer a Google Partner (And Why You Shouldn't Either) (2020). https://www.storegrowers.com/google-partner/
42. Mroke, P.: The Google Monopoly Over Internet Search: A Visually Compelling Analysis (2020). https://www.clearboxseo.com/google-monopoly-internet-search-rand-fishkin-graph-analysis
43. Musleh Al-Sartawi, A.M.A.: E-learning improves accounting education: case of the higher education sector of bahrain. In: Themistocleous, M., Papadaki, M., Kamal, M.M. (eds.) EMCIS 2020. LNBIP, vol. 402, pp. 301–315. Springer, Cham (2020). https://doi.org/10.1007/978-3-030-63396-7_20
44. Naughton, E.: Making Progress on Diversity and Inclusion (2017). https://blog.google/outreach-initiatives/diversity/making-progress-diversity-and-inclusion
45. Omondi, S.: What are the Four Big Tech Companies in the US? (2019). https://www.worldatlas.com/articles/what-are-the-big-four-tech-companies-in-the-us.html
46. Paul, K.: Google is Facing the Biggest Antitrust Case in a Generation. What Could Happen? (2020). https://www.theguardian.com/technology/2020/oct/21/google-antitrust-charges-what-is-next
47. Razzaque, A., Eldabi, T., Chen, W.: Quality decisions from physicians' shared knowledge in virtual communities. Knowl. Manage. Res. Pract. 1-13 (2020). https://doi.org/10.1080/14778238.2020.1788428
48. Razzaque, A., Hamdan, A.: Internet of things for learning styles and learning outcomes improve e-learning: a review of literature. In: Hassanien, A.-E., Azar, A.T., Gaber, T., Oliva, D., Tolba, F.M. (eds.) AICV 2020. AISC, vol. 1153, pp. 783–791. Springer, Cham (2020). https://doi.org/10.1007/978-3-030-44289-7_73
49. Razzaque, A., Hamdan, A.: Role of financial technology fintech: a survey. In: Hassanien, A.-E., Azar, A.T., Gaber, T., Oliva, D., Tolba, F.M. (eds.) AICV 2020. AISC, vol. 1153, pp. 112–117. Springer, Cham (2020). https://doi.org/10.1007/978-3-030-44289-7_11
50. Rochet, J.-C., Tirole, J.: Platform competition in two-sided markets. J. Euro. Econ. Assoc. **1**(4), 990–1029 (2003)
51. Rosenberg, E.: How Google Makes Money (GOOG) (2020). https://www.investopedia.com/articles/investing/020515/business-google.asp
52. Ryan, J.: Formal GDPR Complaint Against Google's Internal Data Free-for-all (2020). https://brave.com/google-internal-data-free-for-all
53. Schwartz, B.: Google Tries to Clarify New Google Partner Program Performance Requirements (2020). https://www.seroundtable.com/google-clarifies-new-partner-program-performance-requirements-29018.html
54. Swant, M.: The World's Most Valuable Brands (2020). https://www.forbes.com/the-worlds-most-valuable-brands/#747736a3119c

55. Tucker, C., Zhang, J.: Growing two-sided networks by advertising the user base: a field experiment. Mark. Sci. **29**(5), 805–814 (2010)
56. Webb, C.L.: Google's IPO: Grate Expectations (2020). https://www.washingtonpost.com/wp-dyn/articles/A14939-2004Aug19.html
57. Yu, B.T.: Potential competition and contracting in innovation. J. Law Econ. **24**(2), 3 (1981)
58. Yurieff, K.: Google Struggles to Increase Workforce Diversity (2017). https://money.cnn.com/2017/06/29/technology/google-diversity-report/index.html

Social Commerce: A Concept Explication

Maryam Husain Almahdi(✉)

Ahlia University, Manama, Bahrain

Abstract. Attention to social commerce has been growing in research and practice in recent years. Many industry articles and academic papers have been dedicated to defining the concept, outlining its components, and highlighting its potential, producing varying and often conflicting findings. This literature review paper contributes to this conversation by introducing a framework of three core themes of social commerce (i.e., social, commercial, and technological). The framework is utilized to offer a tentative definition of social commerce and shed light on what distinguishes it from two closely related concepts: social media and e-commerce. The paper concludes with highlighting persisting gaps in the conceptualization of social commerce and offers recommendations for future research to further our understanding of the concept.

Keywords: Social commerce · Social media · E-commerce · Digital marketing

1 Introduction

Social Commerce (SC) is defined in the literature as "the delivery of e-commerce activities and transactions via the social media environment" [1, p. 6]. SC is a growing phenomenon and novel area of research [2–4]. It is expected to redefine the consumer shopping experience and assume an increasingly important position among competing shopping channels [5, 6]. Indeed, according to eMarketer, social commerce sales in the US are expected to climb to 36.09 billion US dollars in 2021, growing by 34.8% from the previous year [7].

Some SC tools and mechanisms have been around since the 1990s [8], but the term (as it is known today) was introduced in late 2005 in a blog post by Beach and Gupta. The two former Yahoo! Shopping managers used the term 'social commerce' to describe their marketplace's newly added shopping features that facilitate consumer participation and content creation, an example of which is shopping lists that customers can create and share. However, it is only recently that SC has started to attract the attention of marketers and academics [9], particularly because of the growth in popularity and influence of social media technologies [3, 10]. Indeed, social media is taking an increasing role in the daily routines of millions of users, guiding their buying and shopping preferences, along with other aspects of their lives [11].

SC is shaping up to become "one of the greatest opportunities—and challenges—the retail industry has ever faced" [11, p.2]. It, consequently, represents a thriving research area [1, 10], as scholars endeavor to understand the consumers' interactivity with the

A. M. A. Musleh Al-Sartawi et al. (Eds.): EAMMIS 2021, LNNS 239, pp. 398–404, 2021.
https://doi.org/10.1007/978-3-030-77246-8_37

SC platforms as well as with other consumers using them [12]. However, despite its promise, social commerce research is still in its early stages [4], as a consensus is yet to be reached on a precise meaning of the concept, what it involves, and how it differs from related concepts such as e-commerce and social media [13–17].

These limitations in delineating the essence and boundaries of SC make up the main research gap this paper is addressing. It is important to address this gap because limitations in understanding the concept of social commerce translate to limitations in defining the empirical settings in social commerce research. Specifically, research contexts in the SC literature are often chosen with little justification, as researchers seem to take liberties in what type of platforms they regard as SC. These conceptualization shortcomings similarly bleed into limitations in practice that render practitioners uncertain about the potential of SC as a marketing tool or how capitalize on it to understand and satisfy their customers [2, 6, 13, 15, 18].

To address these conceptual shortcomings, a framework of three core themes (i.e., social, commercial, and technological) is proposed in this paper to reconcile the past definitions of SC. The framework is additionally utilized to distinguish social commerce from three related concepts, namely: social media, electronic commerce (e-commerce), and social shopping. This is followed by introducing a tentative definition of social commerce as being the result of the convergence of social media and e-commerce technologies. Finally, gaps in the extant conceptualizations of SC are discussed and recommendations for future research are provided.

2 Three Core Themes of Social Commerce

A close examination of SC definitions in past academic literature [1, 2, 10, 13–15, 19, 20] reveals three recurring themes, namely, social interactions, commercial activities, and technological infrastructures. Indeed, a key theme of SC in prior research is the support of the consumers' social interactions, with an emphasis on their communication and relationships with others in the online environment [1, 13, 21]. A second major theme covers the customers' commercial activities throughout the different stages of their online shopping journey [13, 14, 20]. Finally, the literature sheds light on the technological infrastructure essential in facilitating the social and commercial activities, specifically in terms of being built on and enabled by the Internet and its interactive mechanisms [1, 2, 20]. In the context of SC, the social, commercial, and technological themes are closely interrelated. Researchers explain that the interactive capabilities of the online environment not only support the consumers' online shopping and buying activities [22, 23]; they further facilitate their social interactions as they transform from passive audiences to active communicators [24]. The consumers' online social interactions, an example of which is seeking product recommendations from their online social network, can in turn influence their shopping experiences and buying decisions in SC [5, 13]. These experiences and decisions, whether positive or negative, will consequently guide the consumers' own opinions and reviews that they might share with others online [2, 16, 21], thus repeating the cycle.

In light of this discussion, SC can be described as the fruit of the interrelation between the technological infrastructures, social interactions and commercial activities.

Interestingly, examining the overlapping areas of each two themes presents a useful tool for understanding three concepts closely related to the study of SC, namely, social media, e-commerce, and social shopping. Specifically, social media is observed in the overlap between the technological and social themes, while e-commerce is reflected in the overlap between the commercial and technological themes. Similarly, social shopping is the result of the overlap between the social and commercial themes. An overview of each of these concepts and how they relate to SC is discussed next.

Social Media

Social media is the result of the overlap between the technological and social themes in the framework [25]. It is, therefore, defined as a range of online platforms and applications that facilitate their users' social activities [24]. These social activities, according to Fuchs [25], are manifest in communication, collaboration, and community building. These activities can be observed in the variety of social media platforms available today, from social networking sites (e.g., Facebook, Instagram), ephemeral messaging apps (e.g., Snapchat), video-sharing apps (e.g., TikTok, YouTube), collaborative communities (e.g., Wikipedia), and even virtual gaming worlds (e.g., Second Life).

It is interesting to note that, while being distinctive for enabling social interactions between its users (i.e., interpersonal communication or human-to-human interactivity), social media is equally able to facilitate one-to-many mass communication models (e.g. in the form of online adverts) [25]. Additionally, social media inherently incorporates human-to-website interactive tools and mechanisms (e.g. links, search, and customization) with the aim of engaging its consumers and enhancing their navigation experiences [25].

Electronic Commerce

Conducting commercial exchanges online are other words to describe e-commerce [22]. Some scholars suggest that for an e-commerce activity to be considered as such, a full financial transaction has to have transpired online [21]. However, other researchers consider any and all of the transactional and non-transactional activities related to the consumers' online shopping experiences a part of e-commerce [22].

For the purposes of presenting a more inclusive view of the term, and in line with the general direction in SC research [14], the researcher follows the second path of using e-commerce to involve to "any kind of [online] activity that leads to commercial benefits" [1, p.7]. Accordingly, the concept of e-commerce includes online presences that support commercial exchanges occurring between businesses, between consumers, between businesses and consumers, and between other types of organizations (e.g. governments) and their consumers [21, 22]. Examples are consumer-to-consumer marketplaces (e.g., Etsy), business-to-consumer marketplaces (e.g., Amazon), brand websites (e.g., Dell.com), news websites (e.g., BBC), online banking services (e.g., TSB.com), and online telecommunication services (e.g., O2.com). Notably, while it often enables basic human-to-human interactivity (e.g. reviews, questions), traditional e-commerce largely facilitate human-to-website interactivity and mass communication models [23].

Social Shopping

The combination of commercial activities and social interactions will result in social shopping. Indeed, shopping in its core is a social experience [6], and social shopping can be understood as the social activity that consists of shopping with other people, whether in the real world or online [23]. It is worth highlighting that some authors use the terms 'social shopping' and 'social commerce' interchangeably [26], while others describe its online form (i.e., online shopping with other people) as a subset of SC [27].

3 SC as the Result of the Convergence Between Social Media and E-commerce Technologies

The discussion, thus far, has highlighted the three main themes of SC and how they join forces to shape the concept of SC. Moreover, the three-theme framework has been employed to shed light on concepts of particular importance to the understanding of SC, including social media and e-commerce. According to the framework, social media is represented in the overlap between the technological and social themes, while e-commerce is the result of the overlap between the technological and commercial themes.

It is then reasonable to suggest that both social media and e-commerce can transform into SC if they utilize the appropriate interactive tools to facilitate their consumers' commercial and social interactions, respectively, thus encapsulating all three themes of the framework. Indeed, social media is often viewed as a type of SC [1–3, 6]; particularly when its consumers' social interactions influence the different stages of their buying decision process [2, 13, 18, 28]. This can involve connecting with and purchasing from firms via their social media, carrying out buying and selling activities with other consumers on social platforms, and sharing opinions about products in social communities [14–16]. Similarly, SC is closely related to e-commerce, and some definitions go as far as describing the first as a development of the latter [16]. Indeed, the two are similar in respect to facilitating their consumers' online shopping and buying activities but differ in regard to allowing their social interactions [20, 27]. As discussed earlier, e-commerce does not typically incorporate social experiences to their customers' shopping journeys [1, 16]. However, when they do (e.g., by adding communication, collaboration, and relationship-building tools), they become a type of SC [6, 14, 15, 20, 23].

Based on the previous discussion, SC can be described in terms of the convergence between social media and e-commerce technologies and activities. This is further evident in the fact that SC can support both human-to-human and human-to-website interactivity. Indeed, SC is created when an online platform is made up of interactive mechanisms that facilitate both the consumers' social interactions and shopping and buying activities along the different stages of their buying decision process [1, 13, 14].

4 Limitations in Conceptualizing SC

The versatility of SC as a concept comes with its own challenges, especially in respect of narrowing it down, categorizing it, and drawing a sure line between SC and e-commerce on the one hand, and between SC and social media on the other hand. Indeed, e-commerce websites have capitalized on the sociality of their consumers by enabling reviews and social sharing for many years now, and social platforms been created around brands and products for quite some time (e.g., Facebook commerce) [8, 10, 15, 19, 25]. This poses the question of whether SC has existed all along [10], or if the concept should be reserved to describe platforms that harbor certain levels of sociability and interactivity? For example, while Amazon enables social interactions in the form of customer reviews and questions, it offers very limited personal profiles and no built-in direct messaging options. In contrast, the users on Etsy can create detailed and vivid profiles that include information about their followers, favorite items, groups, and collections in addition to liking and direct messaging options.

Along the same lines are shortcomings relating the categorizations of SC platforms presented in the literature, which are for the most part arbitrary, contradictory, and unscientific. A rare example of an empirical classification of SC was presented by Saundage and Lee [29], who based it on a study of a mere 15 websites and presented two simple categories; pre- and post-transactional SC. Evidently, these two categories do not offer much information on SC types that can be utilized in future research or practice. A more detailed typology was presented in a conference paper by Almahdi et al. [30], but requires further statistical analyses to validate it. Therefore, a major shortcoming in the literature is the ambiguity concerning SC, particularly its precise meaning and its different types [13, 14, 16]. These difficulties in pinning down the concept are reflected in the wide range of empirical settings used in prior SC research, and consequently in the types of respondents and subjects used. SC studies are set in contexts as varied as social networking sites [18, 28], online shopping platforms [35], and websites identified as SC without a clear justification [19, 31]. Several other researchers resort to targeting general respondents who happen to be online shoppers or social media users [23]. Consequently, when the term 'SC' is used in empirical research, it can be referring to one of many concepts, presenting a problem that could affect the reliability of research outcomes [17].

The conceptualization shortcomings are additionally translated into a real-life problem which finds the potential of SC unclear, with marketers uncertain of the future success of their SC investments [2, 13, 16, 18]. On the one hand, a direction of investigation maintains that SC is a promising business model and growing phenomenon [2, 16], and both reports and predicts a note-worthy increase in use and sales [7]. On the other hand, some warn of overhyping the concept [32], and point out reports of exaggerated financial performance [33]. These contradictions are possibly a result of the inconsistency in defining what SC means and the types of activities it involves. Indeed, when disclosing SC statistics, some experts might be referring to sales from actual purchases using social media buy-buttons [33], while others could be including the outcomes of other types of SC activities, such as word-of-mouth and referrals [16, 34].

This paper highlighted attempts of prior research at conceptualizing SC and contributed to the conversation with a three-theme framework (i.e., technological, social, commercial) to synthesize prior definitions of the concept. It discussed how SC relates to social media and e-commerce, and then defined the concept in terms of the convergence between the two, which facilitates different modes of communication. Several conceptual limitations that exist in the literature were stressed, with their implications highlighted.

Based on the discussion in this paper, it is recommended that future researchers continue empirically examining the concept of SC from different perspectives, especially as it is expected to evolve and change with the constant changes in technology and customer behaviors [3]. Additionally, a typology of different SC websites, one that is rooted in theory and conducted systematically, is much needed to shed a clearer light on the concept. Such research efforts will be beneficial in informing both research and practice in digital and social media marketing.

References

1. Liang, T.-P., Turban, E.: Introduction to the special issue social commerce: a research framework for social commerce. Int. J. Electron. Comm. **16**(2), 5–14 (2011)
2. Stephen, A.T., Toubia, O.: Deriving value from social commerce networks. J. Mark. Res. **47**(2), 215–228 (2010)
3. Llewellyn, G.: Social commerce 2021 trends: Tactics and tools to grow your e-commerce strategy through social media. 2020. https://www.smartinsights.com/ecommerce/ecommerce-strategy/social-commerce-trends/. Accessed 27 Jan 2021
4. Busalim, A.H., Ghabban, F.: Customer engagement behaviour on social commerce platforms: an empirical study. Technol. Soc. **64**, (2021)
5. Anderson, M., et al.: Turning like to buy social media emerges as a commerce channel. Booz Company Inc, **2**(1), 102–128 (2011)
6. Cecere, L., The rise of social commerce. Altimeter Res. Rep. (2010)
7. Lipsman, A.: Social Commerce 2021: Media and Commerce Convergence Creates Growth Opportunity for Brands 2021. https://www.emarketer.com/content/social-commerce-2021. Accessed 09 Mar 2021
8. Indvik, L.: The 7 Species of Social Commerce 2013. https://mashable.com/2013/05/10/social-commerce-definition/
9. Baethge, C., Klier, J., Klier, M.: Social commerce—state-of-the-art and future research directions. Electron. Mark. **26**(3), 269–290 (2016). https://doi.org/10.1007/s12525-016-0225-2
10. Lin, X., Li, Y., Wang, X.: Social commerce research: definition, research themes and the trends. Int. J. Inf. Manage/ **37**(3), 190–201 (2017)
11. Howard, R.: Let's Get Social: How Commerce Transform Retail 2016. https://www.linkedin.com/pulse/lets-get-social-how-commerce-transform-retail-robert-howard/. Accessed 9 Mar 2021
12. Wang, C., Zhang, P.: The evolution of social commerce: the people, management, technology, and information dimensions. Commun. Assoc. Inf. Syst. **31**(1), 5 (2012)
13. Yadav, M.S., et al.: Social commerce: a contingency framework for assessing marketing potential. J. Interact. Mark. **27**(4), 311–323 (2013)
14. Shen, J.: Social comparison, social presence, and enjoyment in the acceptance of social shopping websites. J. Electron. Comm. Res. **13**(3), 198–212 (2012)
15. Turban, E.J., Strauss, L.L.: Social Commerce. Springer, Heidelberg (2016)
16. Mullin, S. Social Commerce: What It Is, What It Isn't and Why You Should Care (2016). https://cxl.com/social-commerce/. Accessed 9 Mar 2021
17. Osatuyi, B., et al.: When it comes to Satisfaction… it depends: an empirical examination of social commerce users. Comput. Hum. Behav. **111**, (2020)

18. Li, X., Wang, C., Zhang, Y.: The dilemma of social commerce. Internet Res. (2020)
19. Curty, R.G., Zhang, P.: Social commerce: looking back and forward. Procee. Am. Soc. Inf. Sci. Technol. **48**(1), 1–10 (2011)
20. Huang, Z., Benyoucef, M.: From e-commerce to social commerce: a close look at design features. Electron. Comm. Res. Appl. **12**(4), 246–259 (2013)
21. Laudon, K.C.: E-Commerce 2020–2021: Business, Technology and Society, Global Edition. Upper Saddle River, NJ: Prentice-Hall (2020)
22. Turban, E., et al., Electronic commerce 2018: a managerial and social networks perspective, Springer, Heidelberg (2017)
23. Huang, Z., Benyoucef, M.: User preferences of social features on social commerce websites: an empirical study. Technol. Forecast. Soc. Change **95**, 57–72 (2015)
24. Kietzmann, J.H., et al.: Social media? Get serious! understanding the functional building blocks of social media. Bus. Horiz. **54**(3), 241–251 (2011)
25. Fuchs, C.: An alternative view of privacy on facebook. Information **2**(1), 140–165 (2011)
26. Wang, C.: Social shopping development and perspectives. Studies in Virtual Communities, Blogs, and Modern Social Networking: Measurements, Analysis, and Investigations: Measurements, Analysis, and Investigations, 97 (2013)
27. Marsden, P.: Social Commerce: Monetizing Social Media (2010). https://digitalwellbeing.org/downloads/White_Paper_Social_Commerce_EN.pdf. Accessed 11 Mar 2021
28. Leong, L.-Y., et al.: Predicting the antecedents of trust in social commerce–A hybrid structural equation modeling with neural network approach. J. Bus. Res. **110**, 24–40 (2020)
29. Saundage, D. Lee, C.Y.: Social commerce activities–a taxonomy. in ACIS 2011: Identifying the information systems discipline: In: Proceedings of the 22nd Australasian Conference on Information Systems. ACIS (2011)
30. Almahdi, M., Archer-Brown, C., Panteli, N.: Developing a Typology of Social Commerce Websites: An Exploratory Study, in Academy of Marketing Conference: The Magic in Marketing. Limerick, Ireland (2015)
31. Bugshan, H., Attar, R.W.: Social commerce information sharing and their impact on consumers. Technol. Forecast. Soc. Change **153**, (2020)
32. PYMENTS. RETAILSolving Social Commerce's Conversion Problem 2020. https://www.pymnts.com/news/retail/2020/solving-social-commerce-conversion-problem/. Accessed 10 Mar 2021
33. eMarketer. As a Direct Response Tool, Social Commerce Is Still Lacking 2016. https://www.emarketer.com/Article/Direct-Response-Tool-Social-Commerce-Still-Lacking/1014866. Accessed 09 Mar 2021
34. Amblee, N., Bui, T.: Harnessing the influence of social proof in online shopping: the effect of electronic word of mouth on sales of digital microproducts. Int. J. Electron. Comm. **16**(2), 91–114 (2011)
35. Curty, R.G., Zhang, P.: Website features that gave rise to social commerce: a historical analysis. Electron. Commer. Res. Appl. **12**(4), 260–279 (2013)

Role of Governance, Cybersecurity, and Privacy in AI and IoT

Financial Statements Fraud and Data Mining: A Review

Zakeya Sanad(✉) and Abdalmuttaleb Al-Sartawi

Ahlia University, Manama, Bahrain
Zsanad@ahlia.edu.bh

Abstract. Financial statement fraud has been a serious concern for many investors and other stakeholders, and this motivated many researchers in developing traditional regression analysis in order to detect financial statement fraud. Intelligent financial statements fraud detection systems were continuously developed to uncover any potential for fraud signals which could assist the stakeholders in taking their decisions. This study reviews and identifies the gaps in the previous studies that tested financial statement fraud detection using data mining techniques. The study expands the knowledge of this research topic among researchers, forensic accountants, auditors and regulators and provides organizations with useful information regarding the various types of financial fraud and data mining techniques available.

Keywords: Fraud · Machine learning · Data mining

1 Introduction

[20] classified internal fraud into financial statements fraud and transactions fraud. Financial statement fraud is related to the "intentional misstatement of certain financial values to enhance the appearance of profitability and deceive shareholders or creditors", whereas transaction fraud is the process related to snatching firms' assets.

Financial statements are considered as the main basis of decision-making for many interested parties including investors, creditors, regulators and other stakeholders, however, fraudulent financial statements are a serious concern which could result in severe damages in terms of firms' sustainable development [39, 51], and bankruptcy [2, 14].

Unlike earnings management, [24, 46] defined financial statement fraud as material omissions or misrepresentations resulting from an intentional deviation in the reporting financial information beyond the generally accepted accounting principles or Accounting standards scope. [33] stated that financial statement fraud deals with doctoring financial statements to make firms sound more profitable. [19] added that the purpose of financial statement fraud is to mislead the interested users.

Manipulative financial statements can be done using a number of methods such as recognizing revenue improperly, misrepresenting or the exclusion of financial statement footnotes, understatement or overstatement of assets, expenses or income [1]. All these actions are taken place in order to mislead the interested parties and make them believe

A. M. A. Musleh Al-Sartawi et al. (Eds.): EAMMIS 2021, LNNS 239, pp. 407–414, 2021.
https://doi.org/10.1007/978-3-030-77246-8_38

as if the financial statements were non-fraudulent. As revealed by the Center for Audit Quality, a number of incentives could be behind the engagement in financial statement fraud, for instance, to meet short term targets and expectations, to avoid sharing bad news regarding the firms' financial condition, or simply, for personal incentives [31, 48]. Other reasons could be related to stock performance enhancement, tax reduction, and performance exaggeration [29, 50].

The detection of fraudulent practices is the main responsibility of the external auditors as it needs high level of expert knowledge in the financial reporting [9]. However, this did not prevent the shortcomings in judgment and causing the previous accounting scandals. Some authors questioned the credibility of the auditors due to the conflicts of interest [25, 37].

Accordingly, in response to the previous accounting scandals that resulted mainly from financial statements fraudulent practices, the audit profession realized the weaknesses attributed with the traditional auditing process including manual verifications, inventory counts, and the application of simple ratios [23, 47].

As stated by [8, 42], the traditional audit methods are slow and could provide imprecise outdated audit reports. Furthermore, prior researchers claimed that the traditional auditing method is associated with a number of drawbacks such as spending long time to analyze the huge volumes of data especially in the current digital economy [10, 30, 32].

Moreover, the traditional audit has many limitations for instance; auditors have limited audit capabilities that could help them to reach the highest audit compliance at low costs [8, 40]. Hence, in today's digital economy, auditors need new mechanisms for analyzing big data transactions. However, the recent technological development in the business world raised the importance of adapting technology for fraud and opportunistic earnings management detection purposes [10].

Accordingly, in order to reduce the probability of manipulation and misleading the stakeholders, high attention was given to developing effective financial accounting fraud detection using technological approaches. Having an accurate detection system is essential because investors are better aware of the firms' financial condition, regulators can better practice their oversight role and audit firms practice their audits more effectively and accurately [36, 49].

According to [2], the percentage of studies that undertook investigating financial statement fraud and bank fraud is the highest among the studies that tested fraud types and data mining methods (41 articles out of the 65 reviewed articles) which indicates that financial statement fraud detection has been given a considerable attention in the existed research. Accordingly, it is important to provide a wide-ranging review on the most frequently used financial fraud detection techniques to expand the knowledge of professionals and researchers; hence, this study provides a comprehensive review of the existed studies that discussed data mining techniques within financial statement fraud context, identifies the gaps in the literature and suggests future research on this topic.

2 Literature Review

[4] and [5] were the two main studies that examined the development of earnings manipulation detection models using financial statements. [5, 48] proposed a model that could

categorize fraudulent firms from firms engaged in aggressive accrual-based earnings management.

Few years later, [4] proposed another model that could classify manipulators firms and non-manipulators firms. Previous studies mostly applied [38] model which is a statistical model that apply financial ratios obtained from firms' financial statements and then the M-Score is calculated to uncover the potential areas of opportunistic earnings management [17, 43, 52].

The Certified Public Auditors were required to use the aforementioned model to make sure that financial reports are not opportunistically managed and free from errors [26, 39]. [16] tested the ability of Beneish M-Score models in uncovering fraud and found that the overall Beneish M-Score model could uncover financial fraud effectively.

Unlike the widely used traditional audit where calculating statistics and ratios manually, the modern audit approach integrates sophisticated techniques such as data mining and machine learning techniques [27]. Intelligent financial statements fraud detection systems were continuously developed to uncover any potential for fraud signals which could assist the stakeholders in taking their decisions. As a result, the number of studies that undertook testing intelligent financial statements fraud detection systems is rising.

Prior researchers agreed that machine learning helps in uncovering earnings manipulation practices [27, 45]. The machine learning method includes complicated techniques like data mining that goes in depth with regards to data analytics [17]. [8] highlighted that machine learning help auditors to make reasonable assurance regarding potential earnings management practices within the financial statements. Studies also documented that machine learning is beneficial for justifying firms' accruals [11], selecting boards directors [10], and forecasting firms' stock returns [13].

Frauds are hidden in large amount of data and the traditional fraud detection methods might not be able to fully detect early these practices. According to the previous studies, data mining method application may solve this issue because of it performs well in inputting large amounts data as well as it provides more precise classification and forecasting [6].

In recent years, data mining methods are increasingly applied to detect financial statements fraud. [8] stated that data mining technique is a process of converting data into knowledge. Moreover, data mining made it possible to advance the Computer Assisted Audit Techniques [15] which includes various techniques such as the Support Vector Machines, Decisions Trees, Logic Base Algorithm, Bayesian Naïve Classifier, and Neural Network Based Machine Learning [3, 8, 41].

Although there are various techniques available, researchers concurred that there is no specific technique that could accurately work with all kinds of data sets [3, 22]. Also, [28] suggested that usually the percentage of manipulators is low compared to the non-manipulators which result in having imbalanced data, hence, leading to biased earnings prediction.

[19] stated that compared with the traditional regression analysis method, data mining techniques are more accurate and classified data mining methods into two main functions: top-down hypothesis testing and bottom-up knowledge discovery. Other researchers classified data mining functions as clustering, classification, forecasting and association [44]. [21] stressed that financial statements fraud represent a typical classification problem.

Although researchers were active in testing different data mining methods, it can be clearly noticed that the findings are still inconclusive. [8] used Bayesian Naïve Classifier to improve the decision-making procedure in the earnings manipulation detection. The findings revealed that mathematical models are better functioning than traditional audit. Another study by [28] focused on the enhancement of earnings manipulation detection using again Bayesian Naïve Classifier as type of supervised machine learning. The study findings showed that Beneish model performs better than the manual auditors' method which indicates that manual auditors' method is less sufficient in uncovering earnings manipulation.

[18] recommended using the non-parametric machine learning techniques because it is useful in different accounting predictions. A more recent study by [7] conducted in Taiwan suggested an improved model for to uncover earnings management through applying hybrid machine learning methods. The study results showed that earnings management detection model using elastic net and C5.0 algorithm perform more accurately.

[14] study showed that ensemble methods are better performing than the other machine learning methods in terms of classifying correctly the fraudulent firms. While Bayesian belief networks is considered the best method in selecting the non-fraudulent firms. On the contrary, [35] study findings revealed that random forest performed better than the other machine learning methods and has good robustness results because it processes high-dimensional data and avoid the overfit.

[19] applied a number of data mining techniques including artificial neural network and a support vector machine as a first stage. In the second stage, different methods of decision trees were used which are chi-square automatic interaction detector, C5.0, classification and regression tree, and quick unbiased efficient statistical tree are constructed for classification. The results showed that an artificial neural network processed by classification and regression tree is the best accurate method for financial statement fraud detection. Furthermore, [6] study showed that Chi squared automatic interaction detector model was effective in performing accurate outcomes.

Furthermore, according to [14], text mining method is an important method and has attracted the researchers' attention in recent years to uncover the financial statement fraud practices and this is due to the large amounts of textual data that could include manipulative managerial notes and comments. Besides, non-annual report data are essential for detecting firms engaged in fraudulent practices.

[2] reviewed the previous studies that applied data mining techniques for financial fraud detection during 2004 to 2015 and found that Logistic regression model is the most frequently used data mining method for financial fraud detection followed by neural network, decision tree, vector machine and naïve Bayes.

3 Conclusion and Future Studies

This study aims at understanding the data mining techniques applied for financial statement fraud detection and provides a foundation to future research in this field. The study also highlights the most frequently data mining methods applied in the literature to help organizations in uncovering and choosing the various types of data mining techniques available for fraud detection.

In general, there are still a number of aspects related to fraud detection using data mining methods that have not yet been the subject of research. Some types of fraud and data mining techniques require future study to be fully understood. In particular, limited but rising number of studies tackled the issue of the advanced technological audit methods such as machine learning role in monitoring financial statements fraud practices.

Some researchers argued that the consequences of integrating technology with fraud detection still vague, hence, more studies are needed to explain how the machine learning and new data mining algorithms as an earnings manipulation detection tool could help forensic accountants, auditors and regulators in uncovering the hidden opportunistic practices.

In addition, [34] argued that few studies agreed on data patterns to uncover potential earnings manipulation and this might be due to the fact that studies do not rely on developing theory or expert information. Hence, future studies should take into consideration this issue.

Moreover, little attention was paid to the combination of financial and linguistic data in intelligent financial statement fraud prediction [14], hence, studies should focus on testing different types of data in order to have a comprehensive picture of fraud practices. Besides, prior studies focused on testing the accuracy of data mining methods and did not pay a lot of attention in checking if these methods are easy to interpret [31].

Most of the previous studies applied one or two methods of data mining, future studies should focus on combining different methods and make a comparison between them. [35] suggested to have hybrid financial statement fraud detection models should be developed to make sure that a number of methods are used, financial and non-financial data can be combined, and most importantly, easy to interpret. Also, the majority of the prior studies have applied matching method to match the non-fraudulent firms with the fraudulent firms, hence, other methods should be applied in order to check other possibilities.

Further, [28] suggested the use of supervised machine learning models by regulators because they are more likely to identify high risk firms at early stages, thus, qualitative studies are needed to know exactly the perception and awareness regarding the role of supervised machine learning in the audit field. Studies can also test advanced models for predicting manipulation.

Based on previous studies statistics, studies that investigated the relationship between data mining and financial statement fraud were mostly applied in the United States followed by Taiwan, China and Spain [2]. Therefore, future studies should focus more on other countries to uncover this relationship using different contexts. Also, [33] stressed that previous studies samples were relatively low (i.e. less than one hundred) because sharing sensitive fraudulent financial data in order to be investigated in academic research is not easy for firms. This of course would result in biased data as it does not reflect the real situation, hence, future studies should include samples that are more accurate and reflect real examples of fraud detection [12].

Lastly, it is commonly known that fraud detection is associated with high costs to firms such as systems maintenance [21], however, there is a lack of studies that took into consideration the costs associated with fraud detection using data mining techniques.

References

1. Gupta, M., Sikarwar, T.S.: Modelling credit risk management and bank's profitability. Int. J. Electron. Bank. **2**(2), 170–183 (2020)
2. Erel, I., Stern, L.H., Tan, C., Weisbach, M.S.: Selecting directors using machine learning (No. w24435). National Bureau of Economic Research (2018)
3. Abdulrasool, F.E., Turnbull, S.I.: Exploring security, risk, and compliance driven IT governance model for universities: applied research based on the COBIT framework. Int. J. Electron. Bank. **2**(3), 237–265 (2020)
4. Albashrawi, M.: Detecting financial fraud using data mining techniques: a decade review from 2004 to 2015. J. Data Sci. **14**(3), 553–569 (2016)
5. Anto, S., Chandramathi, S.: Supervised machine learning approaches for medical data set classification: a review. Int. J. Comput. Sci. Technol. **2**(4), 234–240 (2011)
6. Beneish, M.: The detection of earnings manipulation. Financ. Anal. J. **55**, 24–36 (1999)
7. Beneish, M.D.: Detecting GAAP violation: implications for assessing earnings management among firms with extreme performance. J. Account. Public Policy **16**(3), 271–309 (1997)
8. Al-Sartawi, A.: The effect of corporate governance on the performance of the listed companies in the gulf cooperation council countries. Jordan J. Bus. Adm. **11**(3), 705–725 (2015)
9. Chen, S.: Detection of fraudulent financial statements using the hybrid data mining approach. Springerplus **5**(1), 1–16 (2016). https://doi.org/10.1186/s40064-016-1707-6
10. Al-Sartawi, A. (ed.): The big data-driven digital economy: artificial and computational intelligence In: Kacprzyk, J. (ed.) Studies in Computational Intelligence, vol. 974. Springer (2021)
11. Chen, S., Shen, Z.D.: An effective enterprise earnings management detection model for capital market development. J. Econ. Manag. Trade 77–91 (2020)
12. Hogan, C.E., Rezaee, Z., Riley, R., Velury, U.: Financial statement fraud: insights from the academic literature. Auditing J. Pract. Theory **27**(2), 231–252 (2008)
13. Dbouk, B., Zaarour, I.: Financial statements earnings manipulation detection using a layer of machine learning. Int. J. Innov. Manag. Technol. **8**(3), 172–179 (2017)
14. Dyck, A., Morse, A., Zingales, L.: Who blows the whistle on corporate fraud? J. Finance **65**, 2213–2253 (2010). https://doi.org/10.1111/j.1540-6261.2010.01614.x
15. Frankel, R., Jennings, J., Lee, J.: Using unstructured and qualitative disclosures to explain accruals. J. Account. Econ. **62**(2–3), 209–227 (2016)
16. Al-Sartawi, A.: Institutional ownership, social responsibility, corporate governance and online financial disclosure. Int. J. Crit. Account. **10**(3/4), 241–255 (2018)
17. Kotsiantis, S.B.: Supervised machine learning: a review of classification techniques. Informatica **31**(3), 249–268 (2007)
18. Glancy, F.H., Yadav, S.B.: A computational model for financial reporting fraud detection. Decis. Support Syst. **50**(3), 595–601 (2011)
19. Gu, S., Kelly, B., Xiu, D.: Empirical asset pricing via machine learning (No. w25398). National Bureau of Economic Research (2018)
20. Al-Sartawi, A.: Assessing the relationship between information transparency through social media disclosure and firm value. Manag. Account. Rev. **18**(2), 1–20 (2019)
21. Hajek, P., Henriques, R.: Mining corporate annual reports for intelligent detection of financial statement fraud–a comparative study of machine learning methods. Knowl.-Based Syst. **128**, 139–152 (2017)
22. Agarwal, G.K., Medury, Y.: Internal auditor as accounting fraud buster. IUP J. Account. Res. Audit Pract. **13**, 7–29 (2014)
23. Gupta, N.: Influence of demographic variables on synchronisation between customer satisfaction and retail banking channels for customers of public sector banks of India. Int. J. Electron. Bank. **1**(3), 206–219 (2019)

24. Hematfar, M., Hemmati, M.: A comparison of risk-based and traditional auditing and their effect on the quality of audit reports. Int. Res. J. Appl. Basic Sci. **4**(8), 2088–2091 (2013)
25. Al-Sartawi, A.: Ownership structure and intellectual capital: evidence from the GCC countries. Int. J. Learn. Intellect. Cap. **15**(3), 277–291 (2018)
26. Herawati, N.: Application of Beneish M-Score models and data mining to detect financial fraud. Procedia Soc. Behav. Sci. **211**, 924–930 (2015)
27. Al-Sartawi, A.: Information technology governance and cybersecurity at the board level. Int. J. Crit. Infrastruct. **16**(2), 150–161 (2020)
28. Hunt, J., Myers, J., Myers, L.: Improving earnings predictions with machine learning (2019). https://link.springer.com/article/10.1007/s10997-007-9015-8
29. Jan, C.: An effective financial statements fraud detection model for the sustainable development of financial markets: evidence from Taiwan. Sustainability (Basel, Switzerland) **10**(2), 513 (2018)
30. Al-Sartawi, A., Sanad, Z.: Institutional ownership and corporate governance: evidence from Bahrain. Afro-Asian J. Finance Account. **9**(1), 101–115 (2019)
31. Jans, M., Werf, J.M., Lybaert, N., Vanhoof, K.: A business process mining application for internal transaction fraud mitigation. Expert Syst. Appl. **38**, 13351–13359 (2011)
32. Al-Sartawi, A.: Does it pay to be socially responsible? Empirical evidence from the GCC countries. Int. J. Law Manag. **62**(5), 381–394 (2020)
33. Kirkos, E., Spathis, C., Manolopoulos, Y.: Data mining techniques for the detection of fraudulent financial statements. Expert Syst. Appl. **32**(4), 995–1003 (2007)
34. Sun, J.H., Li, Q.H., Huang, K.Y.H.: Predicting financial distress and corporate failure: a review from the state-of-the-art definitions, modeling, sampling, and featuring approaches. Knowl. Based Syst. **57**, 41–56 (2014). https://doi.org/10.1016/j.knosys.2013.12.006
35. Kuenkaikaew, Vasarhelyi, A.M.: The predictive audit framework. Int. J. Digital Account. Res. **13**, 37–71 (2013)
36. Al-Sartawi, A.: Corporate governance and intellectual capital: evidence from gulf cooperation council countries. Acad. Account. Financ. Stud. J. **22**(1), 1–12 (2018)
37. Mohamad Kamal, M.E., Md Salleh, M.F., Ahmad, A.: Detecting financial statement fraud by Malaysian public listed companies: the reliability of the Beneish M-score model. Jurnal Pengurusan **46**, 23–32 (2016)
38. Moore, D., Tetlock, P., Tanlu, L.: Conflicts of interest and the case of auditor independence: moral seduction and strategic issue cycling. Acad. Manag. Rev. **31**, 10–29 (2006)
39. Al-Sartawi, A.: Measuring the level of online financial disclosure in the Gulf Cooperation Council Countries. Corp. Ownersh. Control. **14**(1), 547–558 (2016)
40. Nwoye, U.J., Okoye, E.I., Oraka, A.O.: Beneish model as effective complement to the application of SAS No. 99 in the conduct of audit in Nigeria. Manag. Adm. Sci. Rev. **2**(6), 640–655 (2013)
41. Omar, N., Koya, R.K., Sanusi, Z.M., Shafie, N.A.: Financial statement fraud: a case examination using Beneish model and ratio analysis. Int. J. Trade Econ. Finance **5**(2), 184–186 (2014)
42. Al-Sartawi, A.: Social media disclosure of intellectual capital and firm value. Int. J. Learn. Intellect. Cap. **17**(4), 312–323 (2020)
43. Rahul, K., Seth, N., Kumar, U.D.: Spotting earnings manipulation: using machine learning for financial fraud detection. In: International Conference on Innovative Techniques and Applications of Artificial Intelligence, pp. 343–356 (2018)
44. Musleh Al-Sartawi, A.M.A.: E-learning improves accounting education: case of the higher education sector of Bahrain. In: Themistocleous, M., Papadaki, M., Kamal, M.M. (eds.) Information Systems. EMCIS 2020. Lecture Notes in Business Information Processing, vol. 402. Springer, Cham (2020)

45. Ravisankar, P., Ravi, V., Rao, G.R., Bose, I.: Detection of financial statement fraud and feature selection using data mining techniques. Decis. Support Syst. **50**(2), 491–500 (2011)
46. Sharma, A., Panigrahi, K.P.: A review of financial accounting fraud detection based on data mining techniques. Int. J. Comput. Appl. **39**(1), 37–47 (2012)
47. Al-Sartawi, A.: The effect of the electronic financial reporting on the market value added of the Islamic banks in Gulf cooperation council countries. In: 8th Global Islamic Marketing Conference, 4–6 May. International Islamic Marketing Association, Turkey (2017)
48. Verner. J.: Why you need internal audit at the table. Published on Business Finance (2012). http://businessfinancemag.com
49. West, J., Bhattacharya, M.: Intelligent financial fraud detection: a comprehensive review. Comput. Secur. **57**, 47–66 (2016)
50. Whiting, D.G., Hansen, J.V., McDonald, J.B., Albrecht, C., Albrecht, W.S.: Machine learning methods for detecting patterns of management fraud. Comput. Intell. **28**(4), 505–527 (2012)
51. Yao, J., Zhang, J., Wang, L.: A financial statement fraud detection model based on hybrid data mining methods, p. 57. IEEE (2018)
52. Sanad, Z., Al-Sartawi, A.: Investigating the relationship between corporate governance and internet financial reporting (IFR): evidence from Bahrain bourse. Jordan J. Bus. Adm. **12**(1), 239–269 (2016)

The Impact of Balanced Scorecard on Firm Value: Evidence from Palestine

Renad Frehat(✉) and Saed Al Koni

An-Najah National University, Nablus, Palestine
saed@najah.edu

Abstract. The main purpose of the research is to examine the impact of Balanced Scorecard (BSC) four perspectives (financial, business process, customer, learning and growth) on firm value. Secondary data was collected from the annual reports of industrial corporations listed in the Palestine Stock Exchange (PSE) from the year 2015 to 2018. The behavior of variables were defined using descriptive statistics. Using Multiple Square Regression analysis, the hypotheses were examined. The result of study showed that financial and customer aspects have positive insignificant impact on firm value, aspect of business process has negative insignificant impact on firm value, and aspect of learning and growth has positive significant impact on firm value. The study found that firms with low business processes perform significantly better and are valued higher. Overall, this study concluded that the aspects of the Balanced Scorecard (financial, customers and business process) do not have a significant impact on the value of the firm while the learning and growth aspect has significant effect on the value of the firm.

Keywords: Balanced scorecard · Financial · Business process · Customer · Learning and growth · Perspectives · Firm value

1 Introduction

The Balanced Scorecard (BSC) is one of the modern management tools which was introduced by Kaplan and Norton. BSC, as a strategic tool, improves corporate performance. BSC offers a solution to the weakness and ambiguity in the direction to the old management system, which focuses only on financial performance, but BSC adds other dimensions that transform the strategic plan into actions and results. BSC is considered as a revolution against traditional approaches which rely on financial data and accounting numbers. However, BSC includes financial and non-financial indicators that evaluate the performance of a company and determine its success. After Kaplan and Norton published an article in the Harvard Business Review in 1992 to analyze corporate performance, BSC technique became popular [1].

BSC is an important tool for strategic control, as it allows the organization to evaluate its strategy through four different dimensions, unlike traditional evaluation systems. [2] stated that BSC is one of the measures to evaluate corporate performance through concern for accomplishing both financial and non-financial purposes. The study of [3]

A. M. A. Musleh Al-Sartawi et al. (Eds.): EAMMIS 2021, LNNS 239, pp. 415–429, 2021.
https://doi.org/10.1007/978-3-030-77246-8_39

was conducted with 500 American institutions for testing the effectiveness of BSC as a tool for communication and strategic control, the researchers concluded that BSC is an effective tool for strategic control, managers notice that the information which provided by BSC allows them to take measures that positively affect the strategy of the organization and thus the monitoring capacity of BSC leads to motivating employees to implement the strategy of organization, which leads to success and achieve good results. The study of [4] aims to determine the benefits associated with applying BSC from the point of view of a group of consultants and managers in the Scandinavian countries. The results of the study concluded that the consultants and users of BSC consider that it useful and can be used to achieve a balance between the demands of both shareholders and stakeholders, and that BSC supports the institution's goals and considered as a motivating tool to work and achieve goals. Financial performance assessment is a way of delivering feedback to the management of the company that directs it in formulating its strategies and enhancing its performance level [5]. Companies aim to maximize their value by increasing the market value of their shares instead of focusing only on profit maximization [6]. The goal of maximizing the market value, which is considered as long-term goals, is influenced by several financial and non-financial factors. Corporate value is an economic indicator that represents a business's market value. There are various metrics as a reference for the valuation of the company. Tobin's Q is used by many studies to measure corporate value [7]. Tobin's Q ratio is used to measure corporate performance [8].

There are various studies on BSC and firm value. The study of [9] aimed to test the extent of BSC application as a tool for performance evaluation in the Palestinian listed companies. Both financial and non-financial (customer, internal business process and learning and growth) perspectives of BSC were used. Data obtained through a questionnaire and hypotheses were tested by using descriptive statistics and one sample t-test techniques. The researchers concluded that the majority of Palestinian companies use BSC in the assessment process, either entirely or partly. [10] studied the influence of BSC on dependent variables (ROA and ROE) on a sample of Amman Stock Exchange-listed Jordanian manufacturing firms. The findings showed that the financial performance calculated by ROA and ROE is positively influenced by all dimensions of BSC.

1.1 Research Problem

With the insufficiency of the traditional performance evaluation methods of companies, it was necessary to search for innovative ways through which companies can be evaluated in a manner that includes all financial and non-financial aspects, especially since Palestinian companies are facing accelerating environmental conditions, which requires the use of modern management methods that help the company achieve its strategic goals and its continuity. The research problem is that companies rely exclusively on the financial aspects in assessing their performance without taking into consideration the non-financial factors. Relying only on financial indicators provides insufficient information on the level of performance of the company in general and thus making inaccurate decisions. Traditional performance management ignores non-financial factors, financial data has very limited forecasting ability and it is considered insufficient to measure the performance of companies and enable them to respond to the complexities of customers and the environment [1]. Financial indicators can only reflect the performance of organizations in the past and not reflect the future operating conditions of the company. Therefore, it is imperative to use modern management methods that enable companies to evaluate their performance in all aspects. BSC is considered one of the evaluation tools that contribute to achieving the goals that the company seeks, it includes financial and non-financial measurement indicators.

1.2 Research Importance

This research derives its importance from emphasizing the important role that BSC plays in efficiently evaluating companies and measuring their value. All the financial and non-financial aspects are taken into account, this makes BSC a modern management method that must be followed and applied in companies to help them continue and improve their performance in light of the competitive environment. Moreover, there is a lack of studies in Palestine that have examined the effect of BSC on corporate value. This study is characterized by its reliance on secondary data derived from the annual reports of the industrial companies listed on the Palestine Stock Exchange (PSE) in order to measure the four aspects of BSC, while previous studies rely on questionnaires in order to measure the aspects of BSC. This study is one of the few studies conducted by Palestinian researchers according to the researchers' awareness of the Palestinian environment.

1.3 Research Objectives

The research aims to examine the impact of the four aspects of BSC on the value of industrial corporations listed in the Palestine Stock Exchange (PSE). Firm value is measured by using Tobin's Q model, which is considered as a modern indicator of corporate value. It surpasses traditional measures in expressing the real companies' performance and assisting investors in making their investment decisions. Moreover, the study aims to achieve the following specific objectives to investigate the impact of the financial, customer, internal business process and learning and growth perspectives of BSC on firm value.

1.4 Research Question

The study is mainly designed to answer the following main question: What is the impact of the Balanced Scorecard (BSC) four perspectives (financial, customer, internal business process and learning and growth) on firm value?

The rest of this study is segregated as follows: Section two presents the literature review, hypotheses, and research framework. Research methodology including the source of data, variable measurement, research model, and research technique are discussed in section three. Section four provides findings of this study. Finally, section five discussed the conclusion and recommendations.

2 Literature Review and Hypotheses Development

2.1 Balanced Scorecard

There are many definitions that show the concept of BSC, as some define it as a multi-dimensional framework used to measure strategic performance that combines financial indicators with non-financial indicators consisting of four dimensions which are financial dimension, customer dimension, internal operations dimension, and growth and learning dimension. BSC represents an integrated strategic management framework in terms of cause and effect approach [11]. In addition, it fulfills all of the management requirements such that the organization becomes client oriented, shortens wait time, boosts efficiency, enhances collaboration, and reduces the releasing times of new products. Moreover, BSC reduces information overload by directing managers to focus on a set of more important metrics. It enables the manager to view the business from four main perspectives. First, financial perspective, (what are our shareholders viewed?), contains financial ratios. Second, customer perspective, (how are we viewed by customers?). Third, perspective of internal business, (what do we succeed at?). Fourth, perspective of learning and growth, (will we continue to maximize and generate value?) [12].

[13] indicated that in its first generation, there was a slight correlation between the four dimensions of BSC, though the integration of financial and non-financial indicators had led to the existence of a causality relationship between four dimensions. It was agreed that the design of BSC requires developing a strategic vision for the firm to measure performance. BSC is a set of indicators that provide managers with a comprehensive vision about the work within the organization. It includes financial indicators that reflect the results of actions and decisions in addition to non-financial indicators that include customer satisfaction, internal processes and learning and growth. These non-financial indicators are the main drivers of financial performance. BSC keeps companies looking forward rather than backward. By integrating financial, customer, internal process, innovation, and organizational learning perspectives, BSC helps managers understand the interconnections and thus problem-solving and better decision-making [12]. BSC considers the organization as a value chain, so that each specific indicator of it enters into a relationship of cause and effect that clarifies the meaning of the strategy within the organization, and this makes BSC a tool for strategic management. Each of BSC indicators includes a logical cause and effect relationship that links the desired results from implementing the strategy with the performance drivers that lead towards those results. In addition to BSC, the intellectual capital is also linked to the company's strategy [14]. This is also supported by the finding of [15] study which stated that BSC is used by companies to match critical management processes with the strategy of the company, in addition to its relationship to cost and financial metrics including activity-based costs and quality programs.

The study of [16] presents the essence of BSC by presenting the frame of the card that has been used in more than 20 Japanese companies, it included five basic points that distinguish BSC from other performance measures. First, maintaining a balance between short, medium and long-term goals through diversity in indicators. Second, developing non-financial quantitative measures in addition to financial indicators. Third, reducing ambiguity by retaining quantitative measures. Fourth, Supporting institutional learning by a repetitive hypothesis review cycle. Fifth, enhancing strategic communications between the institution's senior management and employees. The study of [17] aims to identify the extent of BSC for evaluating the performance of the corporations listed on the Palestine Stock Exchange. The questionnaire was distributed to the financial managers of corporations, 37 questionnaires were distributed and 32 questionnaires recovered, the recovery rate was 86.5%. The methodology used in the study is descriptive statistics for testing hypotheses. Its findings indicate that the corporations listed on the Palestine Stock Exchange focused on using the financial and customers satisfaction dimensions in order to evaluate performance. However, it disregarded the internal processes, and learning and growth dimensions in the evaluation process of performance.

2.2 Firm Value

The value of the company is the present value of the expected future cash inflows, discounted at a rate that reflects the company risk [18]. The investor in the stock exchange relies a lot on information that helps him make the right investment decision, as there are metrics for evaluating the financial performance of companies. It is an important axis to know the success or failure of companies. The financial performance of the companies is represented by market value, which is represented by the total value of the shares traded; that is equal share price multiplied by the number of shares. If there is an increase in the company's share price, and despite of this, there is a desire for investors to pay more to obtain shares, then they have confidence in the company and its ability to achieve a higher return for them [19].

The study of [20] stated that one of the most recent indicators that are used to measure the value of the company is Tobin's Q model, which indicates the current assessment of the financial markets about the return on each additional investment dollar [21]. Tobin's Q represents the ratio of the market value of a company's assets to its replacement cost. When the value of Tobin's Q is greater than 1, this indicates efficiency of the company in using its resources [22]. Tobin's Q Ratio is considered a forward-looking market measure of the value of the company [23], it differs from the accounting measures that are related to the past and reflect the company's performance based on historical information. In contrast, market metrics relate to expected future performance and reflect the market's perception of corporate performance [24]. The measures of financial performance include liquidity ratios, profitability ratios, efficiency ratios, leverage ratios, and market ratios. These ratios are considered traditional accounting measures. Recently, modern metrics have been developed to evaluate companies' financial performance based on the market, including market value added (MVA), economic value added (EVA), and Tobin's q model through which the investor can make the right decision. The study of [5] indicates that the evaluation of financial performance is a way to provide the company's management with feedback that helps it to set its strategy and improve its level of performance. The study of [25] recommended the use of Tobin's Q to assess the performance of companies.

2.3 Financial Perspective of Balanced Scorecard and Firm Value

The financial dimension is considered one of the aspects of performance evaluation, the results of this dimension represent indicators for achieving the firm's goals and reaching results by following the strategy. Financial indicators also provide a clear picture of the firm's financial performance and it is a means of ascertaining whether the strategy contributes to improving results and achieving goals [26]. The amount of profit achieved by the company is a measure of efficiency; the greater the profits, the higher financial and operational efficiency of the company's business [27]. The conventional and most used measurement technique is financial metrics. The measurable economic effects of the acts that can be implemented are expressed by fiscal policies. Different metrics such as profitability, return on capital, and return on equity are the emphasis of financial indicators [28]. The financial dimension measures are an important aspect of BSC, especially in the profit-oriented economic units that are concerned with economic results and the implementation of strategies [29]. The financial dimension represents specific goals and objectives and measures to achieve those goals. Measures of the financial dimension such as return on capital, level of operating income, return on assets and return on equity are related to the final results of performance. The study of [30] found that companies prefer to use BSC to evaluate their performance, the most used perspective is financial views, then customer satisfactions, internal business processes, and learning and growth perspectives. The study of [31] tested the degree to which BSC is used to evaluate the performance of banking sector in Nigeria by using the survey method. The study findings found that banks use the perspective of financial performance in order to measure their performance. Accordingly, the first hypothesis of the study can be formulated as follows:

HA1: There is a significant impact of financial perspective of BSC on firm value.

2.4 Customer Perspective of Balanced Scorecard and Firm Value

The purpose of the consumer view is to define the client and market sectors for which the company competes. [15] identified five indicators to measure the customer dimension, which are market share, new customer acquisition, customer retention, customer satisfaction, and customer profitability. The study of [32] found that the use of the four aspects of the Balanced Scorecard has a positive effect on the performance of companies as measured using BSC. The non-financial aspects such as customer satisfaction, internal process design and employee training affect the financial performance of small and medium companies. Within the customer's perspective, the organization seeks to differentiate itself from competitors and retain target customers. This is an important aspect, since it helps the organization to link its internal operations to achieve better results with its customers [15]. The customer side is the core of any business and helps in improving the company's performance in the long run [33]. In this sense, the second hypothesis of the study can be formulated as follows:

HA2: There is a significant impact of the customer perspective of BSC on firm value.

2.5 Business Process Perspective of Balanced Scorecard and Firm Value

The study of [34] stated that the business process dimension depends on identifying the operational processes that have importance in achieving firm's strategic goals. A firm can, through its internal operations, provide value to its customers and satisfy shareholders by achieving good financial results. [35] suggested that both internal business processes, and learning and growth perspectives of BSC had a positive effect on corporate performance, contributing to a growth in the organization's revenue base. In addition, BSC implementation would decrease the cost of product/service per unit. However, the study of [17] found that the corporations listed on the Palestine Stock Exchange did not depend on these two dimensions in the evaluation process of the performance. They rather focused on using financial and customers dimensions in order to evaluate performance. Therefore, the third hypothesis of the study is formulated as follows:

HA3: There is a significant impact of the business process perspective of BSC on firm value.

2.6 Learning and Growth Perspective of Balanced Scorecard and Firm Value

Learning and growth viewpoint describes the corporate infrastructure that ideally suits its strategic priorities and offers "how to get there" steps. There are three aspects of the learning & growth viewpoint; persons, systems and corporate [26]. The dimension of learning and growth is the basis of BSC, it represents the infrastructure of the firm. Kaplan and Norton indicate that organizational learning and growth have three main sources, which are individuals, systems and organizational procedures [34]. [36] indicated that the learning and growth perspective and the customer perspective positively affect the organization's performance, but the perspective of financial and internal processes had a positive insignificant relationship with performance. [10] studied the influence of BSC on performance for a sample of Amman Stock Exchange-listed Jordanian manufacturing firms. The findings showed that the learning and growth viewpoint positively affected performance. The fourth hypothesis of the study is accordingly formulated as follows:

HA4: There is a significant impact of the learning and growth perspective of BSC on firm value.

3 Research Methodology

A methodology is a pivotal part that represents the applied aspect of the study until it achieves its intended objectives, and reaches results. This chapter aims to describe the study methodology, including the population and sample of the study, data sources and tools for their collection, study variables and measurement method for each of them, study model, and statistical methods used in data analysis and hypothesis testing.

Research Population and Sample
The study population comprises all corporations listed on the Palestine Stock Exchange (PSE) which numbered 48 companies divided into five sectors, including industry, services, investment, banks and financial services, and insurance. A sample of 11 industrial

corporations that have published annual reports with available data for the period from 2015 to 2018.

Source of Data Collection
This study depends mainly on the use of secondary sources of data, as they were collected from the annual reports of industrial Palestinian firms released from the year 2015 to 2018. These reports published on the official website of the Palestine Stock Exchange (www.pex.ps).

Research Model
For measuring the impact of BSC four perspectives (financial, business process, customer, and learning and growth) on firm value, the following model was estimated to test the research hypotheses:

$$\mathbf{TbQ = \alpha + \beta 1\ FINBS_{(1,2)} + \beta 2\ CUSBS_{(1,2,3)} + \beta 3\ BUSBS_{(1,2,3)} + \beta 4\ LGRBS + e}$$

Where:

TbQ = TOBIN Q
α is constant, $\beta 1$, $\beta 2$, $\beta 3$, $\beta 4$, are regression coefficients
FINBS = Financial Perspective of Balanced Scorecard
CUSBS = Customer Perspective of Balanced Scorecard
BUSBS = Business Process Perspective of Balanced Scorecard
LGRBS = Learning and Growth Perspective of Balanced Scorecard
e = Error Term, used as a substitute for all other variables that are not used in this model.

Firm value as measured by TOBIN Q is the dependent variable, Balanced Scorecard perspectives are the independent variables. Study's model is shown in the following Fig. 1:

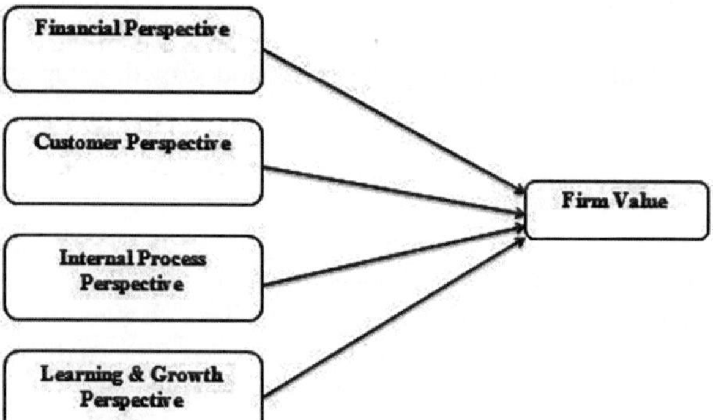

Fig. 1. Research model

3.1 Variables Measurement

The study uses two kinds of variables which are the dependent variable which is firm value and the independent variables which are the four perspectives of BSC.

Independent Variables

Financial Perspective
This study measures financial performance by using return on assets, and return on equity. [37] stated that Return on Assets (ROA) refers to the company's effectiveness in achieving profit through its assets, it represents the institution's ability to invest the assets it owns in generating profits. It is measured by the following equation: ROA = Net Income/Total Assets. Return on Equity (ROE) evaluates the earnings of the company by shareholders' equity investments. ROE reflects the company's financial performance [38]. Companies which achieve profits are able to maintain or increase the market value of their shareholders. An increase in the profitability rates has a positive effect on the share price. The higher the profitability rates, the higher the market value of the company. ROE is measured through dividing net income by total equity: ROE = Net Income/Total Equity.

Customer Perspective
The company seeks to improve the best relationships with customers with a special focus on retaining them, in order to drive sales growth. [15] identified five indicators to measure the customer dimension, which are market share, new customer acquisition, customer retention, customer satisfaction and customer profitability. [39] determined that increased customer loyalty and satisfaction contribute to the company's sales growth. This study measures customer perspective by percentage of sales growth, quality provided by employees to customers achieving sales, and marketing expenses that firm spends in achieving customer satisfaction.

Business Processes Perspective
The study of [26] used inventory turnover in order to measure business processes perspective. Nowadays, time is a crucial factor for new and creative goods and services, so firms must be able to rapidly plan and produce products. This study measures internal business processes by labor productivity, employee efficiency productivity in achieving profit, and bonuses that firms spend in achieving employee satisfaction.

Learning and Growth Perspective
The study of [40] stated that companies should pay attention to employees a significant aspect, since they are really important not only for their necessary duties, but also for their creativity and the development of new ideas. This study measures learning and growth perspective by quality of audit offices and spending on their skills in obtaining sales.

Dependent Variable

Firm Value
The value of the firm represents the dependent variable in this study. A measure of Tobin's Q is used as an indication of the firm's value based on previous studies [41]. This research relied on Tobin's Q model, which is considered one of the modern metrics

for measuring company value. This model can measure the market value of companies based on financial data derived from the financial statements of firms. The use of Tobin's Q model which is market-based measure is appropriate to measure the effect of the BSC on the value of the firm. The value of Q reflects the performance and profitability of the company. If Q > 1, this indicates that the market value of the company is higher than the value of its assets, it reflects good performance and high profitability [42]. In this study, Tobin's Q ratio was calculated according to the following equation:

$$\text{Tobin's Q} = (EMV + DEBT) / TA$$

Where:

EMV: The market value of equity which is measured by (closing price at the end of the fiscal year * the number of shares outstanding)
DEBT: Total Liabilities at the end of the year
TA: Total Assets at the end of the year (Table 1)

Table 1. Dependent and Independent Variables measurement.

Variable name	Measurement
TbQ	Measured by Tobin's Q ratio (EMV + DEBT)/TA
FINBS	Measured by ROE & ROA
CUSBS1	Measured by % Sales Growth (loyal customers)
CUSBS2	Measured by Net sales/Number of employees (quality which provided by the employees to customers in achieving sales)
CUSBS3	Measured by Marketing Expenses/Revenues (marketing expenses that firm spends in achieving customer satisfaction)
BUSBS1	Measured by Total assets/Number of employees (labor productivity)
BUSBS2	Measured by Operating Cash Flow/Total Profits (employee efficiency productivity in achieving profit)
BUSBS3	Measured by Bonuses/Administrative Expenses (bonuses that firm spends in achieving employee satisfaction)
LGRBS	Measured by Professional Fees/Sales Revenues (quality of audit offices and spending on their skills in obtaining sales)

3.2 Research Technique

The effect and relationship between variables are tested by using Pearson correlation coefficient and multiple regression analysis. Therefore, the same techniques will be used in current research to fulfill the purpose of this research, in addition to other statistical methods, such as descriptive analysis.

4 Results

The financial aspect of the balanced scorecard positively affects the value of the firm depending on the figures in Table 2, where the value of t are equal (FINBS1 = .359 & FINBS2 = 1.659), the value of sig. are (FINBS1 = .722 & FINBS2 = .106) as appears in the table. This effect is statistically insignificant because the probability value is more than 5%. Therefore, the financial perspective of the balanced scorecard has a positive effect on the value of the firm. This finding supports and agrees with the findings of study [10, 31] and contradicts [43]. Clients' perspective Balanced Scorecard positively affects the value of industrial companies listed on the Palestine Stock Exchange based on the value of t (CUSBS1 = .742, CUSBS2 = .507 & CUSBS3 = 1.266) and the value of P (CUSBS1 = .463, CUSBS2 = .616 & CUSBS3 = −.214) as shown in the table. This result of the study agrees with the study of [36, 43]. The business process perspective of the balanced scorecard, based on the value of t (BUSBS1 = .000, BUSBS2 = −2.463 & BUSBS3 = −.310) and the value of P (BUSBS1 = 1.000, BUSBS2 = .19 & BUSBS3 = .758), has a non-significantly statistical negative effect on firm value. This indicates that companies with lower business processes have higher value. This finding supports the results of [44] and inconsistent with [36]. Learning and growth balanced scorecard perspective has a positive impact on the value of Palestinian industrial companies, the value of t (LGRBS = 5.079 and the p-value 0.00 according to the table). This effect has statistical significance as its probability value was less than the value of 0.05. Therefore, a balanced scorecard from a learning and growth perspective has a significant impact on the value of the company. This finding is inconsistent with the results of [10, 36].

Table 2 shows the Pearson correlation coefficient between the dependent variable (firm value) and the independent variables (BSC perspectives), the value of (R = 86%), the value of the determination coefficient (R Square = 75%), it is considered excellent measures and fit model. The value of Adjusted R Square is 69%, meaning that the independent variables are explained by 69% of the variance in the dependent variable. Moreover, Durbin-Watson value is 2.055. The ANOVA analysis to test the significance of the regression, the value is 0.000 which is less than 0.05, therefore the regression is significant and there is an effect of the independent variables on the dependent variable. F-value (11.554) and sig. of (0.000) shows that the TOBIN Q regression model is well specified. Moreover, the dependent variable can be predicted through independent variables in this study.

Table 2. Regression results

Variables	T-statistics	Sig
FINBS1	0.359	0.722
FINBS2	1.659	0.106
CUSBS1	0.742	0.463
CUSBS2	0.507	0.616

(continued)

Table 2. (*continued*)

Variables	T-statistics	Sig
CUSBS3	1.266	0.214
BUSBS1	0	1
BUSBS2	−2.463	0.19
BUSBS3	−.310	0.758
LGRBS	5.079	0
R		0.868
R-squared		0.754
Adjusted R-square		0.688
F-Statistic		11.554
Sig		0
Durbin–Watson		2.055

5 Conclusion and Recommendation

This research focuses on an important modern management tool for improving and evaluating the performance of companies, which is Balanced Scorecard (BSC) that was developed to guide the organization to evaluate its strategy through four different dimensions, unlike traditional evaluation systems. The purpose of this study was to examine the impact of BSC on the value of industrial corporations listed in the Palestine Stock Exchange (PSE). BSC includes four aspects which are financial, business process, customers, and learning and growth. Firm value is measured by using Tobin's Q model, which is considered as a modern indicator of corporate performance. It surpasses traditional measures in expressing the real companies' performance and assisting investors in making their investment decisions. The results of study showed that financial and customer aspects have positive insignificant impact on firm value, aspect of business process has negative insignificant impact on firm value, and aspect of learning and growth has positive significant impact on firm value. Therefore, the study rejects the hypotheses of HA1, HA2, HA3 and accepts HA4. Overall, this study concluded that the aspects of BSC (financial, customers and business process) do not have a significant impact on the value of the company while the learning and growth aspect has an important effect on the value of the company.

The study recommends companies to take advantage and activate perspectives of the BSC because it has an impact on the firm value. Moreover, determining performance appraisal measures should be based on the four aspects of BSC in order to evaluate companies in a more accurate and comprehensive manner and improve the effectiveness of performance. There is a need for future research in Palestine in this area, it should focus on studying the factors that facilitate or impede the implementation of BSC. More studies related to impact of BSC on Activity based costing (ABC), profit maximization, and intellectual capital are also needed. The study recommends researchers for using

Tobin's q model to find in firm value in conducting studies. The results of this study have many implications for policy makers in companies in all sectors in terms of evaluating and assessing performance.

References

1. Kaplan, R.S., Norton, D.P.: The Balanced Scorecard–Measures That Drive Performance republished in Harvard Business Review on Measuring Corporate Performance Boston (1992)
2. Al Sawalqa, F., Holloway, D., Alam, M.: Balanced scorecard implementation in Jordan: an initial analysis. Int. J. Electron. Bus. Manag. **9**(3), 196 (2011)
3. Malina, M.A., Selto, F.H.: Communicating and controlling strategy: an empirical study of the effectiveness of the balanced scorecard. J. Manag. Account. Res. **13**(1), 47–90 (2001). https://doi.org/10.2308/jmar.2001.13.1.47
4. Madsen, D.Ø., Stenheim, T.: Perceived benefits of balanced scorecard implementation: some preliminary evidence. Probl. Perspect. Manag. **12**(3), 81–90 (2014)
5. Simionescu, M.: The Competition between London companies regarding their financial performance. J. Competitiveness **8**(2), 5–16 (2016). https://doi.org/10.7441/joc.2016.02.01
6. Berle, A.A., Means, G.C.: The Modern Corporation and Private Property. Transaction Publishers, Piscataway (1991)
7. Yermack, D.: Higher market valuation of companies with a small board of directors. J. Financ. Econ. **40**(2), 185–211 (1996). https://doi.org/10.1016/0304-405x(95)00844-5
8. Chen, K.C., Lee, C.J.: Accounting measures of business performance and Tobin's q theory. J. Acc. Audit. Finance **10**(3), 587–609 (1995). https://doi.org/10.1177/0148558X9501000310
9. Rahhal, A., Darabee, T.: Adoption of balanced scorecard (BSC) in Palestinian corporations (2014)
10. Al Mseden, N.A., Nassar, M.A.: The effect of balanced scorecard (BSC) implementation on the financial performance of the Jordanian companies. In: The International Business and Social Science Research Conference (2015)
11. Speckbacher, G., Bischof, J., Pfeiffer, T.: A descriptive analysis on the implementation of Balanced Scorecards in German-speaking countries. Manag. Account. Res. **14**(4), 361–388 (2003). https://doi.org/10.1016/j.mar.2003.10.001
12. Kaplan, R.S., Norton, D.P.: The balanced scorecard: measures that drive performance. Harv. Bus. Rev. **83**(7), 172 (2005)
13. Cobbold, I., Lawrie, G.: The development of the balanced scorecard as a strategic management tool. Perform. Meas. Assoc. **35**, 37–56 (2002)
14. Mouritsen, J., Thorsgaard Larsen, H., Bukh, P.N.: Dealing with the knowledge economy: intellectual capital versus balanced scorecard. J. Intellect. Cap. **6**(1), 8–27 (2005). https://doi.org/10.1108/14691930510574636
15. Kaplan, R.S., Norton, D.P.: Transforming the balanced scorecard from performance measurement to strategic management: Part II. Account. Horiz. **15**(2), 147–160 (2001). https://doi.org/10.2308/acch.2001.15.2.147
16. Morisawa, T.: Building performance measurement systems with the balanced scorecard approach. NRI Papers **45**, 1–15 (2002)
17. Saad, R.I., Daraghma, Z.M.A.: Using of the balanced scorecard for performance evaluation: empirical evidence from the listed corporations in the Palestine Exchange (PEX). Int. J. Bus. Manag. **11**(3), 215 (2016). https://doi.org/10.5539/ijbm.v11n3p215
18. Tangjitprom, N.: Earnings management and firm value: the role of investor protection and corporate governance. Working Paper (2012)

19. Haryono, U., Iskandar, R.: Corporate social performance and firm value. Int. J. Bus. Manag. Invention **4**(11), 69–75 (2015)
20. Kim, J.Y., Kwak, J., Lee, K.: Estimating Tobin's Q for listed firms in Korea (1980–2005): comparing alternative approaches and an experiment with investment functions. Seoul J. Econ. **28**(1), 1–30 (2015)
21. Tandry, A.Y., Setiawati, L., Setiawan, E.: The effect of CSR disclosure to firm value with earning management as moderating variable: case study of non-financing firms listed at Indonesia Stock Exchange. Int. J. Trade Global Markets **7**(3), 190 (2014). https://doi.org/10.1504/ijtgm.2014.064907
22. Sarun, A.: Corporate governance, earnings quality and firm value: evidence from Malaysia. Doctoral dissertation, Victoria University (2016)
23. Emeka-Nwokeji, N., Osisioma, B.C.: Sustainability disclosures and market value of firms in emerging economy: evidence from Nigeria. Eur. J. Account. Audit. Finance Res. **7**(3), 1–19 (2019)
24. Al Nasser, Z.: The effect of internal corporate governance on earnings quality, firm performance and firm value in GCC Countries. Doctoral dissertation, Trinity College Dublin (2018)
25. Li, M.Z.F., Oum, T.H., Zhang, Y.: Tobin's Q and Airline Performances. Public Works Manag. Policy **9**(1), 51–65 (2004). https://doi.org/10.1177/1087724x04265137
26. Bose, S., Thomas, K.: Applying the balanced scorecard for better performance of intellectual capital. J. Intellect. Cap. **8**(4), 653–665 (2007). https://doi.org/10.1108/14691930710830819
27. Barad, M.M.: A study of liquidity management of Indian steel industry. Doctoral dissertation, Saurashtra University (2010)
28. Al-mawali, H.H., Zainuddin, Y., Ali, N.N.K.: Balanced scorecard (BSC) usage and financial performance of branches in Jordanian banking industry. World Acad. Sci. Eng. Technol. **42**, 1174–1181 (2010)
29. Niven, P.R.: Balanced Scorecard Step-by-Step: Maximizing Performance and Maintaining Results. Wiley, Hoboken (2002)
30. Anand, M., Sahay, B.S., Saha, S.: Balanced scorecard in Indian companies. Vikalpa J. Decis. Makers **30**(2), 11–26 (2005). https://doi.org/10.1177/0256090920050202
31. Mohammed, I.: Investigating the use of the four perspectives of balanced score card (BSC) as technique for assessing performance by Nigerian banks. J. Account. Tax. **7**(4), 62–70 (2015). https://doi.org/10.5897/jat2014.0148
32. Tarurhor, E.M., Osazevbaru, H.O.: The balanced scorecard as a performance management tool for small and medium scale enterprises in Nigeria. AU-eJournal Interdiscip. Res. **4**(1) (2019). ISSN 2408-1906
33. Pineno, C.J.: The balanced scorecard: an incremental approach model to health care management. J. Health Care Finance **28**(4), 69–80 (2002)
34. Kaplan, R.S., Norton, D.P.: Linking the balanced scorecard to strategy. Calif. Manag. Rev. **39**(1), 53–79 (1996). https://doi.org/10.2307/41165876
35. Tibbs, C.Y., Langat, L.K.: Internal process, learning perspective of balance scorecard and organisational performance. Int. J. Econ. Commer. Manag. **4**(11), 458–474 (2016)
36. Fikriani, A.: The effect of balanced scorecard implementation on management performance PT Bank Syariah Mandiri. Al-Uqud: J. Islamic Econ. **3**(1), 18 (2019). https://doi.org/10.26740/al-uqud.v3n1.p18-33
37. Patalas-Maliszewska, J.: Assessing the impact of ERP implementation in the small enterprises. Found. Manag. **4**(2), 51–62 (2012). https://doi.org/10.2478/fman-2013-0010
38. Rashid, A., Kemal, M.U.: Impact of internal (micro) and external (macro) factors on profitability of insurance companies. İktisat Politikası Araştırmaları Dergisi **5**(1), 35–57 (2018)

39. Cao, L., Li, L.: The impact of cross-channel integration on retailers' sales growth. J. Retail. **91**(2), 198–216 (2015). https://doi.org/10.1016/j.jretai.2014.12.005
40. Ivanov, C.-I., Avasilcăi, S.: Measuring the performance of innovation processes: a balanced scorecard perspective. Procedia Soc. Behav. Sci. **109**, 1190–1193 (2014). https://doi.org/10.1016/j.sbspro.2013.12.610
41. Tjia, O., Setiawati, L.: Effect of CSR disclosure to value of the firm: study for banking industry in Indonesia. World J. Soc. Sci. **2**(6), 169–178 (2012)
42. Chung, K.H., Pruitt, S.W.: A simple approximation of Tobin's Q. Financ. Manag. 70–74 (1994)
43. Kala, D., Bagri, S.C.: Balanced scorecard usage and performance of hotels: a study from the tourist state of Uttarakhand, India. Asia-Pac. J. Innov. Hosp. Tour. (APJIHT) **3**(2), 1–21 (2014). https://doi.org/10.7603/s40930-014-0009-8
44. Braam, G.J.M., Nijssen, E.J.: Exploring the antecedents of Balanced Scorecard adoption as a performance measurement and strategic management system (2008)

Corporate Governance and Sustainability of Financial Institutions: A Study of GCC Countries Banks

Samer Al-Waaeli(✉)

University of Delhi, New Delhi, India

Abstract. Corporate governance has become a major concern of all organization to sustain in today's business environment and has received in recent years much attention in its effect on the sustainability of financial institutions in both developed and developing countries (Al-Nasser Mohammed and Mohammed 2017). Especially in the context of strategic decisions which are crucial for the longevity of the institution, this paper primarily focused to highlight the correlation between corporate governance and sustainability of financial institutions as selected banks in GCC countries. Sustainability of financial institutions which was measured by strategy policy development and risk management while the corporate governance attributes included board composition, the board size and audit committee, for the study. The study relies on a descriptive research design with a sample 128 responses from a structured questionnaire. Data has been obtained from 119 out of the 128 respondents from GCC countries board members and analyzed using descriptive statistics and multiple regression analysis. It has been found that there exists a positive correlation between corporate governance and sustainability of financial institutions in GCC countries.

Keywords: Corporate governance · Sustainability · Financial institutions · GCC

JEL Classification: G3 · G30 · G32 · G34 · G38

1 Introduction

"Corporate Governance" can be defined as the concept through which various companies operate in a transparent as well as accountable manner (Alkhamees 2013). As per the Cadbury committee of UK in the year 2002, "Corporate Governance" can be defined as the system by which a firm is directed and controlled. Henceforth, it is clear from the said concept of "Corporate Governance", that the corporate world shall promote transparency as well as accountability for the purpose of fulfilling the expectations of the stakeholders and further protecting their interest through regulations, laws and rules concerning "Corporate Governance" (Aggarwal 2013).

Therefore, when all of the above mentioned four activities are undertaken by a corporate entity and incorporated within its corporate policies, then it can definitely

A. M. A. Musleh Al-Sartawi et al. (Eds.): EAMMIS 2021, LNNS 239, pp. 430–440, 2021.
https://doi.org/10.1007/978-3-030-77246-8_40

achieve higher level of sustainability and accordingly implement the CG principles more effectively (Raut 2019).

Research Questions
Past studies confirm that organizations which focus more on sustainability practices have higher financial performance measures by return on assets, profits before taxation, and cash flow from operations compared to those without such commitments in some activity sectors (Cebeci 2012), Corporate governance from Islamic perspective does not differ much with the conventional definition as it refers to a system by which companies are directed and controlled with a purpose to meet the corporation's objective by protecting all the stakeholders' interest and right (Alhumoudi 2016).

The present study has been envisaged to gather literature and supportive evidence while investigating the following research problem or questions: As can be seen above, the research so far has successfully identified the association between corporate governance and firms' performance as studied by (Kanojia 2018) but the research on corporate governance and its association with the performance of financial institutions in general and with the sustainability of financial institutions is scant. Thus, to fill an existing gap in the literature,the present study poses this main question to be answered by this research.

Does corporate governance have a significant impact on the sustainability of financial institutions in GCC countries?

This question is divided into the following sub-questions.

1. Does corporate governance have a significant impact on strategy development in the financial institutions of GCC countries?
2. Does corporate governance have a significant impact on risk management in the financial institutions of GCC countries?

Second Question
Is there significant difference in "Corporate Governance" of GCC counties financial institutions, this question is divided into the following sub questions:

1. Is there significant difference in strategy and policy development in GCC countries financial institutions?
2. Is there significant difference in risk management in GCC countries financial institutions?

Objectives of the Research
Below are the various objectives which will be achieved by the research through this paper.

1. To examine the impact of corporate governance on strategy development in the financial institutions in GCC countries

2. To investigate the impact of corporate governance on risk management in the financial institutions in GCC countries
3. To provide useful suggestions and policy implications for regulators and policy makers in GCC countries

There is a significant different in "Corporate Governance" of GCC counties financial institutions, this objective is divided into the following sub-objectives: To know if There is a significant difference in board size, board composition, CEO Duality, Board committee and Audit committees of GCC countries financial institutions To study and compare if there is a significant difference in strategy and policy development of GCC countries financial institutions.

Hypotheses

The following section provide the hypothesis formulated to test the data collected in both countries and derive evidence (Bhagat 2008). The hypothesis which has been tested by conducting an in-depth on field survey within the Financial Institutions of Gulf countries: H01a: There is a significant impact of Corporate Governance principles on strategy policy development in GCC countries financial institution.

Research Methodology Used

The study made use of both "Primary and Secondary data" where the former has been collected with the help of a structured questionnaire which has been pretested and validated and the latter made use of the extensive literature survey. Through this methodology, the study attempts to provide for both primary and secondary data analysis where in secondary data has been used for the purpose of analyzing already existing research on the subject matter – board size, board composition, CEO duality, board committee, audit committee, strategy policy development and risk management which is mainly concerned with the analysis of "Corporate Governance" and sustainability as applicable within Gulf countries. However, on the other hand the very aim of primary research had been to collect on field data from respondents which has been further analyzed under this study using regression analysis, for the purpose of establishing clear comparison between the application of "Corporate Governance" norms within GCC region.

- Dependent Variables:
  Below are the dependent variables which will be analyzed within this study as part of the correlation and regression analysis:
  Strategy development and Risk management.
- Independent Variables:
  Independent variables of this study are as follows: Board size and Board composition.

"Corporate Governance" and It is Principles

"Corporate Governance" can be defined as the rules, regulations and laws which are concerned with the operations of corporate entities and how they are controlled as well as regulated (K.N. and C 2019).

Data Collection

Two significant Data Collection methods should be followed under this research will be both primary data as well as secondary data.

1. **Primary Data**
   In this research primary data should be collected by conducting an in-depth survey of the respondents of Financial Institutions using the Likert Scale, which is used within the questionnaire, that has been attached at the end of this research. The questionnaire had seven sections with different number of questions in each segment to adhere to the objective that the study intended to test. The specific sections were included to test the specific objectives like the first section had the questions that referred to board size, the second segment captured the information related to board composition, the third section had questions related CEO duality, the fourth section had questions related to board committee, fifth section was about the different role played by the audit committee, sixth section was about the first dependent variable–strategy policy development and the last section was about risk management.
   Secondary data analysis Secondary **data analysis is collected by analyzing various literary sources like: Books, Journals, Research Papers and Articles Websites etc.**

Data Analysis Instruments

The "reliability and analysis of the data in this study were analyzed using Statistical package for social sciences (SPSS v 16.0) (Al-Matari 2012). Analysis of Moment Structure (SPSS AMOS v.16) which were used to confirm factor structures, and the same was visualized by Tableau Software version 10.0."

Also, Multiple regression analysis is designed to establish the relationship between multiple independent variables and one dependent variable as concluded by (Bachiller 2018). Depicted below is the multiple regression model used in the study. The equations used to establish the relationship was as follows:

$$\mathrm{SPD} = \alpha + \beta_1(\mathrm{BS}) + \beta_2(\mathrm{CEOD}) + \beta_3(\mathrm{BC}) + \beta_4(\mathrm{BCO}) + \beta_5(\mathrm{AC}) + \beta_6(\mathrm{AC}) + \varepsilon$$

$$\mathrm{RM} = \alpha + \beta_1(\mathrm{BS}) + \beta_2(\mathrm{CEOD}) + \beta_3(\mathrm{BC}) + \beta_4(\mathrm{BCO}) + \beta_5(\mathrm{AC}) + \beta_6(\mathrm{AC}) + \varepsilon$$

Where: (SPD) = Strategy policy development, (RM) = Risk Management, (BS) = Board size, (BC) = Board of directors' composition, (CEO D) = CEO Duality (BD) = Board committee, (AC) = Audit committee and α = Constant term representing other factors other than the above "Corporate Governance" which are not defined in the model.

Sample Size

The sample size taken for the purpose of conducting this research is 128 which includes the below-mentioned professionals of GCC bank professionals:

Senior Manager, Board member, Audit Committee Member, Accountant, and Internal Auditor.

Research Tools Used

The above survey was carried out with the help of the questionnaire (referred to in the appendix). In addition, the kind of inquiries made were, for the most part, it was open inquiries.

Qualitative Research: Primary data is used in this research which has been analyzed various statistical tools as (Chepkoech 2015).

Descriptive statistics was carried under this research for further determining the mean as well as maximum or minimum values with respect to each variable, for the purpose of determining the extent to which variable can deviate under this paper.

The Table 1 above showing GCC countries, board size has the maximum value of well and so it can be inferred on the basis of this value that the maximum number of board members within a board GCC region is 25, on the other hand the minimum number of board members within a board in GCC region is 12.67, which means that the minimum no. of members in a broad within the country GCC region is at least 12.

As discussed in the previous chapter the value of Cronbach alpha of more than 0.7 is good enough to conduct the further technical analysis. Scores closer to 1 therefore suggest a greater level of confidence. The table shows the Cronbach alpha's overall group value for the different constructs used in the questionnaire is given in Table 2. The table indicates that all the constructs used has a Cronbach value ranging from 0.8 to 0.9 and that the overall reliability was 0.90, indicating that the questionnaire is very reliable as study by (Author 2017).

The above Table 3 results reflect upon Durbin-Watson correlation value of 1.967, which means that a linear dependence has been recorded in case of corporate governance and no autocorrelation is present in the model strategy policy development since value of DW statistic is close to 2. R squared further reflects that corporate governance did influence performance of financial institutions within this research and that the same may not be impacted by the percentage of factors other than "Corporate governance".

From the above Table 4 ANOVA table, the P-value is 0.00 which means that the model is statistically significant since the P-value is less than 0.005. This depicts the significance of the regression analysis done at 95% confidence level.

The above Table 5 results reflect upon Durbin-Watson correlation value of 1.858, which means that a linear dependence has been recorded in case of corporate governance and no autocorrelation is based in Risk management model since value of DW statistic is close to 2. R squared further reflects that corporate governance did influence performance of financial institutions within this research and that the same may not be impacted by the percentage of factors other than "Corporate Governance".

From the above Table 6 ANOVA table, the P-value is 0.00 which means that the model is statistically significant since the P-value is less than 0.005. This depicts the significance of the regression analysis done at 95% confidence level.

Regression Coefficient Results

From the above Correlation Table 7, it has been found that none of the correlations was

Table 1. Descriptive statistics for explanatory variables descriptive statistics GCC countries

	N	Minimum	Maximum	Mean	Std. deviation	N	Minimum	Maximum	Mean	Std. Deviation
Strategy policy development	89	4.20	21.00	17.71	2.93	119	12.60	21.00	18.16	2.01
Risk management	89	6.20	21.00	16.60	2.51	119	10.80	20.80	16.49	1.64
Board size	89	5.17	25.83	20.74	3.52	119	12.67	25.83	21.16	2.37
Coequality	89	4.20	21.00	16.74	2.74	119	9.60	21.00	17.06	2.07
Board composition	89	4.20	21.00	16.57	2.76	119	12.60	21.00	17.05	1.99
Board committee	89	5.17	25.83	20.78	3.31	119	14.50	25.83	21.31	2.19
Audit committee	89	5.17	25.83	20.90	3.49	119	12.17	25.83	21.09	2.30

Source: Researcher's Output

Table 2. Reliability and validity

	GCC Cronbach's alpha if item deleted
Board size	0.851
CEO duality	0.831
Board composition	0.838
Board committee	0.822
Audit committee	0.834
Strategy policy development	0.839
Risk management	0.871

Source: Researchers output

Table 3. Model summary of SPD for GCC.

Model	R	R square	Adjusted R square	Std. error of the estimate	Durbin-Watson
1	.833[a]	.693	.680	1.11654	1.967

[a]Predictors: (Constant), audit committee, board size, board composition, board committee, CEO duality.
[b]Dependent Variable: strategy policy development.
Source: Researcher's Output

Table 4. ANOVA OF SPD for GCC

Model		Sum of squares	Df	Mean square	F	Sig.
1	Regression	315.573	5	63.115	50.627	.000[b]
	Residual	139.625	112	1.247		
	Total	455.199	117			

[a.]Dependent Variable: strategy policy development.
[b.]Predictors: (Constant), audit committee, board size, board composition, board committee, CEO duality.
Source: Researcher's Output

beyond 5% significance level, thus representing the fact that the independent variables under this research are not serially correlated,

Finding and Conclusion:
On the basis of the primary and secondary research that has been conducted under this study, below are the various findings which can provide a better understanding of the results that has been obtained under this research: The primary research reflects that there is no significant difference between the board size in GCC countries with respect

Table 5. Model summary of RM for GCC

Model	R	R square	Adjusted R square	Std. error of the estimate	Durbin-Watson
1	.424[a]	.180	.143	1.52171	1.858

[a]Predictors: (Constant), audit committee, board size, board composition, board committee, CE duality.
[b]Dependent Variable: risk management.
Source: Researcher's Output

Table 6. ANOVA of RM for GCC

Model		Sum of squares	Df	Mean square	F	Sig.
1	Regression	56.932	5	11.386	4.917	.000[b]
	Residual	259.347	112	2.316		
	Total	316.280	117			

[a]Dependent Variable: risk management.
[b]Predictors: (Constant), audit committee, board size, board composition, board committee, CEO duality.
Source: Researcher's Output

to their Financial Institutions. CEO Duality is found to be prevalent in GCC countries with most significant impact of the same up on "Corporate Governance" measures which is undertaken within the Financial Institutions in GCC regions. It was found that there is no significant difference between board committees which were in existence within the financial Institutions in GCC countries. It was found that there is significant difference within the audit committee of GCC countries.

Recommendations

Below are the suggestions which have been farmed by the researcher, for the purpose of providing various recommendations which can be adopted by Gulf countries, for further enhancing their "Corporate Governance" framework and increasing their sustainability to an extent which required, achievable profitability as well as an ethical code of conduct for performing within their respective countries as well as the global community: First and foremost, it is significant for the said countries to incorporate such "Corporate Governance" norms which could establish adequate transparency and accountability within each corporate firm within the GCC region. It is also important for the Gulf countries to incorporate CEO duality, to enhance the decision-making process and reducing the complexities which are otherwise involved in the same due to divided CEO and chairman seats. However, these decisions taken by the CEO shall always be in conformity to the interest of shareholders of the company.

Table 7. Correlations of GCC countries Banks

		Board size	CEO duality	Board composition	Board committee	Audit committee	Strategy development	Risk management
Board size	Pearson Correlation	1						
	Sig. (2-tailed)							
	N	118						
CEO duality	Pearson Correlation	.678**	1					
	Sig. (2-tailed)	.000						
	N	118	118					
Board composition	Pearson Correlation	.562**	.632**	1				
	Sig. (2-tailed)	.000	.000					
	N	118	118	118				
Board committee	Pearson Correlation	.544**	.736**	.680**	1			
	Sig. (2-tailed)	.000	.000	.000				
	N	118	118	118	118			
Audit committee	Pearson Correlation	.421**	.623**	.532**	.681**	1		
	Sig. (2-tailed)	.000	.000	.000	.000			

(*continued*)

Table 7. *(continued)*

		Board size	CEO duality	Board composition	Board committee	Audit committee	Strategy development	Risk management
	N	118	118	118	118	118		
Strategy development	Pearson Correlation	.426**	.587**	.570**	.803**	.700**	1	
	Sig. (2-tailed)	.000	.000	.000	.000	.000		
	N	118	118	118	118	118	118	
Risk management	Pearson Correlation	.239**	.237**	.264**	.299**	.407**	.161	1
	Sig. (2-tailed)	.009	.010	.004	.001	.000	.082	
	N	118	118	118	118	118	118	118

**Correlation is significant at the 0.01 level (2-tailed).
*Correlation is significant at the 0.05 level (2-tailed).
Source: Researcher's Output

References

Aggarwal, P.: Impact of Corporate Governance on Corporate Financial Performance. [online] Iosrjournals.org. Available at: http://iosrjournals.org/iosr-jbm/papers/Vol13-issue3/A01330105.pdf [Accessed 3 Nov. 2019] (2013)

Alhumoudi, H.Y.: Corporate governance mechanisms and firms' performance: an empirical analysis of firms listed on the Saudi stock exchange. Int. J. Acc. Financ. Rep. **6**(2), 101 (2016). https://doi.org/10.5296/ijafr.v6i2.9192

Alkhamees, A.: The impact of Shari'ah governance practices on Shari'ah compliance in contemporary Islamic finance. J. Bank. Regul. (2013). https://doi.org/10.1057/jbr.2012.12

Al-Matari, Y.A., Al-Swidi, A.K., Bt Fadzil, F.H.: Corporate governance and performance of Saudi Arabia listed companies. Br. J. Arts Soc. Sci. **9**(1), 1–30 (2012)

Al-Nasser Mohammed, S.A.S., Muhammed, J.: The relationship between agency theory, stakeholder theory and *Shariah* supervisory board in Islamic banking. Humanomics **33**(1), 75–83 (2017). https://doi.org/10.1108/H-08-2016-0062

Author, C., Bin Tariq, Y., Abdul Hameed, S., Abbas Jadoon, I., & Ali Jinnah University Islamabad, M.: Corporate Governance in Financial Sector Companies of Pakistan: Current State and Room for Improvement World (2013).

Authors, F.: Corporate governance of Islamic banks a comparative study between GCC and Southeast Asia countries. Int. J. Islamic Middle Eastern Financ. Manage. **7**(3), 346–362 (2014). https://doi.org/10.1108/IMEFM-01-2013-0001

Bachiller, P., Garcia-lacalle, J., Bachiller, P., Garcia-lacalle, J.: Corporate governance in Spanish savings banks and its relationship with financial and social performance. Manage. Decis. **56**(1), (2018). https://doi.org/10.1108/MD-01-2017-0079

Banking, J.: Shari'ah supervision, corporate governance and performance: conventional vs. Islamic banks, pp. 1–39 (n.d.)

Basalighe, A., Khansalar, E.: A review of the relationship between corporate financial performance and the level of related party transactions among listed companies on Tehran stock exchange. Int. J. Econ. Financ. **8**(7), 330 (2016). https://doi.org/10.5539/ijef.v8n7p330

Bhagat, S., Bolton, B.: Corporate governance and firm performance. J. Corp. Finance **15**, 66–79 (2008). https://doi.org/10.1016/j.jcorpfin.2008.03.006

Cebeci, I.: Integrating the social Maslaha into Islamic finance. Acc. Res. J. (2012). https://doi.org/10.1108/10309611211290158

Chepkoech, Z.M., Chenuos, N.K., Kosgei, D.K.: Green procurement strategies as determinants of financial. Int. J. Small Bus. Entrepreneurship Res. **3**(2), 1–15 (2015)

Kanojia, S., Bindra, H.K.: Corporate governance, board of directors and financial performance: evidence from Indian listed banks. Amity Bus. Rev. (2018)

K.N., D., C., S..: The role of corporate governance pratices in selected indian financial institutions. Retrieved 10 November 2019, from http://www.ijbarr.com/downloads/3105201413.pdf (2019)

Raut, S.: Corporate Governance and Sustainability Concepts. Retrieved 16 November 2019, from http://iodglobal.com/Articles/Corporate%20Governance%20and%20Sustainability%20Concepts.pdf (2019)

The Relationship Between Human Resource Practices and Organizational Performance

Jasem Taleb Al-Tarawneh[1(✉)], Mohd Saiful Izwaan Saadon[2], and Aya Naser Maqableh[1]

[1] Doctor of Philosophy Management, Faculty of Business, Universiti Malaysia Terengganu, 21030 Kuala Nerus, Terengganu, Malaysia

[2] Faculty of Business, Economics and Social Development, Universiti Malaysia Terengganu, 21030 Kuala Nerus, Terengganu, Malaysia

Abstract. This study aims to analyses how Jordanian public universities take advantage of the relationship between HRM practices and organizational performance with their combined dimensions. This study is so important for different sectors to choose appropriate HRM practices which help to organizational performance in the organization. As well as, the organizational performance in the Jordanian universities is low comparing with expected results. However, HRM practices integration formulate strong and influential instruments to put a figure on ways Organizational performance. Thus, the problem of this research relates to the gap between the expected organizational performance and the actual organizational performance. The literature review outlined a wide range of relations between HRM practices and it is a positive relationship. The results display gaps in the plane and the implementation of researches. the researchers observed the dominance of positive results, in the studies that seek to evaluate the impact of HRM on performance.

Keywords: Organizational performance · HRM practices · Recruitment and selection · Training and development · Rewards systems · Performance evaluation

1 Introduction

Nowadays institution that follows a successful line in work depends heavily on human capital. In addition to this, human capital is considered a valuable element that may grant benefits to an organization, although human capital alone will not add value to institution t without taking the decisions correctly. Moreover, The Human Resources (HR) of an institution offers the possibility of synergy for sustained competitive advantage, particularly, when it is properly deployed and applied. The traditional HR strategy, as well as the formal system for managing people in the institution, is essentially concerned with transactional and administrative support services. To perform successfully, the roles of business partners and change agents under the HR strategy must be highly knowledgeable, multi-skilled, and be able to acquire core competencies like business knowledge,

A. M. A. Musleh Al-Sartawi et al. (Eds.): EAMMIS 2021, LNNS 239, pp. 441–454, 2021.
https://doi.org/10.1007/978-3-030-77246-8_41

vision strategy, global operating skills, reliability and integrity, internal consulting skills, and many others [3]. Therefore, the human resources management practices affect organizational performance. Whereas, Organizational performance concerning the ideas of effectiveness and efficiency. Because any institution requires to produce the right things by adopting the fewest possible inputs, then just like institution guarantee powerful and strong organizational performance [3]. Commonly, the institution seeks to implement well in different areas: First, they attempt to perform well financially, which seek to accomplish the best return on their investment, further, they must add as much rate as possible in their production operation. Second, they seek to perform well in concepts of the market.in other words, the institution needs to accomplish as much market share as they can, and they should be manufactured a product that is in demand and at a price that enables them to compete strongly on the market [32]. Add to that, companies can manage human resources and foster organizational performance, by having imagined of HR practices working alongside other dimensions. According to [27], these are: Career planning and HR planning, Performance Evaluation, Reward systems, Training, Selection.

On other hand, higher education plays a significant role in various bodies and aspects in the country, such as the national economy and the industrial sector. As well as, the Hashemite Kingdom of Jordan possesses efficient and competent human resource systems that can provide the community with lifelong learning experiences closely related to its current and future needs in response to sustainable economic development and stimulated by preparing educated individuals and skilled workforces. Jordan is witnessing, in general, a qualitative boom in the field of higher education since the early nineties, and the capital Amman in particular, in addition to the major Jordanian cities such as Irbid and Zarqa, contains most of the most important scientific edifices in the Kingdom in most specialties. The private university education sector witnessed a remarkable boom, as Amman was the first in the Middle East. The students who hold a high school diploma are accepted in public or private universities or colleges. Most universities in Jordan implement the American university model based on the credit hours' system, which gives students the flexibility to choose the number of hours and morning or evening hours. There are ten public universities, most of which are affiliated with universities in the United States and the United Kingdom. Jordanian universities attract a large number of Arab foreign students and non-Arabs every year. Thus, this study aimed to investigate the relationship between human resource practices and organizational performance.

2 Problem Statement

An institution's success and competitiveness, in the long term, count on the existence of qualified managers inside it. Many benefits are gotten when institutions strongly measure their treatment of Organizational Performance problems, develop and apply Human Resource Practices plans the problem of the study lies in the different role of HRM practices that followed HRM department, thus reflecting the level of efficiency and effectiveness of these resources and their contribution to the competitive advantage in different [30]. Hence, it is the main concern of the HRM department to explore all possible ways of creating and sustaining organizational performance [1; 33]. Add to that effective HRM practices emphasize on the organizational performance level of the organization and ward off from being sluggish. They also highlight that an organization's efficiency is proven by a robust package of actions and practices pointed at taking advantage of its strengths. In this regard, HRM provides a better trace of support to the higher management before the time of selecting and hiring suitable personnel, managing, and training them to meet up with the requirements needed to get better of organizational performance. The varying nature of humans makes HRM practices a difficult task. Consequently, it is highly significant for organizations to emphasize on strengthen their HRM practices. Thus, attaining better organizational performance needs successful, effective, and efficient exploitation of organizational competencies and resources to create a competitive advantage. HRM practices so on must be expressed and applied by HRM practices specialists with the help of managers to achieve an efficiency of organizational performance. Organizational performance is an important outcome that the management will use to assess the business procedures and the organization's actions. It is significant because a perfect performance would give them confidence in management and thus secure a sustained and money-making organization future [3].

This study is so important for different sectors to choose appropriate HRM practices which help to organizational performance in the organization. As well as, the organizational performance in the Jordanian universities is low comparing with expected results. However, there is a need to increase research opportunities that can enrich the understanding of human resources practices and their relationship to organizational performance in Jordanian universities [4]. As well as the education sector as such is the single most important constituent of the service sector which is very crucial for the economic development of any country. Moreover, the Jordanian education sector is passing through a very critical phase according to Covid-19 because of its need for the unprecedented pace of automation, communication revolution. All these challenges in the economics, social-cultural, technological, psychological environment have important implications on the HRM practices being followed in the education sector as well as on the performance of the Jordanian universities. All the HRM practices like recruitment, training, and development, compensation and rewards, performance appraisal, promotion and empowerment, sharing of decisions, job security benefits, welfare measures, and separations in the form of lay-offs, retrenchment have come under tremendous pressure. As a result, the performance of the education sector has been affected very seriously. In universities, HRM practices integration formulate strong and influential instruments to put a figure on ways Organizational performance. Thus, the problem of this research relates to the gap between the expected organizational performance and the actual organizational performance in the administrative departments of the Jordanian government universities. Thus, this study comes to expand the effort in this area and achieve the most distinguished aim in this study, is there a link occurred in the relationship between human resource practices and organizational performance at Jordanian public universities were selected as part of the higher education sector in Jordan.

3 Research Questions and Research Objectives

Since the success of educational institutions and increase their competitiveness depends on the existence of management excellence, successful and qualified within them, in addition to the ability to provide high quality and efficiency of educational services. Many benefits are gained when educational institutions measure vigorously to address organizational performance problems and apply human resource practice plans. The current research problem lies in the different role of human resources management practices of the Human Resources Management Section, reflecting the level of efficiency and effectiveness of these resources and their contribution to the advantage Competitiveness in the higher education sector. Thus, the main concern of HRM is to maintain organizational performance within educational institutions. Therefore, this research aims to expand the effort in this field and achieve the most distinctive objective in this research to find out if there is a link between the role in the relationship between human resource practices and organizational performance in Jordanian public universities. From this the following main study questions emerge: The main question: What is the relationship between human resource practices and organizational performance at Jordanian public universities?

4 Significance of the Research

The importance of this study comes from the fact that it is based on extrapolation "the Relationship between Human Resource Practices and Organizational Performance". Therefore, the importance of the study divided into the following: Theoretical Significance it will add to the literature to establishing an actual value in the academic research in human resource practices and organizational performance field and encourage other researchers to build up and improve studies, which support this field to achieve the desired goals in the future. From the Practical perspective. the study provides convincing evidence about the direct relationship between human resource practices and organizational performance. Furthermore, the decision-makers who are interested in informational and analytical studies with important information on the relationship between human resource practices and Organizational Performance and will provide assistance to officials and decision-makers in Jordanian universities to identify the relationship between human resource practices and organizational performance, and thus help them in taking appropriate measures to implement these concepts.

5 Literature Review and Previous Studies

Nowadays, Organizations work in an environment that is characterized by uncertainty, unpredictability, and change, which creates several challenges [10]. The modern working environment contains numerous factors, such as increased globalization, speedy technological change, and the increasing need for qualified employees and improved performance. This forces organizations to try to realize the best use of the resources at their disposal to achieve a competitive advantage. Human resources are considered critical factors that contribute to an organization's success [41]. In addition, the appeared need for organizations to change their processes and innovate, as the market requires, the effective response is necessary in order to meet up with the highly competitive market characterized with continuous changing. The Human Resources role, which was primarily represented favorably in administrative worker, personnel data management and legal/legislation requirements, now includes focusing on tangible properties and financial resources. strategically using human resources is required to overcome different business challenges. Managing the human resources of an organization requires the use of different components that play an important role in helping organizations create and maintain the performance they want, as this will affect the attitudes and behaviors of employees [20]. Additionally, some human resource management components can affect employee commitment and motivation. Such as staffing and selection, training and development, performance evaluation, collaboration, and payment and reward [13]. The most important task for many organizations today is to survive the present competitive market through the impact of privatization, liberalization, globalization, competition. Commonly, the institution seeks to implement well in different areas: First, they attempt to perform well financially, which seek to accomplish the best return on their investment, further, they must add as much rate as possible in their production operation. Second, they seek to perform well in concepts of the market.in other words, the institution needs to accomplish as much market share as they can and they should be manufactured a product that is in demand and at a price that enables them to compete strongly on the market [18].

Good HRM can be applied as a magic weapon for organizations, and it is tricky for other rivals to imitate. Consequently, it will share in making a competitive advantage for the organization in regards to human capital [19]. Therefore, this part will shed the light on human resource management, strategy, and practice. HRM begin in England in half of the 1800s in the apprenticeship period and turn into more developed on the industrial revolt at the end of 1800s. In the 19th century, Taylor explained that scientific management and the industrial psychology of employees must be noted. In this case, it was planned that employees should achieve the happiness of the employees with the job and its efficiencies. They could also feel the same by drastic changes in technology, the development of organizations, the increase of associations, and government involvement result by the development of personnel departments in the 1920s. In this stage, personnel administrators were called ‘welfare secretaries’ [19]. The investigators are realizing the concept HRM and personal management in other several ways [38]. Due to the development of and changes in the world of management, the term ‘management’ to HRM was changed. Thus, a new term would seem appropriate to take new thought, concepts of HR, human resource management (HRM) outlined as an asset of united strategies with a political and philosophical underpinning. Accordingly, there are four characteristics of HRM: a specific set of faiths and anticipations, a strategic thrust informing decision about individual management, the central involvement of line managers, and reliance upon a group of levers to characterize the employee relations [7]. As per [25] HRM is generally actuated out of HR systems that bring jointly HR attitudes in an intelligible way. These positions help describe the major values and lead principles that one should take on board when managing the individual. HRM practices include the familiar methods used when managing individuals. HR agendas give HR strategies, policies, and practices to be performed according to plan. Furthermore, the main aim of HRM is to certify that the organization is arable of realizing success by individuals. Moreover, HRM systems can be the substance of organizational abilities that led organizations to pick up and achievement new opportunities [21]. As well as HRM depends on 12 agreement goals as a following: managing individuals, supporting HRM policies, encouraging team-working, empowering employees. Also, developing planned prize strategies, reading a strong customer-first philosophy, improving employee involvement, building greater employee commitment, raising line management accountability, improving the extended role of managers as enablers, and supporting the emergence of a closer group of HR policies, step and systems [7]. Also, it has struggled that managing HR is more problematic than managing capital [32]. Organisations adopt HRM practice to amend the actions, attitudes, and awareness of employees in a way that developed its performance and anticipated outcomes. In addition, what should be considered is that the HRM strategy is not fixed; they change from one company to another. Many scholars defined different practice that is associated with HRM. These include recruitment, selection, training and development, motivation, and Performance Evaluation [20].

6 Recruitment and Selection

Any organisations to be effective its need to catch the best staff by the important and major activity of recruitment. Therefore, recruitment and selection forms as a main part of the main activities depend on HRM: namely, the acquisition, improvement, and reward of staff. It repeatedly shapes a substantial part of the action of HR managers. However, non-specialists, just like line managers, often (for good reason) take recruitment and selection resolution [34]. Recruitment outlined as the operation of searching for and finding candidates for jobs. The resulting group is then interviewed so that the true people can be chosen. According to theory, the recruitment operation is finished off with the reception of applications; in practice, the outcoaching activity boils down to the examination of applications to remove those who are not capable for the work [11]. Likewise, recruitment goals to acquire the correct number of accomplished workers to run into the human resource requirements of organisations. Thus, numerous staffing methods are used to realize this purpose [13]. Recruitment methods can be either internal or external, but most companies tend to emphasise three external methods: advertising, online recruitment, and the use of employment agencies. Each of these has been created to increase the possibility of recruiting gifted employees [7]. The importance of this operation can be seen from the two-point view as a following: 1) the organisation, which depends on the recruitment and selection operation to develop and improve individual performance. 2) The individual, who depends on a good recruitment and selection process. The procedure at large affects safety and trust for individuals within the organisation. Moreover, the importance of the selection procedure is that it is one of the most significant processes in employ policy, and HRM must be organised to dissatisfy applicants looking for the job [26]. The most important object when choosing the individual for the job is the effectiveness. The interviewer must share the individual specification, and then through the interview and testing one can regulate the efficiency of the person therefore determine whether they meet the requirements of the job [7].

7 Training and Development

Due to the many challenge's workers face in the work environment as underappreciated assets, organisations requirement to participate in the training and development of their human capital to get better their abilities and capacities. Further, the motivation behind having training and improvement is to rise and inform the abilities, knowledge, and experiences of an organisation's workers [13]. Training and Development: Training and development is an effort to develop the current and future performance of workers by developing their basic skills. Training is the attempt create by the firms to facilitate the learning operation for its worker [12]. Moreover, the training and development consist of five steps, Needs Analysis, Instruction Design, Validation, Implementation and Evaluation and Follow-up. On other hand, Competencies are applied in development centres, which assist entrant, reinforcement their understanding of the competencies they needed now and in the future. By utilizing this, they can outline their self-directed learning programmes [7]. In addition, training and development rise workers' efficiency and commitment, which in turn improves an organizational performance [42]. The most significant basic components connected with the concept of training and development are culture, training, development, and knowledge Additionally, worker training and development agendas may be activities destined to raise organizational and employee performance in the short run, while job management be a way for the employee and the organisation to raise organizational and employee performance and long-run satisfaction together [11]. Training and development consist of many components that similarly affect one another Trainees must appreciate the need for training; then, it will outcome in a deprived response and operation of working out objectives. Trainers must be accountable for stating the training material in a way that suits trainees and reaches the training objectives. Training Material which must be acceptable and distributed by the trainer to the trainees. And training environment: The environment is a place of training and must be logistically satisfying in terms of its audio and visual instruments [37].

8 Rewards Systems

Contributions made to an organisation by workers and worker achievements should be recognized and reciprocated in some system of reward. A reward system outlined as "a package/systems that take in rewards and benefits, just like holiday leave, medical benefits, transport payments, and performance bonuses" [23, 104]. Furthermore, a reward system is a process used to develop motivation and job engagement by valuing people following their contribution. Also, encouraging productive, discretionary effort by ensuring that people are positive and interested in their jobs, that they are satisfied to work for the organisation and want to continue working for it, and that they take action to accomplish organisational and individual objectives [7]. On other hand, reward systems typically serve many purposes, just like attracting, retaining, and motivating workers [31]. The components of a reward system (such as providing compensation) should be deferential to the major objectives of the system, which are to draw, keep, and inspire workers. In a reward system contain the following components [34] Direct financial rewards, Indirect financial rewards, and Nonfinancial rewards. Reward system is one of the most important parts in organization development and success. Furthermore, the reward system designed according to the employee is an efficient method to raise work motivation. The sustainable type of reward is developed following the organization reward philosophy, strategies [2].

9 Performance Evaluation

Performance Evaluation (PE) is an official system of check and evaluation of worker or team task performance [23, 24]. Moreover, employee performance Evaluations are held out by individuals, if there is an understanding by the worker that the operation has not been lead with most topicality and they are not pleased with the Evaluation system this may outcome in complaints and petitions which in turn, may outcome in damage of management's valuable time to address these matter [8]. PE is the assessment of an individual's performance in a methodical way. An improvement tool adopts for the overall development of the worker and the organization. On another hand, the performance is measured with different factors as job knowledge, quality and quantity of output, initiative, leadership abilities, supervision, dependability, co-operation, judgment, versatility and health. Assessment should be confined to the past as well as potential performance also. Further, PA focused on behaviours as a side of assessment because behaviours do affect job outcomes. Performance concerned with the employee and their performance, typically line managers, to look at their performance and development, also to support they need in their role. It is adopted to assess recent performance and focus on future goals, opportunities, and resources requirement [17]. The performance Evaluation info is vital for worker development since knowledge of feedback is essential to motivate and guide performance developments [12]. On other hand, in performance Evaluations, capabilities are adopted to ensure that performance Evaluations do not simply focus on results, but also reflect the behavioural characteristics of how the work is carried out. Thus, Performance Evaluations are adopted to report personal improvement, development plans, and other learning and development creativities [7].

10 Organizational Performance

Organizational performance is one of the most important variables in the management studies and arguably the most imperative factor of the organizational performance. Though the concept of organizational performance is so common in academic literature, its definition is tricky because of its several moral. Accordingly, there is not a Specific definition of organizational performance. In the '50s, the organizational performance was outlined as the range to which organizations, showed as a social system achieve their objectives. Performance evaluation over this time was focused on work, people and organizational structure. Further, in the 60s and 70s, organizations have started to explained new paths to estimate their performance so organization look to performance as the capability to achieve its environment for accessing and using the limited resources. In the 80s and 90s were marked by the investigation that the similarity of organizational objectives is more complex than at first beheld. Managers start to understand that an organization is effective if it achieves its goals by using a smaller number of resources. Thus, organizational theories that followed confirming the thought of an organization that realize its performance objectives based on the constraints imposed by the limited resources [14].

The concept of performance is one of the most common and used concepts in the field of economics and management of institutions. It has received wide attention by researchers and thinkers, especially in economics, and performance has a broad concept and its contents are renewed with the renewal and change of each component of the institution of different types. Interest and analysis in research and administrative studies. According to Girard, et al. [15] Performance means achievement, implementation, results, and therefore performance is intended to perform business based on a goal or set of objectives considering the consequences of such actions. In terms of management, performance can reflect three general concepts of work, work, and success in performance. Performance is expressed in a rather complex set of indicators, and the term success requires the satisfaction of the internal and external parties of the institution. External recognition of the market [35]. Moreover, performance is defined as: "the result of any activity that includes knowing what to do, who performs, how performance is performed and how performance evaluates?" [29]. According to Bianchi [9]. As well as the organizational performance is the integrated system of the organization's work considering its interaction with the elements of its internal and external environment. It includes: the performance of individuals in their organizational unit, the performance of organizational units within the general policies of the organization, the performance of the organization in the context of the economic, social and cultural environment, It is linked to the efficiency and effectiveness of organizations in achieving their goals and to achieve the aspirations of the visions and goals and values are essential, and hence increased attention to human resources management and improve the level of performance of employees because the success of any organization linked to the level of performance of its members and this will improve the institutional performance of the Organization and carry out its duties to the fullest extent possible [6]. It can be noted that organizational performance reflects the organization's ability and ability to achieve its objectives using available resources efficiently and effectively. Organizational performance defined as a continuous holistic activity that reflects the organization's ability to exploit its capabilities according to certain criteria and criteria, based on its long-term objectives. If the performance results differ from the criteria, the management corrects the procedures to address the imbalance between the expected results and actual results [21]. Also, organizational performance defined as the performance and implementation of the functions and tasks assigned to each component and each element of the institution individually or collectively leading to the completion of the mission of the institution or organization and the achievement of its objectives [16]. Therefore, organizational performance reflects how the organization uses its human and material resources in such a way that it can achieve its objectives and thus is a function of organized indicators and characteristics as well as the expertise of the organization's leaders. Further, [31] investigates the relationship between human resources management practices and the performance of the college of humanities and social sciences of the University of Nairobi, Kenya. The result showed that human resource management practices have a positive relationship with organizational performance.

From the above-mentioned literature and after a deep review of previous studies human resource practices and organizational performance, it can be noted that various studies have explored and researched the impact of human resource practices on many aspects of working life, such as productivity, efficiency, competitive advantage, and organizational performance, whether for employees or managers or the entire organization in different sectors and countries. [40] study aimed to look at (HRM) practices as the right mix in the implementation of (TQM) effectiveness. Such practices. The result represented that recruitment and selection process, teamwork and employee empowerment, training and development, performance Evaluations and compensation have the potential to explore and generate new knowledge for the organizations. The outcomes also recognized the advantages from practices during TQM implementation and they believe this approach could give them a chance to achieve their goals. However, several organizations do not adopt HRM and TQM practices because of the lack of identified reasons and less serious outcomes manifested in the effectiveness of such practices. Furthermore, [36] study examined the impact of HRM practices as well as the factors that affect recruitment and selection and training and development on organizational performance of the Jordanian Public University in the Kingdom of Jordan. The result showed that recruitment, selection, training, and development significantly correlated with the organizational performance of the Jordanian Public University. Furthermore, [22] aimed to examine the relationship between HRM and organizational performance can be explained by the effect of the internal and strategic fit of HRM on the cooperative behaviours of employees. The result showed that cooperation with co-workers is negatively concerning to turn over and positively related to sick leave. [34] study aimed to deals with the mediating role of immaterial satisfaction between substantive HR features and organizational performance. The acquired outcomes offer that HRM practices influence immaterial satisfaction and, satisfaction positively effects on organizational performance. Many studies outcomes show that HR practices training, and information technology have a direct impact on organizational performance. The outcomes found that incentive is positively related to organizational performance but did not moderate the relation between both HR practices and organizational performance [28]. [33] the study aimed to investigate the influence of human resource management (HRM) practices, (recruitment and selection) and organizational performance (OP) through mediation role of ethical climates (ECs) in Nigerian educational agencies. The results revealed strong support for the mediating role of ECs on the relationship between HRM practice (recruitment and selection) and OP.

[5] aimed to investigate the mediating role of knowledge management for organizational performance (OP), and human resource management (HRM) practices and the moderating role of organizational learning (OL) on the relationship between organizational performance and knowledge management capability (KMC). The findings of this study look at a significant effect of HRM practices on knowledge management capabilities, organizational culture, organizational performance, and organizational learning in Saudi Arabia during 2019. [32] aimed to focuses on assessing human resource practices in the manufacturing and service sector institutions in Nepal. This study also attempts to find out the relationship between the adoptions of such practices in their performance. The results suggested that there is a negative association between HRM practices and employee turnover whereas there is a positive association between HRM practices and productivity. Moreover, the result suggested that there is a negative association between HRM practices and employee turnover whereas there is a positive association between HRM practices and productivity.

11 Conclusion

The human resource information systems reveal any changes in the internal and external environment and allow the administration to ready for the changes. It also realizes integration and coordination between the different activities of human resources management and other departments. In addition, the importance of human resource management comes from the appeared need for organizations to change their processes and innovate, as the market requires, an effective response is necessary to meet up with the highly competitive market characterized by continuous changing. As a result, of this, managers have realized that the success of any business enterprise in a complex and changing environment would depend on the effective management of its Human Resources and other departments, and through make valued coordination between them. Additionally, some human resource management components can affect employee commitment and motivation. Such as staffing and selection, training and development, performance evaluation, collaboration, and payment and reward. The most important task for many organizations today is to survive the present competitive market through the impact of privatization, liberalization, globalization, competition. Thus, the organization needs to achieve Organizational performance concerning the ideas of effectiveness and efficiency to achieve competitive advantage. The previous study has suggested connections between HR practices and organisational performance. The present research aims were to analyse how Jordanian public universities take advantage of the relationship between human resource practices and organizational performance with their combined dimensions. Its relation is in the recognition of the importance of individuals in the improvement of resources and abilities and in the making of possible competitive advantages. As well as, some studies have developed a conceptual framework for people management practices (such as recruitment, training, job design, engagement, rewards, and evaluation) that create value for influencing the creation, transfer, and integration of knowledge that forms the basis of a company's competencies. To maintain continuous improvement and development of organizational performance, organizations need to emphasize learning at the level of individuals, teams, and organizations. Human resource practices contribute to the organization's learning and facilitate the development of organizational knowledge. It is believed that higher education institutions need a system of committed human resources practices to improve individual knowledge, skills, and ability, and to stimulate knowledge sharing within projects and the institution, learned from the external environment to facilitate access to organizational knowledge and learning. Therefore, this paper proposes a positive relationship between human resource practices and organizational performance in higher education institutions in general and Jordanian government universities. Thus, according to resource-based theories, the influence of organizational learning on the relationship between human resource practices and organizational performance is assumed. To understand the complex situation between human resource practices, organizational learning, and organizational performance in higher education institutions in general and Jordanian government universities in particular, inductive research must be investigated. As a determination, the study highlights the time structure and the beginning examination, done only from the abstracts. The literature review outlined a wide range of relations between HRM practices and make go measure the relationship between these structures. The results also display gaps in the plane and the implementation of research. With consideration to performance measures, there is a sustainable lack in the measurement of variables, in the integration of more than one dimension of performance, in the comparative analysis between Jordanian public universities, and the temporary analysis of performance. With esteem to HRM, the verified the concomitant use of variables related to HR practices and outcomes, which is not recommended. The researcher also specified the employ of a wide range of variables, a point also noted in the international studies (and that has been the base of criticism since it forces determination on knowledge accumulation. In the analysis of the studies that seek to evaluate the impact of HRM on performance, the researcher observed the dominance of positive results, in line with evidence found in international studies. Moreover, considering the dispersion of variables and the methodological matter, the researcher could not draw more overall conclusions about the Jordanian public universities' reality, suggesting a fruitful field for future research. Considering the study's results and findings, the researcher concludes that it is essential to conduct more research that aims towards filling some of the gaps in this area. It is important to focus on the relationship between Human Resource Practices and Organizational Performance and loyalty towards the university. Moreover, further research is necessary to achieve more diversified samples, thus reaching better generalizability. Further studies are needed to shed light and focus on the impact of Human Resource Practices and dimensions particularly on employee performance, to potentially uncover more thorough results.

References

1. Abu-Mahfouz, S.S.: TQM practices and organizational performance in the manufacturing sector in Jordan mediating role of HRM practices and innovation. J. Manag. Oper. Res. **1**(22), 1–12 (2019)
2. Alkhazali, Z., Aldabbagh, I., Abu-Rumman, A.: TQM potential moderating role to the relationship between HRM practices, Km strategies and organizational performance: the case of Jordanian banks. Acad. Strat. Manag. J. **18**(3), 1–16 (2019)
3. Al-Nsour, M.: Relationship between incentives and organizational performance for employees in the Jordanian Universities. Int. J. Bus. Manag. **7**(1), 78 (2012)
4. Alshammari, A.A.: The impact of human resource management practices, organizational learning, organizational culture and knowledge management capabilities on organizational performance in Saudi organizations: a conceptual framework. Revista Argentina de Clínica Psicológica **29**(4), 714 (2020)
5. Argyris, C.: Integrating the Individual and the Organization. Routledge , London (2017)
6. Awunyo-Vitor, D., Hagan, M.A., Appiah, P.: An assessment of performance appraisal system in savings and loans companies in Ghana: evidence from Kumasi metropolis. J. Facul. Econ. Admin. Sci. **4**(2), 191–206 (2014)
7. Bianchi, M.: Beyond the structural modelling for the analysis of organizational performances in the resilience management. Economia Aziendale Online **10**(1), 63–74 (2019)
8. Edwards, J.: A process view of knowledge management: It is not what you do; it's the way that you do it. Electron. J. Knowl. Manag. **9**(4), 297–306 (2011)
9. El-Ghalayini, Y.: Measuring the Readiness of Managers to Effectively Implement a New Human Resources (HR) Process - An Empirical Study on The International Committee of the Red Cross (ICRC). Department of Business Administration. Faculty of Commerce. Islamic University – Gaza (2013)

10. Fong, C., Ooi, K., Tan, B., Lee, V., Chong, A.: HRM practices and knowledge sharing: an empirical study. Int. J. Manpower **32**(5/6), 704–723 (2011)
11. Hamann, P.M., Schiemann, F., Bellora, L., Guenther, T.W.: Exploring the dimensions of organizational performance: a construct validity study. Organ. Res. Methods **16**(1), 67–87 (2013)
12. Hassan, S.: Impact of HRM practices on employee's performance. Int. J. Acad. Res. Account. Finance Manag. Sci. **6**(1), 15–22 (2016)
13. Inyang, B.J.: Strategic human resource management (SHRM): a paradigm shift for achieving sustained competitive advantage in organisation. Int. Bull. Bus. Admin. **7**, 23–36 (2010)
14. Ivancevich, J.: Human Resource Management. McGraw-Hill/Irwin, New York (2007)
15. Juhdi, N., Pa'wan, F., Hansaram, R.M., Othman, N.A.: HR practices, organizational commitment and turnover intention: a study on employees in Klang Valley, Malaysia. Recent Res. Appl. Econ. 30–36 (2011)
16. Katsikeas, C.S., Morgan, N.A., Leonidou, L.C., Hult, G.T.M.: Assessing performance outcomes in marketing. J. Mark. **80**(2), 1–20 (2016)
17. Lambooij, M., Sanders, K., Koster, F., Zwiers, M.: Human resource practices and organisational performance: can the HRM-performance linkage be explained by the cooperative behaviours of employees? Manag. Revue 223–240 (2006)
18. Lim, L.J.W., Ling, F.: Human resource practices of contractors that lead to job satisfaction of professional staff. Eng. Constr. Archit. Manag. **19**(1), 101–118 (2012)
19. Lobontiu, G., Birle, V., Petrovan, A.: Human resource performance appraisal: ranking the importance of derived information. In: Managing Intellectual Capital and Innovation for Sustainable and Inclusive Society: Managing Intellectual Capital and Innovation; Proceedings of the MakeLearn and TIIM Joint International Conference 2015, pp. 1105–1111. ToKnowPress (2015)
20. Minbaeva, D.B.: HRM practices and MNC knowledge transfer. Personnel Rev. **34**(1), 125–144 (2005)
21. Mohamad, A.A., Lo, M.C., La, M.K.: Human resource practices and organizational performance. Incentives as moderator. J. Acad. Res. Econ. **1**(2) (2009)
22. Mori, N., Golesorkhi, S., Randøy, T., Hermes, N.: Board composition and outreach performance of microfinance institutions: evidence from East Africa. Strat. Change **24**(1), 99–113 (2015)
23. Obeidat, D., Yousef, B., Tawalbeh, H.F., Masa'deh, R.E.: The Relationship between Human Resource Management (HRM) Practices, Total Quality Management (TQM) Practices and Competitive Advantages. Total Quality Management (TQM) Practices and Competitive Advantages (December 12, 2018). Modern Appl. Sci. **12**(11) (2018)
24. Oluoch, J.O.: Influence of best human resource management practices on organizational performance: a case of college of humanities and social sciences university of Nairobi, Kenya. Unpublished Thesis of Master of Arts in Project Planning and Management of the University of Nairobi (2013)
25. Pradhan, G.M.: Impact of human resource management practices on organizational performance (a case of Nepal). Batuk **5**(2), 14–31 (2019)
26. Sabiu, M.S., Ringim, K.J., Mei, T.S., Joarder, M.H.R.: Relationship between human resource management practices, ethical climates and organizational performance, the missing link. PSU Research Review (2019)
27. Sacchetti, S., Tortia, E.C., Lopez Arceiz, F.J.: Human resource management practices and organizational performance. The mediator role of immaterial satisfaction in Italian Social Cooperatives (2016)
28. Saifalislam, K.M., Osman, A., AlQudah, M.K.: Human Resource Management Practices: Influence of recruitment and selection, and training and development on the organizational performance of the Jordanian Public University. Organization **3**, 1–08873 (2014)

29. Tiwari, P., Saxena, K.: Human resource management practices: a comprehensive review. Pakistan Bus. Rev. **9**(2), 669–705 (2012)
30. Usrof, H.J., Elmorsey, R.M.: Relationship between HRM and TQM and its influence on organizational sustainability. Int. J. Acad. Res. Account. Finance Manag. Sci. **6**(2), 21–33 (2016)
31. Vanhala, S., Stavrou, E.: Human resource management practices and the HRM—performance link in public and private sector organizations in three Western societal clusters. Baltic J. Manag. **8**(4), 416–437 (2013)
32. Vlachos, I.P.: The effects of human resource practices on firm growth. Int. J. Bus. Sci. Appl. Manag. **4**(2), 18–34 (2009)
33. Wheelen, L., Hunger, J.: Strategic Management and Business Policy: Toward Global Sustainability. Pearson, New Jersey (2013)
34. Zhou, Y., Zhang, Y., Montoro-Sanchez, A.: Utilitarianism or romanticism: the effect of rewards on employees' innovative behavior. Int. J. Manpower **32**(1), 81–98 (2011)

Women Entrepreneurs (WE): Reshaping and Sustaining the Gulf Cooperation Council (GCC) Economy

Saad Darwish[1](✉), Umair Ahmed[2], Venus D. Bunagan[1], and Sheila M. Sison[1]

[1] Kingdom University, Riffa, Bahrain
{saad.darwish,v.bunagan,s.sison}@ku.edu.bh
[2] Arab Open University, Kuwait, Bahrain

Abstract. This paper focuses on discovering and offering pertinent and attentive evidence on the successes and challenges by women entrepreneurs in GCC. The research aims at reviewing what has been achieved so far in previous literature. The economic changes and development encouraged people to start their own businesses and follow the path of entrepreneurship. These advancements were only possible because of the continuous efforts of the GCC governments. Women are part of this and have gained recently momentum in the world of business in GCC, which reflected positively on the economy. GCC was formed in 1981 and it was inaugurated to promote the economic, social, and cultural cooperation between these Arab Gulf States. This leads us to say that sustainability is supported by women role in creating new businesses.

Keywords: Women Entrepreneurship (WE) · GCC · Sustainability · Economic development · Entrepreneurial ecosystem

1 Introduction

Women entrepreneurship is an essential part of the new emerging markets of the economy. It is emerging at a fast pace, and it is reshaping the economy. This is happening not just in the Gulf countries, but also in other countries of the world. It always played an important role in making the country's economy progression and development [22] see Fig. 1. This world of women entrepreneurship made people come up with innovative and unique ideas, which left the world, look at it with admiration. It has made the markets attain a certain level of maturity, and the young women entrepreneurs are filling it with great initiative and creative ideas.

To follow on this, the World Bank report in 2008 showed that women suffer from some gender barriers, which are not within the business environment that prevent them from starting a business venture. This implies that there are other factors. It concludes the existence of three factors. Women unfavorable attitude towards female entrepreneurship, the high cost of startups and non-positive legislations [14]. Then the report says nearly 1:8 is the ratio of women business owners in a survey of 5,169 firms find it exceedingly

A. M. A. Musleh Al-Sartawi et al. (Eds.): EAMMIS 2021, LNNS 239, pp. 455–466, 2021.
https://doi.org/10.1007/978-3-030-77246-8_42

difficult to invest, and male and female face the same dilemma. However, women face more hurdles in participating in the economy. Research shows that the deprivation and poverty trend decreases as women are educated and prepared for future responsibilities. There is evidence that a country's GDP increases by 3% when 10% girls join school. This will have impact on economic expansion and growth [15]. The global survey by [27] concluded that women are improving, guiding, and affecting how business is performed today. Women business owners must be renowned for who they are, what they do, and how effective they are in influencing the global economy see Table 1 below. In 2013, the GEM surveyed 59 countries targeting women entrepreneurship and to identify how this is influencing the economy [11]. They found out that only Ghana, Nigeria, and Thailand are creating new startups more than men are. However, Brazil, Ecuador, Uganda, and Switzerland the men and women startups are equal. The other 52 countries women are lower than men which means there is male domination on startups. The economies that sponsor and support women will benefit from the creation of new jobs and innovation that will contribute to the Gross Domestic product and their tendency to reinvest in education, their families and community. They are changing families and society, also giving assistances to business growth and expansion [39].

Table 1. Sources: Estes, 1999; NFWBO, 1998; women in business lesotho, 1998; [14, 28]

Women business ownership global facts	
EU	Women in advanced market economies own more than 25% of all businesses, start one-third of new businesses
USA	Women own 38% of all businesses and employ nearly 28 million people
Japan	23% of private firms are established by women
Russia	64% of firms employing more than 10 people are owned by women
China	Women started 25% of the businesses since 1978
Germany	Women have created one-third of the new businesses since 1990 representing more than one million jobs
Hungary	More than 40% of all businesses were started by women since 1990
Poland	38% of all businesses are owned by women
Switzerland	Women own 70% of micro, small, and medium enterprises

Giving some brief stories of successful women entrepreneurs who Influenced the business world gives an impression of how women entrepreneurship functions in the advanced economies. Women today want to be entrepreneurs because it is one of the fastest-growing career paths today [13, 46].

On the other hand, the GCC plays a significant role in the current economic activities for women business ownership. In the past, women who work in the GCC had perceived cultural roles of women in the society as being difficult. They perceived that their main role is to raise children and take care of the family. Presently, this level of thinking and mindset has changed. Women working in GCC countries have continuously been attracting attention and momentum at a very fast pace. There are women who have decided to take part in this movement, and they are changing their perceptions and thinking. They have identified that there are many other ways to make an impact in the world even as they raise children [6]. Raising the children and taking care of house members is not the only way to make an impact on the world. A survey in the GCC on political status of women showed that women in some of the countries have made tremendous progress in political rights as a vote to equality. Governments are also in support of the political rights of women [20] and this increased women participation in workforce Table 2.

Table 2. Comparison of business ownership in GCC. Source [12]

Among adults aged 15 and older

	Bahrain	Kuwait	Saudi Arabia	UAE
Own a business				
Women	3%	2%	2%	2%
Men	11%	7%	10%	7%
Difference (pct. pts.)	-8	-5	-8	-5
Planning to start a business in the next 12 months				
Women	4%	4%	2%	4%
Men	7%	11%	5%	8%
Difference (pct. pts.)	-3	-7	-3	-4

On the other hand, in most of the Gulf countries, men are having the first hold of the family while women have a secondary role. Practically speaking, the number of men entrepreneurs is more than WE. There is always a challenge in balancing the work-life with family life. Although women population contributes to a large part of the economy while working, the norms and the cultural values sometimes does not allow them to work as freely as they want to. This also limits and creates a barrier for them to display their true talents. In every Gulf country, women face similar challenges. However, some notable success cases show the power of women entrepreneurship.

There is evidence that women entrepreneurship has made a great progress and contribution to the GCC economy. Research shows that this contribution has increased by 150% during the last 3 years representing 385 Billion Dollars. In addition, it shows the growth of women business movement to jump from 4% previously to 10% of the total business domain. Furthermore, a good association shows that an increase of 10 percent of women having access to education then the GDP will be better by 3%. This Affluent innovative cluster of women creates new businesses; larger collective openings lead to diverse the economy and contributes to the future development [37, 38] see Fig. 1.

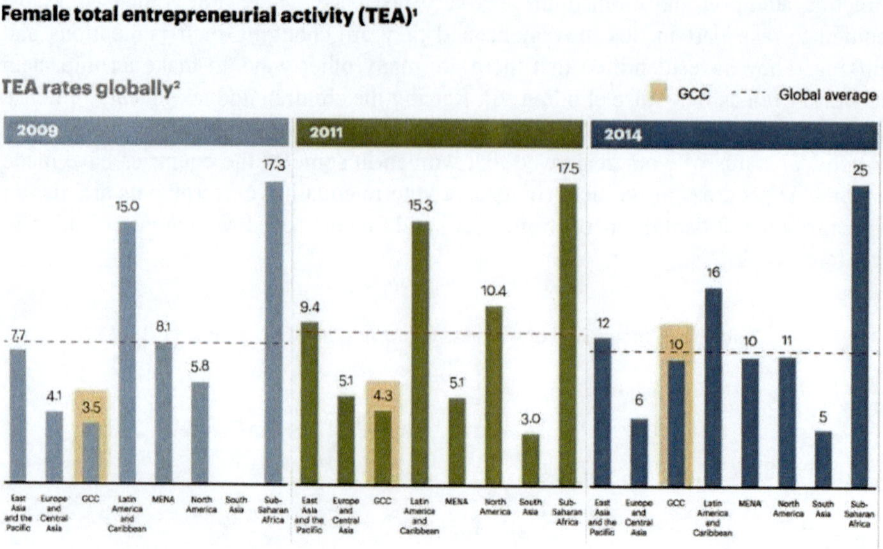

Fig. 1. GEM- Global Entrepreneurship monitor 2009–2014 [23]

2 Literature Review

2.1 The Role of Women Entrepreneurship in Economic Development

Women entrepreneurship in recent years became one of the most important components in the countries for economic development. However, the concept of women entrepreneurship has affected the economies socially as well as culturally. Women entrepreneurial ecosystem benefits the environment of the country while promoting various business ideas, innovations, and of course entrepreneurship. The factors that are affected by women's entrepreneurial ecosystem are government regulations, corporate space, and business environment [43]. Therefore, WE directly relate to government policies as they are made to ease the problems for small businesses. Entrepreneurship is a business where the owners grab the opportunity to make it even bigger and promote it on a large scale.

Another important aspect of entrepreneurship is its effect on the 'ecosystem'. The meaning of ecosystem is not to be taken too literally; the emphasis here is on the effect of women entrepreneurship in a community of independent actors [27, 30]. Women entrepreneurship is not only about what you would like to create, but it is a whole study about the needs and wants of the customers as well as the geographical needs. In recent years, the entrepreneurial ecosystem, academic researchers play a significant role in this whole system as they convey ideas and tips to women who are looking forward to being entrepreneurs in the future [18] (Fig. 2).

Women entrepreneurship has gained major popularity in the past few years. GCC women are taking many inspirations from different entrepreneurs and following their footsteps. Tt also strives for a better economy in the country. This is how you get support from the economy as well [4, 23].

Fig. 2. Entrepreneurial eco system [31]

2.2 Women Entrepreneurship: GCC Highlights

The oil resources make the major economic revenues of GCC countries. Most countries depend on oil resources. However, the decrease in the price of oil has affected the GCC countries. Nevertheless, the GCC countries diversify their economies in order to promote women entrepreneurship within the locals. WE is one of those key components that bring the country's economy to a better extent and increase the competitiveness in GCC. It had to change their economic rules and regulations as the world was on the race of bringing up changes in the business environment and the economic changes [5]. SMEs, on the other hand, are imperative to a thriving economy as they not only contribute roughly one third to the exports but also employ nearly 60% of the local workforce. In addition, it is worthy to note that its contribution to the GDP is well known to be an impressive 50%. The report published by GEM in 2013 had some interesting revelations stating that millennials now formed the largest chunk of women entrepreneurs around the world. People are developing a major interest in this field, and they are working not only because of their passion but also for the improvement of the economy [2] (Fig. 3).

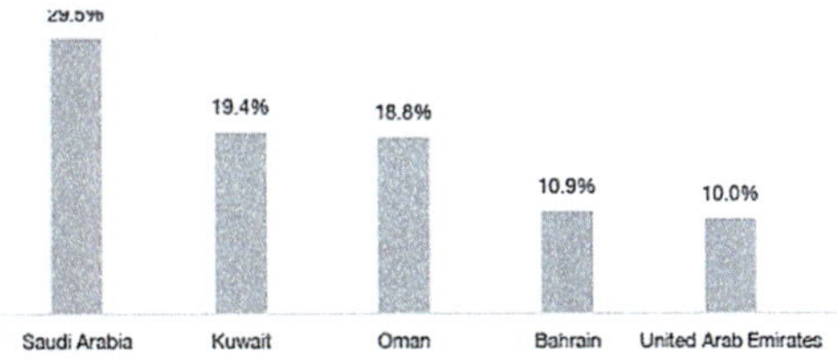

Fig. 3. Youth unemployment rate in GCC [22]

2.3 Women Entrepreneurship in Kuwait

In Kuwait, most of the women have liberties in their life than many other countries socially. A plan was approved for women in Kuwait for their rights in the municipal elections, and the government for the rights of the women passed a bill in National parliament of Kuwait. Kuwait is the most liberal country that considers women's rights. Usually, in most of the Gulf Countries, women are restricted to do any business or to meet anyone who is out of the family. Presently, among the new entrepreneurs, 40% are women and it is the highest percentage so far. However, these results are on a global level. The major portion of this country's economy is dependent on the oil reserves that it actually contains. During the past few years, the prices of oil have been constantly fluctuating, which has increased the chances of danger to the economy. Moreover, other than its dependency on oil, Kuwait aims at diversifying its economy by making use of different products. Recently, they have focused on promoting women entrepreneurship. Kuwait's government has taken the initiative of elevating the private sectors as well as small businesses as they are considered the pillar of the economy [34]. Currently, government is motivating every working woman and acting as a role model for the women looking for inspiration (Gender Gap report, 2012, p.52)

2.4 Women Entrepreneurship in the Kingdom of Bahrain

Bahrain encourages and supports a wide range of projects and individual women entrepreneurs like every other GCC country. Bahrain has also made certain accomplishments in its business and has improved its economy with respect to technological advancements. Many laws have been introduced with the aim to support and empower entrepreneurs. This is achieved in order to keep the economy stable and well established. Just like every other country in the world, Bahrain is bringing up digital economy laws following the rules of GCC [22]. This will indeed bring profits through the processes of e-commerce and many more. Thus, the main aim of the government is to assist women entrepreneurs and let them earn through online means as well. Moreover, to improve the entrepreneurial ecosystem, Bahrain's Ministry of Commerce and Tourism has introduced various centers supporting the entrepreneurial investors. In Bahrain, becoming a woman entrepreneur is more comfortable. Nonetheless, gender discrimination is an argued topic everywhere. Women may encounter problems and hurdles in Bahrain after entering the practical world of entrepreneurship due to the complications of businesses. Many women were asked about their work; and after conducting a proper survey, around 74% of the women work for 48 h a week. Hence, they do not give their families much of their time. This may give some indications of the difficulties facing women because of their family's commitment. However, women entering the field of business have increased their numbers in the private sector. Statistics indicate a higher participation of Bahraini women in the labor force over the period 1991 from 19.4% to 51.3% [8]. There are also a few challenges that women face in the Middle East because of cultural reasons, which eventually create barriers in their way to success [35] (Fig. 4).

A Bahraini Handbag designer known as Sofia Al Asfoor became a successful woman who is now recognized by many international and Arab celebrities. She is an entrepreneur who started her own work using her own name as the brand official name. Her business is not only locally recognized but she had made a name in the international market as well. As stated by Sofia Al Asfoor in one of her interviews: "To succeed, one must be persistent to achieve their goals in life. This involves creating a lifestyle where one physically and mentally strengthens themselves to overcome and embrace the stress and

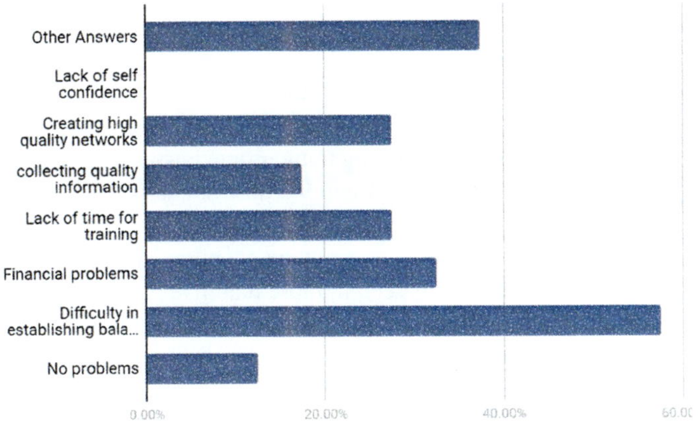

Fig. 4. Problems faced by women entrepreneurs in Bahrain [22]

challenges that they are faced with in today's world. To me, this defines "success" [3, 44]. Another success story is that of Mona Ataya from Bahrain who is a founder and CEO of Mumzworld. She explains what it looks like to face challenges and still succeed in the male-dominant state. She had the thought in mind to provide the people in her state with better baby products. Hence, with this sole purpose and motivation to provide the people with better quality, she started her own work and people look up to her now. She made an example, and she is now a proud winner in a male dominant state [21]. Nada Ala Wai sees another great achievement from Bahrain who is known as the founder of Annada Online. She made this online website in order to sell her own made Scarves, Abayas. In the world of advanced technology, she used the concept effectively in her business. She also thought about the online process for the ease of people. Today, it is a successful business and is considered a role model organization for young entrepreneurs [36].

2.5 Omani Women Entrepreneurship

In Oman, approximately in 1994, women were granted their rights. However, those rights were only applicable to a certain number of dwellers in a specific city. In an election in October, there was a reelection in which two women were selected to serve. Slowly and gradually, this number of elected women increased until 2003 when they had the greatest number of women i.e., seven who are now serving the state council of the GCC. Among these women, two were designated as ministers, four were selected as undersecretaries, and one was chosen as an ambassador to travel all over the world and make the people aware of their newly granted rights in the organization. Oman, like any other country in the gulf, depends on the oil resources for its economy. Its agriculture, as well as fishing sectors, contributes a major chunk to the Omani economy sector. A few years back, the Omani government said that they would be ensuring smooth business activities in every kind of sector. There is a major change in the number of local businesses from the past few years. In 2013, there were about 323 businesses. However, within a time span of a few years in 2015, there are now 4,300 local businesses. This has proved to be a successful portion of the era for entrepreneurs. Wadi accelerator is a program that provides funding for new entrepreneurs and helps these people with investments as well as other activities. In addition, this smart initiative has built up a well-defined ecosystem for entrepreneurs in Oman [32]. Just like any other place, Omani women face some adversities whilst working in their own businesses. Since 50% of the Omani population is women, supporting them would eventually grant a huge amount of profit to the economy. Domestic women of Omani have taken extremely influential steps to innovate their business in order to grow the economy. Their role of stabilizing the economy is remarkable, and this is seen in more than five various industry markets. The industry includes clothing, technology, institutions, and fashion designing. The attempt to establish cooperation with European Union had helped to encourage trade exchange and new investments. This helped in promoting women's entrepreneurship, as new avenues were open for further collaboration [45]. A success story about a graduate known as Bahwan from Harvard Business School once again gives an example of success. She is the CEO and founder of her own company "Cybertek Group". Breaking the hurdle of Stereotypes, Bahwan made her own company. She is known as one of the most influential entrepreneurs and a businessperson in Arab. She had received many awards as Global Young Leader in 2007 and Women of Power Award in 2011 [17].

2.6 Women Entrepreneurship in Saudi Arabia

Saudi Arabia, as stated by many other people, is the land of opportunities. It is one of the most famous countries because of its religious tourism. Many entrepreneurs are taking over the responsibility for the advancement of the economy in order to remove the dependency of the economy from oil and gas. Around 76% of the population is pursuing the idea of entrepreneurship as per the reports of the Global Entrepreneurship Monitor. Many people have been investing in good business startups, which have increased the economic rate since the past few years. It has become a key point for Saudi Arabia to encourage more women entrepreneurs [7]. Many women were highly educated but their stories of success were not shared for various reasons. It is getting better with time, but some issues are still facing women entrepreneurs in these present days. Thus, the government took progressive steps to accelerate the involvement of women in the social, political, and economic life [34]. Some notable stories to inspire women is the story of Al Dossary of Saudi Arabia who is an entrepreneur. She works in a Petrochemical Industry. She is also known for founding a company named "Oriental Promises Trading" [7]. Another outstanding achievement is noted by Rahma Al Ghanem who is an entrepreneur from Saudi Arabia and also a dentist. She has been volunteering in different places and has invested time into growing her profession for benefiting her people. It is not an easy task to support your people while being a woman in a male dominant country. Some women like Rahma still managed to help her people and provide them with the best service possible.

2.7 Women Entrepreneurship in UAE

UAE aims at maintaining the entrepreneurial environment just to ensure their contribution to the economy. The vision is to enhance and support these activities by investments and funding the right amount to the entrepreneurs [34, 42]. Despite the rapid change in women entrepreneurship in the UAE, there are still many challenges that they might face until this date. For example, displaying their ideas and abilities to the fullest might be limited because of various reasons. Taking care of families and business often costs a lot of energy and time, which most women entrepreneurs really do. The culture of the people is sometimes a demotivating factor to many women striving to be entrepreneurs. Many women are also burdened to look out for the family problems, performing household chores, and rearing children. This makes the women miss the time they should have given to their own work. Al Madani who is a fashion designer is another example of a success story. She is a working woman entrepreneur and her continuous work and efforts made her open up her own fashion hub by the name "Rouge Couture". At a very young age of 15, she decided to launch her business career, which became successful because of her constant motivation. Being a woman, she had fashion in mind and wanted to keep it modest. Hence, she went for something unique, which was an Abaya Store. Simplicity and modesty were kept while following a way that no one can easily attain. Today, she is still considered as the youngest member to open up her own business venture. Al Madani has also joined business ventures with the Sharjah Chamber of Commerce & Industry. This has led to immense success for this woman entrepreneur [24, 38]. Another story of success is a woman who not only started her own business as a woman entrepreneur but also made people love her work. She is known by the name Huda Kattan. She came up with an idea of cosmetics. Presently, she has made the world know her name by the brand "Huda Beauty". Her utmost efforts and the support of her two sisters made this venture successful. Currently, she is the most famous Makeup influencer not only in the Gulf but the whole world.

3 Research Methodology and Objectives

The approach adopted for this paper is to follow an extensive review of the current literature to envisage how women entrepreneurship affects the economy. The approach concentrated on discussing the historical development and current trends in WE based on a broad review of the secondary data and literature and own experiences. Through this study, the researcher attempts to formulate a basis for further research by developing measures to help the GCC in following systematically the development of WE. Secondary data is collected from international reports and statistics to reflect on the status of WE. An orderly analogy is explained to achieve the paper's objectives. Some examples are given here to interpret partially the status of WE. This is due to the limitations in stating many other cases of successes for women, as this paper objective does not permit.

4 Conclusion

There appears to be a diverse variety of factors that can help infuse entrepreneurial spirit for sustainable growth and development in the emerging economies like that of GCC. Government support terms to be a significant prospect for enhancing entrepreneurial activities with a particularized focus on women to help encourage them for responsive engagement. Notably, government funding and support initiatives are also essential to be considered for responsive outcomes in this regard [29]. Considering a way forward, there are a few suggestions for boosting entrepreneurship particularly when it comes to women. For example, Ahmed [1] recently underlined that involvement of youth is essential in boosting entrepreneurship and entrepreneurial activities in countries like Bahrain. Accordingly, [19] urged training and development activities for women entrepreneurship. Moreover, at times, they also require additional training keeping in view their limited skills and competencies [9, 40]. Equally, women who are already engaged in the entrepreneurial activities have a notable role to play to boost women entrepreneurship [11]. Above all, to connect with the entity of sustainable growth and development, formal education opportunities is an avenue to focus [10, 41]. Women entrepreneurial ecosystem in GCC countries in today's world is one of the most effective ways to benefit the economy. Entrepreneurship has brought competitiveness within the companies and people from all around the world. However, a major portion of the economy has been significantly impacted in various ways. GCC is very effective to promote women entrepreneurship to boost the economy. All the countries provided financial aid to the new women entrepreneurs in order to encourage them to start their own businesses [33]. Irrespective of the challenges faced by women in the Gulf countries, many women have managed to fight through all the hurdles and barriers just to follow their dreams and achieve success. Some have made an example for the world by balancing the work-life and family life together.

From this paper, it is evident that the GCC has leaped forward in supporting women entrepreneurship. Women's role is conspicuous in achieving sustainable society by entering the world of entrepreneurship. However, there is a need to create an educational system that provides the base for women entrepreneurs to be equipped with the skills and knowledge that make their involvement in the business life as smooth as possible. This can be done from early stages in the educational system. Further, using successful entrepreneurs as role models will inspire potential women entrepreneurs to start their new ventures. Comparing the achievements of other nations, with what is happening in the GCC; we can say there is a steady progress here

References

1. Ahmed, U.: Influencing youth Involvement in entrepreneurship: implications for policy makers in the Kingdom of Bahrain. Ann. Contemp. Dev. Manage. HR (ACDMHR) **1**(3), 39–52 (2019)
2. Al-Zayyat, Y.: What is the GCC? (2017). https://www.aljazeera.com/news
3. Alhudaithy, F.: Exclusive Interview with Sofia Al Asfoor. Retrieved from felwa alhudaithy blog: (2012). http://felwa-alhudaithy.blogspot.com/2012/09/exclusive-interview-with-sofiaa lasfoor.html?m=1
4. Arabia, G.: Arabian Business Global (2019). https://www.arabianbusiness.com/
5. Associates, E.: Business Case for Women's Economic Empowerment in the Arab States Region (2016). https://ergonassociates.net/wp-content/uploads/2017/09/BusinessCase for-Womens-Employment-Arab-States-Report-EN-2016.pdf?x74739
6. Atef, T.M., Al-Balushi, M.: Entrepreneurship as a means for restructuring employment patterns. Tourism Hospitality Res. **15**(2), 73–90 (2014). https://doi.org/10.1177/146735841455 8082

7. Basaffar, A.A., Niehm, L.S., Bosselman, R.: Saudi Arabian women in entrepreneurship: challenges, opportunities and potential. J. Dev. Entrepreneurship **23**(02), 1850013 (2018). https://doi.org/10.1142/s1084946718500139
8. BCCI, Women in Bahrain in the fields of Business. (2015). https://www.bcci.bh/sites/default/files/researchandstudy/eld%20of%20Business_0.pdf
9. Birley, S., Moss, C., Saunders, P.: Do women entrepreneurs require different training? Am. J. Small Bus. **12**(1), 27–36 (1987)
10. Bhardwaj, B.R.: Impact of education and training on performance of women entrepreneurs. J. Entrepreneurship Emerg. Econ (2014)
11. Brush, C.: How Women Entrepreneurs Are Transforming Economies and Communities, (2103). https://www.forbes.com/sites/babson/2013/02/16/how-womenentrepreneursare-transforming-economies-and-communities/#43640ee388c4
12. Bugshan, F.: Lack of Mentors Hinder Women's Entrepreneurship in GCC, Women Express Entrepreneurial treats, But Less Likely to own a Business. (2012). https://news.gallup.com/poll/155204/lack-mentors-may-hinder-women-entrepreneurshipgcc.aspx
13. Bunagan, V., Sison, S.: Promoting Entrepreneurs: Challenges and Success (2020). https://workingcapitalreview.com/2017/04/women-tech-entrepreneurs-need-to-be-valued)
14. Carter and Cannon (1992). https://numerons.files.wordpress.com/2012/04/17women-entrepreneursin-the-globaleconomy.pdf
15. Chamlou, N., Klapper, L., Muzi, S.: The Environment for Womens Entrepreneurship in the Middle East and North Africa (2008). https://doi.org/10.1596/978-0-8213-7495-5report
16. Cherneff, A.: 6 Reasons to Invest in Women, (2012). https://www.huffpost.com/entry/invest-in-women_b_1612911
17. CyberTek (2019). https://www.bahwancybertek.com/#/
18. Darwish, S.: 20 Common Mistakes of Startups, Chapter 17 in the Book Startup, 1st Edition Editors Mohammed Isa & Raoof Hujairi, Grant Thornton, Bahrain, National Library Deposit Number: 224/2019 (2019)
19. Davis, P.J.: The global training deficit: the scarcity of formal and informal professional development opportunities for women entrepreneurs. Ind. Comm. Training (2012)
20. Elkitbi, I.: Women's Political Status in the GCC States: (2008). https://carnegieendowment.org/sada/21229
21. Farah. 100 Tech.co Interview with Mona Ataya of Mumzworld (2017). https://100tech.co/100tech-co-interview-with-mona-ataya-of-mumzworld/
22. Fattahova, N.: Bahrain Invites GCC Entrepreneurs to invest in oil and industrial sectors (2019). https://news.kuwaittimes.net/website/bahrain-invites-gcc-entrepreneurstoinvest-in-oil-and-industrial-sectors/
23. GEM, Global Entrepreneurship Monitor 2018–2019. https://www.middleeast.atkearney.com/web/global-business-policy-council/article?/a/powerwomen-in-arabiashaping-the-path-for-regional-gender-equality
24. Group, A.M.: Profile (2019). http://www.almadanigroup.com/web/
25. Hamdan, A.: Saudi Arabia: higher education reform since 2005 and the implications for women. Educ. Arab World (2017). https://doi.org/10.5040/9781474271035.ch-011
26. Ijamee IJAMEE (2017). http://www.ijamee.info/index.php/IJAMEE
27. Isenberg, D.: What an Entrepreneurship Ecosystem Actually Is (2018). https://hbr.org/2014/05/what-an-entrepreneurial-ecosystem-actually-is
28. Jalbert, E.S.: Women Entrepreneurs in the Global Economy, (2000). https://numerons.files.wordpress.com/2012/04/17women-entrepreneurs-in-the-globaleconomy.pdf
29. Khalid, N., Ahmed, U., Tundikbayeva, B., Ahmed, M.: Entrepreneurship and organizational performance: Empirical insight into the role of entrepreneurial training, culture and government funding across higher education institutions in Pakistan. Manage. Sci. Lett. **9**(5), 755–770 (2019)

30. Lee, M.S.: How entrepreneurship shapes a community: (2014). https://www.crainsdetroit.com/other-voices/how-entrepreneurship-shapes-community
31. Mazzarol, T.: Growing and sustaining entrepreneurial ecosystems: What they are and the role of government policy, White Paper WP01–2014, Small Enterprise Association of Australia and New Zealand (SEAANZ) (2014)
32. Mc Elwee, G.: Women entrepreneurs in Oman: Some barriers to success (2017). https://www.researchgate.net/publication/235286376_Women_entrepreneurs_in_Oman_Some_barriers_to_success
33. Metcalfe, B.D.: Women, management and globalization in the middle East. J. Bus. Ethics **83**(1), 85–100 (2008)
34. Miniaoui, H.A.: Innovation and Entrepreneurship for the growth and diversification of the GCC Economies. Retrieved from MPRA: (2017). https://mpra.ub.uni-muenchen.de/71898/
35. Moghadam, V.M.: Women, structure, and agency in the Middle East: introduction and overview to feminist formations' special issue on women in the Middle East, 9 (2016)
36. Alaa, N.: (2019). https://www.facebook.com/search/top/?q=Nada+Alaa+Eldin
37. Peterson, J.E.: The political status of women in the Arab Gulf States. Middle East J. **43**(1), PP43–50 (2018)
38. PNE Group (n.d). https://www.pne.org/pne-group-insights/rise-in-female-entrepreneurshipgivesgulf-businesses-a-boost/
39. Rao, N.: Women's rights to land and assets: experience of mainstreaming gender in development projects, 4709 m, 2011, UNESCO; Geneva (2017)
40. Roomi, M.A., Harrison, P.: Behind the veil: women-only entrepreneurship training in Pakistan. Int. J. Gender Entrepreneurship (2010)
41. Sajuyigbe, A.S., Fadeyibi, I.O.: Women entrepreneurship and sustainable economic development: Evidence from Nigeria. J. Entrepreneurship, Bus. Econ. **5**(2), 19–46 (2017)
42. Shahnawaz, M.R.: Preliminary Investigation of Emirati Women (2016). https://www.ijstr.org/final-print/aug2015/Preliminary-Investigation-Of-Emirati-
43. WomenEntrepreneurship-In-The-Uae-Motivating-Factors-Challenges-And-Government-Initiatives.pdf Springborg, R. GCC Countries as "Rentier States".Middle East J. **67**(2), 301309 (2013)
44. Step Feed 8 Female Arab Entrepreneurs who made it big (2017). https://stepfeed.com/8-female-arab-entrepreneurs-who-made-it-big-2660
45. Times News Service, NBC hosts roundtable for Omani women entrepreneurs (2018). https://timesofoman.com/article/141458
46. Toledo, R.: The Entrepreneur (2021). https://www.entrepreneur.com/article/364947

The Mediating Role of Employee Engagement Between Quality of Work-Life and Job Performance in Jordanian Commercial Banks _conceptual Paper

Aya Naser Maqableh[1](✉), Khatijah Omar[2], and Jasem Taleb Al-Tarawneh[3]

[1] Faculty of Business, Universiti Malaysia Terengganu, 21030 Kuala Nerus, Terengganu, Malaysia

[2] Faculty of Business, Economics and Social Development, Universiti Malaysia Terengganu, 21030 Kuala Nerus, Terengganu, Malaysia
Khatijah@umt.edu.my

[3] Faculty of Business, Universiti Malaysia Terengganu, 21030 Kuala Nerus, Terengganu, Malaysia

Abstract. Increased job performance is an essential goal for businesses to sustain them market growth. The corporate initiative is geared towards optimizing performance outcomes, taking into account the organizational sense in which performance is generated. This study aims to identify the mediating role of employee engagement between quality of work-life and job performance in Jordanian commercial banks. This paper found that there are indications that the level of engagement is positively associated with job performance in terms of financial benefits, greater client loyalty and better adaptation to the working environment. However, engagement is positively related to finance. Lastly, work engagement has a positive impact on engagement when job demands are high. In addition, work engagement has a positive effect on job performance.

Keywords: Employee engagement · Quality of work-life · Job performance

1 Introduction

The digital economy is a recently-emerging phenomenon of increasing importance given estimates of double-digit annual growth around the world, with particularly strong growth in the global South [34]. The driving forces behind this emergence are economic and political, but they of course also have roots in technological innovation (itself shaped by wider forces). In the 1990s, economic changes were associated mainly with emergence of the Internet, and this remains a foundation for growth of the digital economy. But during the 2000s and 2010s a succession of new information and communication technologies (ICTs) has diffused and underpinned economic change. This includes the embedding of connected sensors into more and more objects (the Internet of things);

A. M. A. Musleh Al-Sartawi et al. (Eds.): EAMMIS 2021, LNNS 239, pp. 467–474, 2021.
https://doi.org/10.1007/978-3-030-77246-8_43

new enduser devices (mobile phones, smartphones, tablets, netbooks, laptops, 3D printers); new digital models (cloud computing, digital platforms, digital services); growing intensity of data usage through spread of big data, data analytics and algorithmic decision-making; and new automation and robotics technologies [24]. That impact can be understood as a disruption of existing economic processes, systems and sectors, reshaping existing consumer behaviour, business interactions and business models [12]. It can also be understood as the emergence of new economic processes, systems and sectors. Within individual sectors, we see this readily reflected in dominance of new firms.The digital economy is unquestionably altering the dynamic of economic growth and business performance. Digital trends such as cloud computing, mobile web services. smart grids and social media are radically changing the business landscape, reshaping the nature of work, the boundaries of enterprises and the responsibilities of business leaders. small and Medium Enterprise (SMEs)are showing their impact on national and regional economic through the world. They have been recognized in both developed and developing countries as an effective instrument for creating employment opportunities with a small amount of capital investment. Increased job performance is an essential goal for businesses to sustain them market growth. As a result, the corporate initiative is geared towards optimizing performance outcomes, taking into account the organizational sense in which performance is generated [8].

Employee performance, in general, refers to behaviors that are relevant to organizational goals and that are under the control of individual employees [23]. The performance derived from it is his main source of joy. Job performance is the feeling of the employee about his job. Yet, it is worthy to mention that all factors need to harmonize and align together to reach the desired progress, success, and sustainability [11].

Most of the banks realized that employees of the bank are the best assets of the firm that they can compete with external as well as internal bank in their sectors [3]. Previous studies evidence that involvement of the employee in their work is the key factor in improving job performance. Quality of work-life in a bank is essential for the smooth running and success of its employees. Quality of work-life helps the employees to feel secure and like they are being thought of and cared for by the bank in which they work. For an effective development of both individual objectives and bank's objective, the bank should provide a good overall working environment considering factors like career opportunities participative management working environment. In this process, it can generate a sense of satisfaction for which on his own capability to achieve simultaneously both individual and bank's objectives. Employees are engaged when organizations have healthy work culture and communication practices, where they can get platforms to express their concerns and opportunities to grow and develop their potential. The level of engagement in employees can be enhanced by identifying its drivers (influential factors) and work on them. The banking sector is a leading segment in the Jordanian economic. Many performance metrics could show how banks manage their resources in Jordan For example; the effectiveness and the ability of the banking sector, represented by the evolution of return on their assets ratio, which have been relatively high compared to other countries in the MENA area despite the higher ratio of non-performing loans in (Algeria, Saudi Arabia, Lebanon, and UAE), which could be attributed to factors that the banking industry have little control over. Accordingly, the paper aims to know the relationship and the mediating relationship between the quality of work-life has a relationship with job performance in Jordanian commercial banks.

2 Literature Review

2.1 Job Performance

Job performance concept expresses the net impact of individual efforts that begin with the capabilities and awareness of the role or tasks which refers to the degree of achievement and completion constituent of the function of the individual tasks [33]. Performance is an integrated system that represents individual performance, it is the key element, and this is due to the human element, it is the active ingredient in performance because of its expertise and capacity to carry out the business [19]. Job performance defined as "the result of three factors together which are the skill, effort, and the nature of working conditions, which include knowledge skills, abilities, and competencies that brought the individual to the organization, the effort is the degree of rush the employee to complete his work, and the nature of working conditions and suitability to facilitate and improve the productivity of the individual [36]. Job performance is considered as an important parameter in the hotels' profession, in such a way which in the last pent etic, new innovative ways of calculation and consideration have invented [6] even for new registered professionals [26]. After a careful detection of international literature, what can be seen is that job performance is directly and strongly related to stress and burnout [18]. Equally, crucial role in shaping the professional performance playing both the employee leadership [28] and the procedure of rational decision making. Of course, we must evaluate the close relationship of employees' job performance with the work schedules, staffing and epidemiological characteristics of the population they serve [6]. Work performance for bank employees can be viewed as the effectiveness of the employee in carrying out their roles and responsibilities related to direct customer service. Four elements to assess job performance of staff banks, first expected bank behavior including courtesy, respect, communication, comfort, responsiveness, teamwork, and professionalism. Second, employees process including assessment, diagnosis implementation, evaluation, and documentation. Third, quality of service and, fourth education [9].

2.2 Quality of Work Life

Quality of work life (QWL) is fundamentally a multidimensional concept that represents the mechanism, which regulates the relationship between individual and their work [16, 21], defines QWL as a "process by which an organisation responds to the employee's needs by developing mechanisms to allow them to share fully in making the decisions that design their lives at work". Likewise, [31] refer to QWL as "employee satisfaction with a variety of needs through resources, activities, and outcomes stemming from participation in the workplace". Furthermore, "QWL encompasses the physical, technological, psychological and social dimension of work corresponding to the ideals of a more humane and healthy organization" [14]. Despite widespread recognition of the concept of QWL, there has been a lack of consensus among management scholars regarding its definition and measurement [22], attributed this fact to a wide range of focus concerning research in the areas of QWL. In general, QWL is a construct that deals with the well-being of employees and are significantly different from job performance.

Bank employees are the largest group work force in financial industry. Without their contribution quality of bank services and customer satisfaction cannot be improved. Therefore, Quality of work life QWL of employees working in tertiary Jordanian commercial bank should be very high. Notably, enhancing the QWL is an all-inclusive course of a process to improve the quality of life of workforces in the facilities and is crucial in any organization to recruit and hold its staff. High quality of work life (QWL) is a crucial issue for service industry facilities to have a qualified, dedicated, and inspired employee. Service provision in the service sector facilities depends on the capacity and capabilities of their manpower [13]. Among different specialties in private service sector settings, employees have a major share among other service providers, and enhancing their work life quality has become a critical issue in private service sector settings [17]. The quality of work life framework shows a difference between results and chances while proving its worth to public policy [25]. Chances have a place with the space for open strategy prompt activity, while it comes about constitute the last inspiration of open arrangement. Governments, as well as countless organizations, usually would contemplate quality of work life and observe their improvement, but the concluding assessment of open approaches must be considered [29].

2.3 Employee Engagement

People with high levels of engagement show positive attitudes towards their jobs and organizations, including job satisfaction and commitment to the company, and they do not frequently shift jobs. Furthermore, those with high work engagement exhibit high learning motivation and proactive behaviors [4], and they work diligently because they enjoy their work even when they are tired, describing fatigue as pleasant because they can associate it with positive achievements [15]. There are indications that the level of engagement is positively associated with job performance in terms of financial benefits, greater client loyalty and better adaptation to the working environment. Empirical studies are also available that indicate that engagement is positively related to finance. For example, engaged employees have been shown to suffer less from depression and stress and to have fewer psychosomatic symptoms [35]. According to [5] this model suggests that job resources are the start of a motivational process that leads to more engagement, which in turn results in higher performance. The model further suggests that job resources become more important when employees are confronted with higher job demands. Job resources and personal resources are independent predictors of work engagement [6], further states that employees with more personal resources are better "able to mobilize their job resources and generally are more engaged in their work" [5] summaries the model of work engagement by stating that job and personal resources have "a positive impact on engagement when job demands are high"; in turn work engagement has a positive effect on job performance.

While this study has offered contributions to the understanding on the relationship between talent management, quality of work-life, organizational climate, employee engagement toward job performance there are a number of limitations that must be acknowledged. This study will focus only on commercial banks for data collection and the sampling of the study only consists of employees of the commercial banks. Thus, the study only managed to evaluate job performance from commercial banks. Thus, it limits the generalizability of the findings. Next, the survey instruments of the study only consist of closed-end questions, thus only a limited amount of information can be gathered from the respondents of the study. Finally, this study will be using a self-reported questionnaire and that might lead to respondents not taking answering the questionnaires seriously.

3 Research Methodology

This paper aims to the mediating role of employee engagement between quality of work-life and job performance in Jordanian commercial banks, while the quantitative method is the most appropriate to be taken. The reason lies in the fact that the current 56 study focuses on research problems in an objective way rather than subjectively. As stated by [27]; and [32] the quantitative strategy works on objective and measure it through the actions and opinions which helps the researcher to describe the data rather than interpret the data. Additionally, the nature of this study is empirical where researchers emphasize fresh data collection by the research problem. Thus, this study is quantitative using a survey design.

By reviewing the reports of the central bank of Jordan (CBJ) there are 25 operating banks, 15 listed in the Amman Stock Exchange (ASE), led by Arab Bank. Also, the number includes regional institutions (e.g., Kuwait National Bank, National Bank of Abu Dhabi, and Egyptian Arab Land Bank), and Western multinationals banks (e.g., Citibank and Standard Chartered). The banking sector history is dating back to the 1940s when Arab Bank moved to Amman. Banking accounted for 18.82% of the GDP as of mid-2018, this justified according to it the most significant economic sectors in Jordan too. The size of the sample was calculated with the help of the table proposed by [30]; and [20]. From the table, population with sample sizes are matched with a 95% level of confidence concerning normally distributed data. Therefore, for a population of 17569 employees, the calculated required size of the sample is 377.

For the methodology section, this paper suggests statistical techniques such as normality tests, multicollinearity test, and EFA using the SPSS. For the measurement model and hypothesis testing using smart PLS. Even though PLS-SEM is methodologically well-established and frequently applied, editors and reviewers often require researchers to justify their choice of the method [10]. The present study used PLS-SEM for several reasons; In real world scenario and application of complex models, modeling of PLS has better advantage hence its suitable and appropriate compared to other models in the literature [1]. The basic assumption guiding the modelling of PLS including the ability to dynamically develop and measure complicated models, make the model to possess the superior qualities of evaluating complex and large models [1]. Hence, this study applied a complex model with several variables. This serves as justification for the adoption of PLS technique in this study.

4 Discussions

Work performance for bank employees can be viewed as the effectiveness of the employee in carrying out their roles and responsibilities related to direct customer service. Four elements to assess job performance of staff banks, first expected bank behavior including courtesy, respect, communication, comfort, responsiveness, teamwork, and professionalism. Second, employees process including assessment, diagnosis implementation, evaluation, and documentation. Third, quality of service and, fourth education. In addition, quality of work life (QWL) is fundamentally a multidimensional concept that represents the mechanism, which regulates the relationship between individual and their work. However, QWL is a construct that deals with the well-being of employees and are significantly different from job performance. Therefore, quality of work life QWL of employees working in tertiary Jordanian commercial bank should be very high. Enhancing the QWL is an all-inclusive course of a process to improve the quality of life of workforces in the facilities and is crucial in any organization to recruit and hold its staff. High quality of work life (QWL) is a crucial issue for service industry facilities to have a qualified, dedicated, and inspired employee. Lastly, job resources are the start of a motivational process that leads to more engagement, which in turn results in higher performance. The model further suggests that job resources become more important when employees are confronted with higher job demands. Job resources and personal resources are independent predictors of work engagement Several works are exploring new opportunities for lifelong learning. Others focus on the digital economy, recruitment potential and a labor market, or answers the question how digital technologies transform human functions in the process of work, improving the quality of working life and business efficiency. The results of the analysis indicate an increase in discussions about the digital economy and its impact on QWL. requiring further development based on a sophisticated approach that would harmoniously involve both objective and subjective aspects of this topic [28].

References

1. Akter, S., Fosso Wamba, S., Dewan, S.: Why PLS-SEM is suitable for complex modelling? An empirical illustration in big data analytics quality. Prod. Plan. Control **28**(11–12), 1011–1021 (2017)
2. Al-Ahmadi, H.: Factors affecting performance of hospital nurses in Riyadh Region, Saudi Arabia. Int. J. Health Care Qual. Assur. **22**(1), 40–54 (2009)
3. Al-dalahmeh, M., Khalaf, R., Obeidat, B.: The effect of employee engagement on organizational performance via the mediating role of job satisfaction: The case of IT employees in Jordanian banking sector. Modern Appl. Sci. **12**(6), 17–43 (2018)
4. Bakker, A.B.: Strategic and proactive approaches to work engagement. Organ. Dyn. **46**(2), 67–75 (2017)
5. Bakker, A.B., Albrecht, S.L., Leiter, M.P.: Key questions regarding work engagement. Eur. J. Work Organ. Psychol. **20**(1), 4–28 (2011)
6. Barker, L.M., Nussbaum, M.A.: Fatigue, performance and the work environment: a survey of registered nurses. J. Adv. Nurs. **67**(6), 1370–1382 (2011)

7. Becton, J.B., Matthews, M.C., Hartley, D.L., Whitaker, L.D.: Using biodata as a predictor of errors, tardiness, policy violations, overall job performance, and turnover among nurses. J. Manage. Organ. **18**(5), 714 (2012)
8. Blackman, D., Buick, F., O'Donnell, M.: Developing High Performance: Performance Management in the Australian Public Service (2018)
9. Carmeli, A., Tishler, A.: The relationships between intangible organizational elements and organizational performance. Strateg. Manage. J. **25**(13), 1257–1278 (2004)
10. Chin, W.W.: How to write up and report PLS analyses. In: Esposito Vinzi, V., Chin, W., Henseler, J., Wang, H. (eds.) Handbook of Partial Least Squares. Handbooks of Computational Statistics. Springer, Berlin, Heidelberg (2010). https://doi.org/10.1007/978-3-540-32827-8_29
11. Crowley-Henry, M., Al Ariss, A.: Talent management of skilled migrants: Propositions and an agenda for future research. Int. J. Hum. Resour. Manage. **29**(13), 2054–2079 (2018)
12. Dahlman, C., Mealy, S., Wermelinger, M.: Harnessing the Digital Economy for Developing Countries, OECD, Paris (2016). http://www.oecd-ilibrary.org/docserver/download/4adffb24-en.pdf.
13. Dargahi, H., Changizi, V., Jazayeri, G.E.: Radiology Employees'Quality of Work Life (2012)
14. Daubermann, D.C., Tonete, V.L.P.: Quality of work life of nurses in primary health care. Acta Paul Enferm. **25**(2), 277–283 (2012)
15. García-Sierra, R., Fernández-Castro, J., Martínez-Zaragoza, F.: Work engagement in nursing: an integrative review of the literature. J. Nurs. Manag. **24**(2), E101–E111 (2016)
16. Hsu, M.Y., Kernohan, G.: Dimensions of hospital nurses' quality of working life. J. Adv. Nurs. **54**(1), 120–131 (2006)
17. Janaabadi, H.: Two effective factors in the staffs performance: Quality of life and quality of working life. Zahedan J. Res. Med. Sci. **13**(suppl 1) (2012)
18. Karimi, O., Daraei, M. R., Farajzadeh, F.: Analyzing the impact of Emotional Intelligence EI on the employees' Quality of Work Life QWL Case Study Central bureaus of Agricultural Bank in Tehran. J. UMP Soc. Sci. Technol. Manage. **3**(2) (2015)
19. Kartikaningdyah, E., Surya, R.O.: The influence of motivation, voice behavior and work engagement of the individual performance at Islamic Bank. J. Appl. Account. Taxation **2**(1), 1–8 (2017)
20. Krejcie, R., Morgan, D.: Determining sample size for research activities. Educ. Psychol. Measur. **30**(3), 607–610 (1970)
21. Lau, R.S.: Quality of work life and performance–An ad hoc investigation of two key elements in the service profit chain model. Int. J. Service Ind. Manage. (2000)
22. Lawler, E.: Measuring the psychological quality of working life: the why and how of it. Qual. Working Life **1**, 123–133 (1975)
23. Moore, B.R., Höhna, S., May, M.R., Rannala, B., Huelsenbeck, J.P.: Critically evaluating the theory and performance of Bayesian analysis of macroevolutionary mixtures. Proc. Natl. Acad. Sci. **113**(34), 9569–9574 (2016)
24. OECD: OECD Digital Economy Outlook 2015, OECD, Paris (2015). http://www.oecd.org/sti/oecddigital-economy-outlook-2015-9789264232440-en.htm.
25. Pawirosumarto, S., Sarjana, P. K., Gunawan, R.: The effect of work environment, leadership style, and organizational culture towards job satisfaction and its implication towards employee performance in Parador Hotels and Resorts, Indonesia. Int. J. Law Manage. (2017)
26. Platis, C., Reklitis, P., Zimeras, S.: Relation between job satisfaction and job performance in healthcare services. Procedia-Soc. Behav. Sci. **175**(1), 480–487 (2015)
27. Rahi, S.: Research design and methods: a systematic review of research paradigms, sampling issues and instruments development. Int. J. Econ. Manage. Sci. **6**(2), 1–5 (2017)

28. Salanova, M., Lorente, L., Chambel, M.J., Martínez, I.M.: Linking transformational leadership to nurses' extra-role performance: the mediating role of self-efficacy and work engagement. J. Adv. Nurs. **67**(10), 2256–2266 (2011)
29. Sari, N.P.R., Bendesa, I.K.G., Antara, M.: The influence of quality of work life on employees' performance with job satisfaction and work motivation as intervening variables in star-rated hotels in Ubud tourism area of Bali. J. Tour. Hosp. Manage. **7**(1), 74–83 (2019)
30. Sekaran, U., Bougie, R.: Research Methods for Business: A Skill Building Approach. Wiley, Chichester (2016)
31. Sirgy, M.J., Efraty, D., Siegel, P., Lee, D.-J.: A new measure of quality of work life (QWL) based on need satisfaction and spillover theories. Soc. Indic. Res. **55**(3), 241–302 (2001)
32. Tashakkori, A., Creswell, J.W.: Exploring the Nature of Research Questions in Mixed Methods Research. Sage Publications Sage CA, Los Angeles (2007)
33. WallapaBoonrod, R.: Quality of working life: perceptions of professional nurses at Phramongkutklao Hospital. J. Med. Assoc. Thai. **92**(1), S7-15 (2009)
34. WEF: Expanding Participation and Boosting Growth: The Infrastructure Needs of the Digital Economy, World Economic Forum, Geneva (2015). www3.weforum.org/docs/WEFUSA_DigitalInfrastructure_Report2015.pdf
35. Welthagen, C., Els, C.: Depressed, not depressed or unsure: prevalence and the relation to well-being across sectors in South Africa. SA J. Ind. Psychol. **38**(1), 57–69 (2012)
36. Yen, H.H.: Perception on ethical climate and individual job performance of bank employees. J. Econ. Dev. **24**(4), 85–98 (2019)

Fifth Industrial Revolution: The Internet of Ideas

A New Perspective on Decision Making and Coordination in Smart City Projects

Basar Atac[1(✉)] and Tarkan Oktay[2]

[1] Project Management Professional Certification (PMP®), Istanbul, Turkey
[2] Istanbul Medeniyet University, Istanbul, Turkey

Abstract. With the use of advanced technologies in smart cities, some particular questions such as which digital tools ought to be used in which projects and which specific areas in smart cities have begun to be further considered. Nevertheless, the projects applied to the smart cities have been carried out in unplanned manners and some resources have been wasted since there is no relevant map that might enable us to determine our path throughout the process unceasingly. Hence, existing technological applications should be strengthened with recent management approaches, and herein a coordination between all stakeholders and city projects should be established. In this article, specific requirements are additionally accentuated and some explicit solutions are decidedly recommended to prevent cities from turning into a 'technology dump' in time. In order to achieve the stated aim, a new assessment map has been created through combining the basic approaches applied separately in the projects of the best smart cities ever known in the world, and then this assessment map has been tested in a smart city, Istanbul. As a consequence, it is alleged that using six different approaches, which are all appraised as substantial, in smart city projects up to the hilt and also envisioning them as whole all during evaluations will provide considerable benefits to all stakeholders of cities.

Keywords: Smart cities · Smart city · Municipality · E-governance · Smart city projects

1 Introduction

Due to the complex structures of city ecosystems, there is a momentous requirement for coordination between infrastructure services such as transportation, water, sewage, energy, health and security, which must be provided by local governments. In order to provide these services effectively and efficiently, digital systems should be benefited during the delivery of such services and these digital systems should communicate with each other. In addition to the necessity for mutual conversation within these systems, the effective use of the data obtained in the provision of these services can be presumed as possible only with the effective use of digital connections.

This article has been derived from the doctoral dissertation entitled "Management of Transition to a Smart City in Local Governments: The Case of the Metropolitan Municipality in Istanbul".

A. M. A. Musleh Al-Sartawi et al. (Eds.): EAMMIS 2021, LNNS 239, pp. 477–489, 2021.
https://doi.org/10.1007/978-3-030-77246-8_44

In addition, not only municipal services, but also electronic commerce structures that emerged as a consequence of technological revolution that begins with the Industry 4.0 are further vitalized by this necessity for connection in smart cities, and local, national and international commercial relations are formed through digital networks. Moreover, these structures of commercial relations, which are inevitably established through internet networks, not only provide commercial connections, but also perform a vigorous role in every sector of life namely in education, health, security, energy and politics especially in the implementation of democracy through e-governance.

As a result, in addition to meeting all these basic requirements, the provision of infrastructure services required for the sustainability of Industry 4.0 is also under the responsibility of smart cities and of who administer them. Since, together with the use of digital technologies in city services, an effective use of data is deemed as the fuel of information societies and data is collected by means of the IoT and the internet both, which require to encounter some basic principles, as well.

Nevertheless, since the fact that certain projects implemented in smart cities might be driven with political concerns, they are not addressed entirely with some other projects and the areas that might palpably convey votes to some politicians during elections are enticingly prioritized. Besides, such a problem is not only related to politicians. Obviously, some private companies tend to invest their resources in areas that they consider that will be profitable at the highest level, so an excess supply ensues in some areas of cities. Such an excess supply can be stabilized only with the efforts in a long period of time, and all these processes result in an inefficient use of limited technological resources. In order to control this situation and to act within a model, some particular approaches have been propounded for the best smart city practices.

Inopportunely, since these approaches are applied fragmentally disorganized despite all needs, they have failed to provide the efficiency targeted. Nonetheless, with the systematic implementation of these approaches, the current overpriced digital technologies can be effectively coordinated, and significant opportunities shall egress for an optimum use of scarce resources.

Among these basic objectives, complying with the criteria of efficiency, meeting the expectations of citizens, ensuring central coordination and applying sustainable flexibility can be shown [1]. In addition to these forthright goals, the criteria for becoming a smart city include having a participatory management, adapting to change, making output-oriented decisions, taking care of cooperation between stakeholders, keeping the digital connection logic in the forefront, being transparent and covering all components, and decision making and used in coordination processes [2].

Herein, our article written as considering the aforesaid basic requirements has formed a road map which local governments can benefit from for a smart city creation process, and make recommendations for the decision-making and coordination approaches that should be applied in due course of following this road map. Since this study concentrated on the management of technical issues by presenting newfangled managerial approaches, it is contrary to the view that the crucial share of the academic literature related to smart cities is consisting of technical issues.

These approaches are repetitively used in academic literature that contains a smart city theory and smart city practices, and they can form the decision-making and coordination approaches in which the article is based. Additionally, they provide a perspective to eliminate the uncertainty in some processes during the evaluation of smart city projects and it is assumed that they will strengthen the possibility of fulfilment in smart city goals if they are used in combination together.

Even though all these approaches are included in the academic literature related to smart cities, the fact that they are examined separately and not partaking of a collective perspective is the terminus a quo in this article. Beginning with this, based on the forestated approaches, the smart city management processes of the Istanbul Metropolitan Municipality have been examined and evaluated on a scale organized and it has been concluded that an effective and efficient management could not be achieved yet.

2 Literature Review

Researchers and organizations that examine the approaches which form the basis of our research, whose implementation during decision-making and coordination, and which are assumed to use sources effectively and efficiently are, all, given in the table below (Table 1):

Table 1. Researchers and institutions examining the approaches

Central coordination approach	T. Aragonez, G. Caetano, A. Blanco-Gonzalez, China Academy of Information and Communications
Citizen-oriented approach	Smart Cities Council, United Nations, World Bank, T.A. Spil, R. Effing, J. Kwast
Holistic approach	K. Kavta, P.K. Yadav, R. Giffinger, The British Standards Institution, J.M. Schleicher
Adaptation approach	F. Fontana, L.G. Anthopoulos, C. Thorne, C. Griffiths
Connected approach	M. Bawa, Smart Cities Council, D. Perslein, European Commission, D. Bennet, D. Perez-Bustamente, M.L. Medrano
Sustainability approach	C. Lohrmann, M. Macagnano, M. Araksinen, H. Ailisto, N.O. Nylund, M. Manfren, A. Townsend

Likewise, a table covers researchers and organizations that have stated the aforesaid approaches in their studies in one hand but could not be included in this study due to the lack of space is presented below, and it is aimed to guide the researchers after us (Table 2):

The researchers and organizations mentioned in both tables have stated analogous views in their several studies included in the academic literature related to smart cities in accordance with the fundamental logic of these approaches. Nevertheless, these approaches have been uttered one by one by everybody, but the subject has been handled in an all-inclusive manner in our doctoral dissertation study. During the implementation,

Table 2. Other researchers studying relevant approaches

Central coordination approach	A. Townsed, L. Mora ve D. Martinez
Citizen-oriented approach	L. Anthopoulos, A. Monzon, A. Caragliu
Holistic approach	A. Monzon, R.P. Dameri, A. Townsend
Adaptation approach	A. Townsend, M. Areksinen
Connected approach	H. Chourabi, T. Nam, T.A. Pardo, R.P. Dameri
Sustainability approach	T.M. Vinod Kumar, C. Del Bo, P. Nijkamp

it was claimed that these approaches should have been followed in a way to create an ecosystem.

3 Research Methodology

This article was derived from a doctoral dissertation [3] and it is aimed to examine and to measure the decision-making and coordination approaches of the municipal staff who administrate Istanbul as a smart city. For this reason, a qualitative research method was used as a research methodology in this study. For an interview, seven people from two different companies were questioned face to face and fifty questions were addressed to them. The data used in the study were collected by open-ended questionnaire method and observation techniques, and the result of the data was compared with scales prepared separately for each decision-making and coordination approach. In addition to these, interviews, correspondence and multidimensional literature reviews were conducted with certain companies outside the Istanbul Metropolitan Municipality. A scale has formed for this comparison was determined based on behavioral patterns which are considered as impartial and so implemented by the best smart cities ever known in the world. The results obtained with the questionnaire were compared with this scale organized and categorized.

4 The Main Objectives of Smart Cities

The fact that five main objectives deliberated in the digital transformation process which have emerged as an outcome of prodigious impact of information and communication technologies, but they have been presumed to be realized through smart cities are, in regular turn, listed as efficiency, effectiveness, sustainability, governance, transparency, creating economic and public value.

Efficiency can be abridged as reaching maximum output with minimum waste in all kinds of production and consumption processes, and it can be described with the example of LED lamps which consume less energy even if they give more light than incandescent lamps.

Effectiveness can be defined as a level of achieving a set of goals for activity, productivity and consumption process. Following the aforesaid example, any use of incandescent lamps in art exhibitions so as to create a diverse upshot on audiences, disregarding the saving feature of LED lamps is a simple instance for this term to comprehend.

Sustainability can be defined as a continuation of economic and social ecosystems that constitute a city under the same conditions with the both systems, despite population growth or changes. Some tools used by smart cities for the sake of sustainability which have been asserted to ensure the sustainability of natural resources are as the following: New business models and innovation, e-governance, cloud computing, internet of things, 5G, fiber optic connection, green transportation, and governance.

Governance and transparency, on the other hand, includes basic principles such as transparency, accountability, participation and equality; subsequently it is a structure that can be summarized as citizens' participation in decision-making processes in administration. In addition to participation methods especially on the internet, public shares of local governments are just a couple of methods of accessibility to all kinds of information and transparency governance.

Creating economic and public value should not be considered only as an efficient and effective use of natural resources for smart cities. Smart cities are also seen as a source of job market that can reach up to 1 trillion USD by 2025 as an economic development engine based on a digital and innovative economy idea to be built on information and communication technologies [4] The idea of creating public value in smart cities is depending upon the understanding of generating an improved life and environmental conditions for people living in smart cities by means of technology and of creating a further prosperous city.

5 The Basic Components to Form Smart City Ecosystems

As a living organism, each city might be imagined as home innately including its own objectives, structures, characteristics, actors, and relationship models between each actor in different structures within the same home. Defining the dynamic relationship between these components and classifying these components systematically is a prerequisite for understanding a city. For these reasons, the smart city ecosystem is divided into six main classes in the literature and has been systematized as Smart People, Smart Governance, Smart Economy, Smart Living, Smart Mobility, and Smart Environment [5, 6].

Although ***smart people*** come to the forefront in the smart city concept due to the natural charm of technology, they are the central element for the establishment and sustainability of a smart city. Therefore, the economic and social potential of a smart city can only be revealed through people who are well-educated by the Industry 4.0 and can use technology-skilled smart city technologies that can easily adapt to global trends and create public value through them [7].

The understanding of citizens' participation in administration, which is the basic motto of the New Public Management, is called as governance in the relevant literature and forms the basis of smart cities as ***smart governance***. Among the factors that support governance, developments in information and communication technologies and the idea that the transformation into smart cities should be from bottom to top in a way that is directed by the citizens can be counted [8].

The concept of e-governance refers to achieving the goals related to governance by using information and communication technologies, especially the internet, and transformation of the relationship between stakeholders, citizens and the state through technology is the main goal [9, 10].

A **smart economy** is the penetration of an innovative and entrepreneurial economy using information and communication technologies into the production, distribution, and consumption functions of the smart city [11].

The fundamental logic of smart economy is depending on innovation and the use of these innovations in commercial activities. Information and communication technologies based on smartphones, e-commerce, virtual currency, electronic banking, and online supply and production management constitute the basis of the smart economy.

Smart Living, as a result of the application of information and communication technologies to public spaces and facilities as well as public services, is called the effective and efficient provision of basic vital needs such as welfare, comfort, health, and safety in city life [12]. The concept of smart living has included many different areas such as municipal services, security, urban transport, social solidarity, healthy food supply chain, improvement of living conditions in houses, distribution of clean water, and prevention of noise pollution [13].

Smart mobility means using green energy and electricity instead of fossil fuels, preferring public vehicles, and promoting the sustainability of natural resources, such as car-sharing and the use of bicycles, is a smart city component, which is why the smart mobility component includes the concept of a green city [14].

In addition to the mobility rate in cities that are expected to triple by 2050, [15] fiber connection schemes based on the distribution of digital information are also considered mobility [16].

Smart environment, cities that cause the 80% of greenhouse gas emissions in the world, increasing traffic, air and water pollution, climate change, excessive energy consumption, uncontrolled waste management, are dealing with problems such as the inability to use natural resources efficiently and recommend the use of information and communication technologies to overcome them. For this reason, it stands out as one of the basic components of smart cities to increase the quality of life of citizens and to reduce carbon emissions, and to use natural resources effectively and efficiently [17].

An efficient and effective environmental policy should be followed, a citizen-oriented living space should be created [18] and innovative inventions such as big data and IOT should be used to solve [19] the problem of getting sick or dying of thousands of people every year due to adverse conditions in city life [20].

6 The Factors Affecting the Smart City Ecosystems

Two elements form an equation in smart cities, the very first one is the adaptability to human and city infrastructure. These factors occur in different forms depending on each country and cultural structure, and any transformation takes a long period of time. And the second one, the structure affecting the equation refers to factors such as financing, business models and technology, and it is among the factors that can change more rapidly according to goals and needs in cities and such factors are more easily affected by external factors. Thus, the correct coordination of relations between all these elements stands out as a factor that can determine whether the investments will be efficacious and it is of special standing to comprehend a smart city ecosystem and to contemplate factors which might affect the ecosystem.

To understand the smart city ecosystem, we can classify the factors affecting the ecosystem in six sections as human factor, governance and institutional factors, funding and financing factor, business models factor, technology factor and structure of the city.

The human factor is the main factor affecting the smart city ecosystem because of the fact that, a smart city is, contrary to the general projections, not only a structure that improves the life quality of its citizens, but also it has a structure that is formed by the knowledge, creativity and ability of its citizens. It is such an influential factor that the power to produce economic value using information and communication technologies is used as an important economic and political weapon today, [21] and the greatest powers of cities and states are made up of educated, innovative thinking and technology-savvy white-collar workers [22].

Governance and institutional factors are among the most prominent factors affecting the smart city ecosystem. The concept of governance expresses a participatory management style and has become the general name of the management style referred to as bottom-up, as information and communication technologies and especially internet-enabled new administrators find a living space in smart cities [23].

Institutional factors are a concept that expresses the general structure of the city and includes dozens of different variables such as management style, relationship models, working and life culture, social and political participation mechanisms [24].

The funding and financing factor is one of the most crucial problems in the academic literature related to smart cities because of the necessity to meet the technological requirements that rise within the dynamic structure of smart cities and which are expensive particularly. Therefore, various funding and financing factors are used to obtain the necessary capital to solve this problem, attract foreign investments, and ensure technological sustainability in smart cities. Especially since the funding includes a limited structure, whichever of these techniques is chosen, the direction of the smart city is shifting towards that side [25].

The business model factor is the supportive component of funding and financing, which is among the necessities for the creation of smart cities. The production service and information flow architecture that is operated by an organization and generates revenue is called as a business model. In terms of smart cities, it refers to the use of all kinds of service layers in which the entire infrastructure of the city, software and hardware, human relations, production and consumption systems, and in short, information and communication technologies, within the business model architecture [26].

The technology factor has paved a way for the idea of smart city and formed it and not only changed the way of workflow as equipping all kinds of physical facilities but also resulted in an emergence of many virtual business models. Therefore, it is the most important factor that has critical effects on the decision-making processes in the public and private sectors [27]. However, contrary to the general opinion about smart cities, the real importance of the technology factor stems from the effective application of technology to human behavior and services, not to the surrounding smart cities [28]. But, the idea that the understanding of identifying the needs correctly and presenting them effectively and efficiently during this process should prevent "technology fetishism", which is widespread everywhere, continues to be important.

The structure of the city is the basis of decisions about smart cities. This ground determines whether the transformation strategy of a smart city will be successful or not, as well as any direction of the determination towards situations and also the decisions to be taken about the form of transformation in the preliminary evaluation phase of smart cities. The factors that determine such decisions include the following: Is it considered part of a national plan that includes other cities within the smart city or as a single smart city? [29]. Is this city an old brownfield city in both developed and developing countries, or a greenfield city built from scratch? [30]. Is priority given to physical or social infrastructure? Are the plans to be implemented in the region selected as a reference sector-based or geographically based? Are the decisions taken from top to bottom or bottom to top? [31].

7 Decision Making and Coordination Approaches in Smart Cities

During the administration and transformation of smart cities, some basic principles have started to emerge in accordance with the New Public Management. These principles are repetitively used in academic literature that contains a smart city theory and smart city practices, and they form decision-making and coordination approaches in which the article is based on. Additively, they provide a perspective to eliminate the uncertainty in some processes during the evaluation of smart city projects.

With a systematic view, these approaches can be termed respectively as the following: Central Coordination Approach, Citizen-oriented Approach, Holistic Approach, Adaptation Approach, Connected Approach, Sustainability Approach. As can be seen in Fig. 1, these approaches should cover all projects of a smart city and be effective on this ecosystem continuously.

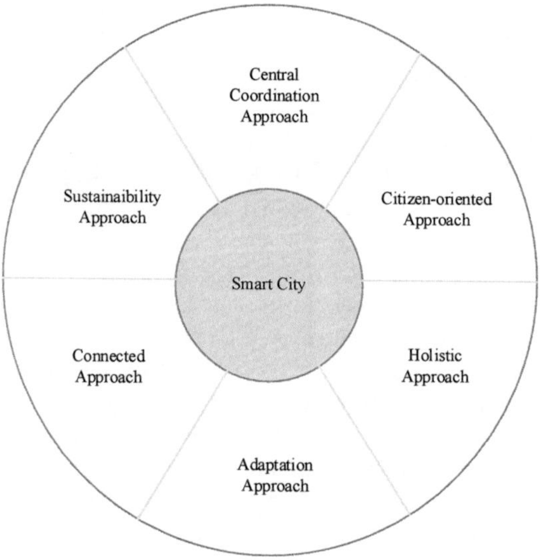

Fig. 1. Decision making and coordination approaches in smart cities

7.1 Central Coordination Approach

Gathering and directing all kinds of activities and resources in an organization under a central coordination so as to achieve basic goals is called coordination. Since, it includes many stakeholders and structures that create value such as energy, service, material resources, digital structures, human resources, business processes, private sector, public sector and non-governmental organizations, which are included in the smart city ecosystem, which is a systematic whole, and due to this diversity, central institution coordination is needed [32]. Therefore, ensuring the balance in the market by managing the decision-making needs that will arise during the smart city process by local governments as a leader and coordinator is also a suitable understanding for the New Public Management [33].

7.2 Citizen-Oriented Approach

The smart city is not just a city where technology is used in all its processes, but a city where technology is used to improve the quality of life of its citizens [34]. For this reason, the idea that not only the technology but also the people using this technology shall be at the center of the smart city should be fortified.

The citizen-oriented thinking is based on human behavior needs, demands, and values, which has a slower changing process compared to technology, [35] as well as human psychology, which is thought to be involved in decision-making processes and more likely to support the implementation of ideas designed by itself [36, 37].

7.3 Holistic Approach

The most common way of thinking in the smart city literature is that technology is almost made into a "fetishism" and everything else is perceived as something of "secondary importance" [38]. The Habitat Program of the United Nations, organized with the awareness of this, ignored the exaggerated interest shown in technology and looked at the events from a holistic city perspective based on the combination of smart people, smart economy, smart governance, smart mobility, smart environment, and smart living components [39, 40].

7.4 Adaptation Approach

If the projects unbased on economic, social, and political structures in societies, organizations of smart cities are not acceptedly recognized by the cultural and structural ecosystem of these cities and their sustainability is not possible. According to this view, the implementation of the projects successfully implemented in a smart city to other smart cities, taking into account the special conditions in that city, is called adaptation approach [41].

Therefore, the economic, social, and physical infrastructure and resources of both cities should be analyzed correctly to adapt projects from different smart cities during the transfer process [42].

7.5 Connected Approach

Based on the idea that smart city structures resemble the human body, the understanding of keeping the human alive by showing the necessary physical responses of the "nervous system" spreading throughout the body expresses the connected approach [43].

The need to communicate between established city networks is the most basic requirement of smart cities. Therefore, the distribution of different types of sensors in the city and the need to transfer the data obtained through different communication protocols such as HTTP, TCP to other systems and the need to continue this uninterrupted is the best example of the connected approach [44, 45].

7.6 Sustainability Approach

Although the idea of sustainability emerged as a sustainable environmental target due to the damages caused by greenhouse gases in its early periods, today it is perceived in three dimensions, including economic, social, and environmental factors [46]. This idea has been brought to the forefront as an approach aimed at increasing the quality of life of stakeholders [47].

To ensure sustainability in smart cities, it has become essential to use innovative approaches of digital technology in all areas related to the city and provides new business opportunities [48, 49]. Finally, the idea of flexible structuring of all management principles, material resources, and infrastructure systems can be put forward as the main factor underlying sustainability in smart cities [50].

8 Discussion and Conclusion

City is a space that might be defined in interdisciplinary respects, including a wide variety of structures within. This is a compelling factor for continuous works of different scientific disciplines not only during the development of technology but also during the joint management with technology in due course of this technology's application to smart cities. It is because of that any access to technology is getting easier day by day thanks to economic and technological developments, but implementing this technology within the framework of scarce resources and social features might involve certain obstacles. Hereby, it is crucial to embark on critical approaches, criteria and perspectives that should be considered as a basis for creating efficacious samples of smart cities. Explicitly both city administrations and technology providers should be aware of requirements to administer smart cities through these approaches.

Consequently, it might be essential to actively apply our six-main component hybrid approach, and firstly, it has set forth as a solution proposal and has possibly added the seventh and eighth approach alongside the others, depending on the probable results to emerge. Secondly, it is significant to make these approaches, which are asserted as a consequence of qualitative research, to be measurable with quantitative methods so as to provide a more effective management. Finally, the development of these approaches on the basis of agile management principles in order to adapt to rapid technological transformation can be envisaged as a new-fangled field that can be examined for other researchers.

References

1. IDC: IDC government insights, smart cities strategies, business strategy: smart city maturity model (2018)
2. The British Standards Institution: PAS 181:2014, smart city framework-guide to establishing strategies for smart cities and communities (2014). https://shop.bsigroup.com/, http://shop.bsigroup.com/upload/267775/PAS%20181%20(2014).pdf
3. Atac, B.: Management of the transition to smart city in local governments: Istanbul Metropolitan Municipality example. Unpublished doctoral dissertation, Istanbul Medeniyet University, Istanbul (2020)
4. Anthopoulos, L.G.: Understanding smart cities: a tool for smart government or an industrial trick? PAIT, vol. 22. Springer, Cham (2017). https://doi.org/10.1007/978-3-319-57015-0
5. European Parliament: Mapping smart cities in the EU (2014)
6. Raj, A., Dwivedi, G.: Smart city: an integrated approach using system dynamics. In: Kar, A.K., Gupta, M.P., Vigneswara Ilavarasan, P., Dwivedi, Y.K. (eds.) Advance in Smart Cities-Smarter People, Governance, and Solutions, pp. 93–104. CRC Press Taylor & Francis Group, Florida (2017)
7. Budhai, D., Proag, V.: Converting Mauritius into a Smart Country. Mauritius University, Institution of Engineers Mauritius, Mauritius (2015)
8. Coe, A., Roy, J., Paquet, G.: E-governance and smart communities. Soc. Sci. Comput. Rev. **19**(1), 80–93 (2001)
9. World Bank: Building blocks of e-government: lessons from developing countries (2004). http://documents.worldbank.org/curated/en/524191468135582219/Building-blocks-of-e-government-lessons-from-developing-countries
10. Nam, T., Pardo, T.A.: Conceptualizing smart city with dimensions of technology, people, and institutions. In: Proceedings of the 12th Annual International Digital Government Research Conference: Digital Government Innovation in Challenging Times. Association for Computing Machinery, New York (2011)
11. Government of Ireland: Building Ireland's smart economy (2008). https://www.lenus.ie, https://www.lenus.ie/handle/10147/76919.
12. Vinod Kumar, T.M., Dahiya, B.: Smart economy in smart cities. In: Vinod, K.T., et al. (eds.) Smart Economy in Smart Cities, pp. 3–76. Springer, Switzerland (2017). https://doi.org/10.1007/978-981-10-1610-3
13. Vienna University of Technology: European smart cities (2019). http://www.smart-cities.eu/model_5.html
14. Zhao, K.: Urban human mobility data mining an overview. In: IEEE International Conference on Big Data (2016)
15. Zipf, A.: Crowdsourcing for individual needs- the case of routing and navigation for mobility-impaired person. In: European Handbook of Crowdsourced Geographic Information, pp. 325–382. Ubiquity Press, London (2016)
16. Orlowski, A., Romanowska, P.: Smart cities concept: smart mobility indicator system. Cybern. Syst. Int. J. **50**, 118–131 (2019)
17. Boone, C.G., Modarres, A.: City and Environment. Temple University Press, Philadelphia (2006)
18. Dameri, R.P.: Smart City Implementation. PI, Springer, Cham (2017). https://doi.org/10.1007/978-3-319-45766-6
19. OECD: Smart sensor networks-technologies and applications for green growth (2009). https://www.oecd.org/sti/44379113.pdf
20. World Health Organization: The health effects of air pollution (2019). https://www.who.int/airpollution/events/expertconsult2019/en/

21. Forest, F.: Roadmap for real world internet applications-socioeconomic scenarios and design recommendations. In: Tselentis, G., et al. (eds.) Towards to Future Internet, pp. 325–334. IOS Press, Amsterdam (2009)
22. United Nations University: Smart cities for sustainable development (2016). https://unu.edu/projects/smart-cities-for-sustainable-development.html#outline
23. Conroy, M.M., Cowley, J.E.: E-participation in planning: an analysis of cities adopting on-line citizen participation tools. Environ. Plann. C Gov. Policy **24**, 371–384 (2006)
24. Deloitte: Smart cities-not just the sum of its parts (2018). https://www2.deloitte.com/xe/en/pages/strategy/articles/smart-cities-where-to-go.html
25. Hamilton, S., Zhu, X.: Funding and financing smart cities. Deloitte Center for Government Insight (2017). https://www2.deloitte.com/global/en/pages/public-sector/articles/smart-cities-funding-and-financing-strategies.html
26. Anthopoulos, L., Fitsilis, P., Ziozias, C.: What is the source of smart city value? A business model analysis. Int. J. Electron. Gov. Res. **12**(2) (2016)
27. Riva Sanseverino, R.: Competitive urban models. In: Riva Sanseverino, E., Riva Sanseverino, R., Vaccaro, V., Zizzo, G. (eds.) Smart Rules for Smart Cities. SSISS, vol. 12, pp. 1–14. Springer, Cham (2014). https://doi.org/10.1007/978-3-319-06422-2_1
28. Wohlers, T.E., Bernier, L.L.: Setting Sail into the Age of Digital Local Government. PAIT, vol. 21. Springer, Boston, MA (2016). https://doi.org/10.1007/978-1-4899-7665-9
29. Angelidou, M.: Smart city policies: a spatial approach. Cities **41**, S3–S11 (2014)
30. Macomber, J.D.: The 4 types of cities and how to prepare them for the future. Harvard Business Review (2016)
31. Brodowicz, D.P., Pospieszny, P., Grzymala, Z.: Eco_Cities: Challenges, Trends and Solutions. Warsaw School of Economics, Warszawa (2015)
32. Aragonez, T., Alves, G.C., Blanco-González, A.: Strategic management of city brands and its influence in smart cities. In: Peris-Ortiz, M., Bennett, D.R., Pérez-Bustamante Yábar, D. (eds.) Sustainable Smart Cities. ITKM, pp. 79–88. Springer, Cham (2017). https://doi.org/10.1007/978-3-319-40895-8_6
33. China Academy of Information and Communications: Comparative Study of Smart Cities in Europe and China 2014. CCERS, Springer, Heidelberg (2016). https://doi.org/10.1007/978-3-662-46867-8
34. Smart Cities Council: Smart Cities Readiness Guide (2015)
35. United Nations-ESCWA: Smart cities: regional perspectives. United Nations-ESCWA, Beirut, 2015
36. World Bank: Public Consultation in the EA Process: A Strategic Approach (1999)
37. Spil, T.A.M., Effing, R., Kwast, J.: Smart city participation: dream or reality? A comparison of participatory strategies from Hamburg, Berlin & Enschede. In: Kar, A.K., et al. (eds.) I3E 2017. LNCS, vol. 10595, pp. 122–134. Springer, Cham (2017). https://doi.org/10.1007/978-3-319-68557-1_12
38. Kavta, K., Yadav, P.K.: Indian smart cities and their financing: a first look. In: Seta, F., Sen, J., Biswas, A., Khare, A. (eds.) From Poverty, Inequality to Smart City. STCEE, pp. 123–141. Springer, Singapore (2017). https://doi.org/10.1007/978-981-10-2141-1_9
39. Giffinger, R.: Smart Cities-Ranking of European Medium-Sized Cities. Vienna University of Technology, Vienna (2007)
40. Schleicher, J.M., et al.: A holistic, interdisciplinary decision support system for sustainable smart city design. In: Alba, E., Chicano, F., Luque, G. (eds.) Smart-CT 2016. LNCS, vol. 9704, pp. 1–10. Springer, Cham (2016). https://doi.org/10.1007/978-3-319-39595-1_1
41. Fontana, F.: The smart city and the creation of local public value. In: Dameri, R.P., Rosenthal-Sabroux, C. (eds.) Smart City. PI, pp. 117–137. Springer, Cham (2014). https://doi.org/10.1007/978-3-319-06160-3_6

42. Thorne, C., Griffiths, C.: Smart, smarter, smartest: redefining our cities. In: Dameri, R.P., Rosenthal-Sabroux, C. (eds.) Smart City. PI, pp. 89–99. Springer, Cham (2014). https://doi.org/10.1007/978-3-319-06160-3_4
43. Bawa, M., Caganova, D., Szilva, I., Spirkova, D.: Importance of internet of things and big data in building smart city and what would be its challenges. In: Leon-Garcia, A., et al. (eds.) SmartCity 360 2015-2016. LNICSSITE, vol. 166, pp. 605–616. Springer, Cham (2016). https://doi.org/10.1007/978-3-319-33681-7_52
44. European Commission: European innovation partnership on smart cities and communities (2013)
45. Bennett, D., Pérez-Bustamante, D., Medrano, M.-L.: Challenges for smart cities in the UK. In: Peris-Ortiz, M., Bennett, D.R., Pérez-Bustamante Yábar, D. (eds.) Sustainable Smart Cities. ITKM, pp. 1–14. Springer, Cham (2017). https://doi.org/10.1007/978-3-319-40895-8_1
46. Lohrmann, C.: Online learning—do MOOCs contribute to the goals of agenda 21: "education for sustainable development"? In: Osburg, T., Lohrmann, C. (eds.) Sustainability in a Digital World. CSEG, pp. 211–224. Springer, Cham (2017). https://doi.org/10.1007/978-3-319-54603-2_18
47. Macagnano, M.: The future of the built environment in the information age: integrating concepts of sustainability, information and communication technology, and architecture. In: Proceedings of the Smart and Sustainable Built Environment (SASBE) Conference. University of Pretoria, Pretoria (2015)
48. Araksinen, M., Ailisto, H., Nylund, N.O.: Smart City. VTT Technical Research Center of Finland Smart City-Research Highlights, Espoo (2015)
49. Manfren, M.: Multi-scale computing for a sustainable built environment. In: Smart Cities-Foundations, Principles, and Applications, pp. 53–98. Wiley, Hoboken (2017)
50. Townsend, A.M.: Smart Cities, 1st edn. Norton Company, New York (2013)

Internet of Things and Its Impact on the Future of Education

Ismail Noori Mseer(✉)

Ahlia University, Manama, Bahrain

Abstract. Internet of Things (IoT) is the change cycle in various parts of our everyday life. IoT advances contrast from past developments as they are universal and urge answers to be wise and self-sufficient Advances in the IoT are a significant key innovation pattern. Ubiquitous sensors and the capacity to overcome any barrier between the actual world and the machine world were seen as the new learning model's calculated system. The deduction behind this incredible change in perspective is the capacity to insert sensors into any article and use Machine-to-Machine (M2M) correspondence to associate billions of items/gadgets to the current Internet foundation. The aggregate of the actual world is coming on the web quickly.

Keywords: Internet of Things · Education · Future of education · Strategies for future classes

1 Introduction

In the coming years, innovation will affect the learning experience from numerous points of view. Internet of Things (IoT) keeps on affirming its significant situation with regards to Data and Communication Technologies and the improvement of society. With the backing of IoT, establishments can improve learning results by giving more well-off learning encounters, improved operational proficiency, and by picking up ongoing, significant knowledge into understudy execution. This investigation aims to discover the capability of IoT in higher schooling and how to boost its advantages and decrease the dangers engaged with it. Further endeavors are essential for delivering the maximum capacity of IoT frameworks and innovations. Accordingly, this paper presents an investigation into the effect of IoT on advanced education, particularly colleges. IoT stands to change drastically how colleges work and improve understudy learning in numerous controls and at any level. It has enormous potential for colleges or some other instructive organizations; if reliable and steady to guarantee boundless and fruitful execution by authority, staff, and understudies. IoT needs advancement where colleges can lead [2]. Scholastics, scientists, and understudies are in an exciting spot to lead the revelation and advancement of IoT frameworks, gadgets, applications, and administrations. This paper also confirms the fate of IoT in advanced education during the following, not many years, which have been offered by several examination associations and ventures. Then again,

A. M. A. Musleh Al-Sartawi et al. (Eds.): EAMMIS 2021, LNNS 239, pp. 490–499, 2021.
https://doi.org/10.1007/978-3-030-77246-8_45

IoT additionally brings enormous difficulties to advanced education. Henceforth, this paper likewise presents the viewpoint on the challenges of IoT in progressive education.

IoT carries colossal difficulties and occasions to advanced education. The remarkable development of all-inclusive processing, creating IoT advancements, for example, distributed computing, and extensive information and investigation are useful in improving the fundamental beliefs of instructing and nature of the examination and building up an IoT society empowering another computerized culture. A portion of the IoT challenges in the advanced education area incorporate Cloud Computing, Instructional Technologies, Mobility Applications, Security and Privacy, Research Computing, Quality and Ethics, and Financing [1]. The IoT can help tackle difficulties across a comprehensive exhibit of subjects, from coordination to the organization to understudy life.

2 Change Numerous Areas

The IoT is an actual worldwide organization that interfaces gadgets, articles, and things to the Internet foundation to impart or cooperate with the interior and the outside climate as delineated in Figure 1 and trade data through the data detecting gadgets per explicit conventions. In this manner, IoT is empowering network for anything and for anybody to be organized far and wide whenever, and anyplace utilizing any organization or any assistance [2] to accomplish the objective of clever recognizing, following, and overseeing things [3]. It is an augmentation and extension of Internet-based organization, which grows the correspondence between human to human (H2H), human to things (H2T) or something to things (T2T) [5] as introduced in Figure. Various examination associations and experts have anticipated the eventual fate of IoT and its likely effect on the Internet: billions of actual gadgets, everywhere in the world, that have computerized sensors and are interrelated by utilizing any organization. As per the examination led by [4], an expected 13.4 billion gadgets were associated in 2015, which speak to more than the complete populace on the earth at that point, and this number is required to increase to 38.5 billion gadgets by 2020 significantly. The chance of collaborating with heaps of regular items associated with the Internet permits an individual to get to unlimited data whenever and anyplace. This vision opens another skyline of thoughts and advancements that is now being considered by research researchers and scholastics.

The IoT vision is grounded in the conviction that the solidly fixed advances in microelectronics, interchanges, and data innovation we have seen in current years will suffer into the anticipated future [5]. Applications for the IoT are now being utilized in areas like medical care and client assistance. Presently, colleges and schools are joining the gathering. A portion of the ways the IoT can profit schooling might be shown, while others are not as self-evident. So this paper will introduce the most significant ramifications for associated gadgets in advanced education and how they could shape learning for the future. So when planning an encounter, institutional pioneers should move toward it first by finding the most significant problem areas. Smoothing out and upgrading offices can help accomplish monetary reserve funds (e.g., reacting to climate occasions, computerizing tasks). Shrewd gadgets can make staff and suppliers aware of administration gear before an issue even presents itself. Smart entryways, locks, and cameras can be utilized to screen and control development in various offices. As more gadgets become associated, grounds pioneers will have the option to extricate significantly more incentive from the persistent stream of information and data, encouraging them to move from a value-based relationship with understudies, personnel, directors, and suppliers to an iterative cycle in which miniature choices can be made on an ongoing premise [6].

The fate of colleges isn't tied in with utilizing and utilizing accessible innovation. It is about how colleges will adjust to the changing requirements of things to come information specialists, the eventual fate of work, and the economy. This paper presents an outline of IoT in advanced education establishments, particularly in colleges, investigates a few arising patterns that are advancing advanced education, and explores the expected effect of IoT and the eventual fate of the IoT in progressive education. Also, investigating some of the IoT challenges concerning the advanced education area.

Essential Components of IoT. The IoT is engaging innovation specialists to create more modest and more moderate remote frameworks that devour less power and can be incorporated into practically any sort of gadget [7]. Three IoT parts empower consistent associations which are: Hardware: comprised of sensors, actuators, and inserted correspondence equipment, Middleware: on interest stockpiling and registering apparatuses for information examination, and Presentation: novel, clear perception and translation apparatuses which can be broadly gotten to on various stages and which can be intended for multiple applications. There are different expected methodologies for presenting low-power interchanges to an IoT hub, going from reason planned conventions, for example, ZigBee, to low-power variations of Bluetooth, Wi-Fi, and NFC. The IoT's accomplishment relies upon the availability of the web and its spread all over [8] Also, the web should be accessible at a sensible expense; and this issue may not be anything but difficult to address in individual nations. New gadgets and the spread of cell phones and applications create colossal information measures, which will keep on expanding.

3 Adjust the Strategies

Advanced Campus System is a significant stage for understudies to get a wide range of data [9]. Innovations are likewise influencing different regions of grounds organization. There is an expanding request for advanced education organizations, particularly, colleges to digitize their substance and exercises and adjust their strategies to permit scholarly and scientists to work efficaciously in a progressive climate [10]. An all-around planned actual ground, totally coordinating innovation, is critical for building the brand of computerized college by improving the understudy insight and giving the proper settings and offices for educating, learning, and exploration. It advances, underpins, and empowers deep-rooted understanding [11]. A computerized college should have the innovation that empowered educating and learning and enable synergistic examination. All new automated dangers can be looked by colleges on the off chance that they contend, anyway few have the vision, adaptability, stages, or appropriate authority, to set up the systems to guarantee that they can enhance, or respond to commercial center Inside an advanced ground, innovation can lessen operational expenses, improve security, and offer devices for analysts, scholastics, understudies and staff. These advantages provide a genuine benefit to college tasks and advancements, the experience of understudies, and analysts. The computerized grounds include two principal parts. Initially, it reuses the

I.T. Service Delivery Platform – start to finish foundation to give network availability, portability, and security for all applications and administrations across the grounds [12].

Furthermore, it incorporates an enormous number of Internet of Things (IoT) applications working over the stage framework to help the college's expert, empower instructing and learning exercises, and improve understudy's insight. As per Cisco – "Digitizing Higher Instruction To upgrade encounters and improve results"- IoT applications vary from typical network applications as they uphold sensors and sensor information instead of clients and client information. IoT applications for the computerized grounds incorporate five fundamental classes: Building Control and Management; Security and Access Control; Video and Information Systems; Location and Attendance Systems; Energy Monitoring and Control. The small organization has a primary task to carry out inside the advanced grounds, subsequently should be intended to meet the high requests of a progressive college. What's more, IoT change frequently the educating and learning in the feet, for instance, the IoT in the preparation of a similar sight, making the showing space, preparing settings, sports scenes, learning residences, eateries, and understudies across the grounds of IoT, making the grounds to prepare understudies whenever and anyplace, can turn into a physical, mental, and abilities spot to obtain a full scope of learning and training. This will lead schools to become universal learning and preparing [7, 16]. They accept three regions require massive speculation and cooperation before an environment can arise to interconnect individuals, spaces, and establishments: Strong information the board and personality controls should be incorporated into the biological system (an unpredictable organization or interconnected framework) from the beginning. The current IoT scene comprises individual arrangements, or "walled gardens," that offer advantages for clients who purchase from a specific item family. Increasing the number of associated items ought not to build the screens or consoles required for design or use [1]. The IoT will influence all aspects of society eventually sooner rather than later. Advanced education establishments, when all is said in done, and colleges specifically, can work across controls and lead the advancement of the IoT innovations, plans of action, morals, and heads of the IoT empowered economy of things to come. For example, college educators of software engineering always coordinate IoT labs to improve IoT advancements. What's more, Informatics College can instruct how to use the sizes of IoT information with TIPPSS. Additionally, they can work with business schools to set and plan IoT courses to make new action plans. Clinical schools can enable the Internet of Medical Things just as, Law schools can show IoT morals, protection, and strategy. As per Zebra innovations, as higher training establishments initiate to create and use arrangements, for example, radio recurrence recognizable proof (RFID) and distributed computing through IoT advances, they will have the option to break down [11].

Moreover, IoT can build the learning experience by accommodating constant and significant experiences into understudy execution. These days, understudies, especially in college, are moving ceaselessly from reading material to innovations, such as tablets and workstations. The high-level e-learning applications permit understudies to learn at their speed and have an indistinguishable learning experience in homerooms and homes [1], which increases move and fulfillment rates just as teachers can convey coordinated guidance and determined understudy evaluations. Also, through IoT innovation, teachers can gather information about understudies' presentation and afterward decide which ones need more consideration and consideration.

This information investigation likewise assists teachers with evolving plans and strategies for future classes. Also, associated gadgets can permit educators to do dynamic study hall. Mediations just as logging participation will be streamlined if understudies have a wearable device that tracks ECG designs. These gadgets can divert an understudy's consideration by surrendering a warm action and exercise to their gadgets. Additionally, EEG sensors can be utilized during courses to screen understudies' intellectual activities [14].

The IoT isn't only an innovation update and improvement inside the business. Yet, it can lead to growing the change to society, including advanced education organizations. IoT will lead the change and change the progressive education foundations. As indicated, IoT will lead to instructive innovation changes, change the schooling, change in educating, change in learning, the executives of progress, trial and pragmatic modifications, changes in grounds, training assets changes, and others. With the improvement of IoT, the planned application in advanced education lies in the three viewpoints: understudies' reformist assessment, the mix of current showing stages, and improvement of instructive middleware [16]. This change gives expanded accommodation to understudies and makes the showing measure more successful for teachers and educators. The stream in associated gadgets and innovation implies that teachers and educators can zero in on the real discovery that is more helpful to the understudies instead of playing out the usual errand. Innovation should "blur out of the spotlight" using articles and administrations that give genuine client profit by availability and can be controlled through voice, signal, or other applicable methods for input [1]. To ensure the data trade in the IoT, existing encryption innovation should be care completely audited [5], The encryption calcul4ations should be quicker and less energy burning-through. Security at both the gadget and organization is essential to the activity of IoT. The creator (The Internet of Everything in Higher Education (CISCO) presents some significant focuses on accomplishing useful IoT in instruction. Improving resource knowledge This vision and understanding give partners a perspective on understudies, association, and monetary resources. This resource insight empowers the association to settle on educated choices to improve understudy information and learning encounters, operational capability, and grounds' security. As per [6] improving resource knowledge, instructive establishments can improve results by adding values in certain regions include Enhanced Learning Experiences and Results, Improved Operational Efficiency, Safer Campus Designs. Also, outside of the homeroom, colleges can utilize associated gadgets to screen their understudies, staff, assets, and hardware at a diminished working expense. Besides, the development of portable innovation and the IoT empower colleges to improve the security of grounds, upgrade admittance to data and applications whenever from anyplace, and monitor primary assets [1]. IoT is changing the understudy learning experience other than offices the board by interfacing individual, information, and things. All learning conditions. IoT can uphold homeroom guidance by improving picking upsetting, upgrade learning assets, improve strategies and procedures of learning, raise the executive's proficiency, and save the executives costs. The assets accessible for learning on gadgets, similar to digital books, are additionally captivating and intelligent. In any case, there is a consistent requirement for innovations for learning measures, for example, rapid remote organizations with the data transfer capacity for real-time sound and video exercises. The article reports that IoT in instruction has enormous potential for advanced education foundations. Be that as it may, three key territories should be routed to guarantee far-reaching and effective reception: Security: IoT's unavoidable advancement will not occur across instructive organizations. The data should be accessible, yet classified, when required, and the data proprietor should conclude who is permitted to get to the data, regardless of whether individuals or associations. The security way to deal with insert gadgets should consider individuals and articles' mysterious organization in both the general population and private areas. Such devices will probably make new connections among individuals and P.C.s. As per the [17], in the following five years, IoT innovation will improve the learning experience unexpectedly. The learning experience will become more virtual, and understudies will burn through information and learning, and study halls will be better prepared for education. At last, knowledge will turn into an astonishing encounter for educators and understudies with information quickening while at the same time bringing novel thoughts and arrangements around the world. Too, understudies are ready for the eventual fate of work and assumptions at the work environment of what's to come. Innovation will consistently have a spot in all instructive orders. IoT additionally has numerous open doors for Science, Technology, Engineering, and Mathematics (STEM) disciplines, for example, Colleges have since quite a while ago understood the capacity of innovation to upset instructing, learning, and appraisal. Besides, innovation disturbance is necessary if a cutting-edge college recognizes its understudy offer, so expanding confirmations, improving maintenance, and conveying wanted results. Yet, planning understudies to be specific for the universe of work is perplexing [14]. It requires a trustworthy academic initiative, admittance to an excellent educational plan and content, and the presentation of understudies to the possible utilization of innovation. With the improvement of IoT, numerous foundations of advanced education have begun to zero in on the connected creation and use of the IoT. Likewise, this endeavor is utilized in college [1]. The Internet has profoundly established itself in schools and colleges, and e-learning has become standard practice in many college frameworks. Even though it's anything but an undeniable application of the IoT, in any case, schooling is on that rundown, and the uses of the IoT in colleges are various, and the suggestions for this are enormous. IoT will take into consideration better operational productivity.

The conversation about embracing IoT should keep harmony between the positive effects and creating mindfulness about the dangers to both protection and security. The two instructors and students should have a superior comprehension of moral issues and the risks of IoT, just as approaches to kill those dangers. Individual and public information should be dealt with unexpectedly, and the person's protection should be regarded and taken into point of view. Data uprightness: the honesty of information should likewise be guaranteed, just as its precision, genuineness, practicality, and culmination [13].

4 Lead the New Innovation

Additionally, the future IoT economy can be molded by specialists and pioneers in advanced education area also, by teaching the understudies [1]. The advancement in education frameworks will improve and lead the new innovation advancements. Along these lines, the progressive education area should work with business and the mechanical regions to shape and construct an IoT-empowered economy's eventual fate. Moreover, advanced education areas, particularly colleges, have to lead the future of IoT advances by planning courses for specialized and business pioneers and encouraging understudies and analysts to assemble new business techniques that influence IoT advances in a multidisciplinary way. In 2016, a workshop directed by IEEE, National Science Foundation (NSF), and Internet, named "Start to finish Trust and Security for the Internet of Things," and followed by an IEEE Experts in Technology and Policy Forum, the specialists and members affirmed that the IoT needs advancement what's more, improvement which analysts, teachers, scholastics, and understudies in colleges and higher schooling area are in a great spot to lead the progress and advancement of IoT gadgets, frameworks, applications, and administrations. Likewise, they underscored that new stages and thoughts across disciplines must be created and found to tackle numerous issues always that we are confronting these days [12]. In expansion, IoT and information examination apparatuses can be utilized to develop and improve adequacy and capabilities nearby and across social orders, to improve data catch, to address security and protection issues, to limit the energy use, and to dissect information and give noteworthy bits of knowledge and comprehension to create and improve wellbeing area. To fabricate this framework sight, working across aptitude sets and trains are required. Besides, scholastics and examination researchers can assemble start to finish TIPPSS answers for the IoT just as they can construct IoT gadgets and administrations with a "protection in profundity" technique, including security at the equipment, programming, firmware, and administration levels. P.C. programming and actual registering. It is anything but difficult to predict how IoT abilities can be utilized in STEM orders, advanced mechanics, and anything having to do with gathering explicit information. It is all in the capability of the IoT [13]. Nonetheless, at last, the educationists should have the option to recognize the privilege innovation and coordinate it appropriately in the homeroom for figuring out how to develop. Albeit principle IoT advancements are so far hazy, the point indeed is that a ton of substance is the result of the new improvement stage. Thinking about the interest of more expert examination, setting the IoT major is relative material and straightforward for graduate understudies. Nonetheless, for college understudies, actually need a broad scope of fundamental courses, so it isn't easy to set IoT major freely like different majors at present. New preparing strategies will be investigated for college understudies. Various universities need to investigate the suitable methodology as per their own qualities. The framework approach, what's more, courses content should be dynamically settled and improved. Since IoT is accomplishing the solidarity of the virtual world and the actual world, numerous new preparing strategies and cross-cutting regions will be created later on [1].

5 Advanced Education

IoT carries colossal difficulties and occasions to advanced education. The exceptional development of pervasive figuring, creating IoT advancements, for example, distributed computing, and extensive information and examination are useful not just in improving the guiding principle of educating and nature of exploration yet, in addition, building up an IoT society and empowering another computerized culture [5]. With expanding on the web degree openings and consistent admittance to an instructional substance in both organized and unstructured organizations, the IoT leads mechanical force into progressive education organizations. IoT is an emotional move in the customary instructional worldview while coordinating more extensive orders, including sociology, to improve the estimation of considerable information accessible from web-based media. A portion of the IoT challenges in higher instruction area include:

Numerous colleges are utilizing crossover cloud as their venture engineering for facilitating IoT applications. The blend of recent college grads, the most technically knowledgeable understudies in the colleges, just as the ascent of tablet and versatile innovation, has opened new techniques to build the viability of significant business engineering, instructional advancements, examination, and learning conditions. With universal figuring, the cloud gives consistent associations and administrations to data innovation administrations [15]. By and by, big business design in numerous advanced education organizations rely upon crossover cloud frameworks with processing stages on private mists, while undertaking and instructional applications steadily move to public mists. Undertaking design in these organizations needs to decrease idleness time in light of the interest for content in instructional advances, the enormous expansion in sound and recordings for directions, and the requirement for dynamic undertaking organizations.

The developing utilization of learning the executive's frameworks LMS like Moodle and Blackboard is making a gigantic measure of organized and unstructured information, for example, sound and video content. Modern electronic schoolrooms outfitted with address catch frameworks and web streaming give an open door for understudies to get to instructional substance on interest whenever. IoT applications are in effect progressively used to coordinate versatile learning applications and for evaluation and reviewing frameworks [11]. The ideal application can help understudies to profit by learning assets, oversee tasks, and work on undertakings. Educators likewise utilize a portion of these applications to show concrete ideas, intricate physical, logical reproductions, and social subjects. The usage of IoT innovations presents new and exceptional security and protection challenges and issues. Tending to these difficulties and problems to guarantee security in IoT gadgets and administrations ought to be a basic need. One of the central rules for IoT is the need to incorporate robust also, reliable protection and security instruments [11].

Advanced education is defenseless against the security; also, the security of the IoT biological system. Despite the fact that there has been a further force to manage the safety of the IoT foundation, there is still no technique to distinguish business hazards related to information breaks. Advanced education areas need to create norms to make sure about IoT applications. As advanced education makes a vast number of future specialists, it needs to grasp IoT stages and frameworks even with the difficulties of IoT financing, developing computerized instructive instructional method, preparing, and interdisciplinary examination. Likewise, IoT applications should draw later on labor force ethically and morally to address network protection issues as society relies more upon IoT applications. Subsequently, a shared technique to wellbeing and security [2].

Advanced education keeps on profiting by IoT joining. As the expense of equipment diminishes, interdisciplinary exploration has picked up energy in the most recent years. Also, with the accessibility of enormous information, much more modest colleges can expand their interdisciplinary examination impression and put in elite processing (HPC), massive information stages, and investigation. STEM instruction has seen the need to distinguish more extensive coordinated effort with IoT biological systems by utilizing sensor advancements, Automated Aerial Vehicles (UAVs), and microcontrollers. Designing labs use sound video advancements, UAV, Raspberry Pi, and open source frameworks (OSS) that are driving developments and upgrading learning measures in designing undertakings [8]. Sociology scientists interested by the plenty of enormous information created by online media and ubiquitous figuring are continually utilizing disseminated processing stages, for example, HPC, GPU groups, Hadoop bunches, and colossal information examination to improve IoT research. The nature of learning both on the web and nearby and the increasing expense of advanced education has been seriously bantered in most recent years. The IoT offers enjoyable occasions to convey computerized courses. Nonetheless, it additionally acquaints difficulties with keeping up the nature of guidance and assessment of understudies' work. IoT instructive applications need apparatuses and innovations for teachers, educators furthermore, established researchers to improve the nature of the examination and address morals issues inside higher schooling [13].

6 Conclusions

The expense of data advances keeps on expanding each year as substance and an application. These application stacks keep on becoming both evenly and vertically on instructional advancements, research figuring, and endeavor innovations. Close by the data innovation, furthermore, lab expenses, most colleges don't have a technique for sharing expenses and distinguishing the aggregate cost of possession for an IoT framework. Advanced education should think of groundbreaking plans to back a data innovation foundation and administrations. With the progression in innovation, for example, Web of Things, colleges can resolve numerous difficulties, for example, monitoring fundamental assets, create admittance to data, assemble more brilliant plans, and plan more secure grounds. IoT frameworks can possibly carry critical qualities to higher instruction by drawing in and propelling the understudies and staff and to speed up learning. The motivation behind this examination was to discover the capability of IoT in advanced education and how to augment its benefits while tending to its difficulties and diminishing the dangers associated with it. Accordingly, our future work will be to zero in on IoT execution in advanced education.

The achievement will be predicated on an open stage that will permit teams to utilize similar benchmark advancements. Teachers should work intimately with governments to guarantee the improvement of IoT in training. Simultaneously, the public authority should safeguard the wellbeing and security of its residents. Education arrangements: the strategies that energize the selection of innovation in the homeroom and its effective coordination into programs are necessary. Such a strategy should incorporate sound change-the board rehearses among instructive foundations to diminish innovation selection boundaries and increment its scale. Proficient advancement programs for teachers should consolidate IoT instruments to energize early reception and assist instructors with creating inventive procedures and suitable instructional methods for the learning climate.

References

1. Aldowah, H., Ghazal, S., Muniandy, B.: Issues and challenges of using e-learning in a Yemeni Public University. Indian J. Sci. Technol. **8**(32), 9 (2015)
2. Friess, P.: Internet of Things: Converging Technologies for Smart Environments and Integrated Ecosystems. River Publishers, Gistrup (2013)
3. Stankovic, J.A.: Research directions for the internet of things. IEEE Internet Things J. **1**, 3–9 (2014)
4. Juniper Research: Internet of Things' connected devices to almost triple to over 38 billion units by 2020 (2015)
5. Agarwal, S., Pati, S.: Study of Internet of Things. Int. J. Sci. Res. Dev. **4**(05), 4 (2016)
6. Ryan, P., Watsonm, R.: Research challenges for the Internet of Things: what role can or play? Systems **5**, 24 (2017). https://doi.org/10.3390/systems5010024
7. Jin, D.: Application of "Internet of Things" in electronic commerce. Int. J. Digit. Content Technol. Appl. (2012)
8. Mukhopadhyay, S.C., Suryadevara, N.K.: Internet of Things: challenges and opportunities. In: Mukhopadhyay, S.C. (ed.) Internet of Things. SSMI, vol. 9, pp. 1–17. Springer, Cham (2014). https://doi.org/10.1007/978-3-319-04223-7_1
9. Han, Z., Wang, J.: Development and research of digital campus system based on android (2014). 1 Sherson, G.: Education and the Digital Campus (1999): p. 9
10. Porter, A., Sherwin, M.: The digital campus the online future for higher education, p. 38 (2013)
11. Aldowah, H., Rehman, S., Ghazal, S., Umar, I.: Internet of Things in higher education: a study on future learning (2017)
12. Zhiqiang, H., Junming, Z.: The application of Internet of Things in education and its trend of development. Mod. Distance Educ. Res. **2**, 19 (2011)
13. Aditya, B., Kumar, P., Doegar, A.: Internet of Things (IoT): IPV implementation for smart vision of the world. Int. J. Adv. Found. Res. **6**(2), 48 (2018)
14. Gubbi, J., et al.: Internet of Things (IoT): a vision, architectural elements, and future directions. Futur. Gener. Comput. Syst. **29**(7), 1645 (2013)
15. Qi, A.Q., Shen, Y.J.: The application of Internet of Things in teaching management system. In: International Conference on Information Technology, Computer Engineering and Management Sciences (ICM). IEEE (2011)
16. Kortuem, G., et al.: Educating the Internet-of-Things generation. Computer **46**(2), 53 (2013)
17. Tan, D.: Engineering technology, engineering education and engineering management. In: Proceedings of the 2014 International Conference on Engineering Technology, Engineering Education and Engineering Management (ETEEEM 2014), Hong Kong, 15–16 November 2014. CRC Press (2014)

Worldwide Artificial Intelligence Studies with a Comparative Perspective: How Ready is Turkey for This Revolution?

İsmail Ermağan(✉)

İstanbul Medeniyet University, Istanbul, Turkey
ismail.ermagan@medeniyet.edu.tr

Abstract. This study focuses on AI in its mission of fundamental dynamics of transformation and the work carried out in this field in the world. Then AI studies in Turkey are analysed and several recommendations are made for a more effective policy there. The main hypothesis of the study is that however R&D investments from Turkey's budget are increased in recent years, these investments are not enough in this time of intense AI revolution and education policies, infrastructure equipment and human capital have not been integrated effectively within the framework of a short-medium-long term plan. Although, some certain progresses are achieved, Turkey is situated in the mid/ranking level in comparison to other developing countries.

Keywords: AI · Industry 4.0 · Turkey · Global AI strategies · Proposals for Turkey's AI strategies

1 Introduction

Along with the 2020s, the world is experiencing a groundbreaking historical break in the industrial field [1]. Consequently, with new technologies, transformations that can bring upheavals in economy, politics, health and many other fields are experienced. In this process, especially AI inventions produce a separate effect [2]. In simple terms, new industrial revolutions are now taking place [3], AI reorders states economically. "Who's ahead, who's behind?", this new economic model based on AI determines the answer to the question.

AI studies, which have undergone ups and downs from the 1950s to the present, entered a revolutionary process in the 2010s with the use of machine learning and artificial neural network techniques. Dozens of states from China to the USA [4], from India to Canada or from Finland to Kenya are making significant investments in this field. The importance of this process, which includes international structures such as the United Nations (BM) and the European Union (EU), where 50 countries have determined their national AI strategies as of 2020 [5], is becoming increasingly important.

Why are AI and deep learning [6] breakthroughs critical? Because by 2030, AI studies will generate a GDP of more than today's Chinese economy ($15 trillion) [7]. In

A. M. A. Musleh Al-Sartawi et al. (Eds.): EAMMIS 2021, LNNS 239, pp. 500–512, 2021.
https://doi.org/10.1007/978-3-030-77246-8_46

this case, what is the policy of Turkey which wants to be among the first ten economies in the world? What steps are taken by Turkey's business and political elite? Dr. Ali Taha Koç, Head of the Digital Transformation Office of the Presidency of Turkey which was established in 2018, sees AI as the one of main pillars of "Digital Turkey" and "National Technology Advance" [8].

In this article, firstly, information on AI (such as its definition and usage areas) is given, and examples of AI policies of successful countries are presented. AI studies of Turkey which wishes to come to an ambitious position in this field, are discussed and various suggestions are subsequently brought for Turkey. The basic argument is: since the transformation of the new economies of the 21st century will take place through AI technologies, the development calculations of the countries that do not eliminate the deficits in both technical and human education dimensions will be surprised. Therefore, Turkey must take more comprehensive plans on the agenda within her politics, business, universities and scholars circles.

2 Literature Review

2.1 Definition and Usage Areas of Artificial Intelligence

The concept of AI, one of the most fundamental components of the process called the Fourth Industrial Revolution. AI, in the most reductionist definition, emerges as a technology developed for computers and robots to think and behave like humans. With the data provided, it can imitate the movements of human beings, the most intelligent creature known, and the system of thinking, "reasoning". The logic of artificial intelligence studies in its simplest form is this: Human imitations, which are independent of human influence, can learn and apply more easily than humans, and the error rate is almost zero [9].

It is realized that artificial intelligence is divided into two as narrow and powerful. Narrow artificial intelligence applications have been created to perform a single task in just one application area. Strong artificial intelligence has human-level intelligence and problem-solving capacity. In AI studies; internet of things, deep learning, machine learning, neural networks are complementary elements. Super artificial intelligence, on the other hand, points to the model that can think and apply itself, which is the landmark of all discussions [10].

Innovations such as accelerating production, decreasing error rates, increasing product quality and decreasing costs with AI applications; indicates the level of competition desired by developed and developing states. These artificial intelligence technologies have now become a shaping element of societies and individuals as well as states. From the Siri application on smartphones to computer games to autonomous cars, from drug production to thief determination, artificial intelligence has become the sought-after blood [11]. The main applications of AI revolution can be seen below (Fig. 1).

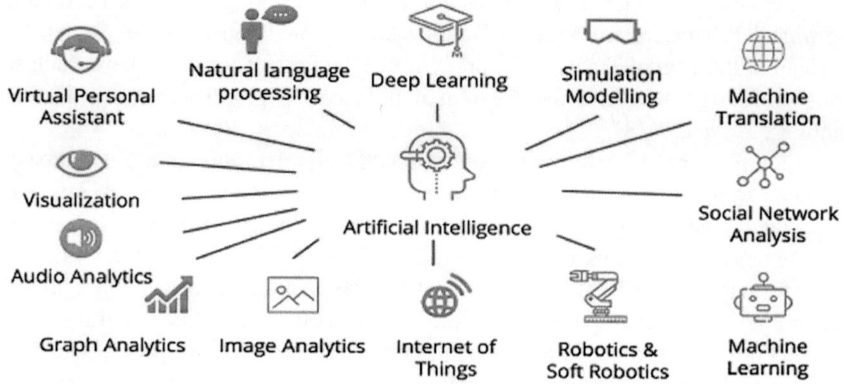

Fig. 1. Applications of artificial intelligence. Reference: Rupali Kamble & Prof. Deepali Shah (2018). "Applications of artificial intelligence in human life." International Journal of Research - Granthaalayah, 6(6), 178–188, here p. 186.

2.2 Artificial Intelligence and Performances of Countries

Most of the R&D investments of governments and private companies involve artificial intelligence works [12, 13]. For example, Russia's president Putin said in 2017 that whoever becomes the leader in AI will become the ruler of the world. Another example is Singapore, an international financial hub, being one of the first countries to announce a national strategy called AI Singapore in 2017. The United Arab Emirates (UAE) is the first country in the world to establish a university on AI in 2020.

AI performances of countries are evaluated by various institutions and organizations such as the UN, EU, World Bank (WB), World Economic Forum, PwC, Stanford University, or Oxford [14]. They also provide some valuable criteria for countries in order to develop successful AI strategies. For example, in the study named "Government AI Readiness Index" prepared by Oxford Insights in 2019, 194 countries were examined in four basic categories (governance; infrastructure and data; skills and education; government and public services). Performance of countries regarding artificial intelligence works; to the required data – from desk research to artificial intelligence strategies, databases such as the number of registered artificial intelligence initiatives in Crunchbase, to performances such as the UN E-Government Development Index – has been ranked/graded by giving a number of points. According to this report, North America is the best performing region on average, while the worst performing regions are Africa and Asia-Pacific. There are large inequalities between countries on a global scale. High-income countries rank higher and better than middle and low-income countries. The top 20 states in the report are Singapore, UK, Germany, USA, Finland, Sweden, Canada, France, Denmark, Japan, Australia, Norway, New Zealand, Netherlands, Italy, Austria, India, Switzerland, UAE, and China. Points of some other countries: Singapore - 9.186 (1st), India - 7.515 (17th), Spain - 6.332 (36th), Turkey - 5.879 (46th). The average of 194 countries is 3.558 [15].

In Oxford's 2020 report, critical changes are seen in the AI performance of countries and, accordingly, in the ranking: "This year, we have changed our approach to try and get to the heart of what makes a government AI ready. We therefore developed three new hypotheses, each of which corresponds to a fundamental pillar of government AI readiness: 1. The Government needs to be willing to adopt AI, and able to adapt and innovate to do so; 2. The Government needs a good supply of AI tools from the technology sector; and 3. These tools need to be built and trained on high quality and representative data, and need the appropriate inf5astructure to be delivered to and used by citizens" [15] (Fig. 2).

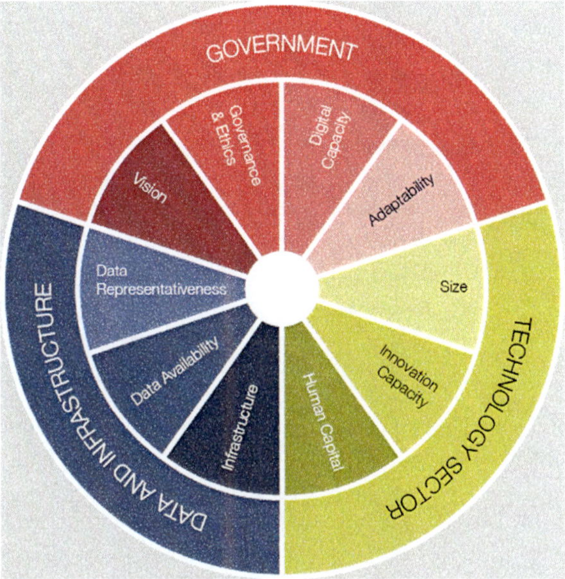

Fig. 2. The pillars and dimensions of the Government AI Readiness Index. Reference: Oxford (2020), Government AI Readiness Index 2020, p. 10.

While the USA (85,479) rose to first place in this report; India (55,983) 40th, and Turkey (46,010) decreased to 67th (Fig. 3).

The average GDP that countries devote to R&D is %2. South Korea doubles the average by devoting %4. Some other countries; Germany %2.5, although a growing trend in Russia and Turkey only %1 [16]. In research conducted in China in 2019, it was determined that a total of 78 billion dollars was invested in 15,916 AI initiatives worldwide at the end of 2018 [17]. Related information can be seen in the graph below (Fig. 4).

Rank	Country	Score
1	United States of America	85.479
2	United Kingdom	81.124
3	Finland	79.238
4	Germany	78.974
5	Sweden	78.772
6	Singapore	78.704
7	Republic of Korea	77.695
8	Denmark	75.618
9	Netherlands	75.297
10	Norway	74.430
11	France	73.767
12	Australia	73.577
13	Japan	73.303
14	Canada	73.158
15	Luxembourg	72.616
16	United Arab Emirates	72.395
17	Estonia	69.922
18	Switzerland	69.219
19	China	69.080
20	Israel	68.825

Fig. 3. Government AI Readiness Index 2020. Reference: Oxford (2020), Government AI Readiness Index 2020, p. 128.

EqualOcean | Landscape of AI fundraising in top 10 countries, by funding event*

Country	United States	United Kingdom	China	Canada	Germany
GNI Per Capita (USD)	62,850	41,330	9,470	44,860	47,450
Number of AI Funding Event	488	131	111	57	45
Amount of AI Funding (USD, Mn)	8,328.77	555.30	3,987.97	311.91	212.29

Country	France	India	Switzerland	Spain	Sweden
GNI Per Capita (USD)	41,070	2,020	83,580	29,450	55.070
Number of AI Funding Event	32	32	21	18	17
Amount of AI Funding (USD, Mn)	585.25	57.17	107.93	56.78	21.99

*The funding event mentioned here happened in 2019, as of July 31.

Source: World Bank, Crunchbase EqualOcean.com

Fig. 4. Landcape of AI fundraising in the top 10 countries, by funding event*. Reference: Yang, Shang (2019). EqualOcean.

2.3 Turkey and Artificial Intelligence

The concept of Industry 4.0 first emerged in 2011 and has been considered as the genesis of a new industrial era. Both the private sector in Turkey and the Turkish government have kept a close eye on the developments in this area. For instance, Turkish Industry and Business Association (TÜSİAD) and BCG Turkey prepared a report titled "Turkey's Capabilities for Digital Transformation of its Industry" in 2017, which analyzed the current situation and offered a roadmap [18].

Turkey has also launched an initiative in the field of AI. Development of a plan to foster cooperation between the public and private sector is on the agenda since 2019. There have been efforts to establish an institute of AI. Universities and some private firms have also undertaken such initiatives. The Cloud Computing and Big Data Research Laboratory project was launched in 2013, and the AI Research Centre at Istanbul Technical University, established in 2018. Two universities, TOBB-ETÜ and Hacettepe, started undergraduate AI Engineering programs. Koç University and İş Bankası established the AI Application and Research Center in 2020. President Erdogan summarized these developments by stating: "The number of research and development centers have reached a thousand while there are close to three-hundred design centers and 81 techno parks. ... The ascension of AI in all areas in the near future is inevitable" [19]. Department of Big Data and Artificial Intelligence Applications have been established under the Presidential Digital Transformation Office, which has a crucial role in Turkey's plans for digital transformation. This department is responsible for ensuring big data and AI is employed actively and commonly in the public by ensuring coordination, developing strategies, and prioritizing the use of AI in public projects.

Microsoft and EY (Ernst & Young) makes the following remarks in their report on the use of AI in the MENA region: Turkey is the leading country in its region in terms of strategic importance, areas of use, awareness, and investments regarding AI technologies. Similarly, 80% of private companies in Turkey handle their AI strategies at the highest levels of management. In Turkey, AI is most commonly used in the areas of prediction, automation, building insight, customization of services, and development of precautionary skills. The rate of use for machine learning is 85% in Turkey as opposed to the 61% in the MENA region [20].

It can be stated that Turkey does not lag far behind the developed countries in AI. Nevertheless, in comparison to countries such as France and Germany [21], there are certain shortcomings despite the increased R&D budgets. Moreover, a platform for the politicians, private sector, universities, and researchers to cooperate could not be developed. As a result, the number of patents remain low. TÜBİTAK, Turkey's leading public agency for science and technology research, have been criticized for leading to brain drain in crucial areas including AI because of its failures in funding projects, innovative training, and development of infrastructures and hardware [22]. Workshops organized by Turkish Artificial Intelligence Initiative (TRAI), which bring together the numerous institutions that work on AI technologies, revealed that private sector is the impetus that drives Turkey's AI revolution. The founder of TRAI Halil Aksu said "There have been a great improvement in the area of artificial intelligence in the last 2–3 years in Turkey. We are going through an artificial intelligence revolution that will transform companies and entire industries. Some are leading this revolution, some are on the front lines and some are at the back, some may remain at the back permanently. What is certain is that this revolution will remain on the top of our agenda in the coming years" [23]. In any case, Turkey still lacks a national strategy for artificial intelligence [24].

"The Data and Infrastructure pillar is Turkey's lowest-ranked, suggesting that this is an area for improvement if the country wants to build AI readiness. […] The National Data Dictionary project, which aims to standardise data systems across government, which may have a beneficial effect on data availability. Turkey ranks 79th in the world in terms of its telecommunications infrastructure according to the 2020 UN eGovernment Survey. However, the country has ambitious plans for its infrastructure in future. It plans to switch to 5G by 2021, and networks have already been trialled in the country." [25] (Fig. 5).

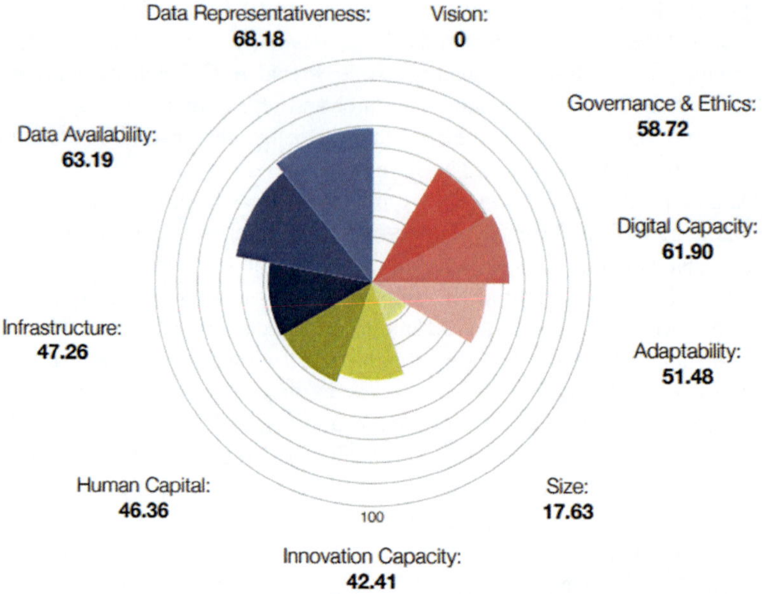

Fig. 5. Turkey's AI Features in the Oxford Report 2020. Reference: Oxford (2020), Government AI Readiness Index 2020, p. 103.

Ultimately, there are problems especially in the following matters: funding needs of Small and Medium-sized Enterprises (SME), data storage and cybersecurity, development of the information infrastructure and its standards, support for entrepreneurialism, identification of key industries for AI improvements, competitiveness of the national AI industry, and the design and organization of rewarding projects.

3 Research Methodology

The principal methodological steps followed in this study are access to scientific resources; analytical analysis of data sets and relationships between data sets; drawing the current situation of countries and finally objective analysis of the main subject with numerical data sets and comparative analysis. Primarily in this research, AI and its field of use has been elucidated and its economic benefits remarked. Subsequently, with a focus on the activities and course of actions of governments, productivity and performance analysis have been carried out. At this stage, through AI national strategy documents and R & N investments overall country vision has been asserted. Finally, the overall situation of Turkey has been explained, and its deficiencies with respect to performance of both top-ranking/developed countries and performance of developing/ascending countries have been discussed.

The drivers of development and/or fields of use of artificial intelligence, its key positioning in the new industry revolution and related activities of the governments have been the subthemes of this this study. Therefore, the study based on both primary and secondary resources in three different languages that are English, German and Turkish.

With regards to analysis of artificial intelligence and Fourth Industrial Revolution, the principal scientific books accepted word wide, reports on preparation and discussion of national roadmaps published by global institutes and/or universities specialists (i.e. MIT, IBM, Harvard, Oxford, Mc Kinsey) and distinguished studies of institutions such as OECD, World Bank, United Nations, EU have been reviewed. Country- based studies of global institutions provide advantages on comparative analysis because of presenting large amount of data sets at the same time. In order to confirmation or falsification of information, official country papers and their web-interpretations have been reviewed.

In the evaluation of Turkey's activities on AI studies of both governmental (i.e. Presidential Digital Transformation office, universities) and civil society entities (i.e. Turkish Industry and Business Association, Digital Turkey Platform or Turkey's Artificial Intelligence Initiative) have been screened.

At this stage, a valuable contribution has been provided both for published documentations and wen sources. For instance, the report of "Artificial Intelligence Strategies and Turkey" published by Ankara Yildirim Bayazit University in May 2020 has a similar objective to this study; that explains AI and countries at a considerable extent, however Ulaşan's recommendations to Turkey here remained quite limited [26].

The information transmitted in this study based on the scientific and statistical/ graphically recourses that increase the credibility of this study and screening of various sources prevents the subjectivity. Overall, such methodology provides window of opportunity for the related literacy in a comparative perspective.

4 Suggestions: How Turkey Would Perform Better on Artificial Intelligence?

Existing research reveal that Turkey has some serious shortcomings despite the positive developments. Suggestions for a more effective process for the development of AI industry can be compiled under two categories.

4.1 Starting Points Before Taking Any Action

4.1.1 Turkey Is A LATECOMER, But It Is Still Possible To Make A Move

First of all, this reality should be accepted. A few institutions that are claimed to work on artificial intelligence cannot bring this industry forward. Turkey's lack of a structured AI policy, especially as a G20 country, is a serious deficiency. While the moves toward Industry 4.0 is underway around the world, Turkey has not reached the third stage of industrial revolution just yet. Thus, work on AI should be carried out swiftly and effectively but also on solid grounds, recognizing the current shortcomings.

4.1.2 Acknowledgement of Limited Resources

Having a GDP of $650 billion, Turkey does not have as strong an economy as its competitors in this race. Therefore, it is not realistic to expect Turkey to invest the amounts that the US, China, or India does. Accordingly, Turkey needs to designate areas of need and distribute the budget for the development of the AI technology through following a preset roadmap and output-efficiency based projects.

4.1.3 Setting-Up Reasonable Goals

China's 2017 plan foresees to match other competitors in the AI industry by 2020, to lead the industry in some areas by 2025, and to become the center of the AI technology development by 2030. Such a plan is within the realm of possibility for China as the country has the necessary technology infrastructure and the economic power. Indeed, the billions of dollars' worth of projects and R&D funded by China proves its commitment to these aims. India, on the other hand, first plans to 'improve and enhance the skills of the Indian workforce', accelerate economic growth and social development, and to export Indian-made AI technologies to the developing world. Clearly, even India, set to become the second largest economy of the world in the near future, makes plans cognizant of its capability. Accordingly, Turkey should also set more realistic short-medium and long-term goals.

4.1.4 Setting-Up Target Industries

Russia is one of the leading powers in military technology, even though its economy is based on natural resource exports and lacks the capabilities, tools, and industrial infrastructure for advanced technology. Therefore, most of its investment to the AI industry is directed toward military technology. China is leading the world in the development of communication technology. USA is the venue for significant improvements in the technology of unmanned aerial vehicles (UAV). Germany, the country where the concept of Industry 4.0 first emerged, is struggling against China's comparative advantages. Similarly, Turkey should also identify its areas of need and prioritize the work in certain industries. Focusing on investments in military technology and the development of new technologies for armed and/or unarmed UAVs seems to be the most reasonable strategy given Turkey's already advanced capabilities in the production of UAVs and its foreign policy goals. Secondly, as automation is increasingly replacing the economic system based on cheap workforce, there will remain no incentive for the transfer production centers to East Asia in the near future. In that case, Turkey can attract production in the industries that it has advantage, and even become a destination for the capital currently invested in East Asia. Therefore, AI investments are crucial for the economy of Turkey. Thirdly, Turkey is quite successful in the healthcare industry. Investments to development of AI technologies in that field has the promise of a hundredfold return within a few years. The potential areas for investment include robot-assisted surgery, fraud detection, and many more. Turkey can improve its position in the healthcare industry even further and expand its already strong healthcare tourism potential through the development and use of artificial intelligence technologies in that area. The main principle here is to act in accordance with the country's potential and needs.

4.2 Actions to be Taken

4.2.1 R&D Funds Should be Created.

The driving force of AI technologies is money, R&D requires high costs. This is the reason why the investments made by the states in this field vary according to their economies; the state who has a lot of money is researching a lot. Leaving aside China and the USA whose investments in this area exceed $ billion; for example, France invests €1.5 billion in the field separately from the EU Commission. 700 million of this amount is allocated for research funds. The driving force of Singapore's $150 million 5-year artificial intelligence plan set in 2017 is research. Therefore, universities and research centers are the places to take the first steps.

4.2.2 Coding Courses in Primary and Secondary Education

Turkey in primary and secondary students can not yet be given adequate training to use computers. There are no fully-fledged computer laboratories that would enable this training to be delivered. If development is to be achieved in this field, it is imperative to lay the foundations of a locomotive population that will carry with the technology.

4.2.3 The Field is not Limited to Engineering

This is another misconception is located in Turkey. AI technologies are also a political, social, and philosophical issue due to their effects. The first AI department is completely based on engineering in Hacettepe. This is a mistake. Politicians, sociologists, philosophers, lawyers, and similar professional groups are needed to work in this field. It is necessary to approach AI technology as a whole, which is expected to affect every aspect of human life.

4.2.4 Cooperation with Developed Countries in the Field

In this context, from countries such as China, the USA, Germany, and South Korea student programs that allow this technology to be moved to Turkey will be effective (programs that will enable students to learn this technology and develop in this field). The coworking of states will pave the way for progress in terms of cost and technology sharing. Good partners must be found in this field.

4.2.5 R &D Should be Incorporated into a State Policy

AI works are becoming a state policy in all developed countries. Saudi Arabia and the UAE have established AI ministries and demonstrated their seriousness in the field. It is now certain that AI technologies are at the top of the factors that will shape the socio-economic world of the future. In Turkey – even if not at the level of creating a ministry – specific appropriations for this area are essential to be added to the budget.

4.2.6 R &D Centers Should be Established

Silicon Valley in the USA, Shenzhen in China, Silicon Wadi in Israel, the whole country in Singapore. It is critical that states have centers that produce technology, pioneer innovations, are based on science and trade, and their efficient management. Taken in terms of Turkey, for sure, it is difficult to think about it in the first place the size of a city or region; However, such centers should be established in places that are suitable for development and expansion, close to big cities (Shenzhen-Hong Kong is adjacent), where trade rules can be stretched for the region over time.

4.2.7 Experts from Other Countries Should be Brought in

Creating an educated and producing locomotive population is a relatively long and planned process. In order to create this population; large funds should be provided to experts who will work in scientific centers, will teach this technology, and will continue to work in Turkey if possible and these experts should be attracted to the country. In addition, reverse migration should be promoted against emigration.

4.2.8 Industry Incentivization Required (On Sectoral Basis)

Technological investments in areas with a certain potential, not yet invested much, in which Turkey was successful should be encouraged and funds should be provided for these investments. When Turkey's geopolitical position and political balances in this position are taken into consideration, it is clear that not having trouble finding new markets. State-sponsored investments should be made to become the artificial intelligence technology provider of the region and to further strengthen the existing strong sectors with artificial intelligence technologies (health, dental, jewelry, casting, spare parts production, military technologies, transportation, logistics).

4.2.9 Ethical Frame and Legal Framework to be Designed

New developing technologies deeply affect the daily life, challenges the conventional and creates room for itself. The reference point here is to produce a frame that allows new technologic developments but as well protects the society.

5 Conclusion

Since 18th century steam engine, discovery of electricity, transistors, computers and internet have been symbols of industrial revolution. Today, artificial intelligence and information technologies designate a new era of industrial discoveries.

The fact that technologically and economically developed countries especially such as the USA, China, Japan, Germany, Israel etc., are interested in the studies on AI both as public and private enterprises and they are in competition with each other, has created a competitive environment around the world. The necessity for the studies to have a systematic within a certain plan has pushed countries to determine strategies and policies. It is unlikely for those countries that did not start yet the investments on artificial intelligence to gain a significant leverage on the future world. Such a revolutionary technology has a potential impact capacity from defence systems to economics, agriculture to textile, culture to education, environment to laws, social media to entertainment and transportation to scientific research.

Herein, the question in Turkey comes up is that: "What Turkey has been doing as a country?" The action plan of such a regional power state as Turkey has crucial importance at its digital transformation stage. It has been observed that mainly the private sector has been the catalyser on development of new technologies, and it requires further show of interest from public sector and academy. Turkey should urgently develop a road map on artificial intelligence and digital transformation. In this study various recommendations have been presented. All in all, the need for a more effective governance involving the decision-makers, private sector, universities, and researchers as well as more innovative projects is clear.

Finally, the king is that naked; the one who does not develop technological infrastructure and not able to improve its human resources in accordance to new trends will face great challenges in their development policies. In this regards, homework's and utilization of the opportunities inevitably exists for Turkey which is fell beyond the level of 2019 according to the reports provides globally analysis.

References

1. Reese, B.: The Fourth Age: Smart Robots, Conscious Computers, and the Future of Humanity. Atria Books; Kaufmann, Timothy & Servatius, Hans-Gerd (Hrsg.) (2020), Das Internet der Dinge und Künstliche Intelligenz als Game Changer. Wege zu einem Management 4.0 und einer digitalen Architektur, Springer; Doleski, O.D. (Hrsg.): Realisierung Utility 4.0 Band 1. Praxis der digitalen Energiewirtschaft von den Grundlagen bis zur Verteilung im Smart Grid, Springer (2020)
2. Jackson, P.C.: Introduction to Artificial Intelligence, Dover Publications: 3rd edition; Russell, Stuart & Norvig, Peter (2015), Artificial Intelligence – A Modern Approach, Pearson: 3rd Edition (2019)
3. Davenport, T.H.: The AI Advantage: How to Put the Artificial Intelligence Revolution to Work. The MIT Press, Cambridge (2019)
4. Lee, K.-F.: AI Superpowers: China, Silicon Valley, and the New World Order Hardcover. Houghton Mifflin Harcourt Company (2018)
5. HolonIQ: The 2020 AI Strategy Landscape (2020). https://www.holoniq.com/notes/50-national-ai-strategies-the-2020-ai-strategy-landscape/. Accessed 16 Dec 2020
6. Goodfellow, L., Bengio, Y., Courville, A.: Deep Learning. The MIT Press, Cambridge (2016)
7. Bughin, J., Seong, J., Manyika, J., Chui, M., Joshi, R.: Modeling the global economic impact of AI. McKinsey Global Institute, p. 2 (2018)
8. BT Haber: Yapay zekâ; 'Dijital Türkiye' ve 'Milli Teknoloji Hamlesi'nin en büyük taşıyıcılarından (2020). https://www.bthaber.com/yapay-zeka-dijital-turkiye-ve-milli-teknoloji-hamlesinin-en-buyuk-tasiyicilarindan/. Accessed 17 Jan 2021

9. IBM: Artificial Intelligence (AI) (2020). https://www.ibm.com/cloud/learn/what-is-artificial-intelligence. Accessed 15 Jan 2021
10. Microsoft Corporation: The Future Computed: Artificial Intelligence and its Role in Society, Independently published (2018)
11. Pannu, A.: Artificial intelligence and its application in different areas. Int. J. Eng. Innov. Technol. (IJEIT) **4**(10), 79–84 (2015)
12. OECD: Hello, World: Artificial Intelligence and its use in the Public Sector. OECD Observatory of Public Sector Innovation (2019)
13. OECD: OECD, AI Policy Observatory. Countries & initiatives overview (2020). https://oecd.ai/countries-and-initiatives. Accessed 08 Jan 2021
14. BM: Developing an artificial intelligence strategy National guide, Economic and Social Commission for Western Asia (ESCWA) (2020). https://www.unescwa.org/sites/www.unescwa.org/files/publications/files/artificial-intelligence-strategy-national-guide-english.pdf, p. iii. Accessed 12 Jan 2021; World Economic Forum: A Framework for Developing a National Artificial Intelligence (2019). http://www3.weforum.org/docs/WEF_National-AI_Strategy.pdf. Accessed 12 Jan 2021; Stanford University: The 2019 AI Index Report; PWC (2018), The macroeconomic impact of AI, pp. 1–75 (2019)
15. Oxford: Government AI Readiness Index 2020, International Development Research Centre, Canada, here p. 7 (2020)
16. OECD: Gross domestic spending on R&D (2021). https://data.oecd.org/rd/gross-domestic-spending-on-r-d.htm. Accessed 27 Jan 2021
17. Yang, S.: USD 18 Billion Rush Into Artificial Intelligence Startups By July 31, 2019. EqualOcean (2019)
18. TÜSİAD & BCG Türkiye: Türkiye'nin Sanayide Dijital Dönüşüm Yetkinliği (2017). https://tusiad.org/tr/yayinlar/raporlar/item/9864-tusiad-bcg-turkiye-nin-sanayide-dijitaldonusum-Yetkinligi. Accessed 14 Jan 2021
19. CBDDO: Cumhurbaşkanımızdan Millî Teknoloji Hamlesi ve Dijital Türkiye Açıklaması (2019). https://cbddo.gov.tr/haberler/4259/cumhurbaskani-erdogan-hgm-atlas-ve-hgm-kure-uygulamalarinin-tanitilmasi-toreni-nde-dijital-gelisimi-vurguladi. Accessed 11 Dec 2020
20. Microsoft Türkiye ve Erns & Young: Orta Doğu ve Afrika'da Yapay Zeka (2019). https://tr.newworldai.com/wp-content/uploads/2019/09/Ortado%C4%9Fu-ve-Afrika-da-Yapay-Zeka.pdf. Accessed 21 Jan 2021
21. European Commission: AI Watch. National strategies on Artificial Intelligence. A European perspective in 2019. Publications Office of the European Union, Luxembourg (2020)
22. Dinler, Z.: Bilimsel Araştırma ve E-Kaynaklar, Ekin Basım Yayın, pp. 10–50 (2019)
23. Hürriyet: Türkiye'de yapay zeka alanında çok ciddi mesafe kat edildi (2019). https://www.hurriyet.com.tr/teknoloji/turkiyede-yapay-zeka-alaninda-cok-ciddi-mesafe-kat-edildi-41211887. Accessed 17 Jan 2021
24. HolonQI: ibid; Oxford, pp. 16–17 (2020)
25. Dutton, T.: An Overview of National AI Strategies. Medium (2018). https://medium.com/politicsai/an-overview-of-national-ai-strategies2a70ec6edfd. Accessed 13 Jan 2021; Aydın, G., Nyadera, I.N., Önder, M.: Strategic Management in Turkey's Public Sector: Reforms and Application Issues. Public Organiz Rev 20, 719–734 (2020); Rao, A.S.: Sizing the prize What's the real value of AI for your business and how can you capitalise? PwC (2017)
26. Ulaşan, F.: Ulusal Yapay Zekâ Strateji Belgeleri ve Değerlendirmeler. In: Demir, İ. (ed.) Yapay Zeka Stratejileri ve Türkiye, Ankara Yıldırım Beyazıt Üniversitesi Uluslararası İlişkiler ve Stratejik Araştırmalar (ULİSA) Enstitüsü, pp. 11–24, here pp. 22–23 (2020)

Technology Readiness as an Antecedent of Technology Acceptance Model: A Meta-analytic Approach

Aygul Donmez-Turan[1] (✉) and Buse Oren[2]

[1] Yildiz Technical University, Istanbul, Turkey
ayturan@yildiz.edu.tr
[2] Galatasaray University, Istanbul, Turkey

Abstract. Technology readiness (TR) is a crucial concept to explain acceptance of an information technology or system by individuals. Based on the prior studies which reported the correlation scores between technology readiness and exogenous variables of technology acceptance model (TAM), we designed a meta-analytic study to attain effect size of TR on perceived usefulness (PU) and perceived ease of use (PEOU). We analyzed correlation scores which reported in 44 articles and conference proceedings with Comprehensive Meta Analysis software program. However, different level correlation scores had been recorded in the research papers, so we conducted six distinct meta-analyses in two stages in order to obtain effect size of TR on PU. We carried out different six meta-analyses in two stages to get effect size of TR on PEOU, as well. As a result, we found that random effect models were appropriate for all effect sizes between the variables of all stages. Findings showed that the average effect size of optimism on PU ($\bar{r} = 0.368$), of optimism on PEOU ($\bar{r} = -0.386$), of innovativeness on PU ($\bar{r} = 0.258$), of innovativeness on PEOU ($\bar{r} = 0.328$) were found positive and medium level. In addition, average effect size of insecurity on PU ($\bar{r} = -0.070$), of insecurity on PEOU ($\bar{r} = -0.096$) were found negative but very small level. However, average effect sizes of discomfort on PU, of discomfort on PEOU were found non-significant. Consequently, average effect size of positive TR on PU ($\bar{r} = 0.355$) and of positive TR on PEOU ($\bar{r} = 0.414$) were found positive and medium level. Nevertheless, average effect size of negative TR on PEOU ($\bar{r} = -0.085$) was found negative but very small level, effect size of negative TR on PU was found non-significant.

Keywords: Technology acceptance model · Technology readiness · Optimism · Innovativeness · Discomfort · Insecurity · Perceived usefulness · Perceived ease of use · Meta-analysis · TAM

1 Introduction

Technology readiness (TR) emerges as a vital concept in pandemic period. Hence, employees have to change their working style and work from their home in this period.

A. M. A. Musleh Al-Sartawi et al. (Eds.): EAMMIS 2021, LNNS 239, pp. 513–522, 2021.
https://doi.org/10.1007/978-3-030-77246-8_47

Employees can adapt themselves easily to those working style, if they have optimistic view towards technology or any other technological systems, if they are openness to new experiences or innovative products, if they feel comfort and secure to use any information system [1]. So main dimensions of TR - optimism, innovativeness, discomfort and insecurity level - determine the acceptance of any technology or system [2]. Furthermore, growing body of literature shows the relationship between TR and acceptance of any technology. Moreover, TR is evaluated as a variable to determine technology acceptance model (TAM) in the literature [3]. On the basis of the literature review, we designed a meta-analytic study to represent the importance of TR on the acceptance any technology or technological transformation.

2 Literature Review

2.1 Technology Acceptance Model

Technology acceptance model (TAM), developed by Davis [4], demonstrates the factors of individual's technology acceptance. In addition, TAM has four main components which are perceived usefulness (PU), perceived ease of use (PEOU), attitude, intention and behavior.

The model focuses on two beliefs which are PU and PEOU. PU is defined as beliefs of individuals on using a certain information technology would improve their performance in their job. On the other hand, PEOU is individuals' belief on the effort that they will make while using a specific information technology [4].

After Davis [1], further theories have been evolved from TAM by other researchers, such as diffusion of innovation [5], information technology innovation [6], unified theory of acceptance and use of technology [7] etc.

Some external variables are used in the literature to explain TAM. Those external variables determine the person's attitude on using the technological product by affecting their PU and PEOU. The variables have a critical role on manipulating the PU and PEOU [8]. These external variables could be negative variables such as resistance or anxiety [9], while they could also be positive variables such as user involvement, technology readiness [10] etc.

2.2 Technology Readiness

TR is an individual's general beliefs on technology, technological systems or services that they use [2]. There are various studies on TR in the literature [1, 3]. However, the most known and most accepted of them is Parasuraman's [2] research. Parasuraman defines TR as an indivdual's tendency to embrace the new technologies. Also he defined four dimensions of TR which are optimism, innovativeness, discomfort and insecurity [2]. Optimism and innovativeness demonstrate the technology readiness, motivators, drivers; on the contrary, discomfort and insecurity demonstrates technology unreadiness, distractors, inhibitors [11]. Therefore, many researchers have evaluated that optimism and innovativeness are the positive dimensions of TR while discomfort and insecurity are negative dimensions of technology readiness [12]. Optimism is an individuals' thought that they enable to control and increase efficiency and effectiveness of their lives with technology. Innovativeness means that an individual is pioneer and leader of using a new technology. Discomfort is an individual's feeling about lack of control over a technological products or services. Insecurity is a individual's distrust of technology and tendency to feel skeptical about its ability to work properly [2].

As mentioned before, TR is one of the external variables of TAM [10] and such external variables have a critical role on manipulating the PU and PEOU [8]. Manipulating directly PU and PEOU, TR in fact influences attitude, intention and usage of a system indirectly. Examining the influence of TR on PU/PEOU on the grounds of the findings in the research paper could be a valuable research for the literature. So we proposes the following hypotheses:

Hypothesis 1: PU is associated with technology readiness dimensions (a. optimism, b. innovativeness, c. discomfort and d. insecurity).
Hypothesis 2: PEOU is associated with technology readiness dimensions (a. optimism, b. innovativeness, c. discomfort and d. insecurity).
Hypothesis 3: PU is associated with (a. positive TR, b. negative TR).
Hypothesis 4: PEOU is associated with (a. positive TR, b. negative TR).

3 Research Methodology

3.1 Literature Search and Criteria of Selection

Literature search was carried out in April 2020 to get appropriate data related to the TAM and TR of individuals towards any information systems. The terms using in the literature search were specified based on the TAM, especially its exogenous variables (PU and PEOU). As mentioned before, to date TAM has evolved and PU dimension has been entitled as relative advantage or performance expectancy, PEOU dimension has been named as complexity or effort expectancy, in further studies [5–7]. Using SCOPUS database, we searched exogenous variables of TAM (PU and PEOU) with TR in 'keywords, title and abstract' of all articles. We typed all terms within the quotation marks and used "or", "and" conjunctions. In addition, literature search was limited to the peer reviewed journal articles, conference proceedings and book chapters with English languages.

Searching terms are typed on the SCOPUS searching engine as follows:

1st searching row: *"perceived usefulness" OR "performance expectancy" OR "relative advantage" OR "perceived ease of use" OR "complexity" OR "effort expectancy"*
2nd searching row: *"technology readiness"*

We reached 163 documents and 79 of them were relevant our research. Scanning all documents whether they had correlation scores between TR and exogenous variables of TAM (PU/PEOU), we attained 44 appropriate ones.

3.2 Coding and Procedures of Meta-analysis

Demographic characteristics (gender rate, countries of the sample), theoretical characteristics (conceptualization and measurement instruments of PU, PEOU TR) and methodological characteristics (correlation scores, means, standard deviations, Cronbach's alpha values of the scales) of the relevant 44 papers were coded.

However, some of the articles had one, some of them two and the others had four different correlation scores which represented the correlation between PU and TR (Similar condition could be seen in the correlation between PEOU and TR). Since TR has four dimensions (optimism, innovativeness, discomfort and insecurity), some of the researches reported correlation scores between each dimensions of TR and PU [2, 13]. Some research evaluated dimensions of optimism and innovativeness under the positive TR, dimensions of discomfort and insecurity under the negative TR latent variables and reported only two correlation scores (between positive TR and PU, between negative TR and PU) [11]. On the other hand, some articles evaluated general TR concept and reported positive TR correlation scores with PU only [3, 14].

Hence the different level correlation scores were reported in the literature, we conducted six distinct meta-analyses in two stages in order to obtain effect size of TR on PU. In addition, we conducted another six meta-analyses in two stages to get effect size of TR on PEOU. In the first stage, we used the correlation scores between PU and each dimensions of TR (optimism, innovativeness, discomfort and insecurity) in four different meta-analyses. In the second stage, we used the correlation scores between PU and positive TR, the scores between PU and negative TR in two distinct meta-analyses (General TR scores recorded as positive TR). Furthermore, the correlation between optimism and PU, the correlation between innovativeness and PU were coded as two different positive TR score; on the contrary, correlation between discomfort and TR, and correlation between insecurity and TR were coded as two negative TR scores. If an article has correlation scores between PU and each dimensions of TR, this article scores were coded as if 4 distinct articles' score. Because it has two positive and two negative correlation scores of TR with PU. Similar stages and analyses were done for TR and PEOU relation. So 12 distinct meta-analyses were conducted in this study.

We applied Copper et al.'s [15] and Lipsey and Wilson's [16] meta-analytic procedures and used correlation coefficient to estimate the effect size. So we attained true correlation and estimated effect sizes and obtained with Comprehensive Meta-analysis package program.

3.2.1 Tests for Outlier and Publication Bias

According to Geyskens et al. [17], outliers in meta-analysis should be controlled with forest and funnel plots. Following this recommendation, we controlled each funnel and forest plots and made decision to exclude one article's correlation coefficient from the meta-analysis conducted between innovativeness and PEOU ($\bar{r}$-fixed = 0.308; $\bar{r}$-random = 0.299). In addition, two articles were excluded from the meta-analysis which was done between insecurity and PEOU ($\bar{r}$-fixed = −0.116; $\bar{r}$-random = −0.045). Those outlier 3 article were removed the data and then the effect size scores were checked. As seen in Table 1, both fixed and random effect models increased after the outlier studies eliminated from the sample of meta-analysis. So those three articles were removed from the study.

As seen in Table 1, Tau's coefficient were non-significant for PU and optimism ($p = 0.10$), PEOU and optimism ($p = 0.18$), PU and innovativeness ($p = 0.29$), PEOU and innovativenes ($p = 0.90$), PU and discomfort ($p = 0.38$), PEOU and discomfort ($p = 0.23$), PU and insecurity ($p = 0.12$), PEOU and insecurity ($p = 0.06$), PU and positive TR (0.71), PEOU and positive TR (0.62), PU and negative TR (0.08), PEOU and negative TR (0.07). Based on those results, we could interpret that there were no publication biases of each meta-analysis in present research.

3.2.2 Q Statistics: Effect Size Variability Across Studies

At the first stage of present study, we conducted four distinct meta-analyses to get the effect sizes of PU and each dimension of TR (optimism, innovativeness, discomfort and insecurity). Table 1 showed significant Q statistics results for optimism and PU ($X^2 = 406.555/df = 20$), innovativeness and PU ($X^2 = 269.171/df = 19$), discomfort and PU ($X^2 = 454.288/df = 20$), insecurity and PU ($X^2 = 190.073/df = 18$). So the random effect model was found suitable for the effect sizes of PU and each dimensions of TR. Finally the random effect size could be reported in this study. The effect size of optimism on PU was positive and significant ($\bar{r} = 0.368$, $p = 0.00$), of innovativeness on PU was positive and significant ($\bar{r} = 0.258$, $p = 0.00$), as well. On the contrary, the effect size of discomfort on PU was non-significant ($\bar{r} = 0.034$, $p = 0.44$), but the effect size of insecurity on PU was negative and significant ($\bar{r} = -0.071$, $p = 0.02$). At the 95% confidence interval, the effect size results for optimism was from 0.293 to 0.438, for innovativeness was from 0.191 to 0.323, for insecurity from −0.133 to −0.010, and the interval was not include the zero point. However, the effect size of discomfort on PU differentiates from −0.053 to 0.121 at the 95% confidence interval, and the interval included the zero point. So hypotheses 1a, 1b, 1d were supported.

After that the same analyses' procedures were applied to get the effect sizes of PEOU and each dimension of TR (optimism, innovativeness, discomfort and insecurity). Table 1 showed significant Q statistics results for optimism and PEOU ($X^2 = 138.440/df = 18$), innovativeness and PEOU ($X^2 = 214{,}125/df = 16$), discomfort and PEOU ($X^2 = 919.915/df = 22$), insecurity and PEOU ($X^2 = 180.482/df = 18$). So the random effect model was found appropriate for the effect sizes of PEOU and each dimensions of TR. Finally the random effect size could be reported in this study. The effect size of optimism on PEOU was positive and significant ($\bar{r} = 0.386$, $p = 0.00$), innovativeness on PEOU was positive and significant ($\bar{r} = 0.328$, $p = 0.00$), as well. On the contrary, the effect size of discomfort on PEOU was non-significant ($\bar{r} = -0.096$, $p = 0.10$), but insecurity on PEOU was negative and significant ($\bar{r} = -0.095$, $p = 0.02$). At the 95% confidence interval, the effect size results for optimism was from 0.340 to 0.430, for innovativeness was from 0.262 to 0.390, for insecurity from −0.155 to −0.034 and the interval was not include the zero point. Nevertheless, the effect size of discomfort and PEOU differentiates from −0.209 to 0.019 at the 95% confidence interval, the interval included the zero point. As a result, hypotheses 2a, 2b, 2d were supported.

Table 1. Meta-analyses results for dimensions of TR, heterogeneity and publication bias

		PU and Optimism	*p(sig.)*	PEOU and Optimism	*p(sig.)*
TR (Optimism)	Sample size	12039		11617	
	Number of effect sizes	21		19	
	Effect Size ($\bar{r}$)				
	Fixed effect model	0.421	0.00	0.404	0.00
	Random effect model*	0.368	0.00	0.386	0.00
	Heterogeneity				
	Q statistics (X^2/df)	406.555/20	0.00	138.440/18	0.00
	Publication Bias (Tau S)	−0.2583	0.10	−0.2222	0.18
	95% CIs - Lower	0.293		0.340	
	95% CIs - Upper	0.438		0.430	
		PU and Innovativeness	*p(sig.)*	PEOU and Innovativeness	*p(sig.)*
TR (Innovativeness)	Sample size	11814		10911	
	Number of effect sizes	20		17	
	Effect Size ($\bar{r}$)				
	Fixed effect model	0.275	0.00	0.330	0.00
	Random effect model*	0.258	0.00	0.328	0.00
	Heterogeneity				
	Q statistics (X^2/df)	269.171/19	0.00	214,125/16	0.00
	Publication Bias (Tau S)	−0.16913	0.29	−0.02206	0.90
	95% CIs - Lower	0.191		0.262	
	95% CIs - Upper	0.323		0.390	

(*continued*)

Table 1. (*continued*)

		PU and Discomfort	*p(sig.)*	PEOU and Discomfort	*p(sig.)*
TR (Discomfort)	Sample size	12219		12641	
	Number of effect sizes	21		23	
	Effect Size ($\bar{r}$)				
	Fixed effect model	−0.006	0.51	−0.211	0.00
	Random effect model	0.034	0.44	−0.096	0.10
	Heterogeneity				
	Q statistics (X^2/df)	454.288/20	0.00	919.915/22	0.00
	Publication Bias (Tau S)	0.13810	0.38	0.17857	0.23
	95% CIs - Lower	−0.053		−0.209	
	95% CIs - Upper	0.121		0.019	
		PU and Insecurity	*p(sig.)*	PEOU and Insecurity	*p(sig.)*
TR (Insecurity)	Sample size	11617		11311	
	Number of effect sizes	19		19	
	Effect Size ($\bar{r}$)				
	Fixed effect model	−0.118	0.00	−0.148	0.00
	Random effect model*	−0.071	0.02	−0.095	0.02
	Heterogeneity				
	Q statistics (X^2/df)	190.073/18	0.00	180.482/18	0.00
	Publication Bias (Tau S)	0.25882	0.12	0.31176	0.06
	95% CIs - Lower	−0.133		−0.155	
	95% CIs - Upper	−0.010		−0.034	

At the second stage of present study, we conducted two different meta-analyses to get the effect size of PU on positive TR, the effect size of PU on negative TR, as well. Table 2 represented significant Q statistics results for positive TR and PU (X^2= 1569,73/df =

Table 2. Meta-analyses results for positive and negative TR, heterogeneity and publication bias

		PU and Positive TR	*p(sig.)*	PEOU and Positive TR	*p(sig.)*
Positive TR	Sample size	31541		30472	
	Number of effect sizes	61		56	
	Effect Size ($\bar{r}$)				
	Fixed effect model	0.368	0.00	0.405	0.00
	Random effect model*	0.355	0.00	0.414	0.00
	Heterogeneity				
	Q statistics (X^2/df)	1569,73/60	0.00	1640.91/55	0.00
	Publication Bias (Tau S)	−0.03173	0.71	0.04545	0.62
	95% CIs - Lower	0.304		0.361	
	95% CIs - Upper	0.405		0.464	
		PU and Negative TR	*p(sig.)*	PEOU and Negative TR	*p(sig.)*
Negative TR	Sample size	27391		28235	
	Number of effect sizes	48		52	
	Effect Size ($\bar{r}$)				
	Fixed effect model	−0.100	0.00	−0.157	0.00
	Random effect model*	−0.064	0.06	−0.083	0.01
	Heterogeneity				
	Q statistics (X^2/df)	1546.55/47	0.00	1427.34/51	0.00
	Publication Bias (Tau S)	0.17125	0.08	0.16843	0.07
	95% CIs - Lower	−0.133		−0.145	
	95% CIs - Upper	0.005		−0.020	

60), negative TR and PU ($X^2 = 1546.55/\ df = 47$). Based on the results, random effect model was founs suitable for the effect sizes of positive TR on PU, and effect sizes of negative TR on PU. Finally the random effect size reported in this study. The effect size of positive TR on PU was positive and significant ($\bar{r} = 0.355$, p = 0.00); on the other hand, the effect size of negative TR on PU was non-significant ($\bar{r} = -0.064$, p = 0.06).

At the 95% confidence interval, the effect size result for positive TR was from 0.304 to 0.405 and the interval was not include the zero point. On the contrary, the effect size result of negative TR differentiates from −0.133 to 0.005, and the interval included the zero point. So hypotheses 3a was supported.

After that the same analyses' procedures were applied to attain the effect size of positive TR on PEOU, and the effect size of negative TR on PEOU. Table 2 represented significant Q statistics results for positive TR and PEOU (X^2 = 1640.91/ df = 55), negative TR and PEOU (X^2 = 1427.34/ df = 51). Based on the results, random effect model was found suitable for the effect sizes of positive TR on PEOU and effect sizes of negative TR and PEOU. Finally the random effect size reported in this study. The effect size of positive TR on PEOU was positive and significant ($\bar{r} = 0.414$, p = 0.00); on the other hand, the effect size of negative TR on PEOU was negative and significant ($\bar{r}$ = -0.081, p = 0.01). At the 95% confidence interval, the effect size results for positive TR and PEOU was from 0.361 to 0.464 and for negative TR and PEOU from −0.145 to −0.020, and the intervals were not include the zero point. So hypotheses 4a and 4b were supported.

4 Conclusion

Based on the prior studies, which reported the correlation scores between TR and exogenous variables of TAM, we designed a meta-analytic study to attain effect size of TR on PU/ PEOU. We analyzed correlation scores which reported in 44 papers with Comprehensive Meta Analysis software program. Different level correlation scores had been recorded in the articles, so we conducted six distinct meta-analyses in two stages in order to obtain effect size of TR on PU. We carried out different six meta-analyses in two stages to get effect size of TR on PEOU, as well. Publication bias was controlled with funnel plot and Kendal Tau's coefficient, so no publication bias was found. After that, Q statistics was evaluated to make decision to relevant model for the effect size estimation.

Findings showed that the average effect sizes of optimism and innovativeness on PU and on PEOU were found positive and medium level. In addition, average effect sizes of insecurity on PU and on PEOU were found negative, but very small level. However, average effect sizes of discomfort on PU and on PEOU were found non-significant. Consequently, average effect sizes of positive TR on PU and on PEOU were found positive and medium level. Nevertheless, average effect size of negative TR on PEOU was found negative but very small level. On the other hand, effect size of negative TR on PU was found non-significant.

Blut and Wang [18] published a meta-analytic research which reported the effect sizes of TR and TAM. However, they did not report the effect sizes of each dimension of TR on TAM variables, they indicated the effect sizes of motivators and inhibitors of TR on TAM variables. Unlike this study, we evaluate each dimensions of TR (optimism, innovativeness, discomfort, insecurity) and TAM in detailed.

References

1. Huy, L.V., Nguyen, P.T.H., Pham, L., Berry, R.: Technology readiness and satisfaction in Vietnam's luxury hotels. Int. J. Manag. Decis. Mak. **18**(2), 183–208 (2019)
2. Parasuraman, A.: Technology Readiness Index (TRI) a multiple-item scale to measure readiness to embrace new technologies. J. Serv. Res. **2**(4), 307–320 (2000)
3. Rejikumar, G., Aswathy Asokan, A., Sreedharan, V.R.: Impact of data-driven decision-making in Lean Six Sigma: an empirical analysis. Total Qual. Manag. Bus. Excell. **31**(3–4), 279–296 (2020)
4. Davis, F.D.: Perceived usefulness, perceived ease of use, and user acceptance of information technology. MIS Q. **13**(3), 319–340 (1989)
5. Rogers, E.M.: Diffusion of Innovations, 3rd edn. The Free Press, New York (1983)
6. Moore, G.C., Benbasat, I.: Development of an instrument to measure the perceptions of adopting an information technology innovation. Inf. Syst. Res. **2**(3), 192–222 (1991)
7. Venkatesh, V., Morris, M.G., Davis, G.B., Davis, F.D.: User acceptance of information technology: toward a unified view. MIS Q. **27**(3), 425–478 (2003)
8. Davis, F.D., Bagozzi, R.P., Warshaw, P.R.: User acceptance of computer technology: a comparison of two theoretical models. Manag. Sci. **35**(8), 982–1003 (1989)
9. Donmez-Turan, A.: Does unified theory of acceptance and use of technology (UTAUT) reduce resistance and anxiety of individuals towards a new system? Kybernetes **49**(5), 1381–1405 (2019)
10. Van Compernolle, M., Buyle, R., Mannens, E., Vanlishout, Z., Vlassenroot, E., Mechant, P.: "Technology readiness and acceptance model" as a predictor for the use intention of data standards in smart cities. Media Commun. **6**(4), 127–139 (2018)
11. Kim, T., Chiu, W.: Consumer acceptance of sports wearable technology: the role of technology readiness. Int. J. Sports Mark. Spons. **20**(1), 109–126 (2019)
12. Khatri, V., Samuel, B.M., Dennis, A.R.: System 1 and System 2 cognition in the decision to adopt and use a new technology. Inf. Manag. **55**(6), 709–724 (2018)
13. Roy, S., Moorthi, Y.L.R.: Technology readiness, perceived ubiquity and M-commerce adoption. J. Res. Interact. Mark. **11**(3), 268–295 (2017)
14. Issaramanoros, E., Khlaisang, J., Pugsee, P.: Auto mechanic students' perceptions and readiness toward mobile learning in Thailand. Int. J. Interact. Mob. Technol. **12**(5), 28–42 (2018)
15. Cooper, H., Hedges, L.V.: The Handbook of Research Synthesis. Russell Sage Foundation, New York (1993)
16. Lipsey, M.W., Wilson, D.B.: Practical Meta-analysis: Applied Social Research Methods Series, vol. 49. SAGE Publications, Londra (2001)
17. Geyskens, I., Krishnan, R., Steenkamp, J.-B., Cunha, P.V.: A review and evaluation of meta-analysis practices in management research. J. Manag. **35**(2), 393–419 (2009)
18. Blut, M., Wang, C.: Technology readiness: a meta-analysis of con-ceptualizations of the construct and its impact on technology usage. J. Acad. Mark. Sci. **48**(4), 649–669 (2020)

E-services and M-services Using IoT: An Assessment of the Kingdom of Bahrain

Magdalena Karolak[1], Anjum Razzaque[2], and Abdalmuttaleb Al-Sartawi[2](✉)

[1] Zayed University, Dubai, United Arab Emirates
[2] Ahlia University, Manama, Bahrain

Abstract. The Kingdom of Bahrain has been reported as the pioneers of the e-government within the Arabian Gulf region. Such particularly is the pride reflective within the healthcare sector. With e-government ranked first among all Arab countries and thirteen worldwide, e-health paved new heights. Its successes further resonated during the COVID-19 era. Various stakeholders in the healthcare sector moved beyond e-services. Thanks to the Internet of Things, have been able to appreciate m-health. It is certain, that the country practiced strategies to assure itself towards a fast-growing Internet users base, and this is particularly evidenced through its citizens and the expat populations. The aim of this study is to explore how the Kingdom of Bahrain was able to succeed in its e-service, e-government, and e-health services to transform from a bureaucratic model to a new IoT based m-health and m-government paradigm. A content analysis of the Kingdom of Bahrain intellectual resources, i.e., their websites as well as journal and conference papers, etc., this chapter extends an overview of how Bahrain was able to transform to an e/m-government and e/m-health paradigm, along with theoretical and practical implication with suggestions for future research are also put forth in this chapter.

Keywords: Bahrain; E-health · M-health · E-government · M-government paradigm · Internet of Things · E-services · M-services

1 Introduction to E-government, M-government, E-health, and M-health

E-Governance is technology that speed up by streamlining the external and internal strategic business processes while connecting numerous nations. There are relevant theories when reviewing e-governance. Such a review to help scholars crafts a unified model of effective e-governance which can systematically blend the pre-requisites for a successful e-governance for any nation. There is empirical evidence driven from the application of e-governance with specific industries, e.g., the healthcare sector [22]. [23] characterizes e-government as legislative data driven by information and communication technologies (ICT) empowering exercises and administrations viable, and straight forward. Electronic (e) government is alluded to an advanced postal of governance for an

A. M. A. Musleh Al-Sartawi et al. (Eds.): EAMMIS 2021, LNNS 239, pp. 523–533, 2021.
https://doi.org/10.1007/978-3-030-77246-8_48

online government platform [17, 18]. A non-Internet foundation utilizes innovations preceding phone, text messages, biometric based proof, television, or radio. There are subtypes of e-government, i.e., m-government (mobile government facilitated on a mobile platform) [27]. In such a case, portable innovations like cell phone interface utilizing the remote accessibility of a local area network via a client to connect to the Web. Holistically, E-government fulfills: (1) virtual collaboration with government stakeholders, (2) upgrading ICT architectures and infrastructures efficiencies, (3) streamlining man and machine interfaces, (4) enhances those interest to participate on social platforms and (5) fortifies administrative protocols.

A generic form of an e-governance architecture, as described by [24] is in reaction for developing an effective e-governance approach as an outcome of what was learnt from the e-business domain. The internet technologies are used within the realm of the e-business domain for delivering a huge set of resolutions that assist in improving knowledge. The E-governance architecture is a model which aids recipient, meaning that an end-user can utilize data for communication purposes, through software such as Microsoft Internet Explorer on a desktop computer or a mobile device, for connecting to the Internet, i.e., the World Wide Web [1]. There are, as well, other relevant software for utilizing like Safari when the end-user is utilizing any device on the Mac OS platform. By applying such devices, the beneficiary links to data communicated application: i.e., the government's Intranet or its extranet. Such a network is then connecting the recipient to a processing system, e.g., any form of a knowledge-enabled systems, an information system, or a decision support system. Such systems, in return, process data, information and/or knowledge; stored within a repository/is also referred as E-governance Architecture Models [26].

[2] describes an e-government model from an e-governance perspective, thereby calling it an E-governance model as illustrated in Fig. 1. This model is built in four phases. The first phase is the information phase where most governments start with a web presence such as an online brochure catering to G2C and G2B to improve democracy and governmental transparency. The second phase is interaction where various applications are considered to simulate interaction between G2C and G2B. Communication and querying methods including emails, search engines, and form downloading assist in this phase. The third phase is transaction where technology is promoting itself by becoming more complex and sophisticated to enhance the value of the government's public (citizens, government body and businesses). Now, a complete transaction is possible using online portal applications to perform procedures such as filing property and income taxes, online voting, or applying for visas. The fourth and final phase is transformation where all information systems are integrated such that G2C and G2B can get all services from a single point.

[18] offer G2C, G2B and G2G solutions for constructing the e-governance model. The G2B solution is similar but focused on businesses. In this phase the internal G2G solution is knowledge-based management using intranet local area networks (LAN). For the interaction phase, G2G and G2B solutions include having the capability to download forms and receive emails from governmental agencies. The G2G is application of email, knowledge management database, and complaints handling tools. In the transformation phase, the G2C solution is a personalized website with integrated personal

account-accessing abilities. With a similar solution, business accounts rather than personal accounts are generated for the G2B in this phase. Finally, the G2G internal solution for this phase is database integration.

[25] described an E-governance Framework, as illustrated by [20]. This framework is made up of 4 parts: (1) Project assessment of sustainability – PAS; (2) e-services development; (3) Re-usable elements; and (4) Inter-operability standards. Since we have understood that e-government is a subset of e-governance, the four parts of this framework are encapsulated by the E-governance Model as depicted in Fig. 1. According to [20], each of the four parts of the E-governance Framework Model collaboratively interacts with each other to work, and each part is also made up of the following: (1) policy makers' re-usable processes, pattern and design standards used in e-governance projects including e-payment and e-filing, (2) developers' standards of protocols and specifications for connecting to services (e.g.: payment online portal or authentication protocol standards) via governmental gateways; and, (3) domain experts' standards pertaining to data.

In E-governance Framework Model [20], the blue book is for policy makers covering the development framework for e-governance with ideas being extracted from the generic information model. The pink book, for developers' community, provides standards and models for developers to construct and deploy e-services. As an example, such standards include information exchange model or information access framework etc. The green book for domain experts holds standards for data for data definition purposes used in governmental services and metadata search framework standards for assistance in easy data search. Project assessment and sustainability empower projects and help to maintain standards. E-service development builds a service through the five illustrated phases. Re-usable elements share and allow previously applied specification, design, pattern, services, codes, and technologies so that there will be no re-designing of the wheel. Interoperability standards are the bases of reference for models and standards for referential and abidance purposes.

A government (G) interatomic with exterior constituents connects with citizens (C), traders/businesses (B), other governmental bodies (G) and workers (E) [18]. Governmental organizations utilize the ICT to portray an allusion of 'electronic' government: referred as e-government, with non-commercial e-government models: (1) G2C: Government-to-Citizen, (2) G2G: Government to Government, (3) G2E: Government to Employees and (4) G2B: Government to Business. Here, G2C is a web entry point to permit those associated to the government [4–6, 40]. G2C is a one-stop-shop for citizens to interact with the services of a government. G2C is also e-democracy: where agents associate with citizens through e-voting. G2B is permits commercial businesses to correspond and trade with the government. G2E or 'e-governance' serves employees for mechanizing workflow, tracking business processes, overseeing human asset and resources [18].

E-government raises admonitive values through online environments where a business communicates with the government's administration environment through the governmental e-services since e-government is a sub-division of e-governance [2]. E-governance is IT-driven operating twenty-four hours daily for the significance of open organizations vital for fulfilling clients' and businesses' needs [32]. Within the interaction stage of an online government various applications mimic intelligence logic between G2C and G2B. This encourages communication through email correspondences, etc.

2 E government and M government: The Case of the Kingdom of Bahrain

The significance of e-government or e-governance, for advancements that are social as well as financial seems evident in Kingdom of Bahrain. The GCC depends on e-government as an online platform for improvement through a distinctive design of financial advancement with various FinTech solutions to facilitate GCC individuals and expats working in Kingdom of Bahrain (Razzaque, Cummings, Karolak, & Hamdan, 2019; Razzaque, Hamdan, & Cummings, 2019). The Kingdom of Bahrain, a tiny island has gotten its oil reserves wiped out. Due to complex populace composition, this nation has been observed to encounter waves of social changes, which have incited specialists working within Kingdom of Bahrain to propose and thus develop solutions which reflect signs of diversifications that lead the change. Such a change has been reported to leading the Kingdom of Bahrain ahead of the other GCC nations. Though in 1932 the Kingdom of Bahrain experienced an oil revelation and then in the 1930s it also reported the gas revolution, but by 1965 the Kingdom of Bahrain's government begun recognizing the dire need for broadening its financial assets [7, 9, 28, 39]. Such a migration that led the Kingdom of Bahrain to steer itself from an oil-based economy to a financial hub of the GCC (Razzaque et al. 2019) started when its specialists began to speculate in various industries, began reengineering their aluminum industry as well as the ship-repair facilities. This is in addition to their financial industry which started cultivating the private sector [15, 34]. Furthermore, since the 1980's the Kingdom of Bahrain got to situate itself as the financial hub within the GCC region [39]; followed by the 2000's when the Kingdom of Bahraini specialists, sought after rearrangements of assist liberalization to diminish the share of the open segment while pulling remote ventures [43].

The crucial adjustments that included in the Kingdom of Bahrain are the privatization of governmental sectors, which facilitated strategies, and the removal of corporate taxes. This is in addition to the outside property and company proprietorships: an initiative that was led without the support of other GCC neighboring countries. The sponsorship of the expat labor was phased out until it gets diminished, allowing laborers to voluntarily transfer their work permits between sponsors without any permission of their present sponsors. Such moves sought the Kingdom of Bahrain to obligate to endeavors which would position itself such that the Kingdom of Bahrain started getting ranked number four regionally and 63rd freest economy worldwide by 2020 [38]. For further clarify, %GDP is depicted in Table 1 for the Kingdom of Bahrain. Oil is still the main asset for the national budgeting since it accounts for more than 60% of the legislative incomes and more than 60% of the added up of the trades [16].

The oil boom revolutionized the Kingdom of Bahraini society, driving it richer. But since then, the richness has gradually declined. At the conclusion of the 1980's the %GDP stood at 4 [44]. It is imperative that the governments of the GCC countries have prevailed with incremental rates [29, 33, 42]. The Kingdom of Bahrain, alongside other

Table 1. The Kingdom of Bahrain's % GDP timeline

% GDP	1990	2009	2019
Agriculture	1	1	9
Natural fuel mining	19	10	18
Production	11	12	15
Economic benefits	19	30	17
Real estate	9	9	5
Other	41	38	8

Source: (Ministry of Industry and Commerce, n.d.; Bahrain Economic Report, 2020; Trading Economics, 2020)

GCC nations, bears the most youthful populations worldwide. The development of the Kingdom of Bahraini population showed increment of nonnatives. [10, 11, 17, 41].

3 E-government Paradigm Shift During the COVID-19 Era

The one thing that the COVID era welcomed the GCC to is the hearted adoption of electronic services. But then this does not mean that the entire credit goes to the COVID-era. Even before the GCC was well advance in this arena. Based on the review of literature mentioned above in this paper, it is evident how serious is the GCC in utilizing e-services. And the review of literature on the acceptance of e-health applications and services, mentioned ahead, puts the Kingdom of Bahrain in the safe zone when it comes to remaining afloat during the COVID crisis. The Kingdom of Bahrain e-government administrations is adjusted with its financial vision of the 2030 where for a steady future advancement endeavoring to supply financial advancements which are free from natural resources dependent assets. The e-government method for Bahrain is the e-government committed for all administrations (Bahrain eGovernment Programme, 2009). The development areas are: (1) E-Government pioneer: Bahrain aims to preserve and improve its regional position as a pioneer committed to advancing its services for its clients. (2) Government administrations: The Kingdom of Bahrain's e-Government procedure should be centered on provisioning client administrations, by electronically enabling all key administrations services. (3) Standard coordinating activities: The Kingdom of Bahrain will extend client fulfillments with government activities through updated customer-centric portals, such that administrations are conveyed clients' needs centrally. And (4) accessible comprehensive society: The Kingdom of Bahrain will guarantee successful conveying of government administrations to all citizens, independent of instruction, nationality, etc. (Table 2).

Table 2. Overview of the Kingdom of Bahrain.bh portal

Analyzed factors	Results
1. The entry is recognized throughout administration firms on governmental websites showing off their services: upheld by appropriate guidelines	Point of get to taking after administrations from the domestic page: pay power & water e-billing; report street blame; issuance of protections arrangements for Ruler Fahad Causeway; instalment of activity negations; ID card reestablishment application and arrangement; legislative schools understudy exam comes about; issuing duplicates of understudy certificates; social occasions; vehicle enlistment reestablishment; Wellbeing Pre-employment restorative arrangement booking benefit; A tremendous cluster of extra administrations is accessible from Commerce and Person tabs; illustrations following of postal administrations; clean complaints; underwriting, accreditation, and approval of scholarly qualifications
2. The Universal Resource Locator (URL) must consist of reminder prompts for the public	YES (www.bahrain.bh was upgraded from e.gov.bh)
3. There ought to be a clear explanation on the Home Page that the entry gives joins to the government offices, and data on how to contact officials	Listed governmental contacts
4. The Domestic Page ought to allow the user to have get to a great location outline as one means of finding required data and services	YES
5. The Domestic Page should allow search empowering end-users to access services	YES
6. The Domestic Page ought to encourage contact with agencies and authorities through directories of road addresses, phone numbers and email addresses	Contacting gets directly referred to government organizations websites
7. Color, format, and web pages plan should be appealing and useful for a user-friendly frontend interface	Website location simple to explore; Locks in see and graphics
8. The online portal ought to give get to e-government shapes and the capacity to execute related administrations online	Forms helped administration services can be requested online

It can be concluded that the Kingdom of Bahrain's e-government reflects the fourth stage of an open organization of change. This entrance offers data, an expanding choice of e-service with interactive portals for citizens to engage with the government personals. Its commitment for brilliance is reflected the accessible client fulfillment interaction which occur between the government and its citizens. An extraordinary consideration is given to outside speculators and sightseers, reflecting innovative financial improvements. The web portal provides services in English and in Arabic. With the issuing of SmartCards for nationals and expats, e-services are accessible, thus reflecting advanced globalization realizing a reality [21, 36].

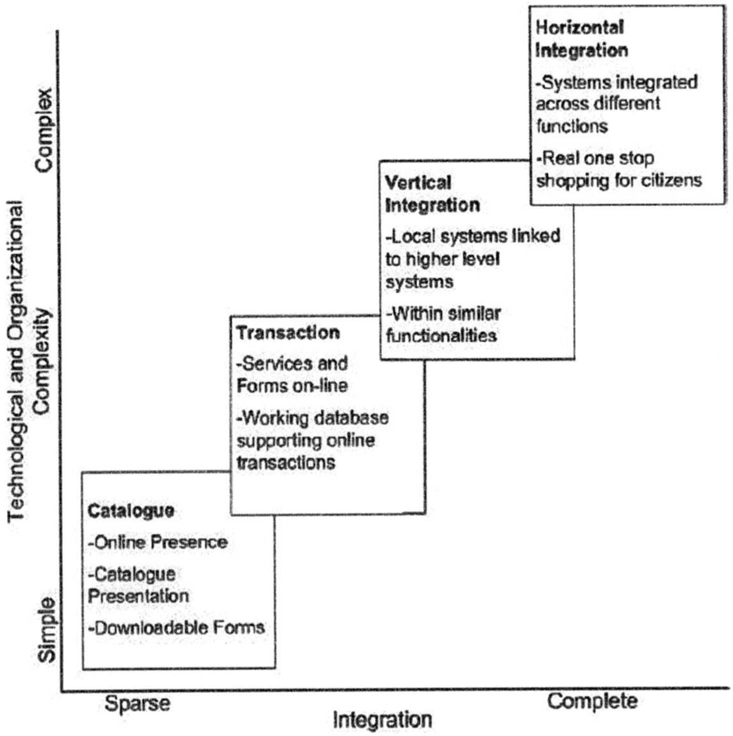

Fig. 1. E-government model of maturity Source: [3].

4 E-health: The Case of the Kingdom of Bahrain

[12, 13] applied the Technology Acceptance Model (TAM) to examine the level of recognition health professionals would be able to have on IT such as the application of Skype and Macedonia within the healthcare context. [13]'s empirical findings revealed that the Perceived Usefulness (PU) and Perceived Ease of Use (PEOU) are vital in recognizing the expecting of the intentions of the nurses and physicians upon using the just-mentioned IT applications. Also, [13, 32]'s model indicated that PEOU was of greater influence upon the acceptance of IT application within the healthcare context, i.e., 63.3%. Furthermore, [35] empirically examined their study in Taiwan to discover crucial factors affecting the physicians' attitudes on accepting Electronic Medical Records (EMRs). By applying a modified version of the TAM model, the empirical results agreed with those of [13] thus suggesting that the highly critical considerations for extremely influencing physicians' intentions for using EMR are the same as [12]'s et al. However, [14] established that the utmost implication of EMR from physicians' perspectives was in the correct entry

and access of data, so to be able to efficiently correspond with the diagnoses, allergies, medications, etc.; hence to mitigate any dangerous circumstances and therefore better management patient care. Yet, [14, 37, 45] exhibited an important notion, i.e., even by the complete positive effect of EMRs to facilitate healthcare performance [9], they remain high resistance and hence a small rate of acceptance among the nurses and doctors. And this is due to certain considerations like provoking of the sense of hard work, the concern that the use of EMR would cause long hours of training, and patients' worries regarding the security concerns when it comes to the application of their private data, the low possible rate of response time, etc. At the end, earlier literature has enthusiastically endorsed the usage of TAM and TAM 2 models within the context of comprehending the rate of acceptance of EMRs by physicians to comprehend their behaviors.

Now what about the Kingdom of Bahrain? EMR is today obtainable and applied in the clinics and hospital in various parts of the world, including the Kingdom of Bahrain. The Kingdom of Bahrain understands the importance to accept EMR. And this is well understood by its physicians and other healthcare stakeholders. The physicians in Bahrain also are aware of the crucial factors which affect to make EMR implementation and integration successful and beneficial [7, 8, 31]. With the improvement in physicians' reliability on medical apps and physicians' ability to use e-textbooks and access various data reserves, for instance: medical calculator, and medications prescriptions via the use of different apps on smartphones it is good news that such practices are well developed and adopted by the physicians of the Kingdom of Bahrain. Nowadays [19, 30]. The notion that healthcare stakeholders utilize app in patient care means that they are making use of the Internet of Things (IoT).

5 Conclusion and Implication

This chapter highlights the awareness that the Kingdom of Bahrain has for the application, implementation, and the significance of the integration of e-government and e-health. The significance of this chapter is the bending of a holistic comprehension of e-government and e-health since these are two portals which form the backbone of any successful society. The fact that the Kingdom of Bahrain has a renowned e-government system ranking it respectfully high in the regional and international arena, goes to show that also its healthcare sector is mature. And this is the reason why the same can be expressed for the Kingdom of Bahrain's e-health initiatives which are not only successful but have been proven to reflect high standards.

For those who would read the review of the literature portrayed in this chapter will be able to understand its theoretical and managerial implications. For the theory standpoint of view the awareness that the Kingdom of Bahrain is diversifying from the utilization of its natural resources for the sake of its income goes to show that this country seems to diversify to become a knowledge economy. Managers who are interested in corresponding with the governmental body and/or with the healthcare stakeholders can comfortably consider that the online portals provided by the Kingdom of Bahrain government withstand an infrastructure which is world-class (Alshaikh et al. 2017).

References

1. Al-Ansari, M.A., Hamdan, A., Razzaque, A., Reyad, S., Sartawi, A.: The moderating role of m-learning activities in the relationship between students' social capital and knowledge sharing. In: 13th International Conference on e-Learning - MCCSIS 2019 (2019)
2. Backus, M.: E-Governance in Developing Countries (2001). http://www.ftpiicd.org/files/research/briefs/brief1.pdf. Accessed 22 Jan 2008
3. Berntzen, L., Olsen, M.G.: Benchmarking e-government - a comparative review of three international benchmarking studies. In: ICDS, pp. 77–82 (2009)
4. Chanana, L.: Outsourcing of G2E Applications – The Next Potential Wave for Pubic Private Partnership Initiatives in India (2007). http://www.iceg.net/2007/books/1/12_376.pdf. Accessed 17 Jan 2008
5. Al-Sartawi, A.: The effect of corporate governance on the performance of the listed companies in the Gulf Cooperation council countries. Jordan J. Bus. Admin. **11**(3), 705–725 (2015)
6. Cheema, G.S.: linking governments and citizens through democratic governance. Public administration and democratic governance: governments serving citizens. In: Proceedings of the 7th Global Forum on Reinventing Government (2007)
7. Building Trust in Government 26–29 June 2007, Vienna, Austria
8. Abdulrasool, F.E., Turnbull, S.I.: Exploring security, risk, and compliance driven IT governance model for universities: applied research based on the COBIT framework. Int. J. Electron. Bank. **2**(3), 237–265 (2020)
9. Durrant, F.: The World Wide Web Enhancing E-government in the Caribbean: an assessment of government portals or gateway websites. IFLA J. **32**, 240 (2006)
10. Gagnon, M.P., Ghandour, E.K., Talla, P.K., Simonyan, D., Godin, G., Labrecque, M., Ouimet, M., Rousseau, M.: Electronic health record acceptance by physicians: testing an integrated theoretical model. J. Biomed. Informat. **48**, 17–27 (2014)
11. Al-Sartawi, A.: Social media disclosure of intellectual capital and firm value. Int. J. Learn. Intellect. Capital **17**(4), 312–323 (2020)
12. Ho, T.K.: Reinventing local governments and the e-government initiative. Public Adm. Rev. **62**(4), 434–444 (2002)
13. Kapiszewski, A.: Arab Labour Migration to the GCC States, IOM (2004)
14. Li, B.: On the barriers to the development of e-government in China. In: Proceedings of the 7th International Conference on Electronic Commerce, XI'an, China, 15–17 August 2005, pp. 549–552 (2005)
15. Al-Sartawi, A.: The effect of the electronic financial reporting on the market value added of the Islamic banks in Gulf Cooperation Council Countries. In: 8th Global Islamic Marketing Conference, 4–6 May. International Islamic Marketing Association, Turkey (2017)
16. Ketikidis, P., Dimitrovski, T., Lazuras, L., Bath, P.A.: Acceptance of health information technology in health professionals: an application of the revised technology acceptance model. Health Inform. J. **18**(2), 124–134 (2012)
17. Lakbala, P., Dindarloo, K.: Physicians' perception and attitude toward electronic medical record, vol. 3, no. 1, 63 (2014). SpringerPlus
18. Al-Sartawi, A., Sanad, Z.: Institutional ownership and corporate governance: evidence from Bahrain. Afro-Asian J Financ. Acc. **9**(1), 101–115 (2019)
19. Musleh Al-Sartawi, A.M.A.: E-learning improves accounting education: case of the higher education sector of Bahrain. In: Themistocleous, M., Papadaki, M., Kamal, M.M. (eds.) Information Systems, EMCIS 2020. Lecture Notes in Business Information Processing, vol. 402. Springer, Cham (2020)
20. Metz, H. Ch.: Persian Gulf States: Bahrain, Kuwait, Oman, Qatar, United Arab Emirates, Federal Research Division, Library of Congress (1993). http://www.mongabay.com/history/united_arab_emirates/united_arab_emirates-introduction.html. Accessed 16 Mar 2011

21. Nakibullah, A., Islam, F.: Effect of government spending on non-oil GDP of Bahrain. J. Asian Econ. **18**(5) (2007). Elsevier
22. OECD: Rethinking e-Government Services: User-Centered Approaches. OECD Publishing (2009)
23. Gupta, N.: Influence of demographic variables on synchronisation between customer satisfaction and retail banking channels for customers of public sector banks of India. Int. J. Electron. Banking **1**(3), 206–219 (2019)
24. Al-Sartawi, A.: Information technology governance and cybersecurity at the board level. Int. J. Crit. Infrastruct. **16**(2), 150–161 (2020)
25. Palvia, S.C.J., Sharma, S.S.: E-Government and E-Governance: Definitions/Domain Framework and Status around the World (2007). http://www.iceg.net/2007/books/1/1_369.pdf. Accessed 17 Jan 2008
26. Payne, K.F.B., Wharrad, H., Watts, K.: Smartphone and medical related App use among medical students and junior doctors in the United Kingdom (UK): a regional survey. BMC Med. Informatics Decis. Making **12**(1), 121 (2012)
27. Al-Sartawi, A.: Does it pay to be socially responsible? Empirical evidence from the GCC countries. Int. . Law Manag. **62**(5), 381–394 (2020)
28. Karolak, M., Razzaque, A.: Bahraini Government and the e-Government initiative: an assessment. In: 2012 Annual Conference of Asian Group for Public Administration (AGPA), Bandos Island Resport, Maldives (2012)
29. Al-Sartawi, A.: Assessing the relationship between information transparency through social media disclosure and firm value. Manag. Account. Rev. **18**(2), 1–20 (2019)
30. Razzaque, A., Eldabi, T.: Link between social capital theory and innovation mediated by knowledge sharing and technology. In: European, Mediterranwan & Middle Eastern Conference on Information Systems 2014, 27–28 Oct, 2014. Qatar University. Qatar (2014)
31. Al-Sartawi, A.: Measuring the level of online financial disclosure in the Gulf Cooperation Council Countries. Corp. Ownership Control. **14**(1), 547–558 (2016)
32. Razzaque, A., Jalal-Karim, A.: Conceptual healthcare knowledge management model for adaptability and interoperability of EHR. In: European, Mediterranean & Middle Eastern Conference on Information Systems (EMCIS 2010), Abu-Dhabi, UAE (2010)
33. Sahraoui, S.: E-Government in the Arabian Gulf 1: Government Transformation vs. Government Automation (2005). http://www.iseing.org/egov/eGOV05/Source%20Files/Papers/CameraReady-16-P.pdf. Accessed 21 Jan 2008
34. Al-Sartawi, A.: Ownership structure and intellectual capital: evidence from the GCC countries. Int. J. Learn. Intellect. Capital **15**(3), 277–291 (2018)
35. Saxena, A., Wadhawa, S.: E-business perspective of E-governance. In: Gupta, M.P. (ed.) Promise of E-governance Operational Challenges, pp. 129–136. Delhi, India (2004)
36. Al-Sartawi, A.: Institutional ownership, social responsibility, corporate governance and online financial disclosure. Int. . Crit. Account. **10**(3/4), 241–255 (2018)
37. Sekir, R., Debnath, S., Bharadwaj, D.: eGovernance Architecture and Standards (2006). http://egovstandards.gov.in/egs/brainstorming-session/egovdef-1.doc/download. Accessed 24 Jan 2008
38. Sanad, Z., Al-Sartawi, A.: Investigating the relationship between Corporate Governance and Internet Financial Reporting (IFR): evidence from Bahrain Bourse. Jordan J. Bus. Admin. **12**(1), 239–269 (2016)
39. Showaiter, A., Razzaque, A., Hamdan, A.: The impact of knowledge management adoption on the government sector's performance: the case of Bahrain. Int. J. Econ. Bus. Res. (2019)
40. Al-Sartawi, A.: Corporate governance and intellectual capital: evidence from Gulf Cooperation Council Countries. Acad. Account. Fin. Stud. J. **22**(1), 1–12 (2018)
41. Trimi, S., Sheng, H.: Emerging trends in M-government. Commun. ACM **51**(5), 53–58 (2008)

42. Al-Sartawi, A.: The Big Data-driven digital economy: artificial and computational intelligence In: Kacprzyk, J. (ed.) Studies in Computational Intelligence, vol. 974. Springer International Publishing (2021)
43. Wilkenson, B., Atti, A.: UNDP microfinance assessment report for Bahrain. United Nations Capital Development Fund (1997). http://www.uncdf.org/english/microfinance/uploads/country_feasibility/bahfinaldb.pdf. Accessed 28 Aug 2007
44. Gupta, M., Sikarwar, T.S.: Modelling credit risk management and bank's profitability. Int. J. Electron. Banking **2**(2), 170–183 (2020)
45. Onn, W.: Arab Political Demography: Population Growth and Natalist Policies. Sussex Academic Press, Brighton (2005)

Author Index

A. M. A. Musleh Al-Sartawi et al. (Eds.): EAMMIS 2021, LNNS 239, pp. 535–536, 2021.
https://doi.org/10.1007/978-3-030-77246-8